Handbook of Therapeutic Antibodies

*Edited by
Stefan Dübel*

1807–2007 Knowledge for Generations

Each generation has its unique needs and aspirations. When Charles Wiley first opened his small printing shop in lower Manhattan in 1807, it was a generation of boundless potential searching for an identity. And we were there, helping to define a new American literary tradition. Over half a century later, in the midst of the Second Industrial Revolution, it was a generation focused on building the future. Once again, we were there, supplying the critical scientific, technical, and engineering knowledge that helped frame the world. Throughout the 20th Century, and into the new millennium, nations began to reach out beyond their own borders and a new international community was born. Wiley was there, expanding its operations around the world to enable a global exchange of ideas, opinions, and know-how.

For 200 years, Wiley has been an integral part of each generation's journey, enabling the flow of information and understanding necessary to meet their needs and fulfill their aspirations. Today, bold new technologies are changing the way we live and learn. Wiley will be there, providing you the must-have knowledge you need to imagine new worlds, new possibilities, and new opportunities.

Generations come and go, but you can always count on Wiley to provide you the knowledge you need, when and where you need it!

William J. Pesce
President and Chief Executive Officer

Peter Booth Wiley
Chairman of the Board

Handbook of Therapeutic Antibodies

Volume II

Edited by
Stefan Dübel

WILEY-VCH Verlag GmbH & Co. KGaA

The Editor

Prof. Dr. Stefan Dübel
Technical University of Braunschweig
Institute of Biochemistry and Biotechnology
Spielmannstr. 7
38106 Braunschweig
Germany

■ All books published by Wiley-VCH are carefully produced. Nevertheless, authors, editors, and publisher do not warrant the information contained in these books, including this book, to be free of errors. Readers are advised to keep in mind that statements, data, drug dosages, illustrations, procedural details or other items may inadvertently be inaccurate.

Library of Congress Card No.:
applied for

British Library Cataloguing-in-Publication Data
A catalogue record for this book is available from the British Library.

Bibliographic information published by the Deutsche Nationalbibliothek
Die Deutsche Nationalbibliothek lists this publication in the Deutsche Nationalbibliografie; detailed bibliographic data are available in the Internet at <http://dnb.d-nb.de>.

© 2007 WILEY-VCH Verlag GmbH & Co. KGaA, Weinheim

All rights reserved (including those of translation into other languages). No part of this book may be reproduced in any form – by photoprinting, microfilm, or any other means – nor transmitted or translated into a machine language without written permission from the publishers. Registered names, trademarks, etc. used in this book, even when not specifically marked as such, are not to be considered unprotected by law.

Cover Design: Schulz Grafik-Design, Fussgönheim

Wiley Bicentennial Logo: Richard J. Pacifico

Composition   SNP Best-set Typesetter Ltd., Hong Kong

Printing   betz-druck GmbH, Darmstadt

Bookbinding   Litges & Dopf GmbH, Heppenheim

Printed in the Federal Republic of Germany
Printed on acid-free paper

ISBN 978-3-527-31453-9

Dedication

To Inge, Hans, Ulrike and Tasso Dübel – the four best things in my life

Stefan Dübel

Contents

Volume II

Overview of Therapeutic Antibodies *XV*

A Greeting by the Editor *XVII*

Foreword *XIX*

List of Authors *XXI*

Part III Beyond IgG – Modified Antibodies

1 Immunoscintigraphy and Radioimmunotherapy *325*
Jason L. J. Dearling and Alexandra Huhalov
1.1 Introduction *325*
1.2 Solid Tumors as Targets for Antibody-Based Therapeutics *326*
1.3 Antigen *326*
1.4 Antibodies as Vehicles for Radionuclide Delivery *327*
1.5 Radioisotope *334*
1.6 Improving Radioimmunoscintigraphy and Therapy *337*
1.7 Summary *338*
References *339*

2 Bispecific Antibodies *345*
Dafne Müller and Roland E. Kontermann
2.1 Introduction *345*
2.2 The Generation of Bispecific Antibodies *346*
2.3 Bispecific Antibodies and Retargeting of Effector Cells *354*
2.4 Bispecific Antibodies and Retargeting of Effector Molecules *361*
2.5 Bispecific Antibodies as Agonists or Antagonists *364*
2.6 Bispecific Antibodies and Somatic Gene Therapy *366*
2.7 Outlook *367*
References *368*

Handbook of Therapeutic Antibodies. Edited by Stefan Dübel
Copyright © 2007 WILEY-VCH Verlag GmbH & Co. KGaA, Weinheim
ISBN 978-3-527-31453-9

3 Immunotoxins and Beyond: Targeted RNases *379*
Susanna M. Rybak and Dianne L. Newton
3.1 Introduction *379*
3.2 Immunotoxins *382*
3.3 Targeted RNases *389*
3.4 Targeted Onconase *394*
3.5 Outlook *399*
References *400*

Part IV Emerging Concepts

4 Automation of Selection and Engineering *413*
Zoltán Konthur
4.1 Introduction *413*
4.2 General Considerations for the Automation of Antibody Generation *415*
4.3 Development of an Antibody Generation Pipeline *417*
4.4 Conclusion *427*
References *428*

5 Emerging Technologies for Antibody Selection *431*
Mingyue He and Michael J. Taussig
5.1 Introduction *431*
5.2 Display Technologies *432*
5.3 Antibody Libraries *433*
5.4 Antibody Selection and Maturation *In Vitro* *435*
5.5 Linking Antibodies to mRNA: Ribosome and mRNA Display *436*
5.6 Advantages of Ribosome Display *436*
5.7 Ribosome Display Systems *437*
5.8 Antibody Generation by Ribosome Display *440*
5.9 Summary *440*
References *441*

6 Emerging Alternative Production Systems *445*
Thomas Jostock
6.1 Introduction *445*
6.2 Production Systems *446*
6.3 Outlook *458*
References *460*

7 Non-Antibody Scaffolds *467*
Markus Fiedler and Arne Skerra
7.1 Introduction *467*
7.2 Motivation for Therapeutic Use of Alternative Binding Proteins *467*

7.3	Single-Domain Immunoglobulins 473	
7.4	Scaffold Proteins Presenting a Contiguous Hypervariable Loop Region 480	
7.5	Scaffold Proteins for Display of Individual Extended Loops 484	
7.6	Scaffold Proteins Providing a Rigid Secondary Structure Interface 488	
7.7	Conclusions and Outlook: Therapeutic Potential and Ongoing Developments 491	
	References 492	
8	**Emerging Therapeutic Concepts I: ADEPT** *501*	
	Surinder K. Sharma, Kerry A. Chester, and Kenneth D. Bagshawe	
8.1	Introduction and Basic Principles of ADEPT 501	
8.2	Preclinical Studies 503	
8.3	Clinical Studies 505	
8.4	Immunogenicity 506	
8.5	Important Considerations/Outlook 508	
	References 509	
9	**Emerging Therapeutic Concepts II: Nanotechnology** *515*	
	Dimiter S. Dimitrov, Igor A. Sidorov, Yang Feng, Ponraj Prabakaran, Michaela A.E. Arndt, Jürgen Krauss, and Susanna M. Rybak	
9.1	Introduction 515	
9.2	Nanoliposomes, Gold Nanoshells, Quantum Dots, and Other Nanoparticles with Biomedical Potential 516	
9.3	Conjugation of Antibodies to Nanoparticles 518	
9.4	Nanoparticle–Antibody Conjugates for the Treatment of Cancer and Other Diseases 520	
9.5	Comparison Between Nanoparticle–Antibody Conjugates and Fusion Proteins for Cancer Diagnosis and Treatment 522	
9.6	Conclusions 526	
9.7	Summary 526	
	References 527	
	Appendix 529	
10	**Emerging Therapeutic Concepts III: Chimeric Immunoglobulin T Cell Receptors, T-Bodies** *533*	
	Thomas Schirrmann and Gabriele Pecher	
10.1	Introduction 533	
10.2	Chimeric Immunoglobulin T-Cell Receptors – "T-Bodies" 534	
10.3	Preclinical Studies 543	
10.4	Therapeutic Considerations 552	

10.5 Perspectives 558
10.6 Conclusions 561
References 561

11 Emerging Therapeutic Concepts IV: Anti-idiotypic Antibodies 573
Peter Fischer and Martina M. Uttenreuther-Fischer
11.1 Introduction 573
11.2 Definition of Anti-idiotypic Antibodies 573
11.3 Anti-idiotypic Antibodies as Autoantigens 575
11.4 Anti-idiotypes as a Tool for Generating Specific Antibodies 575
11.5 Intravenous Immunoglobulin Preparations (IVIG) 575
11.6 Anti-idiotypic Antibodies as Possible Superantigens? 577
11.7 Anti-idiotypic Antibodies in Cancer Therapy 578
11.8 Ab2β Vaccine Trials Mimicking GD_2 580
11.9 Molecular Characterization of the Anti-idiotypic Immune Response of a Relapse-free Cancer Patient 581
11.10 Conclusion 583
References 583

Part V Ongoing Clinical Studies

12 Antibodies in Phase I/II/III: Cancer Therapy 593
P. Markus Deckert
12.1 Introduction 593
12.2 Novel Antibody Constructs 595
12.3 Specific Targeting and Effector Mechanisms 606
12.4 Antigens Without Known Effector Function 627
12.5 Disease-specific Concepts of Unknown Antigen Function and Structure 656
12.6 Summary 656
References 659

13 Antibodies in Phase I/II/III: Targeting TNF 673
Martin H. Holtmann and Markus F. Neurath
13.1 Introduction 673
13.2 Inflammatory Bowel Disease 674
13.3 Pathophysiologic Role of T Cells 675
13.4 Tumor Necrosis Factor-α 675
13.5 TNF Receptors and Signaling 676
13.6 Anti-TNF Antibodies and Fusion Proteins in Clinical Testing 679
13.7 Mechanisms of Action 683
13.8 Other Anti-TNF Biologicals 683
13.9 Other Cytokine-based and Anti-$CD4^+$ T-Cell Approaches 684
13.10 Perspective 685
References 686

Volume I

Overview of Therapeutic Antibodies *XIX*

A Greeting by the Editor *XXI*

Foreword *XXIII*

List of Authors *XXV*

Introduction

1 Therapeutic Antibodies – From Past to Future *3*
   Stefan Dübel

Part I Selecting and Shaping the Antibody Molecule

2 Selection Strategies I: Monoclonal Antibodies *19*
   Gerhard Moldenhauer

3 Selection Strategies II: Antibody Phage Display *45*
   Michael Hust, Lars Toleikis and Stefan Dübel

4 Selection Strategies III: Transgenic Mice *69*
   Marianne Brüggemann, Jennifer A. Smith, Michael J. Osborn, and Xiangang Zou

5 Bioinformatics Tools for Antibody Engineering *95*
   Andrew C.R. Martin and James Allen

6 Molecular Engineering I: Humanization *119*
   José W. Saldanha

7 Molecular Engineering II: Antibody Affinity *145*
   Lorin Roskos, Scott Klakamp, Meina Liang, Rosalin Arends, and Larry Green

8 Molecular Engineering III: Fc Engineering *171*
   Matthias Peipp, Thomas Beyer, Michael Dechant, and Thomas Valerius

Part II The Way into the Clinic *197*

9 Production and Downstream Processing *199*
   Klaus Bergemann, Christian Eckermann, Patrick Garidel, Stefanos Grammatikos, Alexander Jacobi, Hitto Kaufmann, Ralph Kempken, and Sandra Pisch-Heberle

10 **Pharmaceutical Formulation and Clinical Application** *239*
Gabriele Reich

11 **Immunogenicity of Antibody Therapeutics** *267*
Huub Schellekens, Daan Crommelin, and Wim Jiskoot

12 **Regulatory Considerations** *277*
Marjorie A. Shapiro, Patrick G. Swann, and Melanie Hartsough

13 **Intellectual Property Issues** *301*
Michael Braunagel and Rathin C. Das

Volume III

Overview of Therapeutic Antibodies *XXIII*

A Greeting by the Editor *XXV*

Foreword *XXVII*

List of Authors *XXIX*

Volume III Approved Therapeutics

1 **Adalimumab (Humira)** *697*
Hartmut Kupper, Jochen Salfeld, Daniel Tracey, and Joachim R. Kalden

2 **Alemtuzumab (MabCampath)** *733*
Thomas Elter, Andreas Engert, and Michael Hallek

3 **Bevacizumab (Avastin)** *779*
Eduardo Díaz-Rubio, Edith A. Perez, and Guiseppe Giaccone

4 **Cetuximab (Erbitux, C-225)** *813*
Norbert Schleucher and Udo Vanhoefer

5 **Efalizumab (Raptiva)** *827*
Karlheinz Schmitt-Rau and Sigbert Jahn

6 **^{99m}Tc-Fanolesomab (NeutroSpec)** *851*
Christopher J. Palestro, Josephine N. Rini, and Charito Love

7 **Gemtuzumab Ozogamicin (Mylotarg)** *869*
Matthias Peipp and Martin Gramatzki

8 Infliximab (Remicade) *885*
Maria Wiekowski and Christian Antoni

9 Muromonab-CD3 (Orthoclone OKT3) *905*
Harald Becker

10 Natalizumab (Tysabri) *941*
Sebastian Schimrigk and Ralf Gold

11 Omalizumab (Xolair)
Anti-Immunoglobulin E Treatment in Allergic Diseases *951*
Claus Kroegel and Martin Foerster

12 Palivizumab (Synagis) *999*
Alexander C. Schmidt

13 Rituximab (Rituxan) *1033*
Michael Wenger

14 Trastuzumab (Herceptin)
A Treatment for HER2-Positive Breast Cancer *1109*
Paul Ellis

15 Abciximab, Arcitumomab, Basiliximab, Capromab, Cotara, Daclizumab, Edrecolomab, Ibritumomab, Igovomab, Nofetumomab, Satumomab, Sulesomab, Tositumomab, and Votumumab *1131*
Christian Menzel and Stefan Dübel

Index *1149*

Overview of Therapeutic Antibodies

Trade name	FDA name	Chapter	Page
Avastin	bevacizumab	3	781
Bexxar	tositumomab	15	1145
CEA-Scan	arcitumomab	15	1134
Cotara	–	15	1137
Erbitux	cetuximab	4	815
Herceptin	trastuzumab	14	1111
HumaSpect-Tc	votumumab	15	1147
Humira	adalimumab	1	699
Indimacis-125	igovomab	15	1142
LeukoScan	sulesomab	15	1144
Leukosite	alemtuzumab	2	735
MabCampath	alemtuzumab	2	735
Mylotarg	gemtuzumab	7	871
Neutrospec	fanolesomab	6	853
OncoScint	satumomab	15	1143
Oncorad	satumomab	15	1143
Orthoclone	muromonab	9	907
Panorex	edrecolomab	15	1139
ProstaScint	capromab	15	1135
Raptiva	efalizumab	5	829
Remicade	infliximab	8	887
ReoPro	abciximab	15	1133
Rituxan	rituximab	13	1035
Simulect	basiliximab	15	1137
Synagis	palivizumab	12	1001
Tysabri	natalizumab	10	943
Verluma	nofetumomab	15	1142
Xolair	omalizumab	11	953
Zenapax	daclizumab	15	1138
Zevalin	imbritumomab	15	1141

Handbook of Therapeutic Antibodies. Edited by Stefan Dübel
Copyright © 2007 WILEY-VCH Verlag GmbH & Co. KGaA, Weinheim
ISBN 978-3-527-31453-9

A Greeting by the Editor

Today, therapeutic antibodies are essential assets for physicians fighting cancer, inflammation, and infections. These new therapeutic tools are a result of an immense explosion of research sparked by novel methods in gene technology which became available between 1985 and 1995.

This handbook endeavors to present the fascinating story of the tremendous achievements that have been made in strengthening humanity's weapons arsenal against widespread diseases. This story not only includes the scientific and clinical basics, but covers the entire chain of therapeutic antibody production – from downstream processing to Food and Drug Administration approval, galenics – and even critical intellectual property issues.

A significant part is devoted to emerging developments of all aspects of this process, including an article showing that antibodies may only be the first generation of clinically used targeting molecules, making the IgG obsolete in future developments, and novel ideas for alternative therapeutic paradigms.

Finally, approved antibody therapeutics are presented in detail in separate chapters, allowing the clinicians to quickly gain a comprehensive understanding of individual therapeutics.

In such a fast-developing area, it is difficult to keep pace with the rapidly growing information. For example, a PubMed search with "Herceptin" yields more than 1500 citations. Consequently, we have tried to extract the essentials from this vast resource, offering a comprehensive basis of knowledge on all relevant aspects of antibody therapeutics for the researcher, the company expert, and the bedside clinician.

At this point, I express my deep gratitude to all the colleagues who wrote for these books. Without their enthusiasm this project would never have materialized. I would also like to thank Dr Pauly from the publisher's office, who paved the way for this three-volume endeavor, and the biologist Ulrike Dübel – my wife. Both played essential roles in keeping the project on track throughout the organizational labyrinth of its production. The hard work and continuous suggestions of all of these colleagues were crucial in allowing the idea of a comprehensive handbook on therapeutic antibodies to finally become a reality.

Braunschweig, December 2006   Stefan Dübel

Handbook of Therapeutic Antibodies. Edited by Stefan Dübel
Copyright © 2007 WILEY-VCH Verlag GmbH & Co. KGaA, Weinheim
ISBN 978-3-527-31453-9

Foreword

The most characterized class of proteins are the antibodies. After more than a century of intense analysis, antibodies continue to amaze and inspire. This *Handbook of Therapeutic Antibodies* is not just an assembly of articles but rather a state-of-the-art comprehensive compendium, which will appeal to all those interested in antibodies, whether from academia, industry, or the clinic. It is an unrivaled resource which shows how mature the antibody field has become and how precisely the antibody molecule can be manipulated and utilized.

From humble beginnings when the classic monoclonal antibody paper by Kohler and Milstein ended with the line, "such cultures could be valuable for medical and industrial use" to the current Handbook you hold in your hand, the field is still in its relative infancy. As information obtained from clinical studies becomes better understood then further applications will become more streamlined and predictable. This Handbook will go a long way to achieving that goal. With the application of reproducible recombinant DNA methods the antibody molecule has become as plastic and varied as provided by nature. This then takes the focus away from the antibody, which can be easily manipulated, to what the antibody recognizes. Since any type, style, shape, affinity, and form of antibody can be generated, then what the antibody recognizes now becomes important.

All antibodies have one focus, namely, its antigen or more precisely, its epitope. In the realm of antibody applications antigen means "target." The generation of any sort of antibody and/or fragment is now a relatively simple procedure so the focus of this work has shifted to the target, and rightly so. Once a target has been identifed then any type of antibody can be generated to that molecule. Many of the currently US Food and Drug Administration (FDA) approved antibodies were obtained in this manner. If the target is unknown then the focus is on the specificity of the antibody and ultimately the antigen it recognizes.

As the field continues to mature the applications of antibodies will essentially mimic as much of the natural human immune response as possible. In this respect immunotherapy may become immunomanipulation, where the immune system is being manipulated by antibodies. With the success of antibody monotherapy the next phase of clinical applications is the use of antibodies with standard chemotherapy, and preliminary studies suggest the combination of these two modalities is showing a benefit to the patient. When enough antibodies become available then cocktails of antibodies will be formulated for medical use. Since the natural antibody response is an oligoclonal response then cocktails of antibodies can be created by use of various *in vitro* methods to duplicate this in a therapeutic setting. In essence, this will be oligotherapy with a few antibodies. After all, this is what nature does and duplicating this natural immune response may be effective immunotherapy.

And all of this brings us back full circle to where it all starts and ends, the antibody molecule. No matter what version, isotype, form, or combination used the antibody molecule must first be made and shown to be biologically active. Currently, many of the steps and procedures to generate antibodies can be obtained in kit form and therefore are highly reproducible, making the creation of antibodies a straightfoward process. Once the antibody molecule has been generated it must be produced in large scale for clinical and industrial applications. More often than not this means inserting the antibody genes into an expression system compatible with the end use of the antibody (or fragment). Since many of the steps in generating clinically useful antibodies are labor intensive and costly, care must be used to select antibodies with the specificity and activity of interest before they are mass produced. For commercial applications the FDA will be involved so their guidelines must be followed.

Stating the obvious, it would have been nice to have this Handbook series in the late 1970s when I entered the antibody field. It certainly would have made the work a lot easier! And here it is, about 30 years later, and the generation of antibodies has become "handbook easy." In this respect I am envious of those starting out in this field. The recipies are now readily available so the real challenge now is not in making antibodies but rather in the applications of antibodies. It is hoped that this Handbook will provide a bright beacon where others may easily follow and generate antibodies which will improve our health. The immune system works and works well; those using this Handbook will continue to amaze and inspire.

Mark Glassy
Chairman & Professor, The Rajko Medenica Research Foundation, San Diego, CA, USA
Chief Executive Officer, Shantha West, Inc., San Diego, CA, USA
December 2006

List of Authors

James Allen
University College London
Department of Biochemistry and
    Molecular Biology
Gower Street
Darwin Building
London WC1E 6BT
UK

Christian Antoni
Schering-Plough
Clinical Research Allergy/
    Respiratory/Immunology
2015 Galloping Hill Rd
Kenilworth, NJ 07033
USA

Rosalin Arends
Pfizer Inc.
MS 8220-3323
Eastern Point Road
Groten, CT 06339
USA

Michaela A.E. Arndt
Department of Medical Oncology
    and Cancer Research
University of Essen
Hufelandstr. 55
45122 Essen
Germany

Kenneth D. Bagshawe
Department of Oncology
Charing Cross Campus
Imperial College London
Fulham Palace Road
London W6 8RF
UK

Harald Becker
Wetzbach 26 D
64673 Zwingenberg
Germany

Klaus Bergemann
Boehringer Ingelheim Pharma GmbH
    & Co. KG
BioPharmaceuticals
Birkendorfer Str. 65
88397 Biberach a.d. Riss
Germany

Thomas Beyer
University Schleswig-Holstein
Campus Kiel
Division of Nephrology
Schittenhelmstr. 12
24105 Kiel
Germany

Michael Braunagel
Affitech AS
Gaustadalléen 21
0349 Oslo
Norway

Marianne Brüggemann
The Babraham Institute
Protein Technologies Laboratory
Babraham
Cambridge CB22 3AT
UK

Kerry A. Chester
CR UK Targeting & Imaging
    Group
Department of Oncology
Hampstead Campus
UCL, Rowland Hill Street
London NW3 2PF
UK

Daan J.A. Crommelin
Utrecht University
Utrecht Institute for
    Pharmaceutical Sciences (UIPS)
Sorbonnelaan 16
3584 CA Utrecht
The Netherlands

Rathin C. Das
Affitech USA, Inc.
1945 Arsol Grande
Walnut Creek, CA 94595
USA

Jason L.J. Dearling
Royal Free and University College
    Medical School
University College London
Cancer Research UK Targeting &
    Imaging Group
Department of Oncology
Rowland Hill Street
Hampstead Campus
London NW3 2PF
UK

Michael Dechant
University Schleswig-Holstein
Campus Kiel
Division of Nephrology
Schittenhelmstr. 12
24105 Kiel
Germany

Peter Markus Deckert
Charité Universitätsmedizin Berlin
Medical Clinic III – Haematology,
    Oncology and Transfusion Medicine
Campus Benjamin Franklin
Hindenburgdamm 30
12200 Berlin
Germany

Eduardo Díaz-Rubio
Hospital Clínico San Carlos
Medical Oncology Department
28040 Madrid
Spain

Dimiter S. Dimitrov
Protein Interactions Group
Center for Cancer Research
    Nanobiology Program
CCR, NCI-Frederick, NIH
Frederick, MD 21702
USA

Stefan Dübel
Technical University of Braunschweig
Institute of Biochemistry and
    Biotechnology
Spielmannstr. 7
38106 Braunschweig
Germany

Christian Eckermann
Boehringer Ingelheim Pharma GmbH
    & Co. KG
BioPharmaceuticals
Birkendorfer Str. 65
88397 Biberach a.d. Riss
Germany

Paul Ellis
Department of Medical Oncology
Guy's Hospital
Thomas Guy House
St. Thomas Street
London SE1 9RT
UK

Thomas Elter
Department of Hematology and
   Oncology
University of Cologne
Kerpener Str. 62
50937 Köln
Germany

Andreas Engert
Department of Hematology and
   Oncology
University of Cologne
Kerpener Str. 62
50937 Köln
Germany

Yang Feng
Protein Interactions Group
Center for Cancer Research
   Nanobiology Program
CCR, NCI-Frederick, NIH
Frederick, MD 21702
USA

Markus Fiedler
Scil Proteins GmbH
Affilin Discovery
Heinrich-Damerow-Str. 1
06120 Halle an der Saale
Germany

Peter Fischer
Boehringer Ingelheim Pharma
   GmbH & Co. KG
Department of R&D Licensing &
   Information Management
Birkendorfer Str. 65, K41-00-01
88397 Biberach a.d. Riss
Germany

Martin Foerster
Friedrich-Schiller-University
Department of Pneumology and
   Allergy
Medical Clinics I
Erlanger Allee 101
07740 Jena
Germany

Patrick Garidel
Boehringer Ingelheim Pharma GmbH
   & Co. KG
BioPharmaceuticals
Birkendorfer Str. 65
88397 Biberach a.d. Riss
Germany

Guiseppe Giaccone
Vrije Universiteit Medical Center
Department of Medical Oncology
De Boelelaan 1117
1081 HV Amsterdam
The Netherlands

Ralf Gold
Department of Neurology
St. Josef Hospital
Ruhr University Bochum
Gudrunstr. 56
44791 Bochum
Germany

Martin Gramatzki
University of Schleswig-Holstein
Campus Kiel
Division of Stem Cell Transplantation
   and Immunotherapy
Schittenhelmstr. 12
24105 Kiel
Germany

Stefanos Grammatikos
Boehringer Ingelheim Pharma
  GmbH & Co. KG
BioPharmaceuticals
Birkendorfer Str. 65
88397 Biberach a.d. Riss
Germany

Larry Green
Abgenix, Inc.
6701 Kaiser Drive
Fremont, CA 94555
USA

Michael Hallek
Department of Hematology and
  Oncology
University of Cologne
Kerpener Str. 62
50937 Köln
Germany

Melanie Hartsough
Center for Drug Evaluation and
  Research
Food and Drug Administration
Division of Biological Oncology
  Products
10903 New Hampshire Ave.
Silver Spring, MD 20993
USA

Mingyue He
The Babraham Institute
Technology Research Group
Cambridge CB2 4AT
UK

Martin H. Holtmann
Johannes-Gutenberg-University
1st Department of Medicine
Rangenbeckstr. 1
55131 Mainz
Germany

Alexandra Huhalov
Royal Free and University College
  Medical School
University College London
Cancer Research UK Targeting &
  Imaging Group
Department of Oncology
Rowland Hill Street
Hampstead Campus
London NW3 2PF
UK

Michael Hust
Technical University of Braunschweig
Institute of Biochemistry and
  Biotechnology
Spielmannstr. 7
38106 Braunschweig
Germany

Alexander Jacobi
Boehringer Ingelheim Pharma GmbH
  & Co. KG
BioPharmaceuticals
Birkendorfer Str. 65
88397 Biberach a. d. Riss
Germany

Sigbert Jahn
Serono GmbH
Freisinger Str. 5
85716 Unterschleissheim
Germany

Wim Jiskoot
Gorlaeus Laboratories
Leiden/Amsterdam Center for Drug
  Research (LACDR)
Division of Drug Delivery Technology
P.O. Box 9502
2300 RA Leiden
The Netherlands

Thomas Jostock
Novartis Pharma AG
Biotechnology Development
Cell and Process R & D
CH-4002 Basel
Switzerland

Joachim R. Kalden
University of Erlangen-Nürnberg
Medical Clinic III
Rheumatology, Immunology &
 Oncology
Krankenhausstrasse 12
91052 Erlangen
Germany

Hitto Kaufmann
Boehringer Ingelheim Pharma
 GmbH & Co. KG
BioPharmaceuticals
Birkendorfer Str. 65
88397 Biberach a.d. Riss
Germany

Ralph Kempken
Boehringer Ingelheim Pharma
 GmbH & Co. KG
BioPharmaceuticals
Birkendorfer Str. 65
88397 Biberach a.d. Riss
Germany

Scott Klakamp
AstraZeneca Pharmaceuticals LP
24500 Clawiter Road
Hayward, CA 94545
USA

Roland E. Kontermann
University Stuttgart
Institute for Cell Biology and
 Immunology
Allmandring 31
70569 Stuttgart
Germany

Zoltán Konthur
Max Planck Institute for Molecular
 Genetics
Department of Vertebrate Genomics
Ihnestr. 63–73
14195 Berlin
Germany

Jürgen Krauss
Department of Medical Oncology and
 Cancer Research
University of Essen
Hufelandstr. 55
45122 Essen
Germany

Claus Kroegel
Friedrich-Schiller-University
Department of Pneumology and
 Allergy
Medical Clinics I
Erlanger Allee 101
07740 Jena
Germany

Hartmut Kupper
Abbott GmbH & Co. KG
Knollstr. 50
67061 Ludwigshafen
Germany

Meina Liang
AstraZeneca Pharmaceuticals LP
24500 Clawiter Road
Hayward, CA 94545
USA

Charito Love
Long Island Jewish Medical Center
Division of Nuclear Medicine
New Hyde Park, New York
USA

Andrew C.R. Martin
University College London
Department of Biochemistry and
 Molecular Biology
Darwin Building
Gower Street
London WC1E 6BT
UK

Christian Menzel
Technical University of
 Braunschweig
Institute of Biochemistry and
 Biotechnology
Spielmannstr. 7
38106 Braunschweig
Germany

Gerhard Moldenhauer
German Cancer Research Center
Department of Molecular
 Immunology
Tumor Immunology Program
Im Neuenheimer Feld 280
69120 Heidelberg
Germany

Dafne Müller
University Stuttgart
Institute for Cell Biology and
 Immunology
Allmandring 31
70569 Stuttgart
Germany

Markus F. Neurath
Johannes-Gutenberg-University
1st Department of Medicine
Langenbeckstr. 1
55131 Mainz
Germany

Dianne L. Newton
SAIC Frederick, Inc.
Developmental Therapeutics Program
National Cancer Institute at Frederick
Frederick, MD 21702
USA

Michael J. Osborn
The Babraham Institute
Protein Technologies Laboratory
Babraham
Cambridge CB22 3AT
UK

Christopher J. Palestro
Albert Einstein College of Medicine
Bronx, New York
USA
and:
Long Island Jewish Medical Center
New Hyde Park, New York
USA

Gabriele Pecher
Humboldt University Berlin
Medical Clinic for Oncology and
 Hematology
Charité Campus Mitte
Charitéplatz 1
10117 Berlin
Germany

Matthias Peipp
University Schleswig-Holstein
Campus Kiel
Division of Stem Cell Transplantation
 and Immunotherapy
Schittenhelmstr. 12
24105 Kiel
Germany

Edith A. Perez
Mayo Clinic Jacksonville
Division of Hematology/
 Oncology
4500 San Pablo Road
Jacksonville, FL 32224
USA

Sandra Pisch-Heberle
Boehringer Ingelheim Pharma
 GmbH & Co. KG
BioPharmaceuticals
Birkendorfer Str. 65
88397 Biberach a.d. Riss
Germany

Ponraj Prabakaran
Protein Interactions Group
Center for Cancer Research
 Nanobiology Program
CCR, NCI-Frederick
Frederick, MD 21702
USA

Gabriele Reich
Ruprecht-Karls-University
Institute of Pharmacy and
 Molecular Biotechnology
 (IPMB)
Department of Pharmaceutical
 Technology and Pharmacology
Im Neuenheimer Feld 366
69120 Heidelberg
Germany

Josephine N. Rini
Albert Einstein College of
 Medicine
Bronx, New York
USA
And:
Long Island Jewish Medical
 Center
Division of Nuclear Medicine
New Hyde Park, New York
USA

Lorin Roskos
AstraZeneca Pharmaceuticals LP
24500 Clawiter Road
Hayward, CA 94545
USA

Susanna M. Rybak
Bionamomics LLC
411 Walnut Street, #3036
Green Cove Springs, FL 32043
USA

José W. Saldanha
National Institute for Medical
 Research
Division of Mathematical Biology
The Ridgeway
Mill Hill
London NW7 1AA
UK

Jochen Salfeld
Abbott Bioresearch Center
100 Research Drive
Worcester, MA 01605
USA

Huub Schellekens
Utrecht University
Department of Pharmaceutics Sciences
Department of Innovation Studies
Sorbonnelaan 16
3584 CA Utrecht
The Netherlands

Sebastian Schimrigk
Ruhr University Bochum
St. Josef Hospital
Department of Neurology
Gudrunstr. 56
44791 Bochum
Germany

Thomas Schirrmann
Technical University
  Braunschweig
Institute of Biochemistry and
  Biotechnology
Department of Biotechnology
Spielmannstr. 7
38106 Braunschweig
Germany

Norbert Schleucher
Hematolgy and Medical Oncology
Marienkrankenhaus Hamburg
Alfredstr. 9
22087 Hamburg
Germany

Alexander C. Schmidt
National Institutes of Health
National Institute of Allergy and
  Infectious Diseases
Laboratory of Infectious Diseases
50 South Drive, Room 6130
Bethesda, MD 20892
USA
And:
Charité Medical Center at Free
  University and Humboldt
  University Berlin
Center for Perinatal Medicine
  and Pediatrics
Schumannstr 20/21
13353 Berlin
Germany

Karlheinz Schmitt-Rau
Serono GmbH
Freisinger Str. 5
85716 Unterschleissheim
Germany

Marjorie A. Shapiro
Center for Drugs Evaluation and
  Research
Food and Drug Administration
Division of Monoclonal Antibodies
HFD-123
5600 Fishers Lane
Rockville, MD 20872
USA

Surinder K Sharma
CR UK Targeting & Imaging Group
Department of Oncology
Hampstead Campus
UCL, Rowland Hill Street
London NW3 2PF
UK

Igor A. Sidorov
Center for Cancer Research
  Nanobiology Program
CCR, NCI-Frederick, NIH
Protein Interactions Group
P.O. Box B, Miller Drive
Frederick, MD 21702-1201
USA

Arne Skerra
Technical University Munich
Chair for Biological Chemistry
An der Saatzucht 5
85350 Freising-Weihenstephan
Germany

Jennifer A. Smith
The Babraham Institute
Protein Technologies Laboratory
Babraham
Cambridge CB22 3AT
UK

Patrick G. Swann
Center for Drug Evaluation and
  Research
Food and Drug Administration
Division of Monoclonal
  Antibodies
HFD-123
5600 Fishers Lane
Rockville, MD 20872
USA

Michael J. Taussig
The Babraham Institute
Technology Research Group
Cambridge CB2 4AT
UK

Lars Toleikis
RZPD, Deutsches
  Ressourcenzentrum für
  Genomforschung GmbH
Im Neuenheimer Feld 515
69120 Heidelberg
Germany

Daniel Tracey
Abbott Bioresearch Center
100 Research Drive
Worcester, MA 01605
USA

Martina M. Uttenreuther-Fischer
Boehringer Ingelheim Pharma
  GmbH & Co. KG
Department of Medicine
Clinical Research Oncology
Birkendorfer Str. 65
88397 Biberach a.d. Riss

Thomas Valerius
University Schleswig-Holstein
Campus Kiel
Division of Nephrology
Schittenhelmstr. 12
24105 Kiel
Germany

Udo Vanhoefer
Department of Medicine
Hematology and Medical Oncology,
  Gastroenterology and Infectious
  Diseases
Marienkrankenhaus Hamburg
Alfredstr. 9
22087 Hamburg
Germany

Michael Wenger
International Medical Leader
F. Hoffmann-La Roche Ltd.
Bldg. 74/4W
CH-4070 Basel
Switzerland

Maria Wiekowski
Schering-Plough
Clinical Research Allergy/Respiratory/
  Immunology
2015 Galloping Hill Road
Kenilworth, NJ 07033
USA

Xiangang Zou
The Babraham Institute
Protein Technologies Laboratory
Cambridge CB22 3AT
UK

Part III
Beyond IgG – Modified Antibodies

1
Immunoscintigraphy and Radioimmunotherapy

Jason L. J. Dearling and Alexandra Huhalov

1.1
Introduction

This chapter aims to provide an introduction to ways in which antibodies can be modified to optimize their role in the detection and therapy of solid tumors. These modifications include: alterations to the protein structure in order to improve antigen binding and pharmacokinetics; addition of a radionuclide to form a radioimmunoconjugate (RIC); and uses of the resultant RIC. The review is focused on solid tumors which continue to be a challenge to successful therapy. Antibody targeting of hematological cancers is discussed later in this volume (see Chapter 12; for review also see Goldenberg 2001).

Use of the antibody as a vehicle for radionuclide delivery means that the effector function is no longer solely attributed to the interaction of the antibody with the immune system but also the emissions of the radionuclide as it physically decays. The selection of the radionuclide will be influenced by the characteristics of the tumor, the antibody, and also whether the aim is to detect using medical imaging (radioimmunoscintigraphy, RIS) or destroy (radioimmunotherapy, RIT). RIS uses antibody-delivered radionuclides with emissions detectable outside the body combined with medical imaging techniques (principally single photon emission computed tomography (SPECT), or positron emission tomography (PET)) to detect solid tumors, either for diagnosis or to monitor therapy. RIT is based on similar basic principles but has different requirements. The two are closely related and often used in concert, as in prescouting dosimetry in which RIS is used to guide RIT. Localization of the labeled antibody to the tumor is a requirement for the success of RIS and RIT, though it is also a challenge for both techniques due to solid tumor structure. This chapter provides an overview of the progress that has been made in understanding the challenges of RIS and RIT and the optimal tools to improve clinical practise.

Handbook of Therapeutic Antibodies. Edited by Stefan Dübel
Copyright © 2007 WILEY-VCH Verlag GmbH & Co. KGaA, Weinheim
ISBN 978-3-527-31453-9

1.2
Solid Tumors as Targets for Antibody-Based Therapeutics

For both RIS and RIT sufficient radionuclide must be delivered to the tumor but there are many obstacles on the way. Once the antibody has been labeled with an appropriate radionuclide the RIC is introduced into the blood, either as a bolus or infusion. It distributes throughout the body and is retained in the tumor. The vasculature of the tumor is the first obstacle for effective delivery of the RIC as it is inhibited by an inadequate tumor blood supply. A common characteristic of tumor growth is the uncontrolled growth rate of cancerous cells, a consequence of the loss of control of cell behavior. Tumor blood vessels do not keep pace with this increased rate of growth, leading to disorganized vasculature composed of poorly formed vessels with increased fenestrations and poor provision of nutrients to the tumor. Furthermore, tumor blood vessels exhibit a number of unusual structures, ending in blind ends, converging and merging, resulting in vascular arterial shunts, and structurally weak tumor blood vessels can collapse and shut down, temporarily or permanently (Jain 1988; Kallinowski 1996). Drug delivery is seriously affected by these factors. Conversely, the leaky nature of tumor blood vessels aid in tumor uptake, enabling the antibody to leave the vessel lumen and extravasate across the endothelium.

Antibody movement is principally driven by diffusion. Penetration into the tumor mass is limited by the relatively large size of whole antibodies (of the order of 150 kDa) (Jain 1999). Additionally, due to the absence or poor formation of the tumor lymphatic system, interstitial pressure, which is lower at the tumor periphery but elevated at its core, further reduces RIC penetration. The site and size of the tumor can also influence the use and success of RIT. Bulky tumors are less likely to respond to therapy, at least partially due to poor uptake. The size of the tumor is inversely related to the uptake of RIC at the tumor site (Hagan et al. 1986; Pedley et al. 1987; Behr et al. 1997a).

Micrometastases present an ideal target for RIT – they are accessible, a relatively homogeneous dose distribution is achievable throughout the tumor mass, and they might not be detected by current medical imaging techniques, excluding them from external beam radiotherapy (EBRT) (Vogel et al. 1996; Dearling et al. 2005).

1.3
Antigen

The type and pattern of antigen expression in tumor and normal tissue strongly influences the therapy regime and its outcome. The antigen is the feature of the tumor cell, typically a protein expressed at elevated levels on the cancer cell surface, that discriminates it from normal, noncancerous tissues. The antibody interacts with the antigen and is retained at the tumor site. Therefore the antigen must be expressed predominantly on tumor cells and at a level sufficiently high to retain enough radionuclide for detection or therapy and to achieve a good differential between radionuclide at the tumor site and normal tissues. Antibody–antigen complexes formed in the blood are cleared via the liver, so elevated blood levels of shed antigen reduce the maximum amount of radionuclide that can accumulate in the tumor, as well as potentially leading to liver radiotoxicity. Expression of antigen at around 10^4–10^5 molecules per cell is desirable for successful targeting (Goldenberg et al. 1989; Sung et al. 1992). Expression of the antigen homogeneously throughout the tumor aids successful therapy, though the nuclides commonly used for RIT have long emission ranges, accounting for some degree of heterogeneity. Use of interferon has been reported to increase antigen expression, resulting in an increase in RIC tumor uptake (Meredith et al. 1996; Pallela et al. 2000). A cocktail of antibodies binding to different antigens might overcome heterogeneous expression throughout the tumor (Meredith et al. 1996).

From this initial discussion of the tumor as a target it becomes clear that the nature of solid tumor dictates some of the characteristics of successful RICs. Further requirements are made when we consider the clearance of proteins from the body and their use in RIS and RIT. The identification of these characteristics and how they may be achieved is the focus of the following section.

1.4
Antibodies as Vehicles for Radionuclide Delivery

The target of both RIS and RIT is the tumor but there are differences in the requirements of the RIC for optimal results. For RIS the molecule should localize rapidly at a high level to the tumor while clearing from the normal tissues in order to aid discrimination of the disease site. This occurs over a short time-frame, in the order of hours or days. For RIT the time-scale is different (therapy effectively taking place over around a week), and prolonged retention in the blood aids greater uptake in the tumor, resulting in a greater probability of tumor cell death. However, radiotoxicity to early responding, radiosensitive tissues, usually the bone marrow, limits administered dose. Antibody engineering has provided a number of ways in which RIS and RIT are being optimized (Table 1.1).

The development of antibody-based therapeutics was restrained by complex and laborious production and purification techniques until modern methods of genetic manipulation and protein engineering could be brought to bear on the task. With the development of these approaches antibodies could be engineered to have the characteristics of choice. One of the early modifications to the antibody structure was that of size, which can limit penetration into the tumor, but which also influences the rate of clearance of the RIC from the blood. Affinity and avidity of the antibody, properties that can greatly influence its pharmacokinetics and pharmacodynamics, have been optimized for improved tumor retention. Reducing the immunogenicity of the antibody has also been addressed. We will now

Table 1.1 Summary of optimization of protein constructs to act as radionuclide vehicles localizing to tumor deposits for detection (RIS) or therapy (RIT).

Modifiable property	
Size	Large size (e.g. whole IgG, 150 kDa) = longer circulation time, good tumor localization, increased normal tissue toxicity
	Small size (e.g. scFv, 27 kDa) = rapid renal clearance (<60 kDa) and poor tumor uptake
Affinity	Increased affinity improves tumor uptake, reaches a point where other factors become limiting
Avidity	Increased antigen-binding sites improve tumour uptake.
Immunogenicity	Patient immune response accelerates clearance leading to use of humanized or human-derived proteins

discuss these facets of antibody engineering for RIS and RIT of solid tumors in more detail.

1.4.1
Antibody Production

Historically material for antibody-based medical techniques was derived from immunized animals. This source carries several drawbacks, as a large number of animals were involved, producing a diverse range of polyclonal antibodies which required laborious purification. The development of hybridoma technology (Köhler and Milstein 1975) revolutionized the field of antibody research. In this technique, hybridization of myeloma cells with antibody-secreting cells from an immunized animal enables production of monoclonal antibodies raised against the same epitope, or determinant region on the antigen, for research and use. Mouse antibodies have been most commonly used in research, though rat antibodies have higher stability and their producer cells have more favorable yield (Clark et al. 1983).

Further developments in molecular biology have facilitated advances in antibody engineering. Large libraries of antibody sequences have been created, which allows for the selection of genes that encode for an antibody from a diverse repertoire. However, this requires that the protein, which binds to the antigen, and the gene encoding it, which gives the instructions for its structure, are associated. Display technologies in which the appropriate protein and its genetic code are physically linked have been developed to accomplish this. The display technologies employed include use of phage systems, display of the polypeptide on ribosomes, and cell surface systems using bacterial or yeast cells (McCafferty et al. 1990; Maynard and Georgiou 2000).

The field of RIS/RIT is not large, though growing, and consequently production must be efficient and cost-effective. Expression systems to produce these constructs in large quantities have contributed to the rapid developments in this field.

Almost any cell that can produce a protein can produce a usable antibody construct. Consequently both eukaryotic and prokaryotic cells have been used in their large-scale production. Traditionally, mammalian cell lines (e.g. recombinant CHO cell lines) have been employed for the expression of antibodies, but more recently insect cells and plants have been described (Yazaki et al. 2001a; de Graaf et al. 2002; Houdebine 2002; Stoger et al. 2002). Ideally, large amounts of product produced at low cost in a short time-frame are required, leading to the predominace of bacterial and yeast expression systems.

The IgG antibody class is the most widely used in this context. Its long residence time in the circulation is both beneficial and detrimental. While this allows good tumor localization, prolonged radionuclide exposure can result in radiotoxicity and increases immunogenicity. In order to overcome these issues properties including size, affinity, avidity, and immunogenicity have been engineered to improve tumor targeting.

1.4.2
Size

Reduction of the size of the antibody has implications for its circulation half-life, route of clearance from the blood, absolute tumor uptake and tumor penetration.

In order to reduce antibody size, enzymatic digestion can be used. Removal of the Fc region by papain gives two Fab regions, consisting of V_L-C_L and V_H-$C_H 1$ chains linked by disulfide bonds (~55 kDa). Fab fragments exhibit rapid clearance and suboptimal tumor uptake due to their small size and monovalency (Buchegger et al. 1990, 1996; Casey et al. 1999, 2002; Behr et al. 2000). Digestion using pepsin gives an F(ab)$_2$ fragment of around 100 kDa. These fragments have better tumor penetration than whole IgG due to their smaller size. In a clinical comparison (Lane et al. 1994) F(ab)$_2$ fragments localized more rapidly than the parent IgG (8.2% injected activity per kg vs. 4.4% injected activity per kg for the parent IgG, both at 4.25 h, $P < 0.05$) (see also Buchegger et al. 1996). However, the circulation residence times of F(ab)$_2$ fragments are reduced, potentially reducing maximal tumor uptake. This may be a result of the removal of the Fc region. For IgG, the Fc region binds to the FcRn receptor in liver cells, leading to its recycling back into serum. This mechanism contributes to the maintenance of protein levels in the blood.

Protein engineering has employed recombinant technologies to reduce the size of constructs down to the smallest fragment recognizing antigen. Figure 1.1 shows a range of antibody formats. Longer circulation time enabling tumor uptake means that larger antibody molecules will generally have higher absolute localization in the tumor, which is beneficial for RIT, but poorer penetration into the tissue. Longer exposure of the bone marrow to the radionuclide will also result in radiotoxicity, limiting administered dose. While they penetrate more successfully into the tumor, smaller molecules have a shorter circulation time, clearing rapidly from the blood. Thus they achieve higher tumor-to-normal tissue ratios,

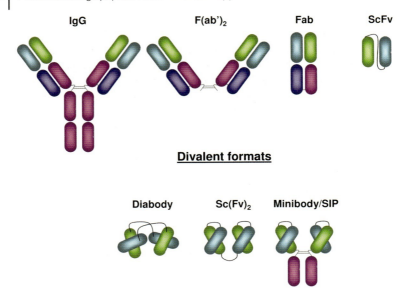

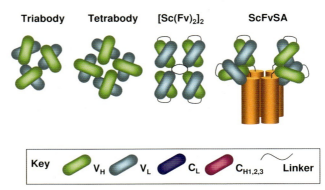

Fig. 1.1 Antibody construct formats. Developments made to optimize performance include varying size and avidity to modify blood clearance and improve tumor localization.

which is advantageous for discrimination of the tumor site from normal tissues, but at the cost of absolute uptake, challenging the sensitivity of RIS and the efficacy of RIT.

One of the smallest antigen-binding constructs used is the single-chain Fv (scFv). It also forms the building blocks of many antibody constructs. Consisting of a V_H and a V_L domain, the scFv has a M_w of ~27 kDa. The V_H and V_L domains, which otherwise dissociate and aggregate, bind either because disulfide bonds have been engineered into the molecule, or because they are tethered by a flexible linker (Huston et al. 1988; Bird et al. 1988) which can be either chemically introduced or engineered into the polypeptide chain. scFvs clear rapidly from the blood, are monovalent and typically achieve low tumor accumulation, limiting their overall performance. While favorable tumor-to-normal tissue (T:N) ratios are obtained, tumor uptake is low (e.g. peaking at 3–5% injected dose per g at 1–4 h) in relation to whole antibodies. Proteins in the blood of <60 kDa are cleared by the kidney, and scFvs show high kidney uptake (Colcher et al. 1990; Milenic et al. 1991; Begent et al. 1996).

scFvs are retained in the glomerulum due to its negatively charged basement membrane. Tubular reabsorption and lysosomal degradation (Tsai et al. 2001) can lead to retention of the radionuclide, increasing radiotoxicity, particularly in the case of radiometals. This effect can be reduced by administration of L-lysine, reducing kidney accumulation of radionuclide (Behr et al. 1995, 1997b).

Increasing the size of the molecule by addition of polyethylene glycol (PEGylation) can decrease blood pool clearance rates, but fails to add otherwise to its performance (Pedley et al. 1994; Francis et al. 1996) and can interfere with antigen binding (Kubetzko et al. 2005). However multimeric antibodies based on scFvs have been produced that have proved advantageous: diabodies, triabodies, and tetrabodies have improved tumor-targeting capabilities. A similar use of scFvs has been investigated in their fusion to larger molecules, including enzymes, cytokines, and proteins with favorable blood maintenance such as human serum albumin (Michael et al. 1996; Bhatia et al. 2000; Cooke et al. 2002; Huhalov and Chester 2004) or into multivalent formats (see Section 1.4.4).

1.4.3
Affinity

The affinity of the antibody for its target antigen, or its affinity, is critical for tumor localization. The antibody/antigen-binding event involves the formation of around 20 noncovalent interactions, in the form of hydrogen bonds and salt bridges between amino acids on the topological surface of the antigen and the antibody-binding site. The kinetics are first order, and stronger or more numerous interactions move the equilibrium to the right, favoring formation of the antibody–antigen complex, reflecting the greater affinity of the two molecules. Often the affinity is quoted in the form of K_D, the dissociation constant, which is equal to k_{off}/k_{on}, where k_{on} is the association of the antibody onto the antigen and k_{off} is the dissociation of the complex.

Increases in affinity result from changes to the binding site of the antibody, specifically of the amino acids that form the noncovalent bonds between the two molecules. These changes might be randomly introduced into a library using error-prone polymerase chain reaction (PCR), complementarity determining region (CDR) walking or chain shuffling (Barbas and Burton 1996; Lantto et al. 2002) while more directed changes require site-directed mutagenesis. Structural knowledge of the antibody, either directly from crystal structures or inferred through comparison with similar molecules, will reveal those amino acids which are involved in binding and could be changed to increase the strength of binding.

Efforts have been made to continually increase the affinity of antibodies for their antigen, in the expectation that this will result in improved targeting. Preclinical *in vivo* studies have provided conflicting evidence of this. Certainly, high affinity is required, and increasing affinity does have a beneficial effect on antibody performance (Colcher et al. 1988; Schlom et al. 1992; Adams et al. 1998) although there are also reports of improvements having no effect (Behr et al. 1997a) on tumor targeting. It is probably the case that there is a threshold above which increases in affinity give no further advantage as other factors become limiting (Sung et al. 1992). Very high-affinity antibodies might pose problems. If the target antigen is not expressed solely on tumor cells then it might be retained in normal tissues, as was encountered in the case of anti-tenascin antibodies, the antigen also being expressed in the liver and spleen. Furthermore, very high-affinity antibodies might hinder tumor penetration of the antibody. Antibody binding to initially available antigen surrounding the blood vessel with very high affinity might not move on to penetrate the tumor. Antibody arriving at the tumor later will not be able to bind to antigen and might rediffuse back into the bloodstream. The "binding site barrier" (van Osdol et al. 1991; Saga et al. 1995) thus could decrease tumor uptake, though the effect is particularly significant when large amounts of protein are administered.

1.4.4
Avidity

The valency of binding of an antibody to its antigen is referred to as the avidity. Increasing the avidity of an antibody construct results in an overall increase in its binding, and its functional affinity (Karush 1970).

Autoradiographic study of radionuclide localiation in preclinical tumors and subsequent analysis of this data using mathematical models has emphasized the importance of valency in addition to affinity (Flynn et al. 2002). Multivalent molecules are retained longer than monovalent ones in radiosensitive (normoxic) regions of the tumor and are more likely to be successful vehicles for RIT.

Monovalent scFv and Fab antibody fragments have been engineered into multimeric constructs in order to improve their tumor retention. Chemical and genetic crosslinkages have been used. Incorporation of a free cysteine residue into the sequence results in crosslinking of two scFv molecules, and this has been shown to improve tumor uptake (Adams et al. 1993, 1995) as does linking two scFv molecules into one polyeptide (Goel et al. 2001a).

Another approach to increase avidity is to take advantage of the dimeric properties of the antibody chains. The V_H chain of one scFv will spontaneously associate with the V_L chain of another. The probability of this occurring can be increased by using a shorter polypeptide linker (e.g. shortening from 15 to 5 amino acids) to tether the V_H and V_L domain of the individual scFv, resulting in inter- rather than intramolecular association of V_H and V_L domains. These divalent formats, also known as diabodies, possess improved pharmacokinetics compared with monovalent molecules (Nielsen et al. 2000). Divalent scFv$_2$ molecules have been demonstrated to have better tumor uptake than scFvs (tumor localization for the scFv$_2$ 3.57% injected dose per g, for scFv 1.25% injected dose per g, at 24 h) (Adams et al. 2006). Their use as RIS agents, when labeled with positron-emitting isotopes, has been investigated preclinically (Robinson et al. 2005). Clinical use of an ^{125}I-labeled diabody raised against the ED-B domain of fibronectin demonstrated tumor localization (Santimaria et al. 2003).

The tandem scFv, composed of two scFvs linked as a single polypeptide chain, again showed improved T:N ratios. When these molecules (sc(Fv)$_2$) spontaneously dimerized to form tetramers ([sc(Fv)$_2$]$_2$) absolute maximal tumor uptake was increased. Tumor uptake (% injected dose per g) for the tetravalent [sc(Fv)$_2$]$_2$ was 21.3 ± 1.3 compared with 9.8 ± 1.3 for the divalent sc(Fv)$_2$ and 17.3 ± 1.1 for the IgG (Goel et al. 2000). Similarly, in a comparison of the ^{177}Lu-labeled [sc(Fv)$_2$]$_2$ and its parental IgG the tumor localization at 8 h was 6.4% injected dose per g and 8.9% injected dose per g respectively (Chauhan et al. 2005). This was surprising in the light of the much more rapid blood clearance of the modified molecule. The $t_{1/2}\alpha$ was 4.40 min for the [sc(Fv)$_2$]$_2$, and 9.5 min for the whole IgG, and $t_{1/2}\beta$ were 375 min and 2193 min, respectively, conferring a significant advantage on the tetravalent form. Kidney accumulation was reduced with use of L-lysine (for further discussion of multivalent molecules see Todorovska et al. 2001; Wittel et al. 2004).

A different approach to forming multimers is to link antibody fragments to molecules that form homomultimers. For example, the minibody is an scFv linked to a human IgG1 C$_H$3 domain, the minibody molecule being bivalent because the C$_H$3 domains dimerize (Hu et al. 1996; Wu et al. 2000). The resultant molecule is similar in size to F(ab')$_2$ fragments (minibody 80 kDa versus F(ab')$_2$ 100 kDa). These molecules gave higher uptake than was achieved with diabody forms of the same scFv in the same xenograft system (Hu et al. 1996; Wu et al. 1996; Yazaki et al. 2001b). The maximal tumor uptake of the minibody at 6 h was 32.9 ± 11.18% injected dose per g while for the diabody it was 10.38 ± 0.81% injected dose per g at 1 h. Relevant for imaging purposes are the greatest tumor-to-blood ratios of 64.89:1 at 48 h for the minibody and 23.23:1 at 24 h for the diabody.

The Small ImmunoProtein (SIP) is a comparable molecule to the minibody, composed of the C$_H$4 region of a human IgE-S2, resulting in a dimeric scFv. This molecule improved on the performance of the diabody form, with maximal tumor uptake at 6 h of 6.14 ± 2.23% injected dose per g compared with 2.47 ± 0.65% injected dose per g, respectively (Borsi et al. 2002). Increasing the size of the (scFv-C$_H$3)$_2$ molecule further, from 80 to 105 kDa, to create the (scFv-C$_H$2-C$_H$3)$_2$ molecule, improved tumor targeting (5.7 ± 0.1% injected dose per g versus 12.2 ± 2.4% injected dose per g, respectively) while reducing kidney uptake (34.0 ± 4.0% injected dose per g versus 13.1 ± 1.5% injected dose per g) (Olafsen et al. 2005).

Other examples of multimers include p53 and streptavidin tetramerization. Tumor uptake of the p53 tetramers was low (4.32 ± 1.94% injected dose per g), and was attributed to dissociation in serum (Willuda et al. 2001). Streptavidin has proved a more successful format (Dubel et al. 1995; Zhang et al. 2003) in terms of tumor targeting, and can also be used in combination with biotin in Pretargeted RIT (PRIT). In this technique a nonradiolabeled antibody is followed by a smaller molecule acting as the radionuclide vehicle.

1.4.5
Immunogenicity

As has been described, the origin of many antibodies under research and development is often rodent cells for genetic material and bacterial or yeast-based systems for production. Systemic administration of such products can lead to immune responses depending on the construct (for a discussion see Mirick et al. 2004). While human anti-mouse antibody (HAMA) responses are directed primarily toward the Fc regions, anti-idiotypic responses can also be encountered. The scale of the response can be increased by rapid introduction of large amounts of protein. The development of immune complexes will lead to faster clearance of RIC from the blood and complicate future treatments. Attempts to minimize or evade this detection and response altogether has led to a range of techniques. The protein may be disguised, or its origin may be changed, leading to chimeric, humanized, and fully human antibodies (Clark 2000). The majority of therapeutic antibodies used in clinical practice are humanized.

A chimeric antibody will have a mixture of sources, for example an human Fc region and the Fab fragment of a rodent antibody. However, it is still possible to elicit a response to the Fab region. Further "humanization" can be achieved through the transfer of the CDR loops, the regions of the rodent antibody bearing the amino acids that form the bonds with the antigen, onto a human antibody (Baca et al. 1997; Rader et al. 1998). In an analogous technique recombinant antibodies are "resurfaced" or "veneered" to provide fewer immunogenic sites (Pederson et al. 1994; Roguska et al. 1994) using structural information. Deimmunizaton, the removal of defined helper T cell epitopes, has also been reported. Fully human antibodies are the goal in this area. They can be produced using transgenic mice (Nagy et al. 2002; O'Connell et al. 2002), and human antibody fragments have been obtained using the phage display system in combination with fully human libaries (Knappik et al. 2000).

1.5
Radioisotope

1.5.1
Selection

Radionuclide choice is related to the aim of the procedure. Different isotopes are suited to imaging or therapy, depending on their properties (Table 1.2). The

Table 1.2 Radionuclides for radioimmunoscintigraphy and radioimmunotherapy.

Radionuclides		Emission energy (MeV)	Physical $t_{1/2}$
Imaging			
γ	^{99m}Tc	0.142	6.01 h
	^{111}In	0.173, 0.247	2.8 days
β$^+$	^{64}Cu	1.675	12.7 h
	^{72}As	1.17	1.1 days
	^{89}Zr	0.9	3.27 days
	^{124}I	1.53	4.18 days
Therapy			
Auger	^{111}In	0.86	2.8 days
	^{125}I	0.179	60.1 days
α	^{211}At	5.980	7.21 h
	^{212}Bi	6.051	1.01 h
	^{225}Ac	5.3–5.8	10 days
β$^-$	^{64}Cu	0.578	12.7 h
	^{67}Cu	0.58	2.58 days
	^{77}As	0.226	1.6 days
	^{90}Y	2.282	2.67 days
	^{131}I	0.606	8.04 days
	^{177}Lu	0.497	6.75 days
	^{186}Re	0.973	3.78 days

Radionuclides commonly used in RIS/RIT, or of potential utility, are shown. Some nuclides have both imaging and therapeutic emissions. Other elements, such as iodine, are represented in both the imaging and therapy sections, recommending them for use in predosing regimes.

emissions of the radionuclide will chiefly guide this selection, but the pharmacokinetics of the antibody construct will also have influence. The half-life of the imaging isotope should be in concordance with the *in vivo* behavior of the antibody. Sensitivity of the imaging procedure will be compromised if too few emissions are detected from the tumor site once the RIC has been allowed to clear from the blood pool and other normal tissues. An ideal range of half-life for the therapeutic isotope is around 1.5–3 times the time taken for the antibody to achieve maximal uptake in the tumor, and mathematical modeling has investigated the relative merits of varying half-lives (Flynn et al. 2002; Howell et al. 1998). This should lead to maximal absorbed dose and dose-rate to the tumor. For both RIS and RIT, administered dose is limited by radiotoxicity to the patient, though other factors, such as dose to medical workers and disposal of waste products should also be borne in mind. As well as procedural considerations practical issues influence this selection, including production, ease of use and cost.

1.5.2
Emissions

Radionuclide emissions will determine use. The two main emissions used for imaging are gamma (γ) and positron (β^+, for positron emission tomography (PET) imaging). There are three main emissions used for therapy: Auger electrons, alpha particles (α) and beta particles (β^-). These therapy emissions vary in their biological effect. For example, Auger emissions are high-energy electrons which are very toxic but with a short path length ($\sim$0.1–5 µm), and are therefore more effective when internalized into the cell (Hofer 1996). Alpha particles are helium nuclei, again with short path lengths. Their lack of availability and short half-lives have limited their application, though they have proved useful in treatment of organ surface tumors when locally injected (so that systemic introduction does not hinder delivery) (McDevitt et al. 1998; Zalutsky and Vaidyanathan 2000) and vascular targeting (Akabani et al. 2002, 2003). Beta-emitting radionuclides are the most widely employed in this area, largely due to the predominance of ^{131}I. Beta emissions have a much longer range than either Auger or alpha articles, but are much less toxic. However their range is more suited to treatment of a small tumor deposit. The problem of heterogeneity of RIC distribution within the tumor, because of both delivery and antigen expression, has been discussed earlier. Use of a radionuclide with long emissions will smooth out this heterogeneity through the cross-fire effect; the radionuclide attached by the antibody to one cell will irradiate neighboring tumor cells whether they express the antigen or not. The emission ranges of commonly used radionuclides such as ^{131}I (0.83 mm) and ^{90}Y (5.2 mm) (Simpkin and Mackie 1990) make them potentially applicable to small deposits and larger bulky tumors, respectively (O'Donoghue, et al. 1995).

1.5.3
Antibody Radiolabeling

Modification of the antibody by addition of the radionuclide may be achieved through either direct or indirect labeling. Direct labeling, mainly through redox reactions, has been described in the use of ^{99m}Tc for RIS and isotopes of iodine for both imaging and therapy, depending on the radionuclide (Greenwood and Hunter 1963). The indirect method of labeling uses a bifunctional chelator to form a bridge between the protein and the radionuclide. The number of additions to the antibody will have an optimal point: too few will limit success of the procedure through insufficient information, while too many might interfere with antigen binding and limit tumor uptake.

1.5.4
Radiobiology

Radioisotope emissions are toxic because they alter biological molecules as they lose their energy. A highly energetic particle of large mass can cause direct damage to important biological structures such as DNA, or because the ionization of water leads to short-lived free radicals, including superoxide (O_2^-), and hyperoxide ($H_2O_2^-$) which are highly reactive and cause indirect damage in the cell. As oxygen is often involved in these reactions it is a potent radiosensitizer. Indirectly caused lesions, which are the main cause of damage by the β^- emitting radionuclides, are greatly reduced under oxygen-lacking (hypoxic) conditions in the tumor, which therefore reduce the success of irradiative therapies (see Thomlinson and Gray 1955; Tannock 1972).

The success of RIT is dependent on the absorbed dose to the tumor and the dose-rate. The dose-rate is important because cells can repair damage at a certain rate, so this must be exceeded in order to have a toxic effect. The time span of RIT is not fixed as with EBRT, but instead depends on the pharmacokinetics and biological half-life of the RIC and the physical half-life of the radioisotope. Once the RIC arrives at the tumor and the rate of cell kill exceeds the rate of proliferation (i.e. there is a net decrease in cell number), therapy has started. RIC uptake at the target site will continue to increase, reach a peak, and then decrease. Once the dose-rate required to overcome cell proliferation is again reached, typically around a week after injection, the therapy has effectively ended.

1.6
Improving Radioimmunoscintigraphy and Therapy

Methods investigated for improving RIS and RIT are generally applicable to both, with some exceptions. For RIS absolute uptake in the tumor is important in order to enable identification and relative tumor localization compared with normal tissues to allow discrimination and sensitivity. For RIT maximizing absolute uptake in the tumor is important, but minimal uptake in normal tissues is desirable in order to reduce radiotoxicity.

The chief methods of modifying RIC pharmacokinetics have already been discussed. Antibody construct size, affinity, avidity, and immunogenicity can all be optimized. Here we will consider ways in which the use of the RICs may be modified in order to improve performance, concentrating on clearing antibodies, dose fractionation and pretargeted RIT.

Use of a clearing agent is applicable to both RIS and RIT. Rapid clearance from the blood decreases radiotoxicity to normal tissues, but also reduces the amount of RIC that can localize to the tumor. A secondary antibody that reacts to the first increases the rate of clearance from the blood (Begent et al. 1982; Pedley et al. 1989). The advantages of this technique include that a greater amount of RIC may be given, and therefore more will localize to the tumor, giving an increased maximal dose-rate and potentially higher overall absorbed dose to the tumor. However, the RIC is only cleared from the vascular space, and exposure of the liver and spleen to radionuclide is increased due to clearance.

Instead of giving a single injection of RIC the dose could be fractionated, that is split into a number of smaller injections over a period of time. This allows for the RIC to clear from normal tissues, allowing repair, but is retained in the tumor, which consequently receives a relatively constant low dose. Using this method a greater total amount of activity can be administered for a similar level of toxicity (Vriesendorp et al. 1993). The timing of the smaller administrations is crucial, and different investigators have reported for (Schlom et al. 1990; Buchsbaum et al. 1995; Goel et al. 2001b) and against fractionation (Pedley et al. 1993) using different regimes (for further discussion see DeNardo et al. 2002). Repeated therapy can reduce the tumor uptake of RIC due to vascular damage (Buchsbaum et al. 1999).

Separate administration of the antibody and radionuclide was first suggested in 1986 (Goodwin et al. 1986a,b). To take the use of the avidin/biotin system as an illustration, the antibody is labeled with avidin before being injected. An antibody construct with a long blood pool residence can be used, increasing tumor uptake. Following use of a clearing agent the radionuclide, attached to biotin, is administered. This is quickly cleared from normal tissues, minimizing toxicity, while delivering therapy to the tumor. Avidin/biotin-based systems have dominated this area, and some clinical applications of RIT use this technology. For example a streptavidin conjugate of the NR-LU-10 pancarcinoma antibody combined with ^{90}Y-DOTA-biotin (a single administration of $110\,\text{mCi}\,\text{m}^{-2}$) achieved a modest overall response rate, with four patients (out of 25) having freedom from progressive disease for 10–20 weeks (Knox et al. 2000).

While the avidin/biotin system has proved the principle it does have disadvantages, major problems being endogenous liver biotin and the immunogenicity of streptavidin. While other systems such as that based around oligonucleotides have been described, the affinity enhanced system (AES) is more relevant to our current survey. This centers upon a bispecific antibody with affinity for the tumor antigen and also for the radionuclide vehicle such as a hapten (Barbet et al. 1998; Schumacher et al. 2001) (for a review of PRIT see Gruaz-Guyon et al. 2005).

1.7
Summary

The ultimate goal of engineering antibodies for RIS and RIT is protein constructs that localize effectively to all tumor sites and clear from normal tissues. In this chapter we have discussed methods to achieve this goal through structural modification. The refinements made to the parent antibody structures have demonstrated that use of antibody-based tumor imaging and therapy has much potential in cancer treatment.

Acknowledgments

The authors wish to thank Dr Kerry Chester for discussions and advice in the preparation of this chapter. Work in this laboratory is supported by programme grants from Cancer Research UK, the Royal Free Cancer Research Trust and NTRAC.

Abbreviations

AES	affinity enhanced system
CDR	complementarity determining region
CHO	Chinese hamster ovary fibroblast
DOTA	a molecule used as an indirect method of antibody radiometal labeling (1,4,7,10-tetraazacyclododecane – N,N',N'',N''' tetraacetic acid)
EBRT	external beam radiotherapy
Fab	fraction antigen binding, after papain digestion of monovalent Ig
Fc	fraction of IgG crystallizable – non-Fab part of IgG
IgG	immunoglobulin class G
K_D	dissociation constant, equal to k_{off}/k_{on}
kDa	kilodalton
M_w	molecular weight
PCR	polymerase chain reaction
PET	positron emission tomography
PRIT	pretargeted RIT
RIC	radioimmunoconjugate
RIS	radioimmunoscintigraphy
RIT	radioimmunotherapy
scFv	single-chain variable fragment
SPECT	single photon emission computed tomography
T:N	ratio of radionuclide localization between tumor and normal tissues
$t_{1/2}\alpha$, $t_{1/2}\beta$	mathematical terms (exponential half-lives) describing the biphasic rate of radiolabeled antibody blood clearance

References

Adams, G.P., McCartney, J.E., Tai, M.S. et al. (1993) Highly specific in vivo tumor targeting by monovalent and divalent forms of 741F8 anti-c-erbB-2 single-chain Fv. *Cancer Res* 53: 4026–4034.

Adams, G.P., McCartney, J.E., Wolf, E.J. et al. (1995) Enhanced tumor specificity of 741F8–1 (sFv')2: an anti-c-erbB-2 single-chain Fv dimer, mediated by stable radioiodine conjugation. *J Nucl Med* 36: 2276–2281.

Adams, G.P., Schier, R., Marshall, K. et al. (1998) Increased affinity leads to improved selective tumor delivery of single-chain Fv antibodies. *Cancer Res* 58: 485–490.

Adams, G.P., Tai, M-S., McCartney, J.E. et al. (2006) Avidity-mediated enhancement of *in vivo* tumor targeting by single-chain Fv dimers. *Clin Cancer Res* 12: 1599–1605.

Akabani, G., McLendon, R.E., Bigner, D.D., Zalutsky, M.R. (2002) Vascular targeted endoradiotherapy of tumors using alpha-particle-emitting compounds: Theoretical analysis. *Int J Radiat Oncol Biol Phys* 54: 1259–1275.

Akabani, G., Kennel, S.J., and Zalutsky, M. R. (2003) Microdosimetric analysis of particle-emitting targeted radiotherapeutics using histological images. *J Nucl Med* 44: 792–805.

Baca, M., Presta, L.G., O'Connor, S.J., Wells, J.A. (1997) Antibody humanization using monovalent phage display. *J Biol Chem* 272: 10678–10684.

Barbas, C.F. III, Burton, D.R. (1996) Selection and evolution of high-affinity human anti-viral antibodies. *Trends Biotechnol* 14: 230–234.

Barbet, J., Peltier, P., Bardet, S. et al. (1998) Radioimmunodetection of medullary thyroid carcinoma using indium-111 bivalent hapten and anti-CEA X anti-DTPA-indium bispecific antibody. *J Nucl Med* 39: 1172–1178.

Begent, R., Keep, P.A., Green, A.J. et al. (1982) Liposomally entrapped second antibody improves tumor imaging with radiolabelled (first) antitumor antibody. *Lancet* 2: 739–742.

Begent, R.H., Verhaar, M.J., Chester, K.A. et al. (1996) Clinical evidence of efficient tumor targeting based on single-chain Fv antibody selected from a combinatorial library. *Nat Med* 2: 979–984.

Behr, T.M., Sharkey, R.M., Juweid, M.E. et al. (1995) Reduction of the renal uptake of radiolabelled monoclonal antibody fragments by cationic amino acids and their derivatives. *Cancer Res* 55: 3825–3834.

Behr, T.M., Sharkey, R.M., Juweid, M.E. et al. (1997a) Variables influencing tumor dosimetry in radioimmunotherapy of CEA-expressing cancers with anti-CEA and anti-mucin monoclonal antibodies. *J Nucl Med* 38: 409–418.

Behr, T.M., Sharkey, R.M., Sgouros, G. et al. (1997b) Overcoming the nephrotoxicity of radiometal-labelled immunoconjugates: improved cancer therapy administered to a mouse model in relation to the internal radiation dosimetry. Cancer 80: 2591–2610.

Behr, T.M., Blumenthal, R.D., Memtsoudis, S. et al. (2000) Cure of metastatic human colonic cancer in mice with radiolabeled monoclonal antibody fragments. *Clin Cancer Res* 6: 4900–4907.

Bird, R.E., Hardman, K.D., Jacobson, J.W. et al. (1988) Single-chain antigen-binding proteins. *Science* 242: 423–426.

Bhatia, J., Sharma, S.K., Chester, K.A. et al. (2000) Catalytic activity of an in vivo tumor targeted anti-CEA scFv:: carboxypeptidase G2 fusion protein. *Int J Cancer* 85: 571–577.

Borsi, L., Balza, E., Bestagno, M. et al. (2002) Selective targeting of tumoral vasculature: comparison of different formats of an antibody (L19) to the ED-B domain of fibronectin. *Int J Cancer* 102: 75–85.

Buchegger, F., Pelegrin, A., Delaloye, B., Bischof-Delaloye, A., and Mach, J.P. (1990) Iodine-131-labelled MAb F(ab')$_2$ fragments are more efficient and less toxic than intact anti-CEA antibodies in radioimmunotherapy of large human colon carcinoma grafted in nude mice. *J Nucl Med* 31: 1035–1044.

Buchegger, F., Gillet, M., Doenz, F. et al. (1996) Biodistribution of anti-CEA F(ab')$_2$ fragments after intra-arterial and intravenous injection in patients with liver metastases due to colorectal carcinoma. *Nucl Med Commun* 17: 500–503.

Buchsbaum, D., Khazaelli, M.B., Liu, T. et al. (1995) Fractionated radioimmunotherapy of human colon carcinoma xenografts with ^{131}I-labeled monoclonal antibody CC49. *Cancer Res* 55: 5881s–5887s.

Buchsbaum, D.J., Khazaelli, M.B., Mayo, M.S., Roberson, P.L. (1999) Comparison of multiple bolus and continuous injections of ^{131}I-labeled CC49 for therapy in a colon cancer xenograft model. *Clin Cancer Res* 5: 3153s–3159s.

Casey, J.L., Pedley, R.B., King, D.J., Green, A.J., Yarranton, G.T., Begent, R.H.J. (1999) Dosimetric evaluation and radioimmunotherapy of anti-tumoour multivalent Fab' fragments. *Br J Cancer* 81: 972–980.

Casey, J.L., Napier, M.P., King, D.J. et al. (2002) Tumour targeting of humanised cross-linked divalent-Fab' antibody fragments: a clinical phase I/II study. *Br J Cancer* 86: 1401–1410.

Chauhan, S.C., Jain, M., Moore, E.D. et al. (2005) Pharmacokinetics and biodistribution of ^{177}Lu-labeled multivalent single-chain Fv construct of the pancarcinoma monoclonal antibody CC49. *Eur J Nucl Med Mol Imaging* 32: 264–273.

Clark, M. (2000) Antibody humanization: a case of the "Emperor's new clothes"? *Immunol Today* 21: 397–402.

Clark, M., Cobbold, S., Hale, G. et al. (1983) Advantages of rat monoclonal antibodies. *Immunol Today* 4: 100–101.

Colcher, D., Minelli, M.F., Roselli, M., Muraro, R., Simpson-Milenic, D., Schlom, J. (1988) Radioimmunolocalisation of human carcinoma xenografts with B72.3 second generation monoclonal antibodies. *Cancer Res* 48: 4597–4603.

Colcher, D., Bird, R., Roselli, M. et al. (1990) In vivo tumor targeting of a recombinant single-chain antigen-binding protein. *J Natl Cancer Inst* 82: 1191–1197.

Cooke, S.P., Pedley, R.B., Boden, R., Begent, R.H., and Chester, K.A. (2002) In vivo tumor delivery of a recombinant single chain Fv::tumor necrosis factor-alpha fusion protein. *Bioconjug Chem* 13: 7–15.

Dearling, J.L.J., Qureshi, U., Whiting, S., Boxer, G.M., Begent, R.H.J., and Pedley, R.B. (2005) Analysis of antibody distribution reveals higher uptake in peripheral intrahepatic deposits than central deposits. *Eur J Nucl Med Mol Imaging* 32: S83.

DeNardo, G.L., Schlom, J., Buchsbaum, D.J. et al. (2002) Rationales, evidence, and design considerations for fractionated radioimmunotherapy. *Cancer* 94: 1332–1348.

Dubel, S., Breitling, F., Kontermann, R. et al. (1995) Bifunctional and multimeric complexes of streptavidin fusd to single chain antibodies (scFv). *J Immunol Methods* 178: 201–209.

Flynn, A.A., Green, A.J., Pedley, R.B. et al. (2002) A model-based approach for the optimization of radioimmunotherapy through antibody design and radionuclide selection. *Cancer* 94: 1249–1257.

Francis, G.E., Delgado, C., Fisher, D., Malik, F., Agrawal, A.K. (1996) Polyethylene glycol modification: relevance of improved methodology to tumour targeting. *J Drug Target* 3: 321–340.

Goel, A., Colcher, D., Baranowska-Kortylewicz, J. et al. (2000) Genetically engineered tetravalent single-chain Fv of the pancarcinoma monoclonal antibody CC49: Improved biodistribution and potential for therapeutic application. *Cancer Res*, 60: 6964–6971.

Goel, A., Colcher, D., Baranowska-Kortylewicz, J. et al. (2001a) 99mTc-labeled divalent and tetravalent CC49 single-chain Fv's: novel imaging agents for rapid *in vivo* localisation of human colon carcinoma. *J Nucl Med* 42: 1519–1527.

Goel, A., Augustine, S., Baranowska-Kortyewicz, J. et al. (2001b) Single-dose versus fractionated radioimmunotherapy of human colon carcinoma xenografts using ^{131}I-labeled multivalent CC49 single-chain Fvs. *Clinical Cancer Res* 7: 175–184.

Goldenberg, A., Masui, H., Divgi, C., Kamrath, H., Pentlow, K., Mendelsohn, J. (1989) Imaging of human tumor xenografts with an indium-111-labelled anti-epidermal growth factor receptor monoclonal antibody. *J Natl Cancer Inst* 81: 1616–1625.

Goldenberg, D.M. (2001) The role of radiolabelled antibodies in the treatment of non-Hodgkin's lymphoma: the coming of age of radioimmunotherapy. *Crit Rev Oncol/Hematol* 39: 195–201.

Goodwin, D.A., Mears, C.F., McTigue, M., David, G.S. (1986a) Monoclonal antibody hapten radiopharmaceutical delivery. *Nucl Med Commun* 7: 569–580.

Goodwin, D.A., Mears, C.F., David, G.F., McTigue, M., McCall, M.J., Frincke, J.M., Stone, M.R. (1986b) Monoclonal antibodies as reversible equilibrium carriers of radiopharmaceuticals. *Int J Rad Appl Instrum B* 13: 383–391.

de Graaf, M., van der Meulen-Muileman, I.H., Pinedo, H.M., Haisma, H.J. (2002) Expression of scFvs and scFv fusion proteins in eukaryotic cells. *Methods Mol Biol* 178: 379–387.

Greenwood, F.C., Hunter, W.M. (1963) The preparation of ^{131}I-labelled human growth hormone of high specific radioactivity. *Biochem J* 89: 116–123.

Gruaz-Guyon, A., Raguin, O., Barbet, J. (2005) Recent advances in pretargeted radioimmunotherapy. *Curr Med Chem* 12: 319–338.

Hagan, P.L., Halpern, S.E., Dillman, R.O. et al. (1986) Tumor size: effect on monoclonal antibody uptake in tumor models. *J Nucl Med* 27: 422–427.

Hofer, K.G. (1996) Biophysical aspects of Auger processes. *Acta Oncol* 35: 789–796.

Houdebine, L.M. (2002) Antibody manufacture in transgenic animals and comparisons with other systems. *Curr Opin Biotechnol* 13: 625–629.

Howell, R.W., Goddu, S.M., and Rao, D.V. (1998) Proliferation and the advantage of longer-lived radionuclides in radioimmunotherapy. *Med Phys* 25: 37–42.

Hu, S., Shively, L., Raubitshek, A., Sherman, M. et al. (1996) Minibody: a novel engineered anti-carcinoebryonic antigen antibody fragment (single-chain Fv-CH3) which exhibits rapid high-level targeting of xenografts. *Cancer Res* 56: 3055–3061.

Huhalov, A., Chester, K.A. (2004) Engineered single chain antibody fragments for radioimmunotherapy. *Q J Nucl Med Mol Imaging* 48: 279–288.

Huston, J.S., Levinson, D., Mudgett-Hunter, M. et al. (1988) Protein engineering of antibody binding sites: recovery of specific activity in an anti-digoxin single-chain Fv analogue produced in *Escherichia coli*. *Proc Nat Acad Sci USA* 85: 5879–5883.

Jain, R.K. (1988) Determinants of tumor blood flow: A review. *Cancer Res* 48: 2641–2658.

Jain, R.K. (1999) Transport of molecules, particles, and cells in solid tumors. *Ann Rev Biomed Eng* 1: 241–263.

Kallinowski, F. (1996) The role of tumor hypoxia for the development of future treatment concepts for locally advanced cancer. *Cancer J* 9: 37–40.

Karush, F. (1970) Affinity and the immune response. *Ann NY Acad Sci* 169: 56–64.

Knappik, A., Ge, L., Honegger, A. et al. (2000) Fully synthetic human combinatorial antibody libraries (HuCAL) based on modular consensus frameworks and CDRs randomised with trinucleotides. *J Mol Biol* 296: 57–86.

Knox, S.J., Goris, M.L., Tempero, M. et al. (2000) Phase II trial of yttrium-90-DOTA-biotin pretargeted by NR-LU-10 antibody/streptavidin in patients with metastatic colon cancer. *Clin Cancer Res* 6: 406–414.

Köhler, G., Milstein, C. (1975) Continuous cultures of fused cells secreting antibody of predefined specificity. *Nature* 256: 495–497.

Kubetzko, S., Sarkar, C.A., Pluckthun, A. (2005) Protein PEGylation decreases observed target association rates via a dual blocking mechanism. *Mol Pharmacol* 68: 1439–1454.

Lane, D.M., Eagle, K.F., Begent, R.H.J. et al. (1994) Radioimmunotherapy of metastatic colorectal tumours with iodine-131-labelled antibody to carcinoembryonic antigen: phase I/II study with comparative biodistribution of intact and F(ab')$_2$ antibodies. *Br J Cancer* 70: 521–525.

Lantto, J., Jirholt, P., Barrios, Y., Ohlin, M. (2002) Chain shuffling to modify properties of recombinant immunoglobulins. *Methods Mol Biol* 178: 303–316.

Maynard, J., Georgiou, G. (2000) Antibody engineering. *Annu Rev Biomed Eng* 2: 339–376.

McCafferty, J., Griffiths, A.D., Winter, G., Chiswell, D.J. (1990) Phage antibodies: Filamentous phage displaying antibody variable domains. *Nature* 348: 552–554.

McDevitt, M.R., Sgouros, G., Finn, R.D. et al. (1998) Radioimmunotherapy with alpha-emitting nuclides. *Eur J Nucl Med* 25: 1341–1351.

Meredith, R.F., Khazaeli, M.B., Plott, W.E. et al. (1996) Phase II study of dual ^{131}I-labelled monoclonal antibody therapy with interferon in patients with metastatic colorectal cancer. *Clin Cancer Res* 2: 1811–1818.

Michael, N.P., Chester, K.A., Melton, R.G. et al. (1996) In vitro and in vivo characterisation of a recombinant carboxypeptidase G2::anti-CEA scFv fusion protein. *Immunotechnology* 2: 47–57.

Milenic, D.E., Yokota, T., Filpula, D.R. et al. (1991) Construction, binding properties, metabolism, and tumor targeting of a single-chain Fv derived from the pancarcinoma monoclonal antibody CC49. *Cancer Res* 51: 6363–6371.

Mirick, G.R., Bradt, B.M., Denardo, S.J., Denardo, G.L. (2004) A review of human anti-globulin antibody (HAGA, HAMA, HACA, HAHA) responses to monoclonal antibodies. Not four letter words. *Q J Nucl Med Mol Imaging* 48: 251–257.

Nagy, Z.A., Hubner, B., Lohning, C. et al. (2002) Fully human, HLA-DR-specific monoclonal antibodies efficiently induce programmed death of malignant lymphoid cells. *Nat Med* 8: 801–807.

Nielsen, U.B., Adams, G.P., Weiner, L.M., and Marks, J.D. (2000) Targeting of bivalent anti-ErbB2 diabody antibody fragments to tumor cells is independent of the intrinsic antibody affinity. *Cancer Res* 60: 6434–6440.

O'Connell, D., Becerril, B., Roy-Burman, A. et al. (2002) Phage versus phagemid libraries for generation of human monoclonal antibodies. *J Mol Biol* 321: 49–56.

O'Donoghue, J.A., Bardies, M., Wheldon, T.E. (1995) Relationships between tumor size and curability for uniformly targeted therapy with beta-emitting radionuclides. *J Nucl Med* 36: 1902–1909.

Olafsen, T., Kenanova, V.E., Sundaresan, G. et al. (2005) Optimizing radiolabelled engineered anti-p185^{HER2} antibody fragments for *in vivo* imaging. *Cancer Res* 65: 5907–5916.

Pallela, V.R., Rao, S.P., Thakur, M.L. (2000) Interferon-alpha-2b immunoconjugate for improving immunoscintigraphy and immunotherapy. *J Nucl Med* 41: 1108–1113.

Pederson, J.T., Henry, A.H., Searle, S.J. et al. (1994) Comparison of surface accessible residues in human and murine immunoglobulin Fv domains. Implication for humanization of murine antibodies. *J Mol Biol* 235: 959–973.

Pedley, R.B., Boden, J., Keep, P.A., Harwood, P.J., Green, A.J., and Begent, R.H.J. (1987) Relationship between tumour size and uptake of radiolabelled anti-CEA in a colon tumour xenograft. *Eur J Nucl Med* 13: 197–202.

Pedley, R.B., Dale, R., Boden, J.A., Begent, R.H.J., Keep, P.A., Green, A.J. (1989) The effect of second antibody clearance on the distribution and dosimetry of radiolabelled anti-CEA antibody in a human colonic tumor xenograft model. *Int J Cancer* 43: 713–718.

Pedley, R.B., Boden, J.A., Boden, R., Dale, R., Begent, R.H. (1993) Comparative radioimmunotherapy using intact or F(ab')2 fragments of ^{131}I anti-CEA antibody in a colonic xenograft model. *Br J Cancer* 68: 69–73.

Pedley, R.B., Boden, J.A., Boden, R. et al. (1994) The potential for enhanced tumour localisation by poly(ethylene glycol) modification of anti-CEA antibody. *Br J Cancer* 70: 1126–1130.

Rader, C., Cheresh, D.A., Barbas, C.F. III. (1998) A phage display approach for rapid antibody humanization: designed combinatorial V gene libraries. *Proc Natl Acad Sci USA* 95 8910–8915.

Robinson, M.K., Doss, M., Shaller, C. et al. (2005) Quantitative immuno-positron emission tomography imaging of HER2-positive tumor xenografts with an iodine-124 labeled anti-HER2 diabody. *Cancer Res* 65: 1471–1478.

Roguska, M.A., Pederson, J.T., Keddy, C.A. et al. (1994) Humanization of murine monoclonal antibodies through variable domain resurfacing. *Proc Natl Acad Sci USA* 91: 969–973.

Saga, T., Neumann, R.D., Heya, T. et al. (1995) Targeting cancer micrometastases with monoclonal antibodies: A binding-site barrier. *Proc Natl Acad Sci USA* 92: 8999–9003.

Santimaria, M., Moscatelli, G., Viale, G.L. et al. (2003) Immunoscintigraphic detection of the ED-B domain of fibronectin, a marker of angiogenesis, in patients with cancer. *Clin Cancer Res* 9: 571–579.

Schlom, J., Molinolo, A., Simpson, J.F., Siler, K., Roselli, M., Hinkle, G., Houchens, D.P., Colcher, D. (1990) Advantage of dose fractionation in monoclonal antibody-directed radioimmunotherapy. *J Natl Cancer Inst* 82: 763–771.

Schlom, J., Eggensperger, D., Colcher, D. et al. (1992) Therapeutic advantage of high-affinity anticarcinoma radioimmunoconjugates. *Cancer Res* 52: 1067–1072.

Schumacher, J., Kaul, S., Klivenyi, G. et al. (2001) Immunoscintigraphy with positron emission tomography: gallium-68 chelate imaging of breast cancer pretargeted with bispecific anti-MUC1/anti-Ga chelate antibodies. *Cancer Res* 61: 3712–3717.

Simpkin, D.J., Mackie, T.R. (1990) EGS4 Monte Carlo determination of the beta dose kernel in water. *Med Phys* 17: 179–186.

Stoger, E., Sack, M., Fischer, R., Christou, P. (2002) Plantibodies: applications, advantages and bottlenecks. *Curr Opin Biotechnol* 13: 161–166.

Sung, C., Shockley, T.R., Morrison, P.F., Dvorak, H.F., Yarmush, M.L., Dedrick, R.L. (1992) Predicted and observed effects of antibody affinity and antigen density on monoclonal antibody uptake in solid tumors. *Cancer Res* 52: 377–384.

Tannock, I.F. (1972) Oxygen diffusion and the distribution of cellular radiosensitivity in tumours. *Br J Radiol* 45: 515–524.

Thomlinson, R.H., Gray, L.H. (1955) The histological structure of some human lung cancers and the possible implications for therapy. *Br J Cancer* 9: 539–549.

Todorovska, A., Roovers, R.C., Dolezal, O., Kortt, A.A., Hoogenboom, H.R., Hudson, P.J. (2001) Design and application of diabodies, triabodies and tetrabodies for cancer imaging. *J Immunol Methods* 248: 47–66.

Tsai, S.W., Li, L., Williams, L.E., Anderson, A.L., Raubitschek, A.A., Shively, J.E. (2001) Metabolism and renal clearance of ^{111}In-labeled DOTA-conjugated antibody fragments. *Bioconjug Chem* 12: 264–270.

van Osdol, W., Fujimori, K., Weinstein, J.N. (1991) An analysis of monoclonal antibody distribution in microscopic tumor nodules: Consequences of a "binding site barrier". *Cancer Res* 51: 4776–4784.

Vogel, C-A., Galmiche, M.C., Westermann, P. et al. (1996) Carcinoembryonic antigen expression, antibody localisation and immunophotodetection of human colon cancer liver metastases in nude mice: A model for radioimmunotherapy. *Int J Cancer* 67: 294–302.

Vriesendorp, H.M., Shao, Y., Blum, J.E., Quadri, S.M., and Williams, J.R. (1993) Fractionated intravenous administration of ^{90}Y-labeled B72.3 GYK-DTPA immunoconjugate in beagle dogs. *Nucl Med Biol* 20: 571–578.

Willuda, J., Kubetzko, S., Waibel, R. et al. (2001) Tumor targeting of mono-, di-, and tetravalent anti-p185(HER-2) mniantibodies multimerized by self-associating peptides. *J Biol Chem* 276: 14385–14392.

Wittel, U.A., Jain, M., Goel, A. et al. (2004) The *in vivo* characteristics of genetically engineered divalent and tetravalent single-chain antibody constructs. *Nucl Med Biol* 32: 157–164.

Wu, A.M., Chen, W., Raubitshek, A. et al. (1996) Tumor localisation of anti-CEA single-chain FVs: improved targeting by non-covalent dimers. *Immunotechnology* 2: 21–36.

Wu, A.M., Yazaki, P.J., Tsai, S. et al. (2000) High-resolution microPET imaging of carcinoembryonic antigen poitive xenografts by using a copper-64-labeled engineered antibody fragment. *Proc Natl Acad Sci USA* 97: 8495–8500.

Yazaki, P.J., Shively, L., Clark, C. et al. (2001a) Mammalian expression and hollow fiber bioreactor production of recombinant anti-CEA diabody and minibody for clinical applications. *J Immunol Methods* 253: 195–208.

Yazaki, P.J., Wu, A.M., Tsai, S.W. et al. (2001b) Tumor targeting of radiometal labeled anti-CEA recombinant T84.66 diabody and t84.66 minibody: comparison to radioiodinated fragments. *Bioconjug Chem* 12: 220–228.

Zalutsky, M.R., Vaidyanathan, G. (2000) Astatine-211-labelled radiotherapeutics: An emerging approach to targeted alpha-particle radiotherapy. *Curr Pharmac Des* 6: 1433–1455.

Zhang, M., Zhang, Z., Garmestani, K. et al. (2003) Pretarget radiotherapy with an anti-CD25 antibody-streptavidin fusion protein was effective in therapy of leukemia/lymphoma xenografts. *Proc Natl Acad Sci USA* 100: 1891–1895.

2
Bispecific Antibodies

Dafne Müller and Roland E. Kontermann

2.1
Introduction

Bispecific antibodies combine the antigen-binding sites of two antibodies within a single molecule. Thus they are able to bind two different epitopes simultaneously, either on the same or on different antigens. Besides applications for diagnostic purposes (e.g. recruiting detectable compounds), bispecific antibodies open up new avenues for therapeutic applications by redirecting potent effector systems to diseased areas or by increasing neutralizing or stimulating activities of antibodies (Fig. 2.1). They are thus able to improve efficacy and selectivity of natural effector functions and to expand effector functions to those not exerted by natural immunoglobulins (Fanger et al. 1992; Cao and Suresh 1998; van Spriel et al. 2000; Cao and Lam 2003).

In the last two decades a variety of different effector functions have been combined with bispecific antibodies, including the retargeting of effector molecules, effector cells, as well as carrier systems and viruses (Fig. 2.1). Thus, bispecific antibodies cover a wide range of applications from the fields of immunotherapy, chemotherapy, radiotherapy, and gene therapy (Cao and Suresh 1998). Developments for clinical applications of bispecific antibodies have mainly focused on the retargeting of different effector cells of the immune system (e.g. to tumor cells), although various other therapeutic strategies have been evaluated. The initial high expectations were not fulfilled, however, for several reasons, as reflected by the fact that as yet no bispecific antibody has been approved. The main problems have been the antibody formats (e.g. whole IgG molecules of murine origin), but also insufficient therapeutic effects. Developments in the field of antibody engineering have resulted in new approaches to improve the efficacy and safety of therapeutic antibodies. This had a substantial impact on the generation of novel bispecific antibody formats and led to a revival of interest in bispecific antibodies (Kufer et al. 2004).

Handbook of Therapeutic Antibodies. Edited by Stefan Dübel
Copyright © 2007 WILEY-VCH Verlag GmbH & Co. KGaA, Weinheim
ISBN 978-3-527-31453-9

2 Bispecific Antibodies

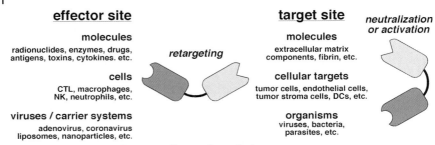

effector site

molecules
radionuclides, enzymes, drugs, antigens, toxins, cytokines. etc.

retargeting

cells
CTL, macrophages, NK, neutrophils, etc.

viruses / carrier systems
adenovirus, coronavirus liposomes, nanoparticles, etc.

target site

molecules
extracellular matrix components, fibrin, etc.

cellular targets
tumor cells, endothelial cells, tumor stroma cells, DCs, etc.

organisms
viruses, bacteria, parasites, etc.

neutralization or activation

Fig. 2.1 Therapeutic applications of bispecific antibodies.

2.2
The Generation of Bispecific Antibodies

Bispecific antibodies are artificial molecules and are not produced by normal B cells. Various methods have been developed to generate them, including chemical conjugation of two antibodies or antibody fragments, fusion of two different antibody-secreting cells to a quadroma, and genetic approaches producing recombinant bispecific antibody molecules.

2.2.1
Somatic Hybridization

Early studies revealed that fusion of a B cell with a myeloma cell line results in a hybrid myeloma cell line (hybridoma) that is not subjected to allelic exclusion (Köhler and Milstein 1975). Subsequently, it was shown that fusion of two antibody-secreting hybridomas results in a quadroma, producing two different heavy chains and two different light chains within one cell, which randomly assemble to immunoglobulin molecules including bispecific molecules (Milstein and Cuello 1983). Besides two hybridomas secreting antibodies of known specificity, a hybridoma can be fused with splenic B cells (e.g. from immunized animals). In this case the resulting quadromas produce a large repertoire of bispecific antibodies from which those with the desired binding properties can be selected (Lloyd and Goldrosen 1991). Fusion of the two antibody-secreting cells can be achieved by standard methods (e.g. using polyethylene glycol (PEG) as fusogenic agent) (Link and Weiner 1993) or electrofusion (Cao et al. 1995; Kreutz et al. 1998). Quadromas with specificities for both antigens are then identified by immunological methods (e.g. by ELISA, flow cytometry, or immunoblotting). The enrichment of quadromas after fusion is facilitated by modified selection protocols that enrich for fused cells. The fusion of two parental cells resistant and sensitive to different chemicals (e.g. HAT/neomycin; HAT/ouabain) allows for a selection for double resistance (Staerz and Bevan 1986; De Lau et al. 1989; Link and Weiner 1993). The necessary resistant parental cells can be generated by growing cells in the presence of increasing concentrations of selection reagent or by introducing resistance genes by genetic means (e.g. by retroviral gene transfer) (De Lau et al. 1989). Alternatively, selection methods based on fluorescence-activated cell sorting (FACS) using parental cells labeled with different fluorescent dyes have been established (Karawajew et al. 1987; Koolwijk et al. 1988; Kreutz et al. 1998).

Although methods to generate quadromas are well established, one of the major limitations of the hybrid hybridoma technology results from the production of a substantial number of nonfunctional molecules containing no or only one active binding site due to random heavy and light chain pairing (De Lau et al. 1991; Smith et al. 1992). On average, only 1 in 10 antibodies produced will be bispecific. Thus, elaborate purification steps are required to obtain homogeneous bispecific antibody preparations.

Interestingly, heterologous pairing of heavy chains with different isotypes or derived from different species can take place, while a preferential and species-restricted heavy–light chain pairing was reported (Corvalan and Smith 1987; Koolwijk et al. 1989; Link and Weiner 1993; Lindhofer et al. 1995). Heterologous heavy chain pairing can help to separate bispecific antibodies from parental or mismatched antibodies (e.g. by ionic exchange chromatography) (Link and Weiner 1993). Studies showed that bispecific antibodies derived from heterologous heavy chain pairings (e.g. mouse IgG1/IgG2a) are still able to bind to human Fc receptors and to recruit complement C1q (Koolwijk et al. 1989, 1991). However, an active Fc region within a bispecific antibody represents a third functional region, which might be undesirable for certain applications. Consequently, several groups have prepared bispecific F(ab')$_2$ molecules by proteolytic cleavage of bispecific antibodies derived from hybrid hybridomas (Fig. 2.2) (Warnaar et al. 1994; Tutt et al. 1995).

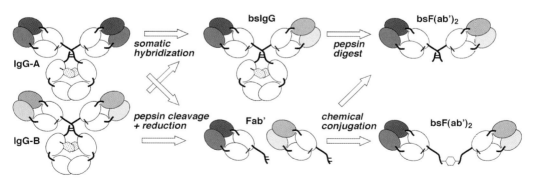

Fig. 2.2 Bispecific antibodies (bsIgG, bsF(ab')$_2$) generated by somatic hybridization or by chemical conjugation of two Fab' fragments. bsIgG, bispecific IgG; bsF(ab')$_2$, bispecific F(ab')$_2$ fragment.

2.2.2
Chemical Conjugation

In 1961 Nisonoff and Rivers described for the first time bispecific antibodies which were generated by oxidation of Fab′ fragments derived from two polyclonal antibody preparations (Nisonoff and Rivers 1961). Although this oxidative reassociation strategy can also be applied to generate bispecific antibodies from monoclonal antibodies, it has the disadvantage of producing a mixture of monospecific and bispecific molecules (Paulus 1985). Improved methods employ homo- or heterobifunctional crosslinking reagents to conjugate two antibodies or antibody fragments such as Fab or Fab′ fragments (Fig. 2.2) (Graziano and Guptill 2004).

Heterobifunctional crosslinkers (e.g. SPDP (N-succinimidyl-3-(2-pyridyldithio)propionate)), introduce in a first reaction free thiol groups at amino groups which then can be used to form disulfide linkages with a second thiol-exposing protein. However, due to the presence of a large number of free amino groups in antibodies and antibody fragments this reaction causes random crosslinking and thus produces heterogeneous populations of crosslinked antibodies (Paulus 1985). This problem can be largely avoided using thiol-reactive homobifunctional reagents. Two such crosslinking reagents, 5,5′-dithiobis(2-nitrobenzoic acid) (DTNB) and o-phenylenedimaleimide (o-PDM), have been primarily used for the generation of bispecific antibodies from Fab′ fragments exposing free thiol groups at the remaining hinge region (Brennan 1986; Glennie et al. 1987). These two reagents differ in the linkage produced between two proteins. While DTNB introduces a disulfide bond identical to that found in natural hinge regions, o-PDM generates more stable thioether bonds but a o-PDM moiety is left attached to the final product with the potential risk of immunogenicity (Fig. 2.3) (Graziano and Guptill 2004). Furthermore, conjugation of two Fab′ fragments with o-PDM requires the presence of an odd number of reactive thiol groups in the maleimidated Fab′ fragment (Fab′-A) due to the dual reactivity of o-PDM.

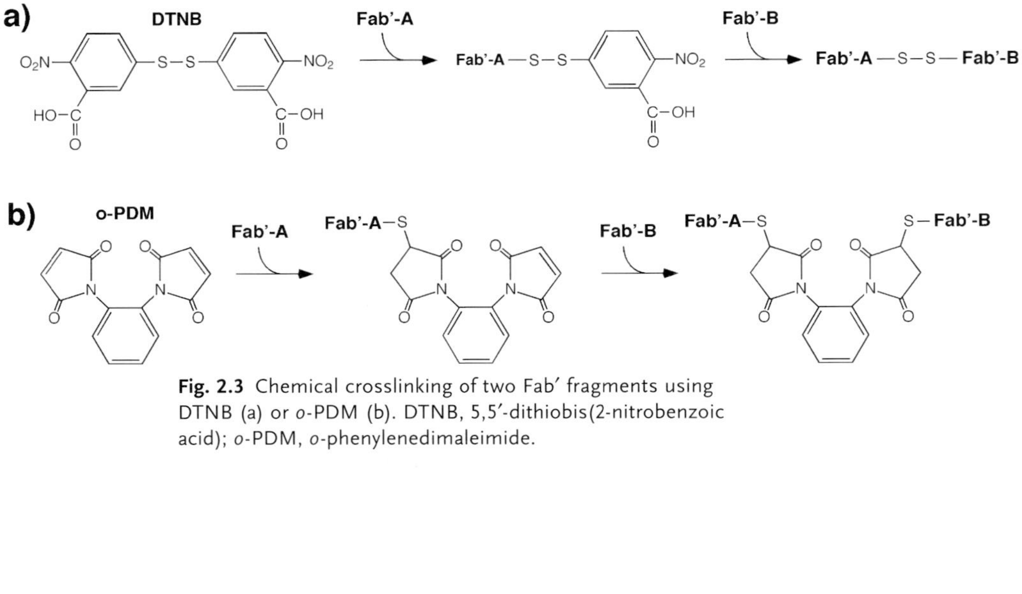

Fig. 2.3 Chemical crosslinking of two Fab′ fragments using DTNB (a) or o-PDM (b). DTNB, 5,5′-dithiobis(2-nitrobenzoic acid); o-PDM, o-phenylenedimaleimide.

Several bispecific F(ab')$_2$ molecules that have entered clinical trials have been generated by the use of DTNB as crosslinking reagent (Keler et al. 1997; Russoniello et al. 1998). Production of such bispecific F(ab')$_2$ molecules is a multistep process starting from two monoclonal antibodies. In the first step Fab' fragments are generated by pepsin digestion and reduction. The reduced first fragment is then incubated with the crosslinking reagent, the intermediate product purified by gel filtration and subsequently incubated with the reduced second fragment. The final product is again purified by gel filtration (Keler et al. 1997).

Problems associated with chemical crosslinking arise from the fact that the hinge regions contain varying numbers of cysteine residues depending on the antibody class and subclass (2–11 in the human IgG subclasses and 1–4 in the murine IgG subclasses). The presence of more than one thiol group may result in intrachain disulfide bond formation, which in the case of DTNB can be avoided by the use of dithiol complexing agents such as arsenite (Brennan et al. 1985). The conjugation of two Fab' fragments derived from different subclasses or species possessing different numbers of free thiols may further leave free reactive thiols or lead to multimeric conjugates (Tutt et al. 1991; Graziano and Guptill 2004). These obstacles can be circumvented using genetically engineered Fab' fragments possessing only one reactive thiol group at the hinge region. This approach also allows for the implementation of humanized or human Fab' molecules leading to bispecific F(ab')$_2$ molecules with reduced immunogenicity (Shalaby et al. 1992).

2.2.3
Recombinant Bispecific Antibody Molecules

Initial clinical trials have revealed several limitations of bispecific IgG molecules derived from monoclonal antibodies (Segal et al. 1999; van Spriel et al. 2000). Besides production problems often leading to heterogeneous antibody preparations, therapeutic efficacy was limited by the induction of a neutralizing immune response against the non-human bispecific antibodies and severe Fc-mediated side effects such as cytokine-release syndrome, thrombocytopenia, and leukopenia.

Genetic engineering offers the possibility of generating novel bispecific antibodies with improved properties, especially for clinical applications. Using DNA from humanized or human antibodies (see Volume I of this book), recombinant antibodies can be generated that are fully or partially human and thus should have reduced immunogenicity compared with bispecific antibodies made from rodent monoclonal antibodies. In addition, various recombinant antibody formats are available which lack the Fc region of normal antibodies and therefore do not induce Fc-mediated side effects. Furthermore, generation of recombinant formats often results in a defined composition that facilitates production.

Over the past 15 years a plethora of different recombinant bispecific antibody molecules have been developed. These formats can be divided into (1) those based on variable domains of immunoglobulins only, (2) those which use constant immunoglobulin domains for heterodimerization, and (3) those which use non-

2.2.3.1 Small Recombinant Bispecific Antibody Formats Derived from the Variable Domain

Single-chain Fv (scFv) fragments are the prototype recombinant antibody molecules containing the complete antigen-binding site of an antibody. ScFv molecules are composed of the variable heavy and light chain domain interconnected by a short peptide sequence of approximately 15 amino acid residues. ScFv fragments can be readily obtained from hybridomas or from other sources, such as combinatorial antibody libraries (e.g. using phage display technology). In addition, they can be easily subjected to affinity maturation and humanization procedures (see Volume I).

Various small bispecific antibody formats can be generated using two scFv fragments with different antigen-binding activities (Kriangkum et al. 2001). The most commonly used formats that have been evaluated for therapeutic applications are bispecific tandem scFv molecules (taFv), bispecific diabodies (Db), and bispecific single-chain diabodies (scDb) as well as several derivatives thereof (Fig. 2.4). All three formats are composed of four variable domains (V_HA, V_LA, V_HB, V_LB) and possess a molecular weight of approximately 60 kDa. They represent the smallest bispecific antibody molecules derived from the entire antigen-binding sites of two antibodies.

Tandem scFv molecules (taFv) are easily generated by connecting two scFv molecules with an additional middle linker sequence (linker M) (Fig. 2.4). They therefore represent a single gene-encoded bispecific antibody format where each scFv unit forms a separate folding entity. Tandem scFv molecules can be expressed using various arrangements of the variable domains: V_HA-V_LA-linker M-V_HB-V_LB, V_LA-V_HA-linker M-V_HB-V_LB, V_LA-V_HA-linker M-V_LB-V_HB, or V_HA-V_LA-linker M-V_LB-V_HB. In all cases, the flanking linkers within the scFv units have a length

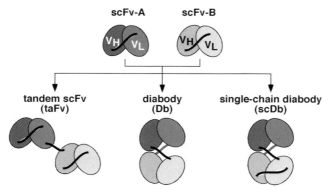

Fig. 2.4 Recombinant bispecific antibodies (tandem scFv, diabody, single-chain diabody) derived from two different scFv molecules (scFv-A, scFv-B).

of 15–20 amino acids to allow for an assembly of the variable heavy and light chain domain of each scFv into an active binding site. In contrast, the middle linkers can be of various length and composition. Examples include a very short Ala$_3$ linker (Brandaõ et al. 2003), a hydrophilic six-residue linker identified by a phage display approach (Korn et al. 2004a), glycine/serine-rich linkers (Kufer et al. 1997; McCall et al. 2001), linkers adopting a helical structure (Hayden et al. 1994), and linkers derived from various natural interconnecting sequences from immunoglobulins or immunoglobulin-like molecules (Grosse-Hovest et al. 2004; Ren-Heidenreich et al. 2004). Most of the described tandem scFv molecules have to be expressed in mammalian cells in order to obtain soluble protein, although several groups meanwhile have demonstrated soluble expression in bacteria (for an overview see: Kontermann 2005). In tandem scFv molecules the two antigen-binding sites are connected in a flexible manner, which might be advantageous for certain applications. However, until now the few direct comparisons (e.g. of tandem scFv and the more rigid single-chain diabody molecules (see below)), have not revealed functional differences in the efficacy of retargeting effector systems to target cells (Korn et al. 2004a).

Diabodies (Db) form when the linker sequence connecting the variable heavy and light chain domains of a scFv fragment is reduced to a length below 8–10 amino acid residues (Holliger et al. 1993). This reduction inhibits assembly of the V_H and V_L domains from one chain and promotes homodimerization of two V_H-V_L chains into a compact bivalent molecule containing two identical active binding sites. The diabody format can also be used to produce bispecific molecules by expressing two chains of the composition V_HA-V_LB and V_HB-V_LA (or V_LA-V_HB and V_LB-V_HA) within the same cell, which assemble into heterodimeric molecules containing two binding sites, one for each antigen (Fig. 2.4) (Holliger et al. 1993). Routinely, five residue linkers (e.g. with the sequence G_4S) are used to connect the variable domains, although other nonrepetitive linkers have been described (Völkel et al. 2001). These bispecific diabody molecules can be expressed in soluble form in bacteria or other systems such as *Pichia pastoris* or mammalian cells (for review see Kontermann 2005). However, the expression of two chains within one cell also leads to the assembly of homodimeric molecules. These molecules are functionally inactive as the V_H and V_L domains forming the antigen-binding sites are derived from two different antibodies. Attempts to improve the heterodimerization of the two chains include the introduction of interchain disulfide bonds (FitzGerald et al. 1997) and knobs-into-holes structures in the V_H-V_L interfaces (Zhu et al. 1997).

Single-chain diabodies (scDb) represent another approach to circumvent the problem of homodimerization observed for expression of bispecific diabodies. In this antibody format the two chains are connected by an additional middle linker (Fig. 2.4). Thus, all variable domains are present in a single polypeptide chain of the composition V_HA-V_LB-linker M-V_HB-V_LA or V_LA-V_HB-linker M-V_LB-V_HA, which assemble into monomeric molecules with a diabody-like structure containing two different antigen-binding sites (Brüsselbach et al. 1999). In the single-chain diabody configuration the two flanking linkers have a length of approximately five residues, whereas the middle linker has the same length used for expression of scFv fragments (i.e. 15–20 residues) (Völkel et al. 2001). All single-chain diabodies analyzed so far could be expressed in bacteria in soluble form (for an overview see: Kontermann 2005). The single-chain diabody format was further modified to generate tetravalent bispecific molecules. This was achieved, for example, by reducing the middle linker to less than 12 amino acid residues, which led to homodimerization of two single-chain diabody chains into a molecule with a molecular weight of approximately 110 kDa (Kipriyanov et al. 1999; Völkel et al. 2001). Such a tetravalent bispecific tandem single-chain diabody directed against CD19 and CD3 mediated improved T-cell cytotoxicity *in vitro* in an autologous system and in combination with a costimulatory anti-CD28 antibody enhanced antitumor effects in an animal model (Cochlovius et al. 2000; Reusch et al. 2004).

For therapeutic applications the pharmacokinetic properties of these small bispecific antibodies are critical. Due to the small size of tandem scFv, diabodies, and single-chain diabodies, these molecules are rapidly cleared from the circulation (Kipriyanov et al. 1999; personal observations). Thus, several strategies are currently pursued to improve their pharmacokinetic properties. One approach is to increase the size of these molecules (e.g. by dimerization as in case of tandem single-chain diabodies, by fusion with other proteins or through PEGylation). Alternatively, direct *in vivo* expression of these molecules can result in high serum concentrations over a prolonged period of time with improved therapeutic efficacy (Blanco et al. 2003). This approach also obviates the need to purify the antibody molecules but requires a safe and efficacious gene transfer system (Kontermann et al. 2002; Sanz et al. 2004).

2.2.3.2 Recombinant Bispecific Antibody Formats Containing Heterodimerization Domains

A variety of other recombinant antibody formats have been developed that use heterodimerization domains for the assembly of bispecific molecules. One strategy applied the knobs-into-holes approach to the immunoglobulin C_H3 domain. Introduction of a knobs-into-holes structure into the C_H3 domain promotes heterodimerization of the immunoglobulin heavy chains or other fusion proteins containing these domains (Fig. 2.5a) (Carter 2001). These knobs-into-holes are generated by replacing in the first domain a small amino acid with a large amino acid and in the second domain at the adjacent position one or two large amino acids with small amino acids (Ridgeway et al. 1996; Atwell et al. 1997). However, although these mutations lead to heterodimerization of the heavy chains of whole IgG molecules, there is no preferential binding of the light chains. Initially, this problem was circumvented by using two parental antibodies possessing identical light chains (Merchant et al. 1998). More recently, single-chain Fv fragments were fused with knobs-into-holes Fc or C_H3 regions (Fig. 2.5a) generating bispecific scFv-Fc or scFv-C_H3 fusion proteins (minibodies) (Shahied et al. 2004; Xie et al. 2005).

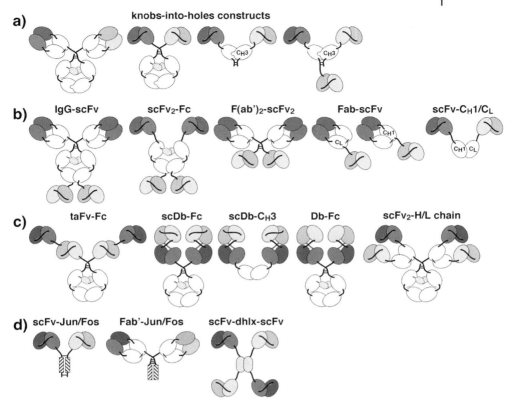

Fig. 2.5 Recombinant bispecific antibodies containing heterodimerization domains. (a) Knobs-into-holes strategies, (b) fusion of scFv fragments to constant immunoglobulin domains, (c) tetravalent and bispecific constructs, (d) fusion of scFv fragments to homo- or heterodimerizing peptides.

Alternatively, the interaction of the C_H1-C_L domains was employed for the generation of bispecific molecules fusing scFv fragments to the C_H1 and the C_L domain (Fig. 2.5b). This results in molecules with a structure similar to that of the above described scFv-C_H3 minibodies (Müller et al. 1998a). In another approach bispecific Fab-scFv fusion proteins were generated by fusing a scFv fragment either to the C-terminus of the C_H1 or the C_L domain of a Fab fragment (Schoonjans et al. 2000; Lu et al. 2002).

Various other bispecific but tetravalent antibody formats have also been developed that differ from the above described molecules in that they contain four antigen-binding sites per molecule (i.e. two for each antigen) (Fig. 2.5b,c). In one approach scFv fragments are fused either to the end of the C_H3 domain of an intact antibody (IgG-scFv) or the hinge region $(F(ab')_2\text{-scFv}_2$ (Coloma and Morrison 1997). Alternatively, the above described small bispecific antibody molecules (tandem scFv, diabody, single-chain diabody, scFv-C_H1/C_L minibodies) were fused to the Fc region or C_H3 domain leading to IgG-like molecules with a molecular weight of 150–200 kDa (Connelly et al. 1998; Alt et al. 1999; Zuo et al. 2000; Lu et al. 2003, 2005; Schneider et al. 2005).

Finally, several approaches have utilized heterodimerizing peptides (e.g. jun-fos leucine zippers) for the generation of recombinant bispecific antibodies (Fig. 2.5d). Fusion of different scFv or Fab' fragments to jun or fos peptides led to peptide-mediated heterodimerization. These dimers can be further stabilized by introducing flanking cysteine residues or using the natural hinge region (Kostelny et al. 1992; de Kruif and Logtenberg 1996). In addition, tetravalent and bispecific antibodies were constructed by linking two scFv fragments with a homodimerizing helix-loop-helix (dhlx) peptide (Müller et al. 1998b).

2.3
Bispecific Antibodies and Retargeting of Effector Cells

Antibodies are able to elicit therapeutic effects by recruiting effector cells of the immune system (e.g. natural killer (NK) cells, monocytes, macrophages, and granulocytes) to target cells, and inducing antibody-dependent cellular cytotoxicity (ADCC) and/or phagocytosis (Brekke and Sandlie 2003). Target cell-bound therapeutic antibodies, e.g. of the IgG1 isotype, are recognized by these effector cells through Fc receptors such as the high-affinity Fcγ receptor I (CD64) and the low-affinity Fcγ receptor III (CD16). However, the induction of ADCC is often limited since these antibodies also bind to non-activating Fc receptors (FcγRIIIb on polymorphonuclear leukocytes (PMNs)), inhibitory Fc receptors (FcγRIIb on monocytes/macrophages) or to Fc receptors on non-cytotoxic cells (FcγRII on platelets and B cells) (Peipp and Valerius 2002). In addition, therapeutic antibodies compete with serum IgGs for binding to the high-affinity Fcγ receptor I (CD64), leading to poor recruitment of monocytes and macrophages (Valerius et al. 1997). Finally, effector cells lacking Fc receptors, such as cytotoxic T lymphocytes, cannot be recruited with conventional antibodies. Developments in the field of bispecific antibodies for therapeutic applications have therefore been focused on the selective retargeting of potent effector cells of the immune system to tumor cells by binding with one arm to a tumor-associated antigen and with the other arm to a trigger molecule on the effector cell. Effector cells retargeted with bispecific antibodies include cytotoxic T lymphocytes, natural killer cells, monocytes and macrophages, as well as PMNs (e.g. neutrophils) by binding to trigger molecules expressed by these cells (Fig. 2.6) (Fanger and Guyre 1991; de Gast et al. 1997; van Spriel et al. 2000). Bispecific antibodies are thus able to elicit local target cell destruction by these effector cells and to circumvent limitations associated with conventional therapeutic antibodies, as described above (van de Winkel et al. 1997).

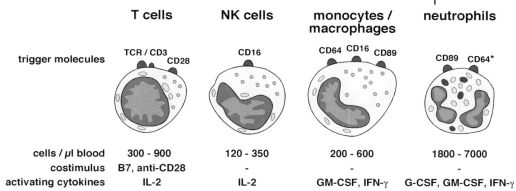

Fig. 2.6 Effector cells, trigger molecules and costimulating/activating molecules. *Upregulated after cytokine activation.

2.3.1
Retargeting of Cytotoxic T Lymphocytes

Cytotoxic T lymphocytes (CTLs) are considered to be the most potent killer cells of the immune system. Under physiological conditions, recognition and killing of a target cell is a highly controlled process involving antigen-specific binding of the T-cell receptor to major histocompatibility complexes (MHCs) on target cells. In order to be fully activated, CTLs need a second stimulus mainly provided by interaction of membrane-bound B7 molecules with CD28 on T lymphocytes during the process of T cell activation by antigen-presenting cells. CTLs can kill target cells by the perforin/granzyme pathways, leading to necrosis and apoptosis and by death receptor-mediated induction of apoptosis (e.g. through FasL/CD95) (Russell and Ley 2002).

Over the last two decades, a large number of bispecific antibodies have been developed for retargeting of CTLs to tumor cells, and more recently also to other cells such as cells of the tumor vasculature and tumor stroma (Molema et al. 2000; Wuest et al. 2001; Korn et al. 2004b; Lum and Davol 2005). Bispecific antibodies used for the retargeting of CTLs have the advantage of bypassing MHC-restricted target-cell recognition by CTLs, a process that is often inadequate due to downregulation or loss of MHC molecules on tumor cells (Bubenik 2003). Most bispecific antibodies developed for the retargeting of CTLs are directed against CD3, which is a multi-subunit complex associated with the T-cell receptor (TCR), but other trigger molecules such as CD2 or the TCR itself have also been evaluated (Fanger et al. 1992). Initially, bispecific antibodies for T-cell retargeting were generated by the hybrid hybridoma technology. Thus, these bispecific antibodies were whole IgG molecules or F(ab')$_2$ fragments of mouse or mouse/rat origin (Table 2.1). While potent tumor cell lysis was observed *in vitro* as well as

Table 2.1 Clinical trials with bispecific antibodies.

Antibody	Specificity	Format	Costimulus	Indication	Status
SHR-1	CD19 × CD3	IgG (rat/mouse)	–	Non-Hodgkin's lymphoma	Phase I
O6A (OKT3 × 6A4)	CD19 × CD3	IgG (mouse)	Anti-CD28	Non-Hodgkin's lymphoma	Phase I
OC/TR	FR × CD3	F(ab')$_2$ (mouse)	–	Ovarian cancer	Phase I/II
Bis-1	EGP2 × CD3	F(ab')$_2$ (mouse)	IL-2	Renal carcinoma	Phase I
OKT3 × Herceptin	HER2 × CD3	IgG (mouse/humanized)	IL-2, GM-CSF	Breast cancer	Phase I
Catumaxomab (removab)[a]	EpCAM × CD3	IgG (rat/mouse)	–	Malignant ascites, NSCLC, ovarian and breast cancer	Phase I–III
Ertumaxomab (rexomun)[a]	HER2/neu × CD3	IgG (rat/mouse)	–	Metastatic breast cancer	Phase II
MT103[b]	CD19 × CD3	Tandem scFv (mouse)	–	Non-Hodgkin's lymphoma	Phase I
rM28	HMW–MAA × CD28	Tandem scFv (mouse)	–	Melanoma	Phase I/II
HRS-3/A9	CD30 × CD16	IgG1 (mouse)	–	Hodgkin's lymphoma	Phase I/II
2B1	Her2/neu × CD16	IgG (mouse)	–	Lung, breast, ovarian, colon, kidney, prostate, pancreas, stomach cancer	Phase I
H22 × Ki-4	CD30 × CD64	F(ab')$_2$ (mouse/humanized)	–	Hodgkin's lymphoma	Phase I
MDX-220[c]	TAG72 × CD64	F(ab')$_2$	G-CSF	Prostate and colon cancer	Phase I/II
MDX-447[c]	EGFR × CD64	F(ab')$_2$ (humanized)	G-CSF	Renal cell carcinoma, head and neck, bladder, kidney, prostate cancer	Phase I/II
MDX-H210[c,d] (IDM-1, Osidem)	HER2/neu × CD64	F(ab')$_2$ (mouse/humanized)	G-CSF, GM-CSF, IFNγ	Renal cell carcinoma, breast, prostate, colon cancer	Phase III

a Fresenius; **b** micromet; **c** Medarex, **d** IDM.

in animal experiments, initial clinical trials faced several problems, despite reported clinical responses (Segal et al. 1999; van Spriel et al. 2000; Withoff et al. 2001). In most of these studies toxicity was observed due to release of inflammatory cytokines (cytokine storm), especially after parenteral administration of the antibodies, which might in part be caused by the presence of a Fc region. In addition, in all studies using bispecific IgG or F(ab')$_2$ molecules of mouse origin, the patients developed human anti-mouse antibody responses (HAMA). It was concluded that the ideal bispecific antibody for retargeting of CTLs, but also other effector cells, should (1) be highly selective for the target cell, (2) retarget relevant effector cells, (3) bind monovalently to the effector cells and activate them only upon binding to target cells, (4) lack a Fc region to avoid Fc-mediated side effects, (5) be human or humanized to avoid a neutralizing antibody response, and (6) be small enough to penetrate tumor tissues but large enough to circulate for a sufficient length of time (Segal et al. 1999).

In the course of progress of recombinant antibody technology, new bispecific antibody formats have been generated in order to address the problems of the HAMA response, Fc-mediated side effects and tumor penetration, but also production of bispecific antibodies with a defined composition. In parallel, much effort has gone into investigation of how to provide appropriate costimulatory signals, considered pivotal in the effective triggering of a cellular immune response. Several bispecific tandem scFv molecules, diabodies and single-chain diabodies with specificity for CD3 and tumor-associated antigens such as CD19, CD20, EpCAM, HER2/neu, and CEA have been reported to efficiently retarget T-cell cytotoxicity to tumor cells and to induce antitumoral activity *in vivo* (for review see Peipp and Valerius 2002; Kontermann 2005). In order to provide a costimulatory signal, anti-CD28 antibodies or recombinant antibody B7 fusion proteins were included in the therapeutic setting (Cochlovius et al. 2000; Holliger et al. 1999). The fusion of an extracellular region of B7 or parts of it to scFv fragments, bivalent diabodies or other antibody fragments directed against a tumor-associated antigen was shown to result in targeted delivery of the costimulatory signal to the target cell (Gerstmayer et al. 1997; Holliger et al. 1999; Biburger et al. 2005).

Interestingly, bispecific antibodies based on anti-CD3 antibody TR66 developed by Lanzavecchia and Scheidegger (1987) and expressed as tandem scFv molecules were shown to be potent T-cell activators even in the absence of a costimulatory signal (Löffler et al. 2000). These bispecific tandem scFv molecules were termed BiTEs (bispecific T cell engagers). *In vitro* studies revealed that efficient target cell lysis is induced by BiTEs at very low antibody concentrations (between 0.01 and 10 ng mL^{-1}) and low effector-to-target cell ratios (E:T = 1:5) (Wolf et al. 2005; Hoffmann et al. 2005). Thus, in addition to the fact that these bispecific antibodies are costimulation independent, they seem to be superior compared with other bispecific antibodies targeting CD3 in respect of the efficacy of tumor cell lysis and the amount of antibody required for T-cell retargeting (Wolf et al. 2005). Several BiTEs directed against different tumor-associated antigens, including CD19 and EpCAM, have been developed and evaluated *in vitro* and in animal experiments (Dreier et al. 2003; Schlereth et al. 2005). Currently, one BiTE directed against CD19 and CD3 is in a clinical phase I trial for the treatment of non-Hodgkin's lymphoma (Table 2.1).

Other studies have shown that binding of certain monoclonal antibodies to CD28 can induce CTL-mediated cytotoxicity without the need for a first activation through the TCR complex (Tacke et al. 1997). These superagonistic antibodies bind to a particular epitope (aa 60–65) on CD28 different from the region that interacts with B7 (aa 99–104) (Lüdher et al. 2003). A non-superagonistic anti-CD28 antibody was recently used to construct a bispecific tandem scFv molecule directed against the melanoma-associated glycoprotein MAPG (HMW-MAA) (Grosse-Hovest et al. 2003). This antibody, r28M, was able to induce efficient tumor cell killing *in vitro* and *in vivo* in a CD3-independent way (Grosse-Hovest et al. 2005). Interestingly, although r28M activates T cells leading to T cell-mediated killing of tumor cells, a contribution to killing of a non-T cell population ($CD56^+$ NK cells) was also observed. It was found that activation of these non-T cells is induced indirectly by rM28 through cytokines secreted by the activated T cells in the presence of antigen-positive target cells (Grosse-Hovest et al. 2005). The observed potency of rM28 *in vivo* might also be due to the large proportion of dimeric molecules in the rM28 preparation (>50%) with a molecular weight of approximately 115 kDa leading to an extended serum half-life compared with monomeric tandem scFv molecules. To produce sufficient amounts for further studies, transgenic cows were generated secreting rM28 at high concentrations into the blood (Grosse-Hovest et al. 2004). Currently, rM28 is being tested in a phase I/II clinical trial for the intralesional treatment of metastatic melanoma and unresectable metastasis (Table 2.1).

Despite the observation that the presence of an Fc part in bispecific antibodies can cause increased toxicity, it was found that intact bispecific hybrid antibodies composed of mouse IgG2a and rat IgG2b and directed against CD3 and a tumor-associated antigen can elicit a strong antitumor response *in vitro* and *in vivo* (Ruf and Lindhofer 2001). It was postulated that these trifunctional bispecific antibodies are able to simultaneously activate T cells via binding to CD3 and accessory cells (NK cells, mononuclear blood cells) through interactions with their Fc region. These activated accessory cells can deliver necessary costimulatory signals to the T cells and can further increase the immune response through phagocytosis of tumor material, leading to long-lasting antitumor immunity (Zeidler et al. 2000, 2001). Two trifunctional bispecific antibodies, removab (catumaxomab) directed against EpCAM and CD3 and rexomun (ertumaxomab) directed against HER2 and CD3 are currently being tested in clinical trials for the treatment of various cancers including malignant ascites, ovarian cancer, and breast cancer (Table 2.1). Results from phase I studies demonstrated a good tolerability after repeated intravenous or intraperitoneal injections and clinical responses (Heiss et al. 2005). The occurrence and involvement of HAMA has not yet been addressed.

2.3.2
Retargeting of Fc Receptor-bearing Effector Cells

A second group of effector cells consists of those naturally recruited by binding to Fc receptors. These include NK cells, monocytes/macrophages and PMNs (e.g. neutrophils). Consequently, bispecific antibodies have been generated for the retargeting of these effector cells by binding to Fc receptors such as FcγRI (CD64), FcγRIII (CD16), and FcαR (CD89) (see Fig. 2.6). NK cell-mediated cytotoxicity is triggered by engagement of the low-affinity Fcγ receptor III (FcγRIIIA = CD16) that is constitutively expressed. Target cell destruction is achieved by mechanisms similar to those observed for CTLs. CD16 is also expressed by monocytes and macrophages, which, in addition, constitutively express the high-affinity FcγRI (CD64) and to some extend the FcαR (CD89). CD89 is the main Fc receptor on PMNs representing the largest effector cell population in the blood. These cells also express CD64 upon induction with interferon γ (IFNγ) and granulocyte colony-stimulating factor (G-CSF). In general, it was found that administration of growth factors (e.g. granulocyte-macrophage colony-stimulating factor (GM-CSF) and G-CSF) or cytokines (IFNγ and IL-2) can lead to further stimulation and proliferation of these effector cells (van Spriel et al. 2000).

Bispecific antibodies directed against Fc receptors are able to extend ADCC to cells normally not or only inefficiently recruited by conventional antibodies, such as PMNs or monocytes and macrophages. These cells are retargeted by binding to CD89 or CD64, respectively, and eliminate target cells directly through cytotoxicity or phagocytosis (Deo et al. 1998; Sundarapandiyan et al. 2001). In addition, since some Fc receptors such as CD64 are also present on antigen-presenting cells, bispecific antibodies can indirectly enhance antitumor immunity by increased antigen presentation (van Spriel et al. 2000).

Although preclinical studies with bispecific antibody-mediated retargeting of NK cells to tumor cells showed promising results *in vivo* (Hombach et al. 1993), first clinical trials were mainly characterized by toxicity (Weiner et al. 1995) and rather limited antitumor responses (Hartmann et al. 1997, 2001). This can be partly attributed to the bispecific antibody format used. Complete murine IgGs are thought to crosslink Fc receptors via the anti-CD16 antigen-binding site and the Fc region, inducing systemic leukocyte activation, characterized by extensive cytokine release. Another reason might be the specificity of most of the anti-CD16 antibodies used for construction, which do not differentiate between the activating (FcγRIIIA on NK cells) and non-activating (FcγRIIIB on PMNs) receptor isoforms (Hombach et al. 1993; Amoroso et al. 1999). Binding of the bispecific antibodies to FcγRIIIB as well as to shedded, soluble CD16 in human plasma might reduce their cytolytic activity unless a large molar excess of bispecific antibodies is applied. In addition, the establishment of the effective effector-to-target cell ratio at the tumor site is another critical point to be considered. Although under physiological conditions the expression pattern of CD64 predisposes it for monocyte/macrophage targeting, most bispecific antibody therapy approaches include growth factor or cytokine treatment to induced CD64 upregulation on PMNs and to increase the effector cell population.

Diverse treatment schedules with IFNγ, G-CSF, or GM-CSF were investigated in clinical studies with MDX-H210 and MDX-447. These bispecific antibodies are partially or fully humanized F(ab')$_2$ molecules specific for CD16 × Her2/neu and CD16 × EGFR, respectively (Table 2.1). Although biological effects such as changes in circulating leukocyte subpopulation cell number and receptor expression, binding of the bispecific antibody to effector cells, enhanced ADCC/phagocytic capacity *in vitro*, cytokine release, and local infiltration of effector cells at the tumor site could be observed, clinical responses remained vague (Curnow 1997; Pullarkat et al. 1999; James et al. 2001; Lewis et al. 2001; Wallace et al. 2001; Repp et al. 2003). Now considerable interest focuses also on CD89 as trigger molecule, because it is constitutively expressed on neutrophils and the main trigger molecule to induce tumor cytolysis by these cells. Retargeting the cytotoxic potential of this large effector subpopulation to tumor cells normally not attacked by this effector mechanism seems feasible at least from *in vitro* studies (Deo et al. 1998).

Comparing a set of bispecific antibodies targeting CD20 or HER-2/neu and Fc receptors CD64, CD16, or CD89 in combination with G-CSF or GM-CSF revealed differences in the retargeting of effector cell cytotoxicity. This was dependent on Fc receptor expression, but also influenced by the growth factors and the tumor antigen involved (Stockmeyer et al. 2001).

Several recombinant bispecific antibodies, such as bispecific tandem scFv molecules directed against CD16 and various tumor-associated antigens (CD19, HER2/neu, HLA class II) have been recently developed for the retargeting of NK cells to tumor cells (McCall et al. 1999; Bruenke et al. 2004, 2005). Increasing binding of bispecific antibody molecules to the tumor-associated antigen, either by using high-affinity binding sites or by increasing the functional affinity with molecules containing two binding sites for the tumor-associated antigen, was shown to enhance antibody-mediated *in vitro* cytotoxicity (McCall et al. 2001; Xie et al. 2003; Shahied et al. 2004). In a further study the combination of bispecific diabodies directed against CD19 × CD3 and CD19 × CD16 demonstrated synergistic antitumor effects in a preclinical model of non-Hodgkin's lymphoma by retargeting different effector cell populations (Kipriyanov et al. 2002). Synergistic antitumor effects in a non-Hodgkin's lymphoma model were also observed by combined therapy with an anti-C19 × anti-CD16 bispecific diabody and the angiogenesis inhibitor thalidomide (Schlenzka et al. 2004). In summary, these findings underline the complexity of this still challenging approach, but also highlight potentials and perspectives of bispecific antibodies for retargeting effector cells to tumor cells.

2.4
Bispecific Antibodies and Retargeting of Effector Molecules

Besides retargeting of effector cells, bispecific antibodies can be applied to redirect effector molecules to other cells or structures associated with diseases. Thus, bispecific antibodies have been explored for the recruitment of a large number of different effector molecules, including radionuclides, drugs, toxins, enzymes, cytokines, complement components, and immunoglobulins (Cao and Lam 2003).

The use of bispecific antibodies circumvents chemical coupling of effector molecules. This might be especially advantageous in cases where chemical modifications may lead to inactivation of the effector molecules or the antibody. Furthermore, bispecific antibodies can be employed for the recruitment of natural effector molecules already present in the organism, such as components of the humoral immune system (Kontermann et al. 1997; Holliger et al. 1997). Importantly, the application of bispecific antibodies allows for an uncoupling of antibody-mediated targeting from delivery of effector molecules. This pretargeting strategy has extensively been studied for radioimmunotherapy (Gruaz-Guyon et al. 2005). In the first step, the bispecific antibody is injected, accumulates in the diseased tissue and unbound antibodies are allowed to clear from circulation and healthy tissues. In the second step, the effector molecule (e.g. a chelated radionuclide) is injected and is recovered by the second binding site of the bispecific antibody at the target site, while unbound effector molecules are rapidly eliminated (Fig. 2.7). Thus, side effects often seen with antibody conjugates can be reduced.

2.4.1
Bispecific Antibodies and Radioimmunotherapy

Radioimmunotherapy (RIT) is based on the selective antibody-mediated delivery of cell-damaging radionuclides into diseased tissues such as tumors. Several

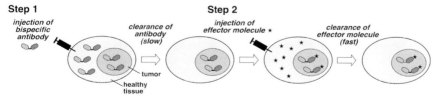

Fig. 2.7 Bispecific antibodies for pretargeting strategies. In the first step, the bispecific antibody is injected and allowed to accumulate in the tumor and to be cleared from circulation and healthy tissues. In the second step the small effector molecule is injected which is retained in the tumor by binding to the bispecific antibodies and is rapidly cleared from nontargeted tissue by renal excretion.

parameters influence efficacy of RIT, including tumor location, size, morphology, physiology and radiosensitivity, physical/chemical properties of the radionuclides and the nature of its radiation (low or high energy transfer), but also pharmacokinetic properties of the antibody (Goldenberg 2003). Several radionuclides are of clinical interest. Currently, mainly ^{131}I and ^{90}Y, both beta-emitters with a half-life of 193 and 64 h, respectively, are used for therapeutic applications, but various other beta-emitters with shorter half-lives (^{188}Rh) or alpha-emitters (^{211}At, ^{225}Ac) have also found increasing interest (Chang et al. 2002; Goldenberg 2003).

One problem associated with the application of antibody–radionuclide conjugates in therapy of solid tumors is an often observed toxicity in normal tissues due to an inappropriate tumor-to-normal tissue ratio. This limits total applicable dose and thus therapeutic efficacy. As described above, a pretargeting approach applying bispecific antibodies can uncouple the slow process of tumor targeting and antibody clearance from a rapid and selective delivery of the radionuclides.

In order to employ bispecific antibodies for RIT they have to bind with one arm to the radionuclide. This can be achieved using antibodies recognizing radionuclides complexed with a chelating agent such as DTPA (diethylenetriaminepentaacetic acid) or DOTA (1,4,7,10-tetra-azacyclododecane-N,N′,N″,N‴-tetraacetic acid) (Fig. 2.8a). This approach was further improved by generating bivalent molecules in respect to the chelating agent (i.e. containing two chelators (Fig. 2.8b)) (Le Doussal et al. 1990). The advantages of these molecules are increased functional affinity (affinity enhancement system) and binding of two radionuclides (e.g. ^{111}In) per molecule. In addition, the aromatic side chain of the tyrosine residue present in these molecules can be used for conjugation of other radionuclides such as ^{131}I (Morandeau et al. 2005). Alternatively, the chelating agent can be coupled to a hapten to which antibodies are available. HSG (histamine-succinyl-glycine) represents such a peptide hapten used for RIT (Fig. 2.8b). In its simplest form a chelating agent such as DOTA is conjugated to one HSG peptide. More advanced systems consist of two HSG peptides, leading again to increased functional affinity, and additional compounds for labeling with radionuclides. Several of these bivalent haptens have been developed which can be used to deliver various radionuclides (e.g. using the tyrosine side-chain of the central core or a chelating agent attached to these molecules) (Morandeau et al. 2005).

Various preclinical studies have shown that the use of the pretargeting approach in RIT results in reduced toxicity and allows for the administration of higher doses with enhanced antitumor effects (Sharkey et al. 2005a). Bispecific antibodies tested so far were directed against CEA, CD20, or the renal cell carcinoma marker G250 (Gautherot et al. 1997; Krannenborg et al. 1998; Sharkey et al. 2005b). Besides bispecific F(ab′)$_2$ molecules generated by chemical coupling, recombinant bispecific molecules (diabodies) have been developed for RIT (DeNardo et al. 2001). Currently a bispecific F(ab′)$_2$ molecule (hMN14 × m734) directed against carcinoembryonic antigen (CEA) and a ^{131}I-labeled di-DTPA molecule is tested in clinical trials for the treatment of patients with various CEA-positive tumors (Kraeber-Bodéré et al. 2003). This bispecific antibody fragment

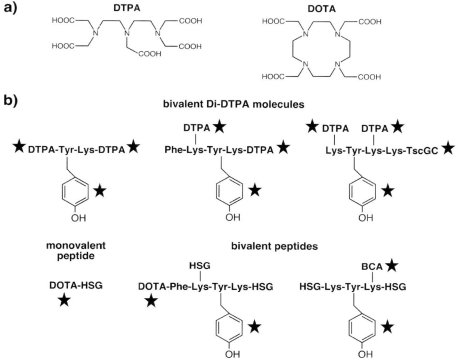

Fig. 2.8 (a) Structures of chelating agents DTPA and DOTA. (b) Examples of various mono- and bivalent haptens developed for pretargeting with bispecific antibodies. Stars indicate possible positions for labeling with radionuclides. DTPA, diethylenetriaminepentaacetic acid; DOTA, 1,4,7,10-tetra-azacyclododecane-N,N′,N″,N‴-tetraacetic acid; HSG, histamine-succinyl-glycine; BCA, bifunctional chelating agent; TscGC, thiosemicarbonylglyoxylcysteinyl.

is derived from a humanized anti-CEA antibody and a mouse anti-hapten antibody and was generated by chemical crosslinking with o-PDM. Thus far, the best results with this antibody have been obtained with a bispecific antibody dose of 40 mg m^{-2} given at 5-day intervals and treatment with doses up to 5.5 GBq in the absence of bone marrow involvement. Although further studies are inevitable to establish safety and efficacy, the encouraging results seen in clinical studies make this a promising approach to improve radioimmunotherapy with monoclonal antibodies.

2.4.2
Bispecific Antibodies and Targeting of Toxins and Drugs

Several proteins and small molecules (e.g. toxins and chemotherapeutic drugs) are known to be potent inhibitors of cell viability, growth, and proliferation.

2 Bispecific Antibodies

Bispecific antibodies have been generated to target such therapeutically useful substances to target cells, with an emphasis on tumor therapy and autoimmune diseases (Embleton et al. 1991; Ferrini et al. 2001; Kuus-Reichel et al. 1995). The advantage is that chemical crosslinking, which might interfere with the activity of the therapeutic molecule and/or the antibody molecule, is not required. By this approach selectivity of the drug as well as target cell sensitivity can be increased (Ford et al. 2001). Toxins targeted with bispecific antibodies include ribosome-inactivating proteins such as saporin, gelonin, and ricin. In addition, several drugs (e.g. the anthracyclin doxorubicin, the antimetabolite drug methotrexate, and the vinca alkaloids vincristine, vindesine, and vinblastine) have been combined with bispecific antibodies for drug delivery (Cao and Lam 2003).

A study with a bispecific antibody generated by the hybrid-hybridoma technology and directed against CEA and doxorubicin (Dox) showed that the antibody can significantly reduce IC_{50} values for Dox with CEA-expressing tumor cell lines and improve inhibition of tumor growth in animal models (Ford et al. 2001). In another study, a remarkable antidotal activity was observed *in vivo* with a bispecific antibody targeting Dox to EGF receptor-expressing tumor cells, while antitumor effects were equal to those of the drug alone (Morelli et al. 1994). Of note, the analysis of a bispecific antibody directed against the IL-2 receptor and vincristine revealed additive and not synergistic effects of the antibody and the vinca alkaloid in the therapy of diabetic mice (Kuus-Reichel et al. 1995). This was attributed to an inactivation of the drug by the antibody, leading to an inaccessibility of the drug to the target cells. This finding indicates that antibodies have to be carefully selected in order to preserve the activity of the drug. Interestingly, synergistic toxic effects were described for the use of two bispecific antibodies, recognizing different nonblocking epitopes on saporin or gelonin redirected to human B cell lymphoma cell lines (Bonardi et al. 1993; Sforzini et al. 1995). Other studies showed that a combination of bispecific antibodies directed against a toxin and two different cell surface markers (e.g. CD22 and CD37 or CD7 and CD38) can also improve cytotoxicity in a synergistic manner (Flavell et al. 1992; French et al. 1995). These findings demonstrate that an increase in functional affinity, either for the drug or the target cell, can enhance therapeutic efficacy.

Currently, there are no data available on comparison of bispecific antibodies and antibody conjugates in a therapeutic preclinical or clinical setting. Initial studies of drug targeting with bispecific antibodies faced the same problems as observed for various other approaches (e.g. related to immunogenicity) (Bonardi et al. 1992). Further studies therefore have to be conducted to show if bispecific antibodies are superior to immunotoxins and antibody–drug conjugates in terms of production, safety, and efficacy.

2.5
Bispecific Antibodies as Agonists or Antagonists

Bispecific antibodies are able to bind to two different epitopes on the same or on different antigens. This dual binding can lead to an increase in functional affinity and can improve the neutralizing or activating potential of antibodies.

Agonistic activities were demonstrated for a bispecific tandem scFv-Fc fusion protein directed against the T-cell antigen CD2. This antibody was shown to be a potent mitogen for T cells, displaying enhanced mitogenic properties compared with the combination of monoclonal antibodies (Connelly et al. 1998). In another study, a bispecific tandem scFv molecule directed against two epitopes on CTLA-4 was even able to convert CTLA-4 from an inhibitor to an activator of T cells. These T cell-activating activities could become clinically useful for boosting immunity (e.g. in vaccination or cancer immunotherapy) (Madrenas et al. 2004).

Bispecific antibodies are also potent antagonists by simultaneous binding to two epitopes on the same or different receptors. Thus, it was shown that a bispecific diabody directed against two epitopes on vascular endothelial growth factor (VEGF) receptor 2 (VEGFR2) efficiently blocked binding of VEGF to its receptor and inhibited VEGF-induced activation of the receptor and mitogenesis of endothelial cells, while neither of the parental scFv fragments showed any inhibitory activity (Lu et al. 1999). In further studies, a bispecific diabody as well as a tetravalent and bispecific diabody – Fc fusion protein (di-diabody) directed against VEGF receptor 2 and 3 blocked binding of VEGF and VEGF-C to their receptors and inhibited activation of both receptors (Lu et al. 2001, 2003; Jimenez et al. 2005). In a similar approach, IgG-like tetravalent and bispecific antibodies (diabody-Fc, scFv$_2$-H/L chain fusion proteins) were generated which bind simultaneously to the epidermal growth factor (EGF) receptor and the insulin-like growth factor (IGF) receptor (Lu et al. 2004). These antibodies blocked binding of EGF and IGF to their receptors and inhibited activation of several signal transduction proteins. In addition, due to the presence of an Fc region in these bispecific molecules they were able to mediate ADCC, which led to growth inhibition of human tumor xenografts *in vivo* (Lu et al. 2005). These studies underline that bispecific antibodies directed against cell surface receptors might be beneficial for therapeutic applications by simultaneously neutralizing two receptors and/or by improving Fc-mediated effector functions.

The approach of simultaneously targeting two essential receptors was also applied to phenotypic knockout by intrabodies. This was shown for a tetravalent and bispecific antibody molecule (scFv$_2$-Fc = intradiabody) directed against VEGFR2 and Tie-2, which was expressed in the endoplasmic reticulum (ER) of endothelial cells by attaching a KDEL retention signal to the C-terminus (Jendreyko et al. 2003). Compared with the ER-targeted parental scFv fragment, expression of the intradiabody resulted in a more efficient and longer lasting surface depletion of the two receptors and strong anti-angiogenic activity in *in vitro* endothelial cell tube formation assays. After adenoviral gene transfer of the bispecific antibody construct by subtumoral injection an efficient inhibition of tumor growth and tumor angiogenesis was observed (Jendreyko et al. 2005).

Further applications of neutralizing bispecific antibodies include the treatment of infectious diseases (e.g. viral infections), as shown with a tetravalent bispecific antibody (tandem scFv-Fc fusion protein) directed against two surface antigens (S and pre-S2) of hepatitis B virus (Park et al. 2000).

2.6
Bispecific Antibodies and Somatic Gene Therapy

The *in vivo* transfer of DNA into living cells offers the possibility of curing monogenic hereditary diseases or expressing a therapeutic protein within the cell (e.g. a suicide protein for cancer therapy). Transfer is accomplished using nonviral carrier systems (liposomes, polymers) or viral vectors (e.g. adenovirus, adeno-associated virus, retroviruses) (El-Aneed 2004; Bartsch et al. 2005; Hendrie and Russell 2005). However, most systems lack specificity for the target cells, which limits efficacy and safety of gene transfer. Targeting to specific cell types and uptake into the cells can be achieved by incorporating ligands such as antibodies or peptides into the surface of the gene transfer vehicle (Walther and Stein 2000). Bispecific antibodies have been developed as adapter molecules to direct viral vectors to target cells, especially in cases where genetic fusion of antibody fragments with viral coat proteins was not successful (e.g. for adenoviral vectors) (Barnett et al. 2002; Everts and Curiel 2004). Bispecific antibodies have the advantage that binding of one arm to a viral coat protein can neutralize the wildtype tropism of the virus, as shown for adenovirus of the serotype 5 and adeno-associated virus (AAV) (Watkins et al. 1997; Bartlett et al. 1999). Thus, bispecific antibodies inhibit binding to natural receptors and allow for a retargeting to a specific receptor on the target cell (Fig. 2.9).

Both bispecific chemical conjugates (F(ab')$_2$) and recombinant bispecific antibodies (tandem scFv, single-chain diabodies) have been evaluated for retargeting

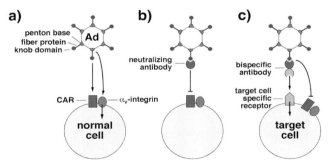

Fig. 2.9 Bispecific antibody-mediated retargeting of adenoviral vectors (serotype 5) to target cells. Wildtype adenovirus (Ad) binds with the fiber knob domain to the coxsackievirus and adenovirus receptor (CAR) on cells. Internalization is mediated by interaction of the penton base with α_v-integrins (a). Antibodies can inhibit adenoviral transduction by neutralizing the binding of the knob domain to CAR (b). Bispecific antibodies against the knob domain and a target cell-specific receptor redirect adenoviruses to new target cells and inhibit transduction of CAR-positive nontarget cells (c).

of recombinant viruses to target cells (Bartlett et al. 1999; Haisma et al. 1999; Kelly et al. 2000; Reynolds et al. 2000; Tillman et al. 2000; Grill et al. 2001; Heideman et al. 2001; Nettelbeck et al. 2001, 2004; Korn et al. 2004a; Würdinger et al. 2005). These approaches include retargeting to tumor cells (e.g. through binding to EGFR, EpCAM, HMW-MAA, TAG-72, and G250), endothelial cells (through binding to endoglin, αv integrins, ACE) and antigen-presenting cells (CD40 binding). In most cases increased and cell type-specific virus transduction was mediated by the antibodies *in vitro*, including transduction of primary tumor cells and spheroids (Grill et al. 2001; Nettelbeck et al. 2004). Importantly, these findings demonstrated that virus uptake is not abolished by the use of bispecific adapter molecules.

A few studies have already shown selective tumor targeting of adenoviral vectors *in vivo*, demonstrating the feasibility of this approach for therapeutic applications. In one study, a bispecific antibody directed against the adenoviral knob domain and angiotensin-converting enzyme (ACE) resulted in targeting of the adenoviruses to pulmonary capillary endothelium as shown by an increase in gene expression in the lung and reduced expression in nontarget organs, especially the liver (Reynolds et al. 2000). In a subsequent study, synergistic effects on selectivity were shown with a combination of transductional and transcriptional targeting using a bispecific antibody directed against ACE and the endothelial cell-specific VEGFR-1 promoter (Reynolds et al. 2001).

The applications of bispecific antibodies for adenoviral gene therapy focus mainly on tumor therapy mediating either the transfer of suicide genes (e.g. thymidine kinase in combination with ganciclovir as prodrug) or transduction by conditionally replicative adenoviruses (CRAds) that replicate only in tumor cells and destroy these cells through lysis (Kanerva and Hemminki 2005). In addition, bispecific antibodies have been developed to redirect other viruses such as coronaviruses (e.g. feline infectious peritonitis virus and felinized murine hepatitis virus), which can kill cancer cells through the formation of syncytia (Würdinger et al. 2005). Targeted adenoviruses have been also applied for vaccination strategies by gene transfer into antigen-presenting cells such as dendritic cells (DCs) (Tillman et al. 2000; de Gruijl et al. 2002). Due to low numbers of coxsackievirus and adenovirus receptor (CAR) on DCs, adenoviral transduction can be drastically improved with bispecific antibodies and can lead to efficient activation of DCs and antigen-specific stimulation of T cells.

2.7
Outlook

As has been extensively shown over the last two decades, bispecific antibodies can extend and improve therapeutic applications of monoclonal antibodies by combining target-specific binding with the recruitment of potent effector mechanisms. Although in recent years successful development and commercialization of monoclonal antibodies and antibody conjugates have come to fore, bispecific antibodies seem to be experiencing a revival. Novel and improved bispecific antibodies generated for example by genetic engineering have already entered clinical trials. The near future will show if these bispecific antibodies will become new, clinically approved therapeutics for treating diseases.

Acknowledgments

We would like to thank Prof. Peter Scheurich (Stuttgart), Dr Jeannette Gerspach (Stuttgart) and Dr Dirk Nettelbeck (Erlangen) for critical reading of the manuscript and helpful suggestions.

References

Alt, M., Müller, R., Kontermann, R.E. (1999) Novel tetravalent and bispecific IgG-like antibody molecules combining single-chain diabodies with the immunoglobulin $\gamma1$ Fc or C_H3 region. *FEBS Lett* 454: 90–94.

Atwell, S., Ridgway, J.B.B., Wells, J.A., Carter, P. (1997) Stable heterodimers from remodelling the domain interface of a homodimer using a phage display library. *J Mol Biol* 270: 26–35.

Amoroso, R.A., Alpaugh, R.K., Barth, M.W., McCall, A.M., Weiner, L.M. (1999) Production and characterization of mice transgenic for the A and B isoforms of human FcγRIII. *Cancer Immunol Immunother* 48: 443–455.

Barnett, B.G., Crews, C.J., Douglas, J.T. (2002) Targeted adenoviral vectors. *Biochim Biophys Acta* 1575: 1–14.

Bartlett, J.S., Kleinschmidt, J., Boucher, R.C., Samulski, R.J. (1999) Targeted adeno-associated virus vector transduction of nonpermissive cells mediated by a bispecific F(ab'γ)2 antibody. *Nat Biotechnol* 17: 181–186.

Bartsch, M., Weeke-Klimp, A.H., Meijer, D.K., Scherphof, G.L., Kamps, J.A. (2005) Cell-specific targeting of lipid-based carriers for ODN and DNA. *J Liposome Res* 15: 59–92.

Biburger, M., Weth, R., Wels, W.S. (2005) A novel bispecific tetravalent antibody fusion protein to target costimulatory activity for T-cell activation to tumor cells overexpressing ErbB2/HER2. *J Mol Biol* 346: 1299–1311.

Blanco, B., Holliger, P., Vile, R.G., Álvarez-Vallina, L. (2003) Induction of human T lymphocyte cytotoxicity and inhibition of tumor growth by tumor-specific d iabody-based molecules secreted from gene-modified bystander cells. *J Immunol* 171: 1070–1077.

Bonardi, M.A., Bell, A., Fench, R.R., Gromo, G., Hamblin, T., Modena, D., Tutt, A.L., Glennie, M.J. (1992) Initial experience in treating human lymphoma with a combination of bispecific antibody and saporin. *Int J Cancer Suppl* 7: 73–77.

Bonardi, M.A., French, R.R., Amlot, P., Gromo, G., Modena, D., Glennie, M.J. (1993) Delivery of saporin to human B-cell lymphoma using bispecific antibody: targeting via CD22 but not CD19, CD37, or immunoglobulin results in efficient killing. *Cancer Res* 53: 3015–3021.

Brandão, J.G., Scheper, R.J., Lougheed, S.M., Curiel, D.T., Tillman, B.W., Gerritsen, W.R., van den Eertwegh, A.J., Pinedo, H.M., Haisma, H.J., de Gruijl, T.D. (2003) CD40-targeted adenoviral gene transfer to dendritic cells through the use of a novel bispecific single-chain Fv antibody enhances cytotoxic T cell activation. *Vaccine* 21: 2268–2272.

Brekke, O.H., Sandlie, I. (2003) Therapeutic antibodies for human diseases at the dawn of the twenty-first century. *Nat Rev Drug Discov* 22: 52–62.

Brennan, M. (1986) A chemical technique for the preparation of bispecific antibodies from Fab fragments of mouse monoclonal IgG1. *BioTechniques* 4: 424–427.

Brennan, M., Davison, P.F., Paulus, H. (1985) Preparation of bispecific antibodies by chemical recombination of monoclonal immunoglobulin G1 fragments. *Science* 229: 81–83.

Bruenke, J., Fischer, B., Barbin, K., Schreiter, K., Wachter, Y., Mahr, K., Titgemeyer, F., Niederweis, M., Peipp, M., Zunino, S., Repp, R., Valerius, T., and Fey, G.H. (2004) A recombinant bispecific single-chain Fv antibody against HLA classII and FcγRIII (CD16) triggers effective lysis of lymphoma cells. *Br J Haematol* 125: 167–179.

Bruenke, J., Barbin, K., Kunert, S., Lang, P., Pfeiffer, M., Stieglmaier, K., Niethammer, D., Stockmeyer, B., Peipp, M., Repp, R., Valerius, T., Fey, G.H. (2005) Effective lysis of lymphoma cells with a stabilised bispecific single-chain Fv antibody against CD19 and FcgammaRIII (CD16). *Br J Haematol* 130: 218–228.

Brüsselbach, S., Korn, T., Völkel, T., Müller, R., Kontermann, R.E. (1999) Enzyme recruitment and tumor cell killing in vitro by a secreted bispecific single-chain diabody. *Tumor Targeting* 4: 115–123.

Bubenik, J. (2003) Tumour MHC class I downregulation and immunotherapy. *Oncol Rep* 10: 2205–2208.

Cao, Y., Lam, L. (2003) Bispecific antibody conjugates in therapeutics. *Adv Drug Deliv Rev* 55: 171–197.

Cao, Y., Suresh, M.R. (1998) Bispecific antibodies as novel bioconjugates. *Bioconjug Chem* 9: 635–644.

Cao, Y., Vinayagamoorthy, T., Noujaim, A.A., Suresh, M.R. (1995) A rapid non-selective method to generate quadromas by microelectrofusion. *J Immunol Methods* 187: 1–7.

Carter, P. (2001) Bispecific human IgG by design. *J Immunol Methods* 248: 7–15.

Chang, C.-H., Sharkey, R.M., Rossi, E.A., Karacay, H., McBridge, W., Hansen, H.J., Chatal, J.-F., Barbet, J., Goldenberg, D.M. (2002) Molecular advances in pretargeting radioimmunotherapy with bispecific antibodies. *Mol Cancer Ther* 1: 553–563.

Cochlovius, B., Kipriyanov, S.M., Stassar, M.J.J.G., Schuhmacher, J., Benner A., Moldenhauer, G., Little, M. (2000) Cure of Burkitt's lymphoma in severe combined immunodeficiency mice by T cells, tetravalent CD3 × CD19 tandem diabody, and CD28 costimulation. *Cancer Res* 60: 4336–4341.

Coloma, M.J., Morrison, S.L. (1997) Design and production of novel tetravalent bispecific antibodies. *Nat Biotechnol* 15: 159–163.

Connelly, R.J., Hayden, M.S., Scholler, J.K., Tsu, T.T., Dupont, B., Ledbetter, J.A., Kanner, S.B. (1998) Mitogenic properties of a bispecific single-chain Fv-Ig fusion generated from CD2-specific mAb to distinct epitopes. *Int Immunol* 10: 1863–1872.

Corvalan, J.R., Smith, W. (1987) Construction and characterisation of a hybrid-hybrid monoclonal antibody recognising both carcinoembryonic antigen (CEA) and vinca alkaloids. *Cancer Immunol Immunother* 24: 127–132.

Curnow, R.T. (1997) Clinical experience with CD64-directed immunotherapy. An overview. *Cancer Immunol Immunother* 45: 210–215.

de Gast, G.C., van de Winkel, J.G.J., Bast, B.E.J.E.G. (1997) Clinical perspectives of bispecific antibodies in cancer. *Cancer Immunol Immunother* 45: 121–123.

de Gruijl, T.D., Luykx-de Bakker, S.A., Tillman, B.W., van den Eertwegh, A.J., Buter, J., Lougheed, S.M., van der Bij, G.J., Safer, A.M., Haisma, H.J., Curiel, D.T., Scheper, R.J., Pinedo, H.M., Gerritsen, W.R. (2002) Prolonged maturation and enhanced transduction of dendritic cells migrated from human skin explants after in situ delivery of CD40-targeted adenoviral vectors. *J Immunol* 169: 5322–5331.

de Kruif, J., Logtenberg, T. (1996) Leucine zipper dimerized bivalent and bispecific scFv antibodies from a semisynthetic antibody phage display library. *J Biol Chem* 271: 7630–7634.

De Lau, W.B., van Loon, A.E., Heije, K., Valerio, D., Bast, B.J. (1989) Production of hybrid hybridomas based on HAT(s)-neomycin double mutants. *J Immunol Methods* 117: 1–8.

De Lau, W.B., Keije, K., Neefjes, J.J., Oosterwegel, M., Rozemüller, E., Blast, B.J. (1991) Absence of preferential homologous

H/L chain association in hybrid hybridomas. *J Immunol* 146: 906–914.

DeNardo, D.G., Xiong, C.Y., Shi, X.B., DeNardo, G.L., DeNardo, S.J. (2001) Anti-HLA-DR/anti-DOTA diabody construction in a modular gene design platform: bispecific antibodies for pretargeted radioimmunotherapy. *Cancer Biother Radiopharm* 16: 525–535.

Deo, Y.M., Sundarapandiyan, K., Keler, T., Wallace, P.K., Graziano, R.F. (1998) Bispecific molecules directed to the Fc receptor for IgA (FcαRI, CD89) and tumor antigens efficiently promote cell-mediated cytotoxicity of tumor targets in whole blood. *J Immunol* 160: 1677–1686.

Dreier, T., Baeuerle, P.A., Fichtner, I., Grün, M., Schlereth, B., Lorenczewski, G., Kufer, P., Lutterbüse, R., Riethmüller, G., Gjorstrup, P., Bargou, R.C. (2003) T cell costimulus-independent and very efficacious inhibition of tumor growth in mice bearing subcutaneous or leukemic human B cell lymphoma xenografts by a CD19-/CD3- bispecific single-chain antibody construct. *J Immunol* 170: 4397–4402.

El-Aneed, A. (2004) An overview of current delivery systems in cancer gene therapy. *J Control Release* 94: 1–14.

Embleton, M.J., Charleston, A., Robins, R. A., Pimm, M.V., Baldwin, R.W. (1991) Recombinant ricin toxin A cytotoxicity against carcinoembryonic antigen expressing tumour cells mediated by a bispecific monoclonal antibody and its potentiation by ricin toxin chain. *Br J Cancer* 63: 670–674.

Everts, M., Curiel, D.T. (2004) Transductional targeting of adenoviral cancer gene therapy. *Curr Gene Ther* 4: 337–346.

Fanger, M.W., Guyre, P.M. (1991) Bispecific antibodies for targeted cellular cytotoxicity. *Trends Biotechnol* 9: 375–380.

Fanger, M.W., Morganelli, P.M., Guyre, P.M. (1992) Bispecific antibodies. *Crit Rev Immunol* 12: 101–124.

Ferrini, S., Sforzini, S., Canevari, S. (2001) Bispecific monoclonal antibodies for the targeting of type I ribosome-inactivating proteins against hematological malignancies. *Methods Mol Biol* 166: 177–192.

Flavell, D.J., Cooper, S., Morland, B., French, R., Flavell, S.U. (1992) Effectiveness of combinations of bispecific antibodies for delivering saporin to human acute T-cell lymphoblastic leuaemia cell lines via CD7 and CD38 as cellular target molecules. *Br J Cancer* 65: 545–551.

FitzGerald, K., Holliger, P., Winter, G. (1997) Improved tumour targeting by disulphide stabilized diabodies expressed in *Pichia pastoris*. *Protein Eng* 10: 1221–1225.

Ford, C.H., Osborne, P.A., Rego, B.G., Mathew, A. (2001) Bispecific antibody targeting of doxorubicin to carcinoembryonic antigen-expressing colon cancer cells lines in vitro and in vivo. *Int J Cancer* 92: 851–855.

French, R.R., Penney, C.A., Browning, A.C., Stirpe, F., Georg, A.J., Glennie, M.J. (1995) Delivery of the ribosome-inactivating protein, gelonin, to lymphoma cells via CD22 and CD38 using bispecific antibodies. *Br J Cancer* 71: 986–994.

Gautherot, E., Bouhou, J., Le Doussal, J.M., Manetti, C., Martin, M., Rouvier, E., Barbet, J. (1997) Therapy for colon carcinoma xenografts with bispecific antibody-targeted, iodine-131-labeled bivalent hapten. *Cancer* 80: 2618–2623.

Gerstmayer, B., Hoffmann, M., Altenschmidt, U., Wels, W. (1997) Costimulation of T-cell proliferation by a chimeric B7-antibody fusion protein. *Cancer Immunol Immunother* 45: 156–158.

Glennie, M.J., McBride, H.M., Worth, A.T., Stevenson, G.T. (1987) Preparation and performance of bispecific F(ab′ γ)2 antibody containing thioether-linked Fab′ γ fragments. *J Immunol* 139: 2367–2375.

Goldenberg, D.M. (2003) Advancing role of radiolabeled antibodies in the therapy of cancer. *Cancer Immunol Immunother* 52: 281–296.

Graziano, R.F., Guptill, P. (2004) Chemical production of bispecific antibodies. *Methods Mol Biol* 283: 71–85.

Grill, J., van Beusechem, V.W., van der Valk, P., Dirven, C.M.F., Leonhart, A., Pherai, D.S., Haisma, H.J., Pinedo, H.M., Curiel, D.T., Gerritsen, W.R. (2001) Combined targeting of adenoviruses to integrins and epidermal growth factor receptors increased gene transfer into primary glioma cells and spheroids. *Clin Cancer Res* 7: 641–650.

Grosse-Hovest, L., Hartlapp, I., Marwan, W., Brem, G., Rammensee, H.-G., Jung, G. (2003) A recombinant bispecific single-chain antibody induces targeted, supraagonistic CD28-stimulation and tumor cell killing. *Eur J Immunol* 33: 1334–1340.

Grosse-Hovest, L., Müller, S., Minoia, R., Wolf, E., Zakhartchenko, V., Wenigerkind, H., Lassnig, C., Besenfelder, U., Müller, M., Lytton, S.D., Jung, G., Brem, G. (2004) Cloned transgenic farm animals produce a bispecific antibody for T cell-mediated tumor cell killing. *Proc Natl Acad Sci USA* 101: 6858–6863.

Grosse-Hovest, L., Wick, W., Minoia, R., Weller, M., Rammensee, H.-G., Brem, G., Jung, G. (2005) Supraagonistic, bispecific single-chain antibody purified from the serum of cloned, transgenic cows induces T-cell-mediated killing of glioblastoma cells *in vitro* and *in vivo*. *Int J Cancer* 117: 1060–1064.

Gruaz-Guyon, A., Raguin, O., Barbet, J. (2005) Recent advances in pretargeted radioimmunotherapy. *Curr Med Chem* 12: 3319–338.

Haisma, H.J., Pinedo, H.M., Rijswijk, A., der Meulen-Muileman, I., Sosnowski, B. A., Ying, W., van Beusechem, V.W., Tillman, B.W., Gerritsen, W.R., Curiel, D. T. (1999) Tumor-specific gene transfer via an adenoviral vector targeted to the pan-carcinoma antigen EpCAM. *Gene Ther* 6: 469–474.

Hartmann, F., Renner, C., Jung, W., Deisting, C., Juwana, M., Eichentopf, B., Kloft, M., Pfreundschuh, M. (1997) Treatment of refractory Hodgkin's disease with an anti-CD16/CD30 bispecific antibody. *Blood* 89 2042–2047.

Hartmann, F., Renner, C., Jung, W., da Costa, L., Tembrink, S., Held, G., Sek, A., König, J., Bauer, S., Kloft, M., Pfreundschuh, M. (2001) Anti-CD16/CD30 bispecific antibody treatment for Hodgkin's disease: role of infusion schedule and costimulation with cytokines. *Clin Cancer Res* 7: 1873–1881.

Hayden, M.S., Linsley, P.S., Gayle, M.A., Bajorath, J., Brady, W.A., Norris, N.A., Fell, H.P., Ledbetter, J.A., Gilliland, L.K. (1994) Single-chain mono- and bispecific antibody derivatives with novel biological properties and antitumour activity from a COS cell transient expression system. *Ther Immunol* 1: 3–15.

Heideman, D.A., Snijders, P.J., Craanen, M. E., Bloemena, E., Meijer, C.J., Meuwissen, S.G., van Beusechem, V.W., Pinedo, H.M., Curiel, D.T., Haisma, H.J., Gerritsen, W.R. (2001) Selective gene delivery toward gastric and esophageal adenocarcinoma cells via EpCAM-targeted adenoviral vectors. *Cancer Gene Ther* 8: 342–351.

Hendrie, P.C., Russell, D.W. (2005) Gene targeting with viral vectors. *Mol Ther* 12: 9–17.

Heiss, M.M., Strohlein, M.A., Jäger, M., Kimmig, R., Burges, A., Schoberth, A., Jauch, K.W., Schildberg, F.W., Lindhofer, H. (2005) Immunotherapy of malignant ascites with trifunctional antibodies. *Int J Cancer* 117: 435–443.

Hoffmann, P., Hofmeister, R., Brischwein, K., Brandl, C., Crommer, S., Bargou, R., Itin, c., Prang, N., Baeuerle, P.A. (2005) Serial killing of tumor cells by cytotoxic T cells redirected with a CD19-/CD3-bispecific single-chain antibody construct. *Int J Cancer* 115: 98–104.

Holliger, P., Prospero, T., Winter, G. (1993) "Diabodies": small bivalent and bispecific antibody fragments. *Proc Natl Acad Sci USA* 90: 6444–6448.

Holliger, P., Wing, M., Pound, J.D., Bohlen, H., Winter, G. (1997) Retargeting serum immunoglobulin with bispecific diabodies. *Nat Biotechnol* 15: 632–636.

Holliger P., Manze, O., Span, M., Hawkins, R., Fleischmann, B., Qianghua, L., Wolf, J., Diehl, V., Cochet, O., Winter, G., Bohlen, H. (1999) Carcinoembryonic antigen (CEA)-specific T-cell activation in colon carcinoma induced by anti-CD3xanti-CEA bispecific diabodies and B7 × anti-CEA bispecific fusion proteins. *Cancer Res* 59: 2909–2916.

Hombach, A., Jung, W., Pohl C., Renner C., Sahin U., Schmits R., Wolf, J., Kapp, U., Diehl, V., Pfreundschuh, M. (1993) A CD16/CD30 bispecific monoclonal antibody induces lysis of Hodgkin's cells by unstimulated natural killer cells in vitro and in vivo. *Int J Cancer* 55: 830–836.

James, N.D., Atherton, P.J., Jones, J., Howie, A.J., Tchekmedyian, S., Curnow, R.T. (2001) A phase II study of the bispecific antibody MDX-H210 (anti-HER2 × CD64) with GM-CSF in HER2+ advanced prostate cancer. *Br J Cancer* 85: 152–156.

Jendreyko, N., Popkov, M., Beerli, R.R., Chung, J., McGavern, D.B., Rader, C., Barbas, C.F. III (2003) Intradiabodies, bispecific, tetravalent antibodies for the simultaneous functional knockout of two cell surface receptors. *J Biol Chem* 278: 47812–4789.

Jendreyko, N., Popkov, M., Rader, C., Barbas, C.F. III (2005) Phenotypic knockout of VEGF-R2 and Tie-2 with an intradiabody reduces tumor growth and angiogenesis in vivo. *Proc Natl Acad Sci USA* 102: 8293–8298.

Jimenez, X., Lu, D., Brennan, L., Persaud, K., Liu, M., Miao, H., Witte, L., Zhu, Z.A. (2005) A recombinant, fully human, bispecific antibody neutralizes the biological activities mediated by both vascular endothelial growth factor receptors 2 and 3. *Mol Cancer Ther* 4: 427–434.

Kanerva, A., Hemminki, A. (2005) Adenoviruses for treatment of cancer. *Ann Med* 37: 33–43.

Karawajew, L., Micheel, B., Behrsing, O., Gaestel, M. (1987) Bispecific antibody-producing hybrid hybridomas selected by a fluorescence activated cell sorter. *J Immunol Methods* 92: 265–270.

Keler, T., Graziano, R.F., Mandal, A., Wallace, P.K., Fisher, J., Guyre, P.M., Fanger, M.W., Deo, Y.M. (1997) Bispecific antibody-dependent cellular cytotoxicity of HER2/neu-overexpressing tumor cells by Fcγ receptor type-I expressing effector cells. *Cancer Res* 57: 4008–4014.

Kelly, F.J., Miller, C.R., Buchsbaum, D.J., Gomez-Navarro, J., Barnes, M.N., Alvarez, R.D., Curiel, D.T. (2000) Selectivity of TAG-72-targeted adenovirus gene transfer to primary ovarian carcinoma cells versus autologous mesothelial cells in vitro. *Clin Cancer Res* 6: 4323–4333.

Kipriyanov, S.M., Moldenhauer, G., Schuhmacher, J., Cochlovius, B., Von der Lieth, C.W., Matys, E.R., Little, M. (1999) Bispecific tandem diabody for tumor therapy with improved antigen binding and pharmacokinetics. *J Mol Biol* 293: 41–56.

Kipriyanov, S.M., Cochlovius, B., Schäfer, H.J., Moldenhauer, G., Bähre, A., Le Gall, F., Knackmuss, S., Little, M. (2002) Synergistic antitumor effect of bispecific CD19 × CD3 and CD19 × CD16 diabodies in a preclinical model of non-Hodgkin's lymphoma. *J Immunol* 169: 137–144.

Köhler, G., Milstein, C. (1975) Continuous cultures of fused cells secreting antibody of predefined specificity. *Nature* 256: 495–497.

Kontermann, R.E. (2005) Recombinant bispecific antibodies for cancer therapy. *Acta Pharmacol Sin* 26: 1–9.

Kontermann, R.E., Wing, M.G., Winter, G. (1997) Complement recruitment using bispecific diabodies. *Nat Biotechnol* 15: 629–631.

Kontermann, R.E., Korn, T., Jérôme, V. (2002) Recombinant adenoviruses for in vivo expression of antibody fragments. In: Welschof M, Kraus J (eds) *Recombinant Antibody Technology for Cancer Therapy: Reviews and Protocols*. Methods in Molecular Medicine. Totowa: Humana Press, pp. 421–433.

Koolwijk, P., Rozemüller, E., Stad, R.K., de Lau, W.B., Blast, B.J. (1988) Enrichment and selection of hybrid hybridomas by percoll density gradient centrifugation and flurescent-activated cell sorting. *Hybridoma* 7: 217–225.

Koolwijk, P., Spierenburg, G.T., Frasa, H., Boot, J.H., van de Winkel, J.H., Bat, B.J. (1989) Interaction between hybrid mouse monoclonal antibodies and the human high-affinity IgG FcR, huFcγRI, on U937. Involvement of only one of the mIgG heavy chains in receptor binding. *J Immunol* 143: 1656–1662.

Koolwijk, P., Boot, J.H., Griep, R., Bast, B.J. (1991) Binding of human complement subcomponent C1q to hybrid mouse monoclonal antibodies. *Mol Immunol* 28: 567–576.

Korn, T., Nettelbeck, D.M., Völkel, T., Müller, R., Kontermann, R.E. (2004a) Recombinant bispecific antibodies for the targeting of adenoviruses to CEA-expressing tumour cells: a comparative analysis of bacterially expressed single-chain diabody and tandem scFv. *J Gene Med* 6: 642–651.

Korn, T., Muller, R., Kontermann, R.E. (2004b) Bispecific single-chain diabody-mediated killing of endoglin-positive endothelial cells by cytotoxic T lymphocytes. *J Immunother* 27: 99–106.

Kostelny, S.A., Cole, M.S., Tso, J.Y. (1992) Formation of a bispecific antibody by the use of leucine zippers. *J Immunol* 148: 1547–1553.

Kraeber-Bodéré, F., Faivre-Chauvet, A., Ferrer, L., Vuillez, F.-P., Brard, P.-Y., Rousseau, C., Resche, I., Devillers, A., Laffont, S., Bardiés, M., Chang, K., Sharkey, R.M., Goldenberg, D.M., Chatal, J.-F., Barbet, J. (2003) Pharmacokinetics and dosimetry studies for optimization of anti-carcinoembryonic antigen × anti-hapten bispecific antibody-mediated pretargeting of iodine-131-hapten in a phase I radioimmunotherapy trial. *Clin Cancer Res* 9: 3973s–3981s.

Krannenborg, M.H., Boerman, O.C., Oosterwijk-Wakka, J.C., de Weijert, M.C., Corstens, F.H., Oosterweijk, E. (1998) Two-step radio-immunotherapy of renal-cell carcinoma xenografts in nuce mice with anit-renal-cell-carcinoma × anti-DTPa bispecific monoclonal antibodies. *Int J Cancer* 75: 74–80.

Kreutz, F.T., Xu, D.Z., Suresh, M.R. (1998) A new method to generate quadromas by electrofusion and FACS sorting. *Hybridoma* 17: 267–273.

Kriangkum, J., Xu, B., Nagata, L.P., Fulton, R.E., Suresh, M.R. (2001) Bispecific and bifunctional single chain recombinant antibodies. *Biomol Eng* 18: 31–40.

Kufer, P., Mack, M., Gruber, R., Lutterbuse, R., Zettl, F., Riethmuller, G. (1997) Construction and biological activity of a recombinant bispecific single-chain antibody designed for therapy of minimal residual colorectal cancer. *Cancer Immunol Immunother* 45 193–197.

Kufer, P., Lutterbuse, R., Baeuerle, P.A. (2004) A revival of bispecific antibodies. *Trends Biotechnol* 22: 38–44.

Kuus-Reichel, K., Knott, C., Sam-Fong, P., Petrella, E., Corvalan, J.R. (1995) Therapy of streptozotocin induced diabetes with a bifunctional antibody that delivers vinca alkaloids to IL-2 receptor positive cells. *Autoimmunity* 22: 173–181.

Lanzavecchia, A., Scheidegger, D. (1987) The use of hybrid hybridomas to target human cytotoxic T lymphocytes. *Eur J Immunol* 17: 105–111.

Le Doussal, J.M., Martin, M., Gruaz-Guyon, A., Gautherot, E., Delaage, M., Barbet, J. (1990) In vitro and in vivo targeting of radiolabeled monovalent and bivalent haptens with dual specificity monoclonal antibody conjugates: enhanced divalent hapten affinity for cell-bound antibody conjugates. *J Nucl Med* 30: 1358–1366.

Lewis, L.D., Cole, B.F., Wallace, P.K., Fisher, J.L., Waugh, M., Guyre, P.M., Fanger M. W., Curnow, R.T., Kaufman, P.A., Ernstoff, M.S. (2001) Pharmacokinenic-pharmacodynamic relationship of the bispecific antibody MDX-H210 when administered in combination with interferon gamma: a multiple-dose phase-I study in patients with advanced cancer which overexpresses HER-2/neu. *J Immunol Methods* 248: 149–165.

Lindhofer, H., Mocika, R., Steipe, B., Thierfelder, S. (1995) Preferential species-restriced heavy/light chain pairing in rat/mouse quadromas. Implications for a single-step purification of bispecific antibodies. *J Immunol* 155: 219–225.

Link, B.K., Weiner, G.J. (1993) Production and characterization of a bispecific IgG capable of inducing T-cell-mediated lysis of malignant B cells. *Blood* 81: 3343–3349.

Lloyd, F. Jr., Goldrosen, M. (1991) The production of a bispecific anti-CEA, anti-hapten (4-amino-phthalate) hybrid-hybridoma. *J Natl Med Assoc* 83: 901–904.

Löffler, A., Kufer, P., Lutterbüse, R., Zettl, F., Daniel, P.T., Schwenkenbecher, J.M., Riethmüller, G., Dörken, B., Bargou, R.C. (2000) A recombinant bispecific single-chain antibody, CD19 × CD3: induces rapid and high lymphoma-directed cytotoxicity by unstimulated T lymphoctes. *Blood* 95: 2089–2103.

Lüdher, F., Huang, Y., Dennehy, K.M., Guntermann, C., Müller, I., Winkler, E., Kerkau, T., Ikemizu, S., Davis, S.J., Hanke, T., Hünig, T. (2003) Topological requirements and signaling properties of T cell-activating, anti-CD28 antibody superagonists. *J Exp Med* 197: 955–966.

Lu, D., Kotanides, H., Jimenez, X., Zhou, Q., Persaud, K., Bohlen, P., Witte, L., Zhu, Z. (1999) Acquired antagonistic activity of a bispecific diabody directed against two different epitopes on vascular endothelial growth factor receptor 2. *J Immunol Methods* 230: 159–171.

Lu, D., Jimenez, X., Zhang, H., Wu, Y., Bohlen, P., Witte, L., Zhu, Z. (2001) Complete inhibition of vascular endothelial growth factor (VEGF) activities with a bifunctional diabody directed against both VEGF kinase receptors, fms-like tyrosine kinase receptor and kinase insert domain-containing receptor. *Cancer Res* 61: 7002–7008.

Lu, D., Jimenez, X., Zhang, H., Bohlen, P., Witte, L., Zhu, Z. (2002) Fab-scFv fusion protein: an efficient approach to production of bispecific antibody fragments. *J Immunol Meth* 267: 213–226.

Lu, D., Jimenez, X., Zhang, H., Atkins, A., Brennan, L., Balderes, P., Bohlen, P., Witte, L., Zhu, Z. (2003) Di-diabody: a novel tetravalent bispecific antibody molecule by design. *J Immunol Methods* 279: 219–232.

Lu, D., Zhang, H., Ludwig, D., Persaud, A., Jimenez, X., Burtrum, D., Balderes, P., Liu, M., Bohlen, P., Witte, L., Zhu, Z. (2004) Simultaneous blockade of both the epidermal growth factor receptor and the insulin-like growth factor receptor signaling pathways in cancer cells with a fully human recombinant bispecific antibody. *J Biol Chem* 279: 2856–2865.

Lu, D., Zhang, H., Koo, H., Tonra, J., Balderes, P., Prewett, M., Corcoran, E., Mangalampalli, V., Bassi, R., Anselma, D., Patel, D., Kang, X., Ludwig, D.L., Hicklin, D.J., Bohlen, P., Witte, L., Zhu, Z. (2005) A fully human recombinant IgG-like bispecific antibody to both the epidermal growth factor receptor and the insulin-like growth factor receptor for enhanced antitumor activity. *J Biol Chem* 280: 19665–19672.

Lum, L.G., Davol, P.A. (2005) Retargeting T cells and immune effector cells with bispecific antibodies. *Cancer Chemother Biol Response Modif* 22: 273–291.

Madrenas, J., Chau, L.A., Teft, W.A., Wu, P.W., Jussif, J., Kasaian, M,. Carreno, B.M., Ling, V. (2004) Conversion of CTLA-4 from inhibitor to activator of T cells with a bispecific tandem single-chain Fv ligand. *J Immunol* 172: 5948–5956.

Marvin, J.S., Zhu, Z. (2005) Recombinant approaches to IgG-like bispecific antibodies. *Acta Pharmacol Sin* 26: 649–658.

McCall, A.M., Adams, G.P., Amoroso, A.R., Nielsen, U.B., Zhang, L., Horak, E., Simmons, H., Schier, R., Marks, J.D., Weiner, L.M. (1999) Isolation and characterization of an anti-CD16 single-chain Fv fragment and construction of an anti-HER2/neu/anti-CD16 bispecific scFv that triggers CD16-dependent tumor cytolysis. *Mol Immunol* 36: 433–445.

McCall, A.M., Shahied, L., Amoroso, A.R., Horak, E.M., Simmons, H.H., Nielson, U., Adams, G.P., Schier, R., Marks, J.D., Weiner, L.M. (2001) Increasing the affinity for tumor antigen enhances bispecific antibody cytotoxicity. *J Immunol* 166: 6112–6117.

Merchant, A.M., Zhu, Z., Yuan, J.Q., Goddard, A., Adams, C.W., Presta, L.G., Carter, P. (1998) An efficient route to human bispecific IgG. *Nat Biotechnol* 16: 677–681.

Milstein, C., Cuello, A.C. (1983) Hybrid hybridomas and their use in immunohistochemistry. *Nature* 305: 537–540.

Molema, G., Kroesen, B.J., Helfrich, W., Meijer, D.K.F., de Leij, L.F.M.H. (2000) The use of bispecific antibodies in tumor cell and tumor vasculature directed immunotherapy. *J Control Release* 64: 229–239.

Morandeau, L., Benoist, E., Loussouarn, A., Quadi, A., Lesaec, P., Mougin, M., Faivre-Chauvet, A., Le Boterff, J., Chatal, J.F., Barbet, J., Gestin, J.F. (2005) Synthesis of new bivalent peptides for applications in the affinity enhancement system. *Bioconjug Chem* 16: 184–193.

Morelli, d., Sardini, A., Villa, E., Menard, S., Colnaghi, M.I., Balsari, A. (1994) Modulation of drug-induced cytotoxicity by a bispecific monoclonal antibody that recognizes the epidermal growth factor receptor and doxorubicin. *Cancer Immunol Immunother* 38: 171–177.

Müller, K.M., Arndt, K.M., Strittmatter, W., Plückthun, A. (1998a) The first constant domain (C_H1 and C_L) of an antibody used as heterodimerization domain for bispecific miniantibodies. *FEBS Lett* 422: 259–264.

Müller, K.M., Arndt, K.M., Plückthun, A. (1998b) A dimeric bispecific miniantibody combines two specificities with avidity. *FEBS Lett* 432: 45–49.

Nettelbeck, D.M., Miller, D.W., Jérôme, V., Zuzarte, M., Watkins, S.J., Hawkins, R.E., Müller, R., Kontermann, R.E. (2001) Targeting of adenovirus to endothelial cells by a bispecific single-chain diabody directed against the adenovirus fiber knob domain and human endoglin (CD105). *Mol Ther* 3: 882–891.

Nettelbeck, D.M., Rivera, A.A., Kupsch, J., Dieckmann, D., Douglas, J.T., Kontermann, R.E., Alemany, R., Curiel, D.T. (2004) Retargeting of adenoviral infection to melanoma: combining genetic ablation of native tropism with a recombinant bispecific single-chain diabody (scDb) adapter that binds to fiber knob and HMWMAA. *Int J Cancer* 108: 136–145.

Nisonoff, A., Rivers, M.M. (1961) Recombination of a mixture of univalent antibody fragments of different specificity. *Arch Biochem Biophys* 93: 460–462.

Park, S.S., Ryu, C.J., Kang, Y.J., Kashmiri, S.V., Hong, H.J. (2000) Generation and characterization of a novel tetravalent bispecific antibody that binds to hepatitis B virus surface antigens. *Mol Immunol* 37: 1123–1130.

Paulus, H. (1985) Preparation and biomedical applications of bispecific antibodies. *Behring Inst Mitt* 78: 118–132.

Peipp, M., Valerius, T. (2002) Bispecific antibodies targeting cancer cells. *Biochem Soc Trans* 30: 507–511.

Plückthun, A., Pack, P. (1997) New protein engineering approaches to multivalent and bispecific antibody fragments. *Immunotechnology* 3: 83–105.

Pullarkat, V., Deo, Y., Link, J., Spears, L., Marty, V., Curnow, R., Groshen, S., Gee, C., Weber, J.S. (1999) A phase I study of a HER2/neu bispecific antibody with granulocyte-colony-stimulating factor in patients with metastatic breast cancer that overexpresses HER2/neu. *Cancer Immunol Immunother* 48: 9–21.

Ren-Heidenreich, L., Davol, P.A., Kouttab, N.M., Elfenbein, G.J., Lum, L.G. (2004) Redirected T-cell cytotoxicity to epithelial cell adhesion molecule-overexpressing adenocarcinomas by a novel recombinant antibody, E3Bi, in vitro and in an animal model. *Cancer* 100: 1095–1103.

Repp, R., van Ojik, H.H., Valerius, T., Groenewegen, G., Wieland, G., Oetzel, C., Stockmeyer, B., Becker, W., Eisenhut, M., Steininger, H., Deo, Y.M., Blijham, G.H., Kalden, J.R., van de Winkel, J.G.J., Gramatzki, M. (2003) Phase I clinical trial of the bispecific antibody MDX-H210 (anti-FcγRIxanti-HER-2/neu) in combination with Filgrastim (G-CSF) for treatment of advanced breast cancer. *Br J Cancer* 89: 2234–2243.

Reusch, U., Le Gall, F., Hensel, M., Moldenhauer, G., Ho, A.D., Little, M., Kipriyanov, S.M. (2004) Effect of tetravalent bispecific CD19 × CD3 recombinant antibody construct and CD28 costimulation on lysis of malignant B cells from patients with chronic lymphocytic leukaemia by autologous T cells. *Int J Cancer* 112: 509.

Reynolds, P.N., Zinn, K.R., Gavrilyuk, V.D., Balyasnikova, I.V., Rogers, B.E., Buchsbaum, D.J., Wang, M.H., Miletich, D.J., Grizzle, W.E., Douglas, J.T., Danilov, S.M., Curiel, D.T. (2000) A targetable, injectable adenoviral vector for selective gene delivery to pulmonary endothelium in vivo. *Mol Ther* 2: 562–578.

Reynolds, P.N., Nicklin, S.A., Kaliberova, L., Boatman, B.G., Grizzle, W.E., Balyasnikova, I.V., Baker, A.H., Danilov, S.M., Curiel, D.T. (2001) Combined transductional and transcriptional targeting improves the specificity of transgene expression in vivo. *Nat Biotechnol* 19: 838–842.

Ridgeway, J.B., Presta, L.G., Carter, P. (1996) "Knobs-into-holes" engineering of antibody CH3 domains for heavy chain heterodimerization. *Protein Eng* 9: 617–621.

Ruf, P., Lindhofer, H. (2001) Induction of a long-lasting antitumor immunity by a trifunctional bispecific antibody. *Blood* 98: 2526–2534.

Russell, J.H., Ley, T.J. (2002) Lymphocyte-mediated cytotoxicity. *Annu Rev Immunol* 20: 323–370.

Russoniello, C., Soomasundaram, C., Schlom, J., Deo, Y.M., Keler, T. (1998) Characterization of a novel bispecific antibody that mediates Fcγ receptor type-I-dependent killing of tumor-associated glycoprotein-72-expressing tumor cells. *Clin Cancer Res* 4: 2237–2243.

Sanz, L., Blanco, B., Àlvarez-Vallina, L. (2004) Antibodies and gene therapy: teaching old "magic bullets" new tricks. *Trends Immunol* 25: 85–91.

Schlenzka, J., Moehler, T.M., Kipriyanov, S.M., Kornacker, M., Benner, A., Bahre, A., Stassar, M.J., Schafer, H.J., Little, M., Goldschmidt, H., Cochlovius, B. (2004) Combined effect of recombinant CD19 × CD16 diabody and thalidomide in a preclinical model of human B cell lymphoma. *Anticancer Drugs* 15: 915–919.

Schlereth, B., Fichtner, I., Lorenczewski, G., Kleindienst, P., Brischwein, K., da Silva, A., Kufer, P., Lutterbüse, R., Juhnhahn, I., Kasimir-Bauer, S., Wimberger, P., Kimmig, R., Baeuerle, P.A. (2005) Eradication of tumors from a human colon cancer cell line and from ovarian cancer metastases in immunodeficient mice by a single-chain Ep-CAM-/CD3-bispecific antibody construct. *Cancer Res* 65: 2882–2889.

Schneider, M.A., Bruhl, H., Wechselberger, A., Cihak, J., Stangassinger, M., Schlondorff, D., Mack, M. (2005) In vitro and in vivo properties of a dimeric bispecific single-chain antibody IgG-fusion protein for depletion of CCR2(+) target cells in mice. *Eur J Immunol* 35: 987–995.

Schoonjans, R., Willems, A., Schoonooghe, S., Fiers, W., Grooten, J., Mertens, N. (2000) Fab chains as an efficient heterodimerization scaffold for the production of recombinant bispecific and trispecific antibody derivatives. *J Immunol* 165: 7050–7057.

Segal, D.M., Weiner, G.J., Weiner, L.M. (1999) Bispecific antibodies in cancer therapy. *Curr Opin Immunol* 11: 558–562.

Sforzini, S., Bolognesi, A., Meazza, R., Marciano, S., Casalini, P., Durkop, H., Tazzari, P.L., Stein, H., Stirpe, F., Ferrini, S. (1995) Differential sensitivity of CD30+ neoplastic cells to gelonin delivered by anti-CD30/anti-gelonin bispecific antibodies. *Br J Haematol* 90: 572–577.

Shahied, L.S., Tang, Y., Alpaugh, R.K., Somer, R., Greenspon, D., Weiner, L.M. (2004) Bispecific minibodies targeting (HER2/neu and CD16 exhbit improved tumor lysis when placed in a divalent tumor antigen binding format. *J Biol Chem* 279: 53907–53914.

Shalaby, M.R., Shepard, H.M., Presta, L., Rodrgues, M.L., Beverley, P.C.L., Feldmann, M., Carter, P. (1992) Development of humanized bispecific antibodies reactive with cytotoxic lymphocytes and tumor cells overexpressing the HER2 protoncogene. *J Exp Med* 175: 217–225.

Sharkey, R.M., Karracay, H., Cardillo, T.M., Chang, C.H., McBride, W.J., Rossi, E.A., Horak, I.D., Goldenberg, D.M. (2005a) Improving the delivery of radionuclides for imaging and therapy of cancer using pretargeting methods. *Clin Cancer Res* 11: 7109s–7121s.

Sharkey, R.M., Karracay, H., Chang, C.H., McBride, W.J., Horak, I.D., Goldenberg, D.M. (2005b) Improved therapy of non-Hodgkin's lymphoma xenografts using radionuclides pretargeted with a new anti-CD20 bispecific antibody. *Leukemia* 19: 1064–1069.

Smith, W., Jarrett, A.L., Beattie, R.E., Corvalan, J.R. (1992) Immunoglobulins secreted by a hybrid-hybridoma: analysis of chain assemblies. *Hybridoma* 11: 87–98.

Staerz, U.D., Bevan, M.J. (1986) Hybrid hybridoma producing a bispecific monoclonal antibody that can focus effector T-cell activity. *Proc Natl Acad Sci USA* 83: 1453–1457.

Stockmeyer, B., Elsässer, D., Dechant, M., Repp, R., Gramatzki, M., Glennie, M.J., van de Winkel, J.G.J., Valerius, T. (2001) Mechanisms of G-CSF- or GM-CSF-stimulated tumor cell killing by Fc receptor-directed bispecific antibodies. *J Immunol Methods* 248: 103–111.

Sundarapandiyan, K., Keler, T., Behnke, D., Engert, A., Barth, S., Matthey, B., Deo, Y.M., Graziano, R.F. (2001) Bispecific antibody-mediated destruction of Hodgkin's lymphoma cells. *J Immunol Methods* 248: 113–123.

Tacke, M., Hanke, G., Hanke, T., Hunig, T. (1997) CD28-mediated induction of proliferation in resting T cells in vitro and in vivo without engagement of the T cell receptor: evidence for functionally distinct forms of CD28. *Eur J Immunol* 27: 239–247.

Tillman, B.W., Hayes, T.L., de Gruijl, T.D., Douglas, J.T., Curiel, D.T. (2000) Adenoviral vectors targeted to CD40 enhance the efficacy of dendritic cell-based vaccination against human papillomavirus 16-induced tumor cells in a murine model. *Cancer Res* 60: 5456–5463.

Tutt, A., Greenman, J., Stevenson, G.T., Glennie, M.J. (1991) Bispecific F(ab'γ)3 antibody derivatives for redirecting unprimed cytotoxic T cells. *Eur J Immunol* 21: 1351–1358.

Tutt, A.L., Reid, R., Wilkins, B.S., Glennie, M.J. (1995) Activation and preferential expansion of rat cytotoxic (CD8) T cells in vitro and in vivo with a bispecific (anti-TRC α/β × anti-CD2) F(ab')2 antibody. *J Immunol* 155: 2960–2971.

Valerius, T., Würflein, D., Stockmeyer, B., Repp, R., Kalden, J.R., Gramatzki, M. (1997) Activated neutrophils as effector cells for bispecific antibodies. *Cancer Immunol Immunother* 45: 142–145.

van de Winkel, J.G.J., Bast, B., de Gast, G.C. (1997) Immunotherapeutic potential of bispecific antibodies. *Immunol Today* 12: 562–564.

van Spriel, A.B., van Ojik, H.H., van de Winkel, J.G.J. (2000) Immunotherapeutic perspective for bispecific antibodies. *Immunol Today* 21: 391–396.

Völkel, T., Korn, T., Bach, M., Müller, R., Kontermann, R.E. (2001) Optimized linker sequences for the expression of monomeridc and dimeric bispecific single-chain diabodies. *Protein Eng* 14: 815–823.

Wallace, P.K., Kaufman, P.A., Lewis, L.D., Keler, T., Givan, A.L., Fisher, J.L., Waugh, M.G., Wahner, A.E., Guyre, P.M., Fanger, M.W., Ernstoff, M.S. (2001) Bispecific antibody-targeted phagocytosis of HER-2/neu expressing tumor cells by myeloid cells activated in vivo. *J Immunol Methods* 248: 167–182.

Walther, W., Stein, U. (2000) Viral vectors for gene transfer: a review of theiruse in the treatment of human diseases. *Drugs* 60: 249–271.

Warnaar, S.O., de Paus, V., Lardenoije, R., Machielse, B.N., de Graaf, J., Bregonje, M., van Haarlem, H. (1994) Purification of bispecific F(ab')$_2$ from murine trinoma OC/TR with specificity for CD3 and ovarian cancer. *Hybridoma* 13: 519–526.

Watkins, S.J., Mesyanzhinov, V.V., Kurochkina, L.P., Hawkins, R.E. (1997) The "adenobody" approach to viral targeting: specific and enhanced adenoviral gene delivery. *Gene Ther* 4: 1004–1012.

Weiner, L.M., Clark, J.I., Davey, M., Li, W.S., Garcia de Palazzo, I., Ring, D.B., Alpaugh, R.K. (1995) Phase I trial of 2B1: a bispecific monoclonal antibody targeting c-erbB-2 and FcγRIII. *Cancer Res* 55: 4586–4593.

Withoff, S., Helfrich, W., de Leij, L.F.M.H., Molema, G. (2001) Bi-sepcific antibody therapy for the treatment of cancer. *Curr Opin Mol Ther* 3: 53–62.

Wolf, E., Hofmeister, R., Kufer, P., Schlereth, B., Baeuerle, P.A. (2005) BiTEs: bispecific antibody constructs with unique anti-tumor activity. *Drug Discov Today* 10: 1237–1244.

Würdinger, T., Verheije, M.H., Raaben, M., Bosch, B.J., de Haan, C.A., van Beusechem, V.W., Rottier, P.J., Gerritsen, W.R. (2005) Targeting non-human coronaviruses to human cancer cells using a bispecific single-chain antibody. *Gene Ther* 12: 1394–1404.

Wuest, T., Moosmayer, D., Pfizenmaier, K. (2001) Construction of a bispecific single chain antibody for recruitment of cytotoxic T cells to the tumour stroma associated antigen fibroblast activation protein. *J Biotechnol* 92: 159–168.

Zeidler, R., Mysliwietz, J., Csanady, M., Walz, A., Ziegler, I., Schmitt, B., Wollenberg, B., Lindhofer, H. (2000) The Fc-region of a new class of intact bispecific antibody mediates activation of accessory cells and NK cells and induces direct phagocytosis of tumour cells. *Br J Cancer* 83: 261–266.

Zeidler, R., Mayer, A., Gires, O., Schmitt, B., Mack, B.M., Lindhofer, H., Wollenberg, B., Walz, A. (2001) TNFα contributes to the antitumor activity of a bspecific, trifunctional antibody. *AntiCancer Res* 21: 3499–3503.

Xie, Z., Shi, M., Feng, J., Yu, M., Sun, Y., Shen, B., Guo, N. (2003) A trivalent anti-erbB2/anti-CD16 bispecific antibody retargeting NK cells against human breast cancer cells. *Biochem Biophys Res Commun* 311: 307–312.

Xie, Z., Guo, N., Yu, M., Hu, M., Shen, B. (2005) A new format of bispecific antibody: highly efficient heterodimerization, expression and tumor cell lysis. *J Immunol Methods* 296: 95–101.

Zhu, Z., Presta, L.G., Zapata, G., Carter, P. (1997) Remodeling domain interfaces to enhance heterodimer formation. *Protein Sci* 6: 781–788.

Zuo, Z., Jimenez, X., Witte, L., Zhu Z. (2000) An efficient route to the production of an IgG-like bispecific antibody. *Protein Eng* 13: 361–367.

3
Immunotoxins and Beyond: Targeted RNases

Susanna M. Rybak and Dianne L. Newton

3.1
Introduction

Antibodies linked to cytotoxic compounds were designed to kill diseased cells while sparing normal ones. The benefits of this "smart drug" approach has great appeal and attempts at therapy with antibodies conjugated to drugs began in the 1950s (Mathe et al. 1958). A major milestone in the evolution of this therapeutic strategy was the development of murine monoclonal antibodies (Köhler and Milstein 1975). The increased specificity further encouraged targeting cancer-associated antigens expressed at high levels on tumor cells and at low levels on non-essential or easily renewable normal cells. Chemicals, radioisotopes, and toxins have been linked to antibodies to produce reagents that progressed from experimental studies to clinical trials (Milenic 2002; Wu and Senter 2005). Initially, standard chemotherapeutic drugs chemically linked to antibodies were widely explored as targeted cancer therapeutics (Blattler 1996). A major problem with those drug conjugates was the low molar cytotoxicity of the drugs, resulting in little improvement in efficacy or nonspecific toxicity over the unconjugated drug. This led to the coupling of chemical toxins that were several fold more potent than the chemotherapeutic drugs (reviewed in Payne 2003; Lambert 2005). The humanized anti-CD33 antibody calicheamicin conjugate (Mylotarg) is of this novel drug type, and has been approved for the treatment of CD33-positive acute myeloid leukemia (Hamann et al. 2002). In addition to giving proof of concept of the power of armed antibody targeting, the approval of Mylotarg demonstrates that the process of linking an antibody to another molecule to create a new drug is not too complex to be commercialized. Mylotarg is also the result of improvements in antibodies, such as humanization, that occurred through developments in technologies that solved some of the problems associated with murine monoclonal antibodies.

Handbook of Therapeutic Antibodies. Edited by Stefan Dübel
Copyright © 2007 WILEY-VCH Verlag GmbH & Co. KGaA, Weinheim
ISBN 978-3-527-31453-9

3.1.1
Targeted Drug Architechture

The progression from murine to human monoclonal antibodies as well as the engineering of novel antibody forms and fragments are coupled to the recent rapid advances in antibody targeted therapeutics (reviewed in Carter 2001; Hoogenboom 2005). Recombinant DNA technology afforded the opportunity to design fusion proteins for specific applications by altering the features of the antibody or enzyme domains. For instance, the nature of the antibody used, IgG, F(ab')$_2$, Fab, or scFv, will depend on the application intended. Natural antibodies as well as F(ab')$_2$ fragments are bivalent and usually bind polyvalent cellular antigens with higher affinities than the monovalent Fab or single-chain Fv (scFv) fragments (Crothers and Metzger 1972). Generally it is recognized that high-affinity antibodies are preferable, yet valency and affinity have to be balanced against factors such as size which affect tumor penetration and plasma clearance rates. Comparative studies show that small antibody fragments clear from the blood faster (Colcher et al. 1990; Milenic et al. 1991; King et al. 1994). The disadvantage is that more fusion enzyme might have to be administered to attain adequate tumor uptake since rapid clearance decreases the latter parameter. Yet, due to the smaller size scFvs exhibit better tumor penetration and are more evenly distributed throughout the tumor compared with intact IgGs (Yokota et al. 1992).

Therefore, for many applications, only the antigen-binding domains of the antibody are required. In scFv analogs the two variable domains are coupled by peptide linkers (Bird et al. 1988; Huston et al. 1988) (reviewed in Huston et al. 1996, 1991). The linker needs to be of sufficient length to bridge the distance between the C-terminus of the first V domain and the N-terminus of the second V domain. Several studies have examined the characteristics of scFvs that are affected by the length and composition of the peptide linker (Whitlow et al. 1993; Desplancq et al. 1994; Alfthan et al. 1995; Kortt et al. 1997). Overall, the nature of the linker can affect affinity, dimerization, and aggregation of the scFv. Indeed, the length of the linker can be adjusted to prevent the V_H and V_L domains on the same chain from pairing with each other (Holliger et al. 1993). This can be exploited to create bispecific-binding proteins. Linkers should not interfere with the association of the two domains. Glycine provides flexibility to the (GGGGS)$_3$ peptide linker originally used by Huston et al. (1988). It is also devoid of charged and hydrophobic residues that might interact with the V domain surfaces and interfere either with the binding of these domains to each other or with the binding of the scFv to the antigen (Huston et al. 1991). Recently, the introduction of an interchain disulfide bond has been used to link V_H-V_L domains after genetically modifying each domain to introduce opposing cysteine residues (Glockshuber et al. 1990; Brinkmann et al. 1993; Rodrigues et al. 1995). The advantages of disulfide-linked Fvs (dsFv) compared with scFvs include enhanced serum stability, decreased tendency to aggregate, increased production, similar or increased antigen binding and improved antitumor activity in animals (reviewed in Reiter and Pastan 1996). Disadvantages include the requirement of additional protein engineering and experiments to investigate possible effects of the introduced disulfide bond on the affinity of the Fv. Finally, in designing the Fv binding unit, the choice of the V region order may be important, that is, V_H-linker-V_L or V_L-linker-V_H. Both the affinity and secretion level of the protein can be influenced by the V domain order (reviewed in Huston et al. 1993).

In some of the studies described in this chapter chimeric or humanized antibodies are being used in the antibody domain in an attempt to reduce immunogenicity. Chimeric antibodies are built by incorporating entire murine variable regions within human constant regions while in humanized antibodies the only murine sequences are in the CDRs (complementarity determining regions), (reviewed in Jolliffe 1993).

All of this fundamental research contributed to the evolution of immunotoxin and targeted RNase architecture described in Sections 3.2 and 3.3.

3.1.2
Targeted Drug Strategies

Different strategies have been adopted for ultimately delivering an effector function to the target antigen. In direct targeting the effector moiety itself is the drug and is transported to the target by the antibody. Most approved and experimental antibody drugs directly target cancer cells and this review will focus on the application of direct antibody targeting to cancer therapy. However, other pathologies such as thrombolysis, dissolving clots formed inside blood vessels (Bode et al. 1985), viral and autoimmune diseases are amenable to direct antibody-targeted therapies as well. References to those studies can be found in Bolognesi and Polito (2004). Indirect targeting is another approach to increase the availability of the drug to its target (Bagshawe et al. 1988; Senter et al. 1988). The antibody–enzyme combination (drug activator) is bound to the tumor cell surface. When cleared from the circulation a pro-drug is administered that will be activated by the antibody enzyme complex. An advantage of the indirect approach for cancer (pro-drug therapy) is that the antibody–enzyme does not have to translocate into the cell, which is an inefficient process. Thus, noninternalizing antigens that may be overexpressed on tumor cells become available for cancer therapy. Also, the activated drug can diffuse to nearby antigen-negative tumor cells. This helps to circumvent the problem of antigen heterogeneity encountered in direct targeting strategies. Excellent reviews by Bagshawe et al. (2004) and Wa (2004) bring pro-drug therapy up to date, and a comparison between indirect and direct antibody targeting is provided by Rybak and Newton (1999a).

The following sections of this chapter examine advances in antibody-targeted proteins. Understanding that effectors had to be extremely potent led to conjugation of antibodies to very toxic plant and bacterial proteins (Pastan et al. 1986; Vitetta et al. 1987). The resulting hybrid proteins were named "immunotoxins." Immunotoxins cause potent cell killing *in vitro* and impressive results in murine models of cancer. In human patients both first-generation chemical conjugates

(Rybak and Youle 1991) and more recent derivations (Frankel et al. 2000) cause toxic side effects and immune responses that limit achievable therapeutic regimens. Yet, some successful clinical results spur continued refinement of these molecules as well as new approaches that are different from, but build on, the broad platform of immunotoxin technologies. For example, antibody-targeted ribonucleases (Section 3.3) may be a natural solution to managing the immunogenicity and toxicity of antibody-targeted proteins.

3.2
Immunotoxins

3.2.1
Diphtheria Toxin-based Immunotoxins

Toxins from plants and bacteria have evolved to kill cells (Fitzgerald 1996). Thus they have evolved structurally to withstand proteolytic degradation and susceptibility to intracellular inhibitors. They target elements in a cell that lead to cell death by inhibiting protein synthesis. The use of extremely toxic proteins isolated from plants and bacteria coupled to monoclonal antibodies to generate drugs that specifically kill target cells was inspired, in part, by the theory that the catalytic activity of extremely toxic poisons would be potent enough to kill tumor cells in spite of low levels of tumor uptake of antibody conjugates in humans. Section 3.2 focuses on the evolution of immunotoxins containing the three most studied toxins.

Diphtheria toxin (DT) is one of the bacterial toxins extensively studied and used in the construction of immunotoxins. It is secreted from *Corynebacterium diphtheria* as a single polypeptide chain consisting of three domains: the N-terminal catalytic domain, a translocation domain, and a C-terminal sequence that mediates binding to the cell surface. Originally the entire DT toxin was coupled to antibodies (Moolten et al. 1975). These immunotoxins were very potent but demonstrated a high level of nonspecific toxicity due to binding to nontarget cells. Understanding the DT structure led to substitution of the native cell binding domain with antibodies and other cell binding molecules. For example, the enzymatically active A chain of diphtheria toxin was coupled to the alternate receptor-binding domain of placental lactogen (Chang et al. 1977) but this did not make a functional immunotoxin. Studies that followed showed that the A chain of DT lacked efficient membrane protein translocation function to form a potent immunotoxin (Colombatti et al. 1986). This led to genetic manipulation of the DT molecule to create deletion mutants such as DAB_{389} that deleted portions of the DT-binding domain (Williams et al. 1990) and site-specific DT mutants such as CRM107 (Greenfield et al. 1987) and CRM9 (Hu and Holmes 1987) that changed critical amino acid residues necessary for native toxin binding. Both of the improved toxin variants retained critical B chain sequences necessary for translocation to the cytosol and have been used to make successful immunotoxins.

Human-transferrin (Tf) chemically conjugated to a receptor-binding mutant of DT, CRM107 (Johnson et al. 1989) has progressed to phase III clinical trials for patients with brain cancer. Phase I clinical trial results demonstrated that Tf-CRM107 (delivered via a high-flow convection method utilizing stereotactically placed catheters) produced tumor response in patients with malignant brain tumors refractory to conventional therapy without severe neurological or systemic toxicity (Laske et al. 1997). In a phase II study, Tf-CRM107 treatment resulted in complete and partial tumor responses without severe toxicity in 35% of the evaluable patients (Weaver and Laske 2003). An important consideration with immunotoxin treatment is delivery of the drug. The success, in part, of Tf-CRM107 was the introduction of an intratumoral infusion under positive pressure. Direct administration to tumors increases the drug concentration and decreases systemic toxicity. Also, more treatment cycles can be given because the high concentration of the drug in the compartment overcomes possible inhibition by induced anti-drug antibodies.

Unfortunately the majority of solid tumors are not amenable to direct intratumoral administration and require systemic infusion. In this regard a cytokine DT fusion protein has received approval from the US Food and Drug Administration for the treatment of relapsed or refractory cutaneous T-cell lymphoma. $DAB_{389}IL-2$ (ONTAK, Ligand Pharmaceuticals) administered, as an intravenous infusion, is specifically targeted to cells expressing high-affinity IL-2 receptors. ONTAK is noted for its importance in establishing the clinical (Eklund and Kuzel 2005) and commercial relevance of immunotoxins. The experience gained by administering a targeted toxin to human patients is generating more confidence in their use. This, in turn, encourages the development of other toxin conjugates and fusion proteins. Although these two DT immunotoxins have made the most progress to date, many other ligands have been chosen for preclinical and clinical development of DT conjugates. A detailed analysis can be found in Frankel et al. (2002).

Anti-CD3 DT immunotoxins are extremely potent because crosslinking the T-cell receptor leads to rapid endocytosis (Youle et al. 1986). They have advanced from chemical conjugates with DT and DTA (Colombatti et al. 1986) to sophisticated fusion proteins (Table 3.1). A chemical conjugate (anti-CD3-CRM9) of an anti-rhesus CD3 antibody and CRM9, a DT-binding site mutant, significantly reduced lymph node T cells in monkeys (Neville et al. 1996). This encouraged the expression and purification of a recombinant anti-CD3 DT-scFv fusion protein using the scFv made from the anti-human CD3 antibody UCHT1 (Thompson et al. 1995). However, the binding affinity of the monovalent fusion protein was significantly decreased compared with the parental UCHT1 antibody. Problems with N-terminal heterogeneity were reported for a similar anti-human CD3 DT single-chain fusion protein, DT389-scFv (Hexham et al. 2001). Additional problems with toxicity, yield, and purity of anti-CD3 scFv fusion proteins spurred the

Table 3.1 Molecular evolution of immunotoxins.

Immunotoxin	Target	Status	References
Mab-DT[a]	SV40 antigens	Experimental	Moolten et al. 1975
UCHT1-DT	CD3	Experimental	Colombatti et al. 1986
UCHT1-DTA[a]	CD3	Experimental	Colombatti et al. 1986
Anti-CD3-CRM9[b]	CD3	Primate study	Neville et al. 1996
DT390-scFv(UCHT1)[c]	CD3	Experimental	Thompson et al. 1995
DT390-bisFv(G4S)[d]	CD3	Experimental	Thompson et al. 2001
Bic3[d]	CD3	Experimental	Vallera et al. 2005
Tf-CRM107[b]	Tf receptor[b]	Phase III	Weaver and Laske 2003
DAB$_{389}$IL2, ONTAK[c]	IL-2 receptor	FDA, approved	Eklund and Kuzel 2005
PE-anti-TAC[a]	CD25	Phase I, halted	Pastan 2003
Anti-Tac(Fv)-PE38[c]	CD25	Phase I	Kreitman et al. 2000
IL4(38–37)-PE38KDEL[c]	IL-4 receptor	Experimental, Phase I	Garland et al. 2005
B3-PE38, LMB1	LeY antigen	Phase I	Pai et al. 1996
BR96scFv-PE40 SGN-10[c]	LeY	Phase I	Posey et al. 2002
RFB4(ds)-PE38, BL22[e]	CD22	Phase I	Kreitman et al. 2005
HA22[e]	CD22	Experimental	Ho et al. 2004
HA22 R490A[e]	CD22	Experimental	Bang et al. 2005
UCHT-1-R[a]	CD3	GVHD, *ex vivo*	Filipovich et al. 1987
Anti-CD25-RTA	CD25	GVHD, *ex vivo*	Solomon et al. 2005
Anti-CD5-RTA[a]	CD5	GVHD, phase I	Hertler et al. 1989
Anti-B4-Blocked ricin	CD19	Phase I, phase II	O'Toole et al. 1998
RFB4-dgA[a]	CD22	Phase I	Amlot et al. 1993
IgG-HD37-dgA + IgG-RFB4-dgA	CD19 + CD22	Phase I	Messmann et al. 2000
IgG-RFB4-SMPT-dgA	CD22	Phase I	Sausville et al. 1995
RFB4-rRTA	CD22	Experimental	Smallshaw et al. 2003

a DT, entire diphtheria toxin molecule; DTA, A chain of DT; R, entire ricin; RTA, A chain of ricin; dgA, deglycosylated ricin A chain; PE, entire *Pseudomonas* toxin molecule.
b CRM9, CRM107, binding site mutants of DT; Tf, transferrin.
c DT390, DT389, DT deletion mutants; PE38, PE40; PE deletion mutants; PE38KDEL, PE38 with endoplasmic reticulum retention signal; scFv, single-chain antibody.
d Bivalent DT fusion proteins.
e BL22, disulfide stabilized single-chain fusion protein; HA22, improved scFv in BL22; HA22R490A, improved PE38 in HA22.

development of novel solutions. A bivalent anti-CD3 single-chain DT was constructed with linkers that reduced aggregation and was expressed with new refolding methods to improve yield (Vallera et al. 2005). The novel bivalent fusion protein was effective in a murine model of established human T-cell leukemia without toxicity to the mice seen with the monovalent fusion protein. Another study combined two anti-CD3 single-chain antibodies to create a single-chain bivalent fusion protein (Thompson et al. 2001). Aggregation and yield was markedly reduced by optimizing culture conditions in *Pichia pastoris* (Woo et al. 2006). These new anti-T-cell immunotoxins could prove useful for the treatment of malignant and autoimmune T-cell disorders. Moreover, they illustrate that imaginative solutions can be found to practical problems as antibody-targeted immunotoxins evolve from chemical conjugates to more complex recombinant antibody toxin fusion proteins.

3.2.2
Pseudomonas Exotoxin-based Immunotoxins

Pseudomonas exotoxin A (PE) secreted by the pathogen *Pseudomonas aeruginosa* is another bacterial toxin under extensive evaluation for immunotoxin use. It has a molecular mass of 60 kDa and the same enzymatic activity as DT (Collier 1988). Both catalyze the ADP-ribosylation of elongation factor 2, thus inhibiting protein synthesis and killing the cell. Immunotoxins built with PE also changed as the different domains of the toxin were understood. An early clinical trial of an anti-CD25 (anti-TAC) PE chemical conjugate had to be halted due to severe liver toxicity caused by nonspecific cell binding of the holotoxin (Pastan 2003). Thus truncated PE molecules (PE38, PE40) were engineered to delete the native toxin cell binding activities while retaining the translocation and enzymatic functions. Notably, a chemical conjugate of modified PE (PE38) linked to the anti-Lewis Y (Le^Y) antibody B3 (LMB-1) was the first immunotoxin to show objective tumor responses in patients with solid tumors (Pai et al. 1996). LMB-1 was supplanted by recombinant immunotoxins in which the Fv portion of the antibody was expressed as an scFv directed against CD25 genetically fused to a truncated toxin (Chaudhary et al. 1989). A slightly smaller molecule (Anti-Tac (Fv)-PE38) was administered to patients with hematologic malignancies in a phase I clinical trial (Kreitman et al. 2000).

Although remissions occurred in some patients, new variations continue to be sought for PE immunotoxins targeting blood cancers. One example is IL-4(38–37)-PE38KDEL that combines targeting the IL-4 receptor with PE38 fused to KDEL, to facilitate transport to the endoplasmic reticulum (Kay et al. 2005). The IL-4-PE immunotoxin did not show any objective responses in patients with advanced solid tumors, possibly due to high levels of neutralizing human antitoxin antibodies (Garland et al. 2005). However, treating patients with more accessible tumor in the vasculature may be more successful since some responses were seen with intratumoral administration.

Single-chain PE fusion proteins have also been designed to try to improve on the activity shown in solid cancers by LMB-1. The anti-Le^Y BR96 scFv-PE40 fusion protein (SGN-10, Seattle Genetics) was evaluated in a phase I trial in patients with advanced solid carcinomas (Posey et al. 2002). Even the smaller scFv-PE fusion elicited an antitoxin antibody response in humans and modest vascular leak syndrome was manifested.

Further iterations in the design of PE single-chain fusion proteins have increased the stability of the scFv, allowing for more efficient delivery of the recombinant immunotoxins (detailed in FitzGerald et al. 2004; Pastan 2003). One of the most exciting of the PE immunotoxins is BL22. BL22 is a fusion protein comprising the cloned variable domains of murine anti-CD22 antibody RFB4, joined by an engineered disulfide bond and fused to PE38. The results of a phase I clinical trial are most impressive for patients with hairy cell leukemia (HCL) (Kreitman et al. 2005). There were 19 complete remissions and six partial responses in patients with HCL. Hairy cells express significantly higher levels of the CD22 receptor compared with CD22-positive tumor cells that were among the other the B-cell malignancies treated. Improved versions of BL22 designed to effectively target malignant cells that express fewer CD22-positive sites are currently being designed using hot spot mutagenesis and phage display (Salvatore et al. 2002). HA22 is a mutant of BL22 with mutations in heavy-chain CDR3 and it was more cytotoxic than BL22 (Ho et al. 2004). HA22 (R490A) is an even more improved version of HA22. It was fused to PE38 that has a mutation located in the catalytic domain (III) of the immunotoxin HA22. This resulted in increased cytotoxic and antitumor activity but without increased toxicity to mice (Bang et al. 2005).

In light of the promising results with PE, numerous studies are reporting new treatment possibilities for pancreatic cancer (Bruell et al. 2005) and mesothelioma (Li et al. 2004). Additionally, Proxinium (MacDonald and Glover 2005) a single-chain anti-Ep-CAM antibody PE38 fusion protein (Viventia Biotech) targets an epithelial cell adhesion molecule. It is being administered intratumorally in a phase I clinical trial for the treatment of squamous cell carcinoma of the head and neck. Like the transferrin-DT chemical conjugate described in Section 3.2.1 (Tf-CRM107), Proxinium is generally well tolerated without the systemic toxicities seen for this type of toxin with intravenous infusions. A detailed compilation of PE clinical trial results can be found in the reviews by Frankel et al. (2000, 2003).

3.2.3
Plant-based Immunotoxins

Ribosome-inactivating proteins (RIPs) from plants inactivate protein synthesis and kill cells by cleaving a single N-glycosidic bond of the 28S RNA of ribosomes. They have a rich history in medicinal and immunotoxin use (Olsnes 2004; Stirpe 2004). One class of RIPs are single-chain (A chain) enzymatically active proteins of about 30 kDa. A second class of these toxins consists of an A chain covalently linked to a B chain. The B chain binds the toxic A chain to the cell surface and aids in the translocation of the toxin into the cell. Although many plant toxins are suitable as immunotoxins, the toxin ricin produced by the castor bean plant (*Ricinus communis*) has been used most extensively (Thrush et al. 1996; Vallera 1988).

Similar to results with the bacterial toxins (Sections 3.2.1 and 3.2.2), a high level of nonspecific toxicity results from use of the intact toxin due to cell binding sites on the toxin. For this reason intact ricin conjugates were used *ex vivo* to deplete T cells to prevent graft-versus-host disease (GVHD) and to circumvent problems of graft failure/rejection in patients needing bone marrow transplantation. The goal was to remove specific undesirable cell types from the bone marrow before infusion into the patient without damaging the hematopoietic stem cells. Both the CD3 and CD5 antigens were targeted in such a clinical trial in 1987 using a cocktail of three ricin conjugates (Filipovich et al. 1987). While *ex vivo* T-cell depletion of the graft can prevent development of GVHD it can also lead to a delay in immune reconstitution as well as an increase in opportunistic infections and recurrence of the cancer. An approach that enables a selective depletion of the donor T cells that causes GVHD while preserving anticancer and antimicrobial functions would be optimal. Recently, a study demonstrated that an anti-CD25 ricin A chain immunotoxin, which reacts with a cell surface activation antigen (IL-2 alpha subunit), might more selectively deplete T cells (Solomon et al. 2005).

Most ricin immunotoxin clinical trials have explored using the A chain of ricin. As described in Sections 3.2.1 and 3.2.2, bacterial toxins were found to need portions of the B chain to aid in translocation. Separating the A and B chains of ricin decreases the nonspecific binding but also the translocation activity of the B chain across the cell membrane. Therefore, the internalization capacity of the antigen is critical for A chain toxins. Blocked ricin is an altered ricin derivative that has its nonspecific binding eliminated by chemically blocking the galactose-binding domains of the B chain (Lambert et al. 1991). Anti-B4-blocked ricin (Anti-B4-bR) is an immunotoxin composed of the murine anti-B4 monoclonal antibody and "blocked ricin." The anti-B4 antibody is directed against the CD19 antigen expressed on more than 95% of normal and neoplastic B cells. Anti-B4-bR is extremely potent due to preservation of the translocation function. Early clinical trials looked promising (Rybak and Youle 1991; Grossbard et al. 1992). Toxicities were manageable and objective responses were obtained. Clinical results with blocked ricin were reviewed in O'Toole et al. (1998) and compared with results using ricin A chain and PE in (Gottstein et al. 1994). To date they are still being tested in phase I and II clinical trials. Serious complications from vascular leak syndrome resulted in the death of a patient in a phase II trial of a blocked ricin in patients with small cell lung cancer (Fidias et al. 2002). Nearly all the patients developed human antibodies against both components of the immunotoxin.

Recently a long-term follow-up of results in patients with chronic lymphocytic leukemia (CLL) treated with anti-B4-blocked ricin was published (Tsimberidou et al. 2003). No patients achieved an objective response. Although found to have an acceptable safety profile in some patients, it was immunogenic even in patients who had previously received immunosuppressive chemotherapy.

Early clinical trials with ricin immunotoxins were done in patients with non-Hodgkin's lymphoma, chronic B- and T-cell lymphocytic leukemia, breast cancer, colon cancer, and melanoma (reviewed in Rybak and Youle 1991). Blood-borne malignancies such as B- and T-cell cancers are thought to be more amenable to immunotoxin therapy because the tumor cells are accessible to the treatment. The CD5 antigen is widespread on cancer cells and was used as a target antigen with ricin A chain (Hertler et al. 1989). It does not mediate potent entry of the

A chain, perhaps explaining disappointing results. Early clinical trials illuminated facets of ricin immunotoxin therapy in humans related to toxicity and immunogenicity (Hertler 1988). Modifications to reduce toxicities included deglycosylated A chain (dgA) constructs to reduce hepatic uptake and new linkers (SMPT) to increase plasma stability. Even so, toxicity not predicted from animal studies persisted, notably vascular leak syndrome (VLS) (Sausville et al. 1995). Although associated with most immunotoxin clinical trials, VLS is particularly a problem for ricin-containing immunotoxins. VLS is characterized by weight gain, edema, serum albumin decrease, and pulmonary edema (Amlot et al. 1993; Sausville et al. 1995; Messmann et al. 2000). Changes in the ricin molecule were made in an attempt to address VLS problems. Recombinant ricin A chains were engineered with mutations in amino acid sequences common to proteins known to cause vascular leak. One mutant was comparable to unmodified ricin A chain except that it did not cause VLS in mice at the same dose and, when conjugated to RFB4 (RFB4-rRTA), was more effective in xenografted immunodeficient mice (Smallshaw et al. 2003). If mice develop VLS due to ricin these results may be predictive of an improved safety profile for ricin-based immunotoxins.

Stability has been another problem with ricin A chain conjugates. Patients were treated with a 1:1 mixture of anti-CD22 RFB4-dgA and anti-CD19 HD37 dgA (Combotox) (Messmann et al. 2000). Patients with 50 or more circulating tumor cells per mm^3 in the peripheral blood tolerated all doses without major toxicities. However those patients with less than 50 circulating tumor cells per mm^3 in the peripheral blood experienced unpredictable toxicity that included two deaths related to the immunotoxin. It was thought that biochemical heterogeneity and/or aggregation of the HD37-dgA preparation may have played a role in the clinical toxicities encountered.

Although ricin is the major plant toxin developed as an immunotoxin, it should be noted that pokeweed antiviral protein, gelonin, momordin, and saporin are among other plant toxins most commonly used to construct immunotoxins. Numerous preclinical (Bolognesi and Polito 2004) and clinical studies have been described and have compared these toxins (Frankel et al. 2000). New candidate plant toxins continue to be isolated and characterized as potent immunotoxins (Bolognesi et al. 2000). Though all these toxins exhibit some difference in levels of toxicity to animals and cell-killing potency, none of them are in clinical trials as natural product drugs; none of them are homologs of human proteins.

The evolution of antibodies, toxins, and consequently immunotoxins described in Section 3.2 is due to advances in recombinant DNA technology and antibody engineering. Representative examples of the changes in immunotoxin design and structure are listed in Table 3.1. Immunotoxins have advanced from heterogeneous chemical conjugates using the whole toxin, to using only catalytic A chains, to sophisticated mutations of the bacterial toxins to decrease nonspecific binding. This was coupled with linkage to increasingly sophisticated recombinant antibody designs. Improvements in ricin immunotoxins relied more on chemical methods to solve problems, with the exception of mutated ricin A chain to reduce VLS. Though the science is impressive there have been few major breakthroughs in clinical use of immunotoxins more than 30 years after their introduction in the 1970s. The most promising areas appear to be in compartmentalized treatment such as Tf-CRM107 administered directly to brain tumors, *ex vivo* T-cell depletion, blood-borne malignancies, and treatment of minimal residual disease in solid tumors where the need for tumor penetration is obviated. Prophylactic tactics are being sought to try to manage the immunogenicity and toxicity. Finally combination trials with other therapeutics may yield better results in the future.

3.3
Targeted RNases

3.3.1
Background

To avoid the problems of immunogenicity and toxicity alluded to in Section 3.2, members of the pancreatic RNase A family have been proposed as possible alternatives to plant and bacterial toxins in the construction of immunotoxins (Rybak and Newton 1999b). Since these small extracellular proteins normally reside in the plasma and tissues of humans, they and their homologs could be expected to cause fewer problems when reinfused into human patients and, in fact, they have been safely administered to humans (Aleksandrowicz 1958; Mikulski et al. 2002). Furthermore, they were well tolerated immunologically in humans (Glukhov et al. 1976; Mikulski et al. 2002).

Moreover, numerous reports link RNases and antitumor activity. In 1955 bovine pancreatic RNase A injected into tumor-bearing mice was reported to impede tumor growth (Ledoux 1955a,b). Thus investigations into the clinical use of RNase A were stimulated and it was used in human clinical trials for the treatment of leukemia. Patients with chronic myelocytic leukemia were given daily subcutaneous injections of 0.5–1 mg of the bovine enzyme and were reported to have a decrease in spleen size and show general improvement (Aleksandrowicz 1958). A dimeric member of this protein superfamily, bovine seminal RNase (BS-RNase), was shown to possess antitumor activity *in vitro* and *in vivo* (Matousek 1973; Laccetti et al. 1994; Soucek et al. 1996; Pouckova et al. 1998). Though antitumor properties have been long associated with diverse members of the pancreatic RNase family, the potency, tumor specificity, and reproducibility in various tumor systems did not encourage serious clinical development in the light of other advances in chemotherapy.

In the early 1990s it was shown that by covalently linking the RNase protein to antibodies or fusing the RNase gene to genes encoding cell-binding ligands, human and mammalian RNases could become potent and specific cytotoxic agents directed towards tumor cells (Rybak et al. 1991, 1992). Subsequently, other studies have confirmed these observations, providing a substantial body of experimental and preclinical studies in support of targeted RNases. Both chemical conjugates and recombinant fusion proteins consisting of various targeting domains linked to RNase proteins or fused to RNase genes have been made (Table 3.2).

3.3.2
Targeted Human and Mammalian RNases

Initially the transferrin (Tf) receptor was targeted with bovine pancreatic RNase (Rybak et al. 1991, 1993; Newton et al. 1992). The Tf receptor is an integral membrane glycoprotein that binds and internalizes Tf-iron complexes into cells. It is widely expressed on both tumor and normal cells, but the number of Tf receptors is coupled to growth rate (Trowbridge and Omary 1981). Most tumor cells proliferate more rapidly than normal cells and express more Tf receptors, thus this antigen was shown to have potential for tumor targeting (Trowbridge and Domingo 1981; Trowbridge 1988). Effector proteins linked to Tf are rapidly internalized by receptor-mediated endocytosis, an important criteria for generation of potent selective cell-killing agents (Taetle et al. 1986).

A complication of targeting the Tf receptor is that anti-Tf antibody drug conjugates cross the blood–brain barrier (Friden et al. 1991) and Tf receptors are present on the luminal side of capillary endothelial cells (Jefferies et al. 1984). Thus systemic administration of an antibody linked to a toxin could target brain endothelium unless the endothelial cells were not sensitive to killing by the toxic protein. However, as described in Section 3.2.1, interstitial infusion of a Tf-DT conjugate (Tf-CRM107) in patients with brain tumors was shown to be possible without serious toxicities, presumably because the conjugate did not interact with Tf receptors inside capillaries (Laske 1995). Tf-CRM107 has advanced to phase III trials, showing that the Tf receptor can be targeted successfully. In that regard, an anti-Tf receptor antibody RNase conjugate was directly compared with the same antibody conjugated to ricin. The RNase conjugate was 1000-fold less cytotoxic *in vitro* than the same antibody conjugated to ricin A chain but equally effective *in vivo* in a solid flank model of brain cancer (Newton et al. 1992). Thus, cell culture cytotoxicity assays may not be reflective of true *in vivo* antitumor efficacy for RNase-based targeted compounds. Transferrin was also conjugated to two human RNases (Suzuki et al. 1999).

RNase fusion proteins have been constructed with growth factors like epidermal growth factor (EGF) (Jinno et al. 1996, 2002; Psarras et al. 1998; Suwa et al. 1999; Yoon et al. 1999), fibroblast growth factor (FGF) (Futami et al. 1999; Hayashida et al. 2005; Tanaka et al. 1998), cytokines such as IL-2 (Psarras et al. 2000), peptide hormones (Gho and Chae 1999), and ligands (Huhn et al. 2001) as the targeting agent (Table 3.2). Some of these fusion proteins are fully humanized since both domains are built from human proteins. Herceptin (trastuzumab), a humanized antibody against the receptor tyrosine kinase ErbB2, has been approved for therapy of metastatic breast cancer, but efforts to increase its efficacy are being sought (Carter 2001). A fully human antitumor RNase targeting the

3.3 Targeted RNases

Table 3.2 Targeted human and mammalian RNases.

Name	RNase	Ligand	Structure	Cell surface target	Biological target	Cytotoxicity in vitro IC_{50}[a] ($nmol L^{-1}$)	Refs
LHRH-RNase	Bovine	Human	Hormone fusion	LHRH receptor	Carcinoma	500–700	Gho and Chae 1999
Tf-RNase[b]	Bovine	Human	Ligand conjugate	Tf receptor	Leukemia	40	Rybak et al. 1991
Anti-TFR-RNase[b]	Bovine	Murine	Mab conjugate	Tf receptor	Leukemia	20	Newton et al. 1992
H17-BSRNase	Bovine	Murine	scFv fusion[b]	hPLAP[c]	Carcinoma	4–400	Deonarain and Epenetos 1998
hpRNase scFv	Human	Chimeric	scFv fusion	Tf receptor	Carcinoma	5–90	Zewe et al. 1997
hpRNase-IL2	Human	Human	Cytokine fusion	IL2 receptor	T-lymphocytes	20	Psarras et al. 2000
RNase-FGF	Human	Human	GF fusion[b]	FGR receptor	Endothelium	100	Tanaka et al. 1998[e]
RNase-EGF	Human	Human	GF fusion	EGF receptor	Carcinoma	300–600	Jinno et al. 1996[e]
ECP-EGF	Human	Human	GF-fusion	EGF receptor	Carcinoma	150	Jinno et al. 2002
TfhRNase[d]	Human	Human	Ligand conjugate	Tf receptor	Glioma	2[d]	Suzuki et al. 1999
hERB-hRNase	Human	Human	scFv Fusion	ErbB-2 receptor	Carcinoma	10–60	Lorenzo et al. 2004
EDNscFv	Human	Chimeric	scFv fusion	Tf receptor	Carcinoma	1–20	Newton et al. 1994
Tf-EDN[d]	Human	Human	Ligand conjugate	Tf receptor	Glioma	5[d]	Suzuki et al. 1999
CH2Ang	Human	Chimeric	Fab fusion	Tf receptor	Carcinoma	0.05	Rybak et al. 1992
AngscFv	Human	Chimeric	scFv fusion	Tf receptor	Carcinoma	4–200	Newton et al. 1996
Ang-EGF	Human	Human	GF fusion	EGF receptor	Carcinoma	10	Yoon et al. 1999
Ang-CD30	Human	Human	Ligand fusion	CD30	Lymphoma	0.24	Huhn et al. 2001
scFv-ANG	Human	Humanized	scFv fusion	CD22	Lymphoma	60	Krauss et al. 2005
Dimeric scFvANG	Human	Humanized	scFv fusion	CD22	Lymphoma	74	Arndt et al. 2005

a Concentration required to inhibit protein synthesis 50%.
b Tf, transferrin; TFR, transferrin receptor; scFv, single-chain antibody; GF, growth factor.
c Tumor-associated isoform of human placental alkaline phosphatase.
d Assayed in the presence of retinoic acid to enhance cytotoxicity; resistant to intracellular RNase inhibitor.
e Multiple studies listed in Section 3.2.

ErbB-2 receptor induced a dramatic reduction in tumor volume in mice bearing an ErbB-2-positive tumor (Lorenzo et al. 2004).

Bovine seminal RNase (BSRNase) was targeted against the tumor-associated human hPLAP that is an isoform of placental alkaline phosphatase present on solid carcinomas such as ovarian and testicular as well as on some bladder and head and neck cancers (Epenetos et al. 1984). The scFv (Savage et al. 1993) was constructed from the H17E2 antibody (Travers and Bodmer 1984) against hPLAP. The scFv was found to localize to human xenografts in a murine model of human cancer more rapidly than the IgG form (Deonarain and Epenetos 1998). H17-BSRNase exhibited a wide range of cytotoxicity against tumor cell lines (Table 3.2).

3.3.3
RNase Fusion Protein Architecture

Genetically engineered RNase fusion proteins have been constructed with several different architectures (Fig. 3.1). In one variation the 5′ region of the human angiogenin ribonuclease gene (ANG) was fused to the 3′ region of the C_H2 domain (Rybak et al. 1992) of a chimeric anti-human Tf receptor antibody, E6 (Hoogenboom et al. 1990). This fusion protein (CH2ANG) displayed activity at picomolar concentrations. This construct was 4 logs more toxic to antigen-expressing cells than were chemical RNase conjugates (Table 3.2). The major problem with CH2ANG was that myeloma cell expression yielded only 1–5 ng mL^{-1}, which made purification difficult. To increase yield, single-chain Fv antibody fusions were constructed with three different human RNases: eosinophil-derived neurotoxin (EDN) (Newton et al. 1994), human pancreatic RNase (hpRNase) (Zewe et al. 1997), or ANG (Newton et al. 1996b; Krauss et al. 2005). Stable active RNase-scFvs were expressed as insoluble protein in inclusion bodies (Newton et al. 1996b) and later in transient mammalian cell expression systems (Krauss et al. 2005) or as proteins secreted from E. coli (Arndt et al. 2005). Another RNase single-chain fusion protein, H17-BSRNase (Deonarain and Epenetos 1994, 1995) was designed using BS-RNase to allow RNase dimerization since this RNase itself dimerizes by virtue of two intersubunit disulfide bonds and an exchanged N-terminus (D'Alessio et al. 1991). H17-BSRNase was cytotoxic to target cells despite being directly attached to the V_L domain of the antibody without an intervening spacer peptide. Altogether the scFv antibody fusion proteins shown in Fig. 3.1 express activity in the nanomolar range irrespective of the RNase, orientation, or linkers used. The extremely potent activity of CH2ANG may be due to possible bivalency of the Fab fragment or may reflect its homogeneous composition resulting in a more perfectly formed fusion protein with full retention of enzymatic and binding activities.

Humanized and human antibodies are being used in antibody-mediated therapeutics to reduce the immune response. Although allergic reactions can occur upon retreatment of patients that have developed antibodies against the reagent, the major effect of the induced antibodies appears to be on the half-life of the

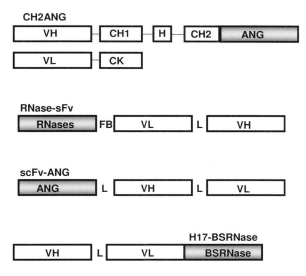

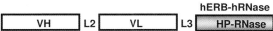

Fig. 3.1 Configurations of recombinant antibody RNase fusion proteins. CH2ANG, the gene for ANG was fused to the 3' end of a chimeric antihuman TfR receptor (E6) antibody (Hoogenboom et al. 1990). It is shown as a Fab enzyme but the possibility of dimerization to a F(ab')$_2$ enzyme existed because of the hinge region and the C_H2 domain (Rybak et al. 1992). Three human RNases – ANG, EDN, and hpRNase – were expressed as fusion proteins with a chimeric scFv derived from E6 in the configuration shown for RNase scFv. The RNases were separated from the V_L by the FB peptide (residues 48–60 of staphylococcal protein A (AKKLNDAQAPKSD))'. The V_L and V_H domains were joined by the flexible linker, L, (GGGGS)$_3$ originally described by Huston et al. (1988). This configuration was found to be optimal for RNase scFvs expressed as insoluble protein from inclusion bodies (Newton et al. 1996a,b). scFv ANG, targeted by a humanized single-chain antibody with specificity to the CD22 antigen, was produced from transiently transfected mammalian Chinese hamster ovary cells (Krauss et al. 2005). H17-BSRNase, the gene for bovine seminal Rnase, was fused to a murine scFv directed against tumor-associated hPLAP (Deonarain and Epenetos 1998). hERB-hRNase is directed to the ErbB-2 receptor. L2, the 15-residue junction peptide SS(G$_4$S)$_2$GGS; L3, the peptide AAASGGPEGGS connecting the scFv and the RNase (Lorenzo et al. 2004).

protein (shortened), which impedes targeting and tumor uptake (reviewed in Rybak et al. 1991; Khazaeli et al. 1994). Thus far it appears that the immune response is attenuated in patients that receive humanized antibodies (Carter 2001; Khazaeli et al. 1994; Stephens et al. 1995), emphasizing the importance of attaching less immunogenic effectors such as RNases to these antibodies. The studies described in Section 3.3 show that the concept of using targeted human and mammalian RNases has been well established in experimental studies but

3.4
Targeted Onconase

3.4.1
Background

Studies with human and mammalian RNases have converged with those of a new direction in cancer research. Sequencing an anticancer protein from frog eggs showed that it belonged to the RNase A superfamily (Ardelt et al. 1991). This new RNase, named Onconase (ranpirnase, Alfacell Corporation), was originally isolated from *R. pipiens* oocytes by following cytotoxic activity against cancer cells *in vitro* (Darzynkiewicz et al. 1988; Mikulski et al. 1990a,b, 1992a,b; Lee et al. 2000) and *in vivo* (Mikulski et al. 1990a,b). Phase I and phase I/II clinical trials of Onconase as a single therapeutic agent in patients with a variety of solid tumors have been completed (Mikulski et al. 1993, 2002; Vogelzang et al. 2001) and have progressed to phase III clinical trials in the United States and Europe for the treatment of malignant mesothelioma (Vogelzang et al. 2000). Although Onconase is an amphibian protein, there have been few problems associated with repeated administration in humans (Mikulski et al. 2002). Recently it was demonstrated that Onconase could selectively enhance activation-induced (e.g. PHA- or mixed lymphocyte reaction-induced) apoptosis of peripheral blood lymphocytes at nanomolar concentrations (Halicka et al. 2002). These results could partly explain an apparent lack of significant adverse immunological reactions observed in Onconase-treated patients.

Since Onconase is currently in clinical trials, understanding the mechanism underlying its antitumor properties is important. Unlike intracellular pancreatic-type ribonucleases that hydrolyze cellular RNAs as a general metabolic function, Onconase targets a specific intracellular RNA that causes changes in intracellular signaling. Onconase binds to the surface of a tumor cell through an as yet unidentified receptor(s) (Wu et al. 1993). Binding is saturable, correlates with cytotoxicity and routing to the cytosol occurs via the Golgi apparatus (Wu et al. 1995). The potent cell killing activity of Onconase is due to damage to tRNA (Lin et al. 1994; Iordanov et al. 2000; Saxena et al. 2002) that causes a physiologically relevant death signal in mammalian cells (Iordanov et al. 2000). The caspase cascade is activated, resulting in apoptosis (Iordanov et al. 2000; Grabarek et al. 2002). However, unlike DNA damage and apoptosis (Benhattar et al. 1996), Onconase-induced apoptosis through tRNA damage is not affected by the absence of a functional p53 protein (Iordanov et al. 2000), an advantage to targeting RNA with Onconase.

Another postulated mechanism of Onconase is that it acts as an intracellular catalyst for the generation of interfering RNAs (RNAi) that could also trigger apoptosis depending upon the microenvironment of the cell (Ardelt et al. 2002). Although apoptosis seems to be a general result of Onconase treatment in all tumor cells tested thus far, specific mechanisms may differ in varying cell types. For example treatment of Jurkat leukemia cells by Onconase corresponds to altered nucleocytoplasmic distribution and reduced expression of the transcription factor NF-κB (Tasi et al. 2004).

Onconase was submitted to the Cancer Drug Discovery and Development Program of the National Cancer Institute (Monks et al. 1991; Grever et al. 1992). The patterns of cell sensitivity of Onconase to all the other agents in the NCI screen database were computed using COMPARE software (Paull et al. 1989). Interestingly, a significant correlation coefficient was found with bleomycin (Abraham et al. 2003), a chemotherapeutic agent now thought to act, in part, by cleaving tRNA. Since it acts on RNA, not DNA, the efficacy of Onconase can be increased in combination therapy with standard DNA damaging chemotherapeutic agents (Lee et al. 2003; Mikulski et al. 1990a,b, 1992a,b; Vasandani et al. 1999) even in the presence of the mdr1 form of multidrug resistance (Rybak et al. 1996). Increased sensitivity to drugs may be due, in part, to the ability of Onconase to decrease tumor interstitial fluid pressure that is an impediment to drug delivery (Lee et al. 2000). Further information on the structure and therapeutic potential of Onconase is found in a recent review (Saxena et al. 2003).

3.4.2
Onconase Conjugates

Attachment of Onconase to targeting ligands markedly improves its specificity and potency (Rybak and Newton 1999b). Chemical conjugates of Onconase to date are listed in Table 3.3. Analogous to Tf-linked toxins (Raso and Basala 1984) and other RNases (Table 3.2) Onconase linked to an anti-Tf receptor antibody was markedly more cytotoxic to human K562 erythroleukemia cells than unconjugated Onconase (IC_{50} 130 versus 5000 nmol L^{-1}, respectively (Table 3.3).

Onconase was conjugated to MRK16, an anti-P-glycoprotein (Pgp) monoclonal antibody (Newton et al. 1996a). A multidrug-resistant phenotype is caused by expression of the Pgp membrane protein encoded by the MDR1 gene (*ABCB1*) (reviewed in Endicott and Ling 1989; Fork and Hait 1990; Ambudkar et al. 1999). The interaction of MRK16-Onconase conjugates with vincristine (VCR) against parental and multidrug-resistant (MDR), Pgp-expressing, human colon carcinoma cells was investigated *in vitro* and *in vivo*. Both reducible disulfide and nonreducible thioether-linked MRK16-Onconase conjugates were more cytotoxic on the Pgp-expressing HT29^{mdr1} cells than on the Pgp-negative parental (HT29^{par}) cells (IC_{50s}, disulfide conjugate 20 nmol L^{-1} versus 130 nmol L^{-1}; thioether conjugate, 100 nmol L^{-1} versus 300 nmol L^{-1}, HT-29^{mdr1} and HT-29^{par} respectively) (Table 3.3). As expected, MRK16-Onconase conjugates with the reducible disulfide bond were more potent, presumably since the Onconase could more easily separate from the antibody and enter the cytosol. Immunofluorescent studies demonstrated that the enhanced toxicity of Onconase conjugates was also related to

Table 3.3 In vitro and in vivo efficacy of Onconase and Onconase chemical conjugates.

Onconase or conjugate	Cell Line[a]	IC$_{50}$ (nmol L^{-1})[b]	% ILS[b]	Refs
Onconase	K562	5000	ND	Rybak et al. 1993
	HT-29^{par}	6000	0	Newton et al. 1996a
	HT-29^{mdr1}	6000	0	Newton et al. 1996a
	Daudi	1200	0	Newton et al. 2001
Anti-TfR-Onconase[c]	K562	130	ND	Rybak et al. 1993
MRK-16-Onconase[d] (DS)[e]	HT-29^{par}	130	ND	Newton et al. 1996a
MRK-16-Onconase[d] (Thio)[f]	HT-29^{par}	300	ND	Newton et al. 1996a
MRK-16-Onconase[d] (DS)[e]	HT-29^{mdr1}	20	93	Newton et al. 1996a
MRK-16-Onconase[d] (Thio)[f]	HT-29^{mdr1}	100	75	Newton et al. 1996a
LL2-Onconase	Daudi	0.02	44	Newton et al. 2001
RFB4-Onconase[g]	Daudi	0.02	153	
LL2-EDN[g]	Daudi	>100	ND	
LL2-hRNase[g]	Daudi	>100	ND	

a The cell lines are as follows: K562, human erythroleukemia; Daudi, human lymphoma; HT-29^{par}, human colon carcinoma; HT-29^{mdr1}, human multidrug-resistant colon carcinoma.
b IC$_{50}$, concentration of RNase or conjugate required to inhibit protein synthesis by 50%; ILS, increase in lifespan.
c Anti-TfR, 5E9 antibody against the human transferrin receptor.
d MRK-16, an anti-P-glycoprotein (Pgp) monoclonal antibody
e DS, disulfide-linked chemical conjugate.
f Thio, thioether-linked chemical conjugate.
g D.L.N., unpublished data.

enhanced internalization of Onconase by the antibody (Newton et al. 1996b). The MRK16 antibody reacts with a portion of the Pgp that functions as an ATP-dependent drug efflux pump not thought to internalize (Hamada and Tsuruo 1986). Thus enhanced internalization by MRK16 most likely reflects the ability of the antibody to concentrate Onconase at the cell surface, where it is able to self-internalize as previously reported (Wu et al. 1993). The implications are that Onconase may be effectively targeted with non-internalizing antibodies. These results were surprising because plant and bacterial toxins must be linked to ligands that efficiently internalize them to the cytosol such as the rapidly internalized 5E9 antibody (Johnson 1991).

Moreover, MRK16-Onconase conjugates sensitized VCR-resistant human colon cancer cells to VCR and this correlated with increased levels of VCR in these cells (Newton et al. 1996a,b). This resulted in an increase in the median survival times of athymic nude mice given transplants of VCR-resistant tumor cells by 40 days, demonstrating that reversal of drug resistance was not an *in vitro* phenomenon (Newton et al. 1996a,b). These results suggest that MRK16-Onconase conjugates could exhibit a dual role: (1) they could directly kill drug-resistant cells; (2) they could decrease resistance to a chemotherapeutic drug such as VCR.

CD22, a B lymphocyte-restricted member of the immunoglobulin superfamily, is a member of the sialoadhesin family of adhesion molecules that include sialoadhesin and myelin-associated glycoprotein (Kelm et al. 1994). Sialoadhesin and CD22 mediate cellular interactions by recognizing specific cell surface sialylated glycoconjugates (Nath et al. 1995). Binding of CD22 to glycoconjugates on neighboring cells alters signaling through the membrane immunoglobulin of B cells by binding cytosolic proteins (Doody et al. 1996). CD22 is an attractive molecular target because of its restricted expression; it is not exposed on embryonic stem or pre-B cells, nor is it normally shed from the surface of antigen-bearing cells (Li et al. 1989). Moreover, it is highly expressed on B cells in non-Hodgkin's lymphoma. A murine anti-CD22 monoclonal antibody (LL2, originally designated EPB-2 (Pawlak-Byczkowska et al. 1989)) was developed for imaging and treatment of non-Hodgkin's lymphoma. LL2 has a highly restricted specificity; it does not cross-react with peripheral blood cells, including the blood's normal B cells, yet is reactive with virtually all cases of non-Hodgkin's lymphoma (Stein et al. 1993).

RFB4, another murine IgG1 antibody against CD22, originally characterized by Campana et al. (1985), was shown to exhibit B-cell specificity that would be favorable for constructing immunotoxins (Li et al. 1989) since it rapidly internalizes (Shih et al. 1994; Shan and Press 1995). Onconase has been conjugated to both LL2 (Newton et al. 2001) and RFB4 (D.L.N., unpublished results). Covalently linking Onconase to these anti-CD22 antibodies increased its cytotoxicity thousands of times (Onconase vs. anti-CD22 Onconase conjugates, IC_{50}s 1200 and 0.02 nmol L^{-1}, respectively) (Table 3.3). Surprisingly, LL2 conjugated to two different human RNases that were successfully conjugated to other antibodies (Table 3.2) did not kill human Daudi cells. To date, all antibodies active with human RNases were also active when conjugated to Onconase. These results imply that human RNases may have a more restricted use than Onconase in targeted therapies.

Anti-CD22 antibodies have been developed to target B-cell lymphomas with plant and bacterial toxins such as derivatives of PE (Kreitman et al. 1993; Mansfield et al. 1996, 1997), deglycoslyated ricin A chain (dgA) (Ghetie et al. 1991, 1992; van Horssen et al. 1996, 1999), and ribosomal inactivating proteins (Flavell et al. 1997; Bolognesi et al. 1998). Some of these anti-CD22 immunotoxins are listed and compared with anti-CD22 Onconase conjugates in Table 3.4. The *in vitro* potency and specificity of anti-CD22 Onconase on Daudi cells is comparable with anti-CD22 immunotoxin conjugates that also kill lymphoma cells in the picomolar range.

Toxins evolved to kill cells and evade host defense mechanisms. Harnessing that extreme toxicity to specifically kill pathological cells is the goal in the immunotoxin field. Though many changes to the toxin molecule were made to try to decrease the nonspecific effects of toxins, severe nonspecific toxicities continue to hamper the clinical effectiveness of these compounds. For instance, all immunotoxins used to date cause VLS to some degree in humans, but the most severe cases have been observed in those patients treated with immunotoxins containing ricin A chain (Frankel et al. 2000). In contrast, VLS has not been associated with Onconase in the clinic (Mikulski et al. 1993, 2002). Moreover, unlike RFB4-ricin

Table 3.4 Comparison of anti-CD22 targeted agents.

Targeted agent	In vitro potency IC_{50} (nmol L^{-1})[a]	In vivo schedule	In vivo toxicity LD_{50} (mg/kg)[a]	Refs
LL2-PE38KDEL[b]	0.01–0.03	ip[c] QD × 4	6	Kreitman et al. 1993
LL2-onconase	0.02	ip QD × 5	350	Newton et al. 2001
RFB4-PE35	0.005	iv QD[c] × 4	5	Mansfield et al. 1996
RFB4(dsFv)PE38	0.005	iv QOD[c] × 3	5	Mansfield et al. 1997
RFB4-dgA[b]	0.001	ip bolus	14	Ghetie et al. 1991
RFB4-onconase[d]	0.02	ip QD × 5	550	
OM124-saporin	0.005	Q3D × 3	0.5	Bolognesi et al. 1998
OM124-momordin	0.009	Q3D × 3	1.75	Bolognesi et al. 1998

a IC_{50}, 50% inhibitory concentration; LD_{50}, 50% lethal dose.
b PE, *Pseudomonas* toxin; dgA, deglycosylated ricin A chain.
c ip, intraperitoneally; iv, intravenously; QD, every day; QOD, every other day; Q3D, every third day.
d D.L.N., unpublished data

A chain, the Onconase conjugate does not damage endothelial cells in culture (D.L.N., unpublished results) when tested in the *in vitro* assay described to be predictive for VLS (Soler-Rodriguez et al. 1993). Onconase is not a toxin. It has been administered to humans and was predicted to cause less toxicity when conjugated to antibodies. As shown in Table 3.4, both anti-CD22 Onconase conjugates are markedly less lethal to mice than immunotoxins. In fact, RFB4(dsFv)PE38, the construct that has caused complete remissions in some hairy cell leukemia patients (Kreitman et al. 2005) is more than 10 times as lethal to mice than RFB4-Onconase.

Anti-CD22 Onconase conjugates are also effective in murine models of cancer. Treatment schedules were designed to assess tumor prevention as well as treatment of minimal and more advanced cancer with both LL2- (Newton et al. 2001) and RFB4-Onconase conjugates (D.L.N., unpublished results). In all of the experiments the conjugate was always effective in significantly increasing the survival of tumor-bearing mice (the increase in lifespan ranged from 40 to 200% over that of mock-treated mice). Again this potency and specificity is comparable to anti-CD22 immunotoxins made with plant and bacterial toxins.

Perhaps as important, considering the lethal toxicities associated with ricin A chain immunotoxin aggregation (Messmann et al. 2000), formulation studies show that both LL2- (Newton et al. 2001) and RFB4-Onconase are very stable. No aggregation either visibly or by HPLC analysis was noted even after the thawing of RFB4-onconase samples stored at −20°C or −70°C (D.L.N., unpublished results). Taken together, targeted Onconase looks promising even though each version has been a first-generation chemical conjugate composed of a heterogeneous mix of molecules. The DNA for Onconase has been cloned, paving the way for the creation of more sophisticated immunofusions as described in Sections

3.2.1, 3.2.2, 3.3.2, and 3.3.3. Yet, because of possible production problems due to misfolded recombinant proteins and poor yield, a new derivative of Onconase has been made to allow site-specific chemical conjugation. (K. Shogen, personal communication). Future generations of targeted Onconase will draw on a variety of possible solutions to enable the best drug for each application.

3.5
Outlook

Immunotoxins have entered the arsenal of anticancer drugs. Problems in clinical use have been identified and attempts to improve the therapeutic index of immunotoxins are being developed. It is widely anticipated that they have the potential to be effective anticancer agents. Many different laboratories have now shown that targeted RNases effectively kill cancer cells in preclinical studies and because they are not toxins naturally display a large therapeutic index. One of these, Onconase, is currently in phase III clinical trials for mesothelioma. Its specificity and potency, particularly against non-Hodgkin's lymphoma, is improved by antibody targeting. Comparisons to classical immunotoxins show that targeted Onconase is effective in preclinical models, causes less nonspecific toxicity in mice and has favorable formulation properties. Anti-CD22 Onconase (RN321) is being developed by the National Cancer Institute in collaboration with Alfacell, Corp., Bloomfield, NJ, USA and is expected to enter clinical trials shortly.

Acknowledgments

Dr Jacob V. Maizel is gratefully acknowledged for expert help in the preparation of this manuscript. Our sincere thanks to Dr Wojciech Ardelt for his interest and helpful comments. This work has been funded in whole or in part with federal funds from the National Cancer Institute, National Institutes of Health, under Contract No. NO1-CO-12400. The content of this publication does not necessarily reflect the views or policies of the Department of Health and Human Services, nor does mention of trade names, commercial products, or organization imply endorsement by the US Government.

Abbreviations

RNase	ribonuclease
scFv	single-chain antibody
dsFv	disulfide linked single-chain antibody
DT	diphtheria toxin
PE	*Pseudomonas* exotoxin A
Le^Y	Lewis Y antigen

RIPs   ribosome inactivating proteins
dgA   deglycosylated ricin A chain
Tf   transferrin
EGF   epidermal growth factor
FGF   fibroblast growth factor
ILS   increased lifespan
i.p.   intraperitoneally
i.v.   intravenously
VLS   vascular leak syndrome

References

Abraham, A.T., Lin, J.J., Newton, D.L., Rybak, S., Hecht, S.M. (2003) RNA cleavage and inhibition of protein synthesis by bleomycin. *Chem Biol* 10: 45–52.

Aleksandrowicz, J. (1958) Intracutaneous ribonuclease in chronic myelocytic leukemia. *Lancet* 2: 420.

Alfthan, K., Takkinen, K., Sizmann, D., Soderlund, H., Teeri, T.T. (1995) Properties of a single-chain antibody containing different linker peptides. *Protein Eng* 8: 725–731.

Ambudkar, S.V., Dey, S., Hrycyna, C.A., Ramachandra, M., Pastan, I., Gottesman, M.M. (1999) Biochemical, cellular, and pharmacological aspects of the multidrug transporter. *Annu Rev Pharmacol Toxicol* 39: 361–398.

Amlot, P.L., Stone, M.J., Cunningham, D., Fay, J., Newman, J., Collins, R., May, R., McCarthy, M., Richardson, J., Ghetie, V., Ramilo, O., Thorpe, P.E., Uhr, J.W., Vitetta, E.S. (1993) A phase I study of an anti-CD22-deglycosylated ricin A chain immunotoxin in the treatment of B-cell lymphomas resistant to conventional therapy. *Blood*. 82: 2624–2633.

Ardelt, W., Mikulski, S.M., Shogen, K. (1991) Amino acid sequence of an anti-tumor protein from *Rana pipieno* ocytes and early embryos. *J Biol Chem* 266: 245–251.

Ardelt, B., Ardelt, W., Darzynkiewicz, Z. (2002) Cytotoxic ribonucleases and RNA interferencde (RNAi). *Cell Cycle* 2: 22–24.

Arndt, M.A., Krauss, J., Vu, B.K., Newton, D.L., Rybak, S.M. (2005) A dimeric angiogenin immunofusion protein mediates selective toxicity towards CD22+ tumor cells. *J Immunother* 28: 245–251.

Bagshawe, K.D., Springer, C.J., Searle, F., Antoniw, P., Sharma, S.K., Melton, R.G., Sherwood, R.F. (1988) A cytoxic agent can be generated selectively at cancer sites. *Br J Cancer* 58: 700–703.

Bagshawe, K.D., Sharma, S.K., Begent, R.H. (2004) Antibody-directed enzyme prodrug therapy (ADEPT) for cancer. *Expert Opin Biol Ther* 4: 1777–1789.

Bang, S., Nagata, S., Onda, M., Kreitman, R.J., Pastan, I. (2005) HA22 (R490A) is a recombinant immunotoxin with increased antitumor activity without an increase in animal toxicity. *Clin Cancer Res* 11: 1545–1550.

Benhattar, J., Cerottini, J.P., Saraga, E., Metthez, G., Givel, J.C. (1996) p53 mutations as a possible predictor of response to chemotherapy in metastatic colorectal carcinomas. *Int J Cancer* 69: 190–192.

Bird, R.E., Hardman, K.D., Jacobson, J.W., Johnson, S., Kaufman, B.M., Lee, S.M., Lee, T., Pope, S.H., Riordan, G.S., Whitlow, M. (1988) Single-chain antigen- binding proteins. *Science* 242: 423–426.

Blattler, W.A., Chari, R.V.J., Lambert, J.M. (1996) Immunoconjugates. In: *Cancer Therapeutics: Experimental and Clinical Agents*. Teicher, B. (ed.), Humana Press Inc., Totowa, NJ, 371–394.

Bode, C., Matsueda, G.R., Hui, K.Y., Haber, E. (1985) Antibody-directed urokinase: a specific fibrinolytic agent. *Science* 229: 765–767.

Bolognesi, A., Polito, L. (2004) Immunotoxins and other conjugates: Preclinical Studies. *Mini-Rev Med Chem* 4: 563–583.

Bolognesi, A., Tazzari, P., Olivieri, F., Polito, L., Lemoli, R., Terenzi, A., Pasqualucci, L., Falini, B., Stirpe, F. (1998) Evaluation of immunotoxins containing single-chain ribosome-inactivating proteins and an anti-CD22 monoclonal antibody (OM124): in vitro and in vivo studies. *Br J Haematol* 101: 179–188.

Bolognesi, A., Polito, L., Tazzari, P.L., Lemoli, R.M., Lubelli, C., Fogli, M., Boon, L., deBoer, M., Stirpe, F. (2000) In vitro anti-tumour activity of anti-CD80 and anti-CD86 immunotoxins containing type 1 ribosome-inactivating proteins. *Br J Haematol* 110: 351–361.

Brinkmann, U., Reiter, Y., Jung, S.H., Lee, B., Pastan, I. (1993) A recombinant immunotoxin containing a disulfide-stabilized Fv fragment. *Proc Natl Acad Sci USA* 90: 7538–7542.

Bruell, D., Bruns, C.J., Yezhelyev, M., Huhn, M., Muller, J., Ischenko, I., Fischer, R., Finnern, R., Jauch, K.W., Barth, S. (2005) Recombinant anti-EGFR immunotoxin 425(scFv)-ETA demonstrates anti-tumor activity against disseminated human pancreatic cancer in nude mice. *Int J Mol Med* 15: 305–313.

Campana, D., Janossy, G., Bofill, M., Trejdosiewicz, L.K., Ma, D., Hoffbrand, A.V., Mason, D.Y., Lebacq, A.M., Forster, H.K. (1985) Human B cell development. I. Phenotypic differences of B lymphocytes in the bone marrow and peripheral lymphoid tissue. *J Immunol* 134: 1524–1530.

Carter, P. (2001) Improving the efficacy of antibody-based cancer therapies. *Nat Rev Cancer* 1: 118–129.

Chang, T.M., Dazord, A., Neville, D.M. Jr. (1977) Artificial hybrid protein containing a toxic protein fragment and a cell membrane receptor-binding moiety in a disulfide conjugate. II. Biochemical and biologic properties of diphtheria toxin fragment A-S-S-human placental lactogen. *J Biol Chem* 252: 1515–1522.

Chaudhary, V.K., Queen, C., Junghans, R.P., Waldmann, T.A., FitzGerald, D.J., Pastan, I. (1989) A recombinant immunotoxin consisting of two antibody variable domains fused to *Pseudomonas* exotoxin. *Nature* 339: 394–397.

Colcher, D., Bird, R., Roselli, M., Hardman, K.D., Johnson, S., Pope, S., Dodd, S.W., Pantoliano, M.W., Milenic, D.E., Schlom, J. (1990) In vivo tumor targeting of a recombinant single-chain antigen-binding protein. *J Natl Cancer Inst* 82: 1191–1197.

Collier, R.J. (1988) Activity relationships in diptheria toxin and pseudomonas aeruginosa exotoxin A. In: *Immunotoxins*, Frankel, A.E., (ed.), Kulwer Academic Publishers, Boston/Dorcrecht/Lancaster. 25–35.

Colombatti, M., Greenfield, L., Youle, R. (1986) Cloned fragment of diphtheria toxin linked to T cell-specific antibody identifies regions of B chain active in cell entry. *J Biol Chem* 261: 3030–3035.

Crothers, D.M., Metzger, H. (1972) The influence of polyvalency on the binding properties of antibodies. *Immunochemistry* 9: 341–357.

D'Alessio, G., Donato, A.D., Parente, A., Piccoli, R. (1991) Seminal RNase: a unique member of the ribonuclease superfamily. *Trends Biochem Sci* 16: 104–106.

Darzynkiewicz, Z., Carter, S.P., Mikulski, S.M., Ardelt, W.J., Shogen, K. (1988) Cytostatic and cytotoxic effects of Pannon (P-30 protein) a novel anti-cancer agent. *Cell Tissue Kinet* 21: 169–182.

Deonarain, M.P., Epenetos, A.A. (1994) Targeting enzymes for cancer therapy: old enzymes in new roles. *Br J Cancer* 70: 786–794.

Deonarain, M.P., Epenetos, A.A. (1995) Construction, refolding and cytotoxicity of a single chain Fv-seminal ribonuclease fusion protein expressed in *Escherichia coli*. *Tumor Targeting* 1: 177–182.

Deonarain, M.P., Epenetos, A.A. (1998) Design, characterization and antitumor cytotoxicity of a panel of recombinant, mammalian ribonuclease-based immunotoxins. *Br J Cancer* 77: 537–546.

Desplancq, D., King, D.J., Lawson, A.D.G., Mountain, A. (1994) Multimerization behaviour of single chain Fv variants for the tumour-binding antibody B72.3. *Protein Eng* 7: 1027–1033.

Doody, G.M., Dempsey, P.W., Fearon, D.T. (1996) Activation of B lymphocytes: integrating signals from CD19: CD22 and FcgRIIb1. *Curr Opin Immunol* 8: 378–382.

Eklund, J.W., Kuzel, T.M. (2005) Denileukin diftitox: a concise clinical review. *Expert Rev Anticancer Ther* 5: 33–38.

Endicott, J.A., Ling, V. (1989) The biochemistry of multidrug resistance. *Annu Rev Biochem* 58: 137–171.

Epenetos, A.A., Travers, P., Gatter, K.C., Oliver, R.D., Mason, D.Y., Bodmer, W.F. (1984) An immunohistological study of testicular germ cell tumours using two different monoclonal antibodies against placental alkaline phosphatase. *Br J Cancer* 49: 11–15.

Fidias, P., Grossbard, M., Lynch Jr., T.J. (2002) A phase II study of the immunotoxin N901-blocked ricin in small-cell lung cancer. *Clin Lung Cancer* 3: 219–222.

Filipovich, A., Vallera, D., Youle, R., Haake, R., Blazar, B., Arthur, D., Neville, D., Ramsay, N., McGlave, P., Kersey, J. (1987) Graft-versus-host-disease prevention in allogeneic bone marrow transplantation from histocompatible siblings. *Transplantation* 44: 62–69.

FitzGerald, D. (1996) Why toxins. *Semin Cancer Biol* 7: 87–95.

FitzGerald, D.J., Kreitman, R., Wilson, W., Squires, D., Pastan, I. (2004) Recombinant immunotoxins for treating cancer. *Int J Med Microbiol* 293: 577–582.

Flavell, D.J., Noss, A., Pulford, D.A.F., Ling, N., Flavell, S.U. (1997) Systemic therapy with 3BIT, a triple combination cocktail of anti-CD19: -CD22: and CD38-saporin immunotoxins, is curative of human B-cell lymphoma in severe combined immunodeficient mice. *Cancer Res* 57: 4824–4829.

Fork, J.M., Hait, W.N. (1990) Pharmacology of drugs that alter multidrug resistance in cancer. *Pharmcol Rev* 42: 155–192.

Frankel, A.E., Kreitman, R.J., Sausville, E.A. (2000) Targeted toxins. *Clin Cancer Res* 6: 326–334.

Frankel, A.E., Powell, B.L., Lilly, M.B. (2002) Diphtheria toxin conjugate therapy of cancer. *Cancer Chemother Biol Response Modif* 20: 301–303.

Frankel, A.E., Neville, D.M., Bugge, T.A., Kreitman, R.J., Leppla, S.H. (2003) Immunotoxin therapy of hematologic malignancies. *Semin Oncol* 30: 545–547.

Friden, P.M., Walus, L.R., Musso, G.F., Taylor, M.A., Malfroy, B., Starzyk, R.M. (1991) Anti-transferrin receptor antibody and antibody-drug conjugates cross the blood-brain barrier. *Proc Natl Acad Sci USA* 88: 4771–4775.

Futami, J., Seno, M., Ueda, M., Tada, H., Yamada, H. (1999) Inhibition of cell growth by a fused protein of human ribonuclease 1 and human basic fibroblast growth factor. *Protein Eng* 12: 1013–1019.

Garland, L., Gitliz, B., Ebbinghaus, S., Pan, H., deHaan, H., Puri, R.K., VonHoff, D., Figlin, R. (2005) Phase 1 trial of intravenous IL-4 pseudomonas exotoxin protein (NBI-3001) in patients with advanced solid tumors that express the IL-4 receptor. *J Immunother* 28: 376–381.

Ghetie, M.A., Richardson, J., Tucker, T., Jones, D., Uhr, J.W., Vitetta, E.S. (1991) Antitumor activity of Fab' and IgG-anti-CD22 immunotoxins in disseminated human B lymphoma grown in mice with severe combined immunodeficiency disease effect on tumor cells in extranodal sites. *Cancer Res* 51: 5876–5880.

Ghetie, M.A., Tucker, K., Richardson, J., Uhr, J.W., Vitetta, E.S. (1992) The anti-tumor activity of an anti-CD22 immunotoxin in SCID mice with disseminated Daudi lymphoma is enhanced by either an anti-CD19 antibody or an anti-CD19 immunotoxin. *Blood* 84: 702–707.

Gho, Y.S., Chae, C.B. (1999) Proliferation of LHRH-receptor positive human prostate and breast tumor cells. *Mol Cells* 9: 31–36.

Glockshuber, R., Malia, M., Pfitzinger, I., Pluckthun, A. (1990) A comparison of strategies to stabilize immunoglobulin Fv-fragments. *Biochemistry* 29: 1362–1367.

Glukhov, B.N., Jerusalimsky, A.P., Canter, V.M., Salganik, R.I. (1976) Ribonuclease treatment of tick-borne encephalitis. *Arch Neurol* 33: 598–603.

Gottstein, C., Winkler, U., Bohlen, H., Diehl, V., Engert, A. (1994) Immunotoxins: is there a clinical value? *Ann Oncol* 5: 97–103.

Grabarek, J., Ardelt, B., Du, L., Darzynkiewicz, Z. (2002) Activation of caspases and serine proteases during apoptosis induced by Onconase (Ranpirnase). *Exp Cell Res* 278: 61–71.

Greenfield, L., Johnson, V., Youle, R. (1987) Mutations in diphtheria toxin separate binding from entry and amplify immunotoxin selectivity. *Science* 238: 536–539.

Grever, M.R., Schepartz, S.A., Chabner, B.A. (1992) The National Cancer Institute: Cancer drug discovery and development program. *Semin Oncol* 19: 622–638.

Grossbard, M.L., Freedman, A.S., Ritz, J., Coral, F., Goldmacher, V.S., Eliseo, L., spector, N., Dear, K., Lambert, J.M., Blattner W.A. (1992) Serotherapy of B-cell neoplasms with anti-B4-blocked ricin: a phase I trial of daily bolus infusion. *Blood* 79: 576–585.

Halicka, D.H., Pozarowski, P., Ita, M., Ardelt, W., Mikulski, S.M., Shogen, K., Darzynkiewicz, Z. (2002) Enhancement of activation-induced apoptosis of lymphocytes by the cytotoxic ribonuclease Onconase (ranpirnase). *Int J Oncol* 21: 1245–1250.

Hamada, H., Tsuruo, T. (1986) Functional role for the 170 to 180-kDa glycoprotein specific to drug resistant tumor cells as revealed by monoclonal antibodies. *Proc Natl Acad Sci USA* 83: 7785–7789.

Hamann, P.R., Hinman, L.M., Hollander, I., Beyer, C.F., Lindh, D., Holcomb, R., Hallett, W., Tsou, H.R., Upeslacis, J., Shochat, D., Mountain, A., Flowers, D.A., Bernstein, I. (2002) Gemtuzumab ozogamicin, a potent and selective anti-CD33 antibody-calicheamicin conjugate for treatment of acute myeloid leukemia. *Bioconjug Chem* 13: 47–58.

Hayashida, T., Ueda, M., Aiura, K., Tada, H., Onizuka, M., Yamada, M.S.H., Kitajima, M. (2005) Anti-angiogenic effect of an insertional fusion protein of human basic fibroblast growth factor and ribonuclease-1. *Protein Eng Des Sel* 18: 321–327.

Hertler, A.A., Schlossman, D.M., Borowicz, M.J., Poplack, D.G., Frankel, A.E. (1988) A phase I study of T101-ricin A chain immunotoxins in refractory chronic lymphocytic leukemia. *J Biol Res Mod* 7: 97–113.

Hertler, A.A., Schlossman, D.M., Borowicz, M.J., Laurent, G., Jansen, F.K., Schmidt, C., Frankel, A.E. (1989) An anti-CD5 immunotoxin for chronic lymphocytic leukemia: Enhancement of cytotoxicity with human serum albumin –monensin. *Int J Cancer* 43: 215–219.

Hexham, J.M., King, V., Dudas, D., Graff, P., Mahnke, M., Wang, Y.K., Goetschy, J.F., Plattner, D., Zurini, M., Bitsch, F., Lake, P., Digan, M.E. (2001) Optimization of the anti-(human CD3) immunotoxin DT389-scFv (UCHT1) N-terminal sequence to yield a homogeneous protein. *Biotechnol Appl Biochem* 34: 183–187.

Ho, M., Kreitman, R.J., Onda, M., Pastan, I. (2004) In vitro antibody evolution targeting germline hot spots to increase activity of an anti-CD22 immunotoxin. *J Biol Chem* 280: 607–617.

Holliger, P., Prospero, T., Winter, G. (1993) "Diabodies": Small bivalent and bispecific antibody fragments. *Proc Natl Acad Sci USA* 90: 6444–6448.

Hoogenboom, H.R. (2005) Selecting and screening antibody libraries. *Nat Biotechnol* 23: 1105–1116.

Hoogenboom, H.R., Raus, J.C.M., Volckaert, G. (1990) Cloning and expression of a chimeric antibody directed against the human transferrin receptor. *J Immunol* 144: 3211–3217.

Hu, V.W., Holmes, R.K. (1987) Single mutation in the A domain of diphtheria toxin results in a protein with altered membrane insertion behaviou. *Biochem Biophys Res Commun* 902: 24–30.

Huhn, M., Sasse, S., Tur, M.K., Matthey, B., Schinkothe, T., Rybak, S.M., Barth, S., Engert, A. (2001) Human angiogenin fused to human CD30 ligand (Ang-CD30L) exhibits specific cytotoxicity against CD30-positive lymphoma. *Cancer Res* 61: 8737–8742.

Huston, J.S., Levinson, D., Mudgett-Hunter, M., Tai, M.S., Novotny, J., Margolies, M.N., Ridge, R.J., Bruccoleri, R.E., Haber, E., Crea, R., Oppermann, H. (1988) Protein engineering of antibody binding sites: Recovery of specific activity in an anti-digoxin single-chain Fv analogue produced in *Escherichia coli*. *Proc Natl Acad Sci USA* 85: 5879–5883.

Huston, J.S., Mudgett-Hunter, M., Tai, M.S., McCartney, J., Warren, F., Haber, E., Oppermann, H. (1991) Protein engineering of single-chain Fv analogs and fusion proteins. *Methods Enzymol* 203: 46–88.

Huston, J.S., McCartney, J., Tai, M.S., Mottola-Hartshorn, C., Jin, D., Warren, F., Keck, P., Oppermann, H. (1993) Medical applications of single-chain antibodies. *Int Rev Immunol* 10: 195–217.

Huston, J.S., Margolies, M.N., Haber, E. (1996) Antibody binding sites. *Adv Protein Chem* 49: 329–450.

Iordanov, M.S., Ryabinina, O.P., Wong, J., Dinh, T.-H., Newton, D.L., Rybak, S.M., Magun, B.E. (2000) Molecular determinants of programmed cell death induced by the cytotoxic ribonuclease Onconase: Evidence for cytotoxic mechanisms different from inhibition of protein synthesis. *Cancer Res* 60: 1983–1994.

Jefferies, W.A., Brandon, M.R., Hunt, S.V., Williams, A.F., Gatter, K.C., Mason, D.Y. (1984) Transferrin receptors on endothelium of brain capillaries. *Nature* 312: 162–163.

Jinno, H., Ueda, M., Ozawa, S., Ikeda, T., Enomoto, K., Psarras, K., Kitajima, M., Yamada, H., Seno, M. (1996) Epidermal growth factor receptor-dependent cytotoxicity for human squamous carcinoma cell lines of a conjugate composed of human EGF and RNase 1. *Life Sci* 58: 1901–1908.

Jinno, H., Ueda, M., Ozawa, S., Ikeda, T., Kitajima, M., Maeda, T., Seno, M. (2002) The cytotoxicity of a conjugate composed of human epidermal growth factor and eosinophil cationic protein. *Anticancer Res* 22: 4141–4145.

Johnson, V.G., Wrobel, C., Wilson, D., Zovickian, J., Greenfield, L., Oldfield, E.H., Youle, R. (1989) Improved tumor-specific immunotoxins in the treatment of CNS and leptmeningeal neoplasia. *J Neurosurg* 70: 240–248.

Johnson, V.G., Youle, R.J. (1991) Intracellular routing and membrane translocation of diphtheria toxin and ricin. In: *Intracellular trafficking of proteins*, Steer, C.J., Hanover, J.A. (eds.), Cambridge University Press, Cambridge. 183–225.

Jolliffe, L.K. (1993) Humanized antibodies: enhancing therapeutic utility through antibody engineering. *Int Rev Immunol* 10: 241–250.

Kay, N.E., Bone, N.D., Lee, Y.K., Jelinek, D.F., Leland, P., Battle, T.E., Frank, D.A., Puri, R.K. (2005) A recombinant IL-4-*Pseudomonas* exotoxin inhibits protein sythesis and overcomes resistance in human CLL B cells. *Leuk Res* 9: 1009–1018.

Kelm, S., Pelz, A., Schauer, R., Filbin, M.T., Tang, S., deBellard, M.-E., Schnaar, R.L., Mahoney, J.A., Hartnell, A., Bradfield, P., Crocker, P.R. (1994) Sialoadhesin, myelin-associate glycoprotein and CD22 define a new family of sialic acid-dependent adhesion molecules of the immunoglobulin superfamily. *Curr Biol* 4: 965–972.

Khazaeli, M.B., Conry, R.M., LoBuglio, A.F. (1994) Human immune response to monoclonal antibodies. *J Immunother* 15: 42–52.

King, D.J., Turner, A., Farnsworth, A.P.H., Adair, J.R., Owens, R.J., Pedley, R.B., Baldock, D., Proudfoot, K.A., Lawson, A.D.G., Beeley, N.R.A., Millar, K., Millican, T.A., Boyce, B.A., Antoniw, P., Mountain, A., Begent, R.H.J., Shochat, D., Yarranton, G.T. (1994) Improved tumor targeting with chemically cross-linked recombinant antibody fragments. *Cancer Res* 54: 6176–6185.

Köhler, G., Milstein, C. (1975) Continous cultures of fused cells secreting antibody of predefined specificity. *Nature* 256: 495–497.

Kortt, A.A., Lah, M., Oddie, G.W., Gruen, C.L., Burns, J.E., Pearce, L.A., Atwell, J.L., McCoy, A.J., Howlett, G.J., Metzger, D.W., Webster, R.G., Hudson, P.J. (1997) Single-chain Fv fragments of anti-neuraminidase antibody NC10 containing five- and ten-residue linkers form dimers and with zero-residue linker a trimer. *Protein Eng* 10: 423–433.

Krauss, J., Arndt, M.A.E., Vu, B.K., Newton, D.L., Rybak, S.M. (2005) Targeting malignant B-cell lymphoma with a humanized anti-CD22 scFv-angiogenin immunoenzyme. *Br J Haematol* 128: 602–609.

Kreitman, R.J., Hansen, H.J., Jones, A.L., FitzGerald, D.J.P., Goldenberg, D.M., Pastan, I. (1993) Pseudomonas Exotoxin-based immunotoxins containing the antibody LL2 or LL2-Fab' induce regression of subcutaneous human B-cell lymphoma in mice. *Cancer Res* 53: 819–825.

Kreitman, R.J., Wilson, W.H., White, J.D., Stetler-Stevenson, M., Jaffe, E.S., Giardina, S., Waldmann, T.A., Pastan, I. (2000) Phase I trial of recombinant immunotoxin anti-Tac(Fv)-PE38 (LMB-2) in patients with hematologic malignancies. *J Clin Oncol* 18: 1622–1636.

Kreitman, R.J., Squires, D.R., Stetler-Stevenson, M., Noel, P., FitzGerald, D.J.P., Wilson, W.H., Pastan, I. (2005) Phase I trial of recombinant immunotoxin RFB4(dsFv)-PAE38 (BL22) in patients with B-cell malignancies. *J Clin Oncol* 23: 6719–6729.

Laccetti, P., Spalletti-Cernia, D., Portella, G., DeCorato, P., D'Alessio, G., Vecchio, G. (1994) Seminal RNase inhibits tumor growth and reduces the metastatic potential of Lewis lung carcinoma. *Cancer Res* 54: 4253–4256.

Lambert, J.M. (2005) Drug conjugated monoclonal antibodies for the treatment of cancer. *Curr Opin Pharmacol* 5: 543–549.

Lambert, J.M., McIntyre, G., Gauthier, M.N., Zullo, D., Rao, V., Steeves, R.M., Goldmacher, V.S., Blattler, W.A. (1991) The galactdose-binding sites of the cytotoxic lectin ricin can be chemically blocked in high yield with reactive ligands prepared by chyemical modificatgion of glycopeptides containing triantennaly N-linked oligosaccharides. *Biochemistry* 30: 3234–3247.

Laske, D., Oldfield, E., Youle, R. (1995) Immunotoxins for brain tumor therapy. In: *Proc Fourth Int Symp Immunotoxins.*, Myrtle Beach.

Laske, D.W., Youle, R.J., Oldfield, E.H. (1997) Tumor regression with regional distribution of the targeted toxin TF-CRM107 in patients with malignant brain tumors. *Nat Med* 3: 1362–1368.

Ledoux, L. (1955a) Action of ribonuclease on certain ascites tumours. *Nature* 175: 258–259.

Ledoux, L. (1955b) Action of ribonuclease on two solid tumors *in vivo*. *Nature* 176: 36–37.

Lee, I., Lee, Y.H., Mikulski, S.M., Lee, J., Covone, K., Shogen, K. (2000) Tumoricidal effects of onconase on various tumors. *J Surg Oncol* 73: 164–171.

Lee, I., Lee, Y.H., Mikulski, S.M., Shogen, K. (2003) Effect of onconase +/− tamoxifen on ASPC-1 human pancreatic tumors in nude mice. *Adv Exp Med Biol* 530: 187–196.

Li, J.L., Shen, G.L., Ghetie, M.A., May, R.D., Till, M., Ghetie, V., Urh, J.W., Janossy, G., Thorpe, P.E., Amlot, P., Vitetta, E.S. (1989) The epitope specificity and tissue reactivity of four murine monoclonal anti-CD22 antibodies. *Cell Immunol* 118: 85–99.

Li, Q., Verschraegen, C.F., Mendoza, J., Hassan, R. (2004) Cytotoxic activity of the recombinant anti-mesothelin immunotoxin, SS1(dsFv)PE38: towards tumor cell oines established from asccites of patients with peritoneal mesothelioma. *Anticancer Res* 24: 1327–1336.

Lin, J.J., Newton, D.L., Mikulski, S.M., Kung, H.F., Youle, R.J., Rybak, S.M. (1994) Characterization of the mechanism of cellular and cell free protein synthesis inhibition by an anti-tumor ribonuclease. *Biochem Biophys Res Commun* 204: 156–162.

Lorenzo, C.D., Arciello, A., Cozzolino, R., Palmer, D.B., Laccedtti, P., Piccoli, R., D'Alessio, G. (2004) A fully human antitumor immunoRNase selective for ErbB-2-positive carcinomas. *Cancer Res* 64: 4870–4874.

MacDonald, G.C., Glover, N. (2005) Effective tumor targeting: Strategies for the delivery of armed antibodies. *Curr Opin Drug Dis Dev* 8: 177–183.

Mansfield, E., Pastan, I., FitzGerald, D.J. (1996) Characterization of RFB4-Pseudomonas exotoxin A immunotoxins tareted to CD22 on B-cell malignancies. *Bioconjugate Chem* 7: 557–563.

Mansfield, E., Amlot, P., Pastan, I., FitzGerald, D. (1997) Recombinant RFB4 immunotoxins exhibit potent cytotoxic activity for CD22-bearing cells and tumors. *Blood* 1997: 2020–2026.

Mathe, G., Loc, T., Bernard, J. (1958) Effet sur la leucemie 1210 de la souris d'un combinasion par diazotation d'A methopterine et de -globulines de hamsters portteurs de cette leucemie par heterogreffe. *CR Acad Sci* 246: 1626–1628.

Matousek, J. (1973) The effect of bovine seminal ribonuclease (AS RNase) on cells of Crocker tumour in mice. *Experientia* 29: 858–859.

Messmann, R.A., Vitetta, E.S., Headlee, D., Senderowicz, A.M., Figg, W.D., Schindler, J., Michiel, D.F., Creekmore, S., Steinberg, S.M., Kohler, D., Jaffe, E.S., Stetler-Stevenson, M., Chen, H., Ghetie, V., Sausville, E.A. (2000) A phase 1 study of combination therapy with immunotoxins IgG-HD37-deglycosylated Ricin A chain (dgA) and IgG-RFB4-dgA (combotox) in patients with refractory CD19 (+), CD22 (−) B cell lymphoma. *Clin Cancer Res* 6: 1302–1313.

Mikulski, S.M., Ardelt, W., Shogen, K., Bernstein, E.H., Menduke, H. (1990a) Striking increase of survival of mice bearing M109 Madison Carcinoma treated with a novel protein from amphibian embryos. *J Natl Cancer Inst* 82: 151–153.

Mikulski, S.M., Viera, A., Ardelt, W., Menduke, H., Shogen, K. (1990b) Tamoxifen and trifluoroperazine (Stelazine) potentiate cytostatic/cytotoxic effects of P-30 protein, a novel protein possessing anti-tumor activity. *Cell Tissue Kinet* 23: 237–246.

Mikulski, S.M., Viera, A., Darzynkiewicz, A., Shogen, K. (1992a) Synergism between a novel amphibian oocyte ribonuclease and lovastatin in inducing cytostatic and cytotoxic effects in human lung and pancreatic carcinoma cell lines. *Br J Cancer* 66: 304–310.

Mikulski, S.M., Viera, A., Shogen, K. (1992b) In vitro synergism between a novel amphibian oocytic ribonuclease (Onconase) and tamoxifen, lovastatin and cisplatin in human OVCAR-3 ovarian carcinoma cell line. *Int J Oncol* 1: 779–785.

Mikulski, S.M., Grossman, A.M., Carter, P.W., Shogen, K., Costanzi, J.J. (1993) Phase 1 human clinical trial of ONCONASE (P-30 protein) administered intravenously on a weekly schedule in cancer patients with solid tumors. *Int J Oncol* 3: 57–64.

Mikulski, S., Costanzi, J., Vogelzang, N., McCachres, S., Taub, R., Chun, H., Mittelman, A., Panella, T., Puccio, C., Fine, R., Shogen, K. (2002) Phase II trial of a single weekly intravenous dose of ranpirnase in patients with unresectable malignant mesothelioma. *J Clin Oncol* 20: 274–281.

Milenic, D.E. (2002) Monoclonal antibody-based therapy strategies: Providing options for the cancer patient. *Curr Pharm Des* 8: 1749–1764.

Milenic, D.E., Yokota, T., Filpula, D.R., Finkelman, M.A.J., Dodd, S.W., Wood, J.F., Whitlow, M., Snoy, P., Schlom, J. (1991) Construction, binding properties, metabolism, and tumor targeting of a single-chain Fv derived from pancarcinoma monoclonal antibody CC49. *Cancer Res* 51: 6363–6371.

Monks, A., Scudiero, D., Skehan, P., Shoemaker, R., Paull, K., Vistica, D., Hose, C., Langley, J., Cronise, P., Vaigro-Wolff, A., Gray-Goodrich, M., Campbell, H., Mayo, J., Boyd, M. (1991) Feasibility of a high-flux anticancer drug screen using a diverse panel of cultured human tumor cell lines. *J Natl Cancer Inst* 83: 757–766.

Moolten, F.L., Capparell, N.J., Zajdel, S.H., Cooperband, S.R. (1975) Antitumor effects of antibody-diphtheria toxin conjugates. II. Immunotherapy with conjugates directed against tumor antigens induced by simian virus 40. *J Natl Cancer Inst* 55: 473–477.

Nath, D., vanderMerwe, P.A., Kelm, S., Bradfield, P., Crocker, P.R. (1995) The amino-terminal immunoglobulin-like domain of sialoadhesin contains the sialic acid binding site. *J Biol Chem* 270: 26184–26191.

Neville, D.M., Scharff, J., Hu, H.Z., Rigaut, K., Shiloach, J., Slingerland, W., Jonker, M. (1996) A new reagent for the induction of T-cell depletion, anti-CD3-CRM9. *J Immunother Emphasis Tumor Immunol* 19: 85–92.

Newton, D.L., Ilercil, O., Laske, D.W., Oldfield, E., Rybak, S.M., Youle, R.J. (1992) Cytotoxic ribonuclease chimeras: Targeted tumoricidal activity *in vitro* and *in vivo*. *J Biol Chem* 267: 19572–19578.

Newton, D.L., Nicholls, P.J., Rybak, S.M., Youle, R.J. (1994) Expression and characterization of recombinant human eosinophil-derived neurotoxin and eosinophil-derived neurotoxin-anti-transferrin receptor sFv. *J Biol Chem* 269: 26739–26745.

Newton, D.L., Pearson, J.W., Xue, Y., Smith, M.R., Fogler, W.E., Mikulski, S.M., Alvord, W.G., Kung, H.F., Longo, D.L., Rybak, S.M. (1996a) Anti-tumor ribonuclease combined with or conjugated to monoclonal antibody MRK16: overcomes multidrug resistance to vincristine in vitro and in vivo. *Int J Oncol* 8: 1095–1104.

Newton, D.L., Xue, Y., Olson, K.A., Fett, J.W., Rybak, S.M. (1996b) Angiogenin single-chain immunofusions: Influence of peptide linkers and spacers between fusion protein domains. *Biochemistry* 35: 545–553.

Newton, D.L., Hansen, H.J., Mikulski, S.M., Goldenberg, D.M., Rybak, S.M. (2001) Potent and specific antitumor activity of an anti-CD22-targeted cytotoxic ribonuclease: potential for the treatment of non-Hodgkin lymphoma. *Blood* 97: 528–535.

O'Toole, J.E., Esseltine, D., Lynch, T.J., Lambert, J.M., Grossbard, M.L. (1998) Clinical trials with blocked ricin immunotoxins. *Curr Top Microbiol Immunol* 234: 35–56.

Olsnes, S. (2004) The history of ricin, abrin and related toxins. *Toxicon* 44: 361–370.

Pai, L.H., Wittes, R., Setser, A., Willingham, M.C., Pastan, I. (1996) Treatment of advanced solid tumors with immunotoxin LMB-1: an antibody linked to *Pseudomonas* exotoxin. *Nat Med* 2: 350–353.

Pastan, I. (2003) Immunotoxins containing *Pseudomonas* exotoxin A: a short history. *Cancer Immunol Immunother* 52: 328–341.

Pastan, I., Willingham, M., FitzGerald, D. (1986) Immunotoxins. *Cell* 47: 641–648.

Paull, K.D., Shoemaker, R.H., Hodes, L. et al. (1989) Display and analysis of patterns of differential actitivy of drugs against human tumor cell lines: development of a mean graph and COMPARE algorithm. *J Natl Cancer Inst* 81: 1088–1092.

Pawlak-Byczkowska, E.J., Hansen, H.J., Dion, A.S., Goldenberg, D.M. (1989) Two new monoclonal antibodies, EPB-1 and EPB-2 reactive with human lymphoma. *Cancer Res* 49: 4568–4577.

Payne, G. (2003) Progress in immunoconjugate cancer therapeutics. *Cancer Cell* 3: 207–212.

Posey, J.A., Khazaelu, M.B., Bookman, M.A., Nowrouzi, A., Grizzle, W.E., thornton, J., Carey, D.E., Lorenz, J.M., Sing, A.P., Siegall, C.B., LoBuglio, A.F., Saleh, M.N. (2002) A phase I trial of the single-chain immunotoxin SGN-10 (BR96 sFv-PE40) in patients with advanced solid tumors. *Clin Cancer Res* 10: 3092–3099.

Pouckova, P., Soucek, J., Jelinek, J., Zadinova, M., Hlouskova, D., Polivkova, J., Navratil, L., Cinatl, J., Matousek, J. (1998) Antitumor action of bovine seminal ribonuclease. Cytostatic effect on human melanoma and mouse seminoma. *Neoplasma* 45: 30–34.

Psarras, K., Ueda, M., Yamamura, T., Ozawa, S., Kitajima, M., Aiso, S., Komatsu, S., Seno, M. (1998) Human pancreatic RNase1-human epidermal growth factor fusion: an entirely human immunotoxin analog with cytotoxic properties against squamous cell carcinomas. *Protein Eng* 11: 1285–1292.

Psarras, K., Ueda, M., Tanabe, M., Kitajima, M., Aiso, A., Komatsu, S., Seno, M. (2000) Targeting activated lymphocytes with an entirely human immunotoxin analogue: human pancreatic RNase1-human IL-2 fusion. *Cytokine* 12: 786–790.

Raso, V., Basala, M. (1984) A highly cytotoxic human transferrin-ricin A chain conjugate used to select receptor-modified cells. *J Biol Chem* 259: 1143–1149.

Reiter, Y., Pastan, I. (1996) Antibody engineering of recombinant Fv immunotoxins for improved targeting of cancer: Disulfide-stabilized Fv immunotoxins. *Clin Cancer Res* 2: 245–252.

Rodrigues, M.L., Presta, L.G., Kotts, C.E., Wirth, C., Mordenti, J., Osaka, G., Wong, W.L.T., Nuijens, A., Blackburn, B., Carter, P. (1995) Development of a humanized disulfide-stabilized anti-p185 HER2 Fv-β-lactamase fusion protein for activation of a cephalosporin doxorubicin prodrug. *Cancer Res* 55: 63–70.

Rybak, S.M., Newton, D.L. (1999) Natural and engineered cytotoxic ribonucleases: Therapeutic potential. *Exp Cell Res* 253: 325–335.

Rybak, S.M., Newton, D.L. (1999a) Immunoezymes. In: Antibody Fusion Proteins. eds. Chamow, S.M., Ashkenazi,

A., John Wiley & Sons, New York, N.Y.: 53–110.

Rybak, S.M., Youle, R.J. (1991) Clinical use of immunotoxins: monoclonal antibodies conjugated to protein toxins. *Immunol Allergy Clin N Am* 11: 359–380.

Rybak, S.M., Saxena, S.K., Ackerman, E.J., Youle, R.J. (1991) Cytotoxic potential of ribonuclease and ribonuclease hybrid proteins. *J Biol Chem* 266: 21202–21207.

Rybak, S.M., Hoogenboom, H.R., Meade, H.M., Raus, J.C., Schwartz, D., Youle, R.J. (1992) Humanization of immuntoxins. *Proc Natl Acad Sci USA* 89: 3165–3169.

Rybak, S.M., Newton, D.L., Mikulski, S.M., Viera, A., Youle, R.J. (1993) Cytotoxic Onconase and ribonuclease A chimeras: Comparison and *in vitro* characterization. *Drug Del* 1: 3–10.

Rybak, S.M., Pearson, J.W., Fogler, W.F., Volker, K., Spence, S.E., Newton, D.L., Mikulski, S.M., Ardelt, W., Riggs, C.W., Kung, H.F., Longo, D.L. (1996) Enhancement of vincristine cytotoxicity in drug-resistant cells by simultaneous treatment with Onconase, an antitumor ribonuclease. *J Natl Cancer Inst* 88: 747–753.

Salvatore, G., Beers, R., Kreitman, R.J., Pastan, I. (2002) Improved cytotoxic activity dtowards cell lines and fresh leukemia cells of a mutant anti-CD22 immunotoxin obtained by antibody phage display. *Clin Cancer Res* 8: 942–944.

Sausville, E.A., Headlee, D., Stetler-Stevenson, M., Jaffe, E.S., Solomon, D., Figg, W.D., Herdt, J., Kopp, W.C., Rager, H., Steinberg, S.M., Gethie, V., Shindler, J., Uhr, J., Wittes, R.E., Vitetta, E.S. (1995) Continuous infusion of the anti-CD22 immunotoxin IgG-RFB4-SMPT-dgA in patients with B-cell lymphoma: a phase I study. *Blood* 85: 3457–3465.

Savage, P., Rowlinson-Busza, G., Verhoeyen, M., Spooner, R.A., So, A., Windust, J., Davis, P.J., Epenetos, A.A. (1993) Construction, characterisation and kinetics of a single chain antibody recognising the tumour-associated antigen placental alkaline phosphatase. *Br J Cancer* 68: 738–742.

Saxena, S., Sirdeshmukh, R., Ardelt, W., Mikulski, S.M., Shogen, k., Youle, R.J. (2002) Entry into cells and selective degradation of tRNAs by a cytotoxic member of the RNase A family. *J Biol Chem* 277: 15142–15146.

Saxena, S.K., Shogen, K., Ardelt, W. (2003) Onconase and its therapeutic potential. *Lab Med* 34: 380–387.

Senter, P.D., Saulnier, M.G., Schreiber, G.J., Hirschberg, D.L., Brown, J.P., Hellstrom, I., Hellstrom, K.E. (1988) Anti-tumor effects of antibody-alkaline phosphatase conjugates in combination with etoposide phosphate. *Proc Natl Acad Sci USA* 85: 4842–4846.

Shan, D., Press, O.W. (1995) Constitutive endocytosis and degradation of CD22 by human B cells. *J Immunol* 154: 4466–4475.

Shih, L.B., Lu, H.H.A., Xuan, H., Goldenberg, D.M. (1994) Internalization and intracellular processing of an anti-B-cell lymphoma monoclonal antibody, LL2. *Int J Cancer* 56: 538–545.

Smallshaw, J.E., Ghetie, V., Rizo, J., Fulmer, J.R., Trahan, L.L., Ghetie, M.A., Vitetta, E.S. (2003) Genetic engineering of an immunotoxin to eliminate pulmonary vascular leak. *Nat Biotechnol* 21: 387–391.

Soler-Rodriguez, A.M., Ghetie, M.-A., Oppenheimer-Marks, N., Uhr, J.W., Vitetta, E.S. (1993) Ricin A-chain and ricin A-chain immunotoxins rqpidly damage human endothelial cells: Implications for vascular leak syndrome. *Exp Cell Res* 206: 227–234.

Solomon, S.R., Mielke, S., Savani, B.N., Montero, A., Wisch, L., Childs, R., Hensel, N., Schindler, J., Ghetie, V., Leitman, S.F., Mai, T., Carter, C.S., Kurlander, R., Read, E.J., Vitetta, E.S., Barrett, A.J. (2005) Selective depletionn of allorective donor lymphocytes-a novel method to reduce the severity of graft-versus-host disease in older patients undergoing matched sibling donor stem cell transplantation. *Blood* 106: 1123–1129.

Soucek, J., Pouckova, P., Matousek, J., Stockbauer, P., Dostal, J., Zadinova, M. (1996) Antitumor action of bovine seminal ribonuclease. *Neoplasma* 43: 335–340.

Stein, R., Belisle, E., Hansen, H.J., Goldenberg, D.M. (1993) Epitope specificity of the anti-B-cell lymphoma monoclonal antibody, LL2. *Cancer Immunol Immunother* 37: 293–298.

Stephens, S., Emtage, S., Vetterlein, O., Chaplin, L., Bebbington, C., Nesbitt, A., Sopwith, M., Athwal, D., Novak, C., Bodmer, M. (1995) Comprehensive pharmacokinetics of a humanized antibody and analysis of residual anti-idiotypic responses. *Immunology* 85: 668–674.

Stirpe, F. (2004) Ribosome-inactivating proteins. *Toxicon* 44: 371–383.

Suwa, T., Ueda, M., Jinno, H., Ozawa, S., Kitagawa, Y., Ando, N., Kitajima, M. (1999) Epidermal growth factor receptor-dependent cytotoxic effect of anti-EGFR antibody-ribonuclease conjugate on human cancer cells. *Anticancer Res* 19: 4161–4165.

Suzuki, M., Saxena, S.K., Boix, E., Prill, R.J., Vasandani, V.M., Ladner, J.E., Sung, C., Youle, R.J. (1999) Engineering receptor-mediated cytotoxicity into human ribonucleases by steric blockade of inhibitor interaction. *Nat Biotechnol* 17: 265–270.

Taetle, R., Castagnola, J., Mendelsohn, J. (1986) Mechanisms of growth inhibition by anti-transferrin receptor monoclonal antibodies. *Cancer Res* 46: 1759–1763.

Tanaka, M., Ozawa, S., Ando, N., Kitagawa, Y., Otani, Y., Seno, M., Futami, J., Ueda, M., Kitajima, M. (1998) Inhibitory effects of recombinant human RNase-FGF fused protein on angiogenesis and tumor growth. *Proc Am Assoc Cancer Res* 40: 68.

Tasi, S.Y., Ardelt, B., Hsieh, T., Darzynkiewicz, Z., Shogen, K., Wu, J.M. (2004) Treatment of Jurkat acute T-lymphocytic leukemia cells by Onconase is accompanied by an altered nucleocytoplasmic distribution and reduced expression of transcription factor NF-kB. *Int J Oncol* 25: 1745–1752.

Thompson, J., Hu, H., Scharff, J., Neville, D.M. Jr. (1995) An anti-CD3 single-chain immunotoxin with a truncated diphtheria toxin avoids inhibition by pre-existing antibodies in human blood. *J Biol Chem* 270: 28037–28041.

Thompson, J., Stavrou, S., Weetall, M., Hexham, J.M., Digan, M.E., Wang, Z., Woo, J.H., Yu, Y., A Mathias, Liu, Y., Ma, S., Gordienko, I., Lake, P., Neville, D.M. Jr. (2001) Improved binding of a bivalent single-chain immunotoxin results in increased efficacy for in vivo T-cell depletion. *Protein Eng* 14: 1035–1041.

Thrush, G.R., lark, L.R., Clinchy, B.C., Vitetta, E.S. (1996) Immunotoxins: an update. *Annu Rev Immunol* 14: 4971.

Travers, P., Bodmer, W. (1984) Preparation and characterization of monoclonal antibodies against placental alkaline phosphatase and other human trophoblast-associated determinants. *Int J Cancer* 33: 633–641.

Trowbridge, I.S. (1988) Transferrin receptor as a potential therapeutic agent. *Prog Allergy* 45: 121–146.

Trowbridge, I.S., Domingo, D.L. (1981) Anti-transferrin receptor monoclonal antibody and toxin-antibody conjugates affect growth of human tumor cells. *Nature* 294: 171–173.

Trowbridge, I.S., Omary, M.B. (1981) Human cell surface glycoprotein related to cell proliferation is the receptor for transferrin. *Proc Natl Acad Sci USA* 78: 3039–3043.

Tsimberidou, A.M., Giles, F.J., Kantarjian, H.M., Keating, M.J., O'Brien, S.M. (2003) Anti-B4 blocked ricin post chemotherapy in patients with chronic lymphocytic leukemia-long term follow-up of a monoclonal antibody-based approach to residual disease. *Leukemia Lymphoma* 44: 1719–1725.

Vallera, D.A., Myers, D.E. (1988) Immunotoxins containing ricin. In: *Immunotoxins*, Frankel, A.E. (ed.), Kluwer Academic Publishers, Boston, Dordrecht, Lancaster. 141–159.

Vallera, D.A., Todhunter, D., Kuroki, D.W., Shu, Y., Sicheneder, A., Panoskaltix-Mortari, A., Vallera, V.D., Chen, H. (2005) Molecular modification of a recombinant, bivalent anti-human CD3 immunotoxin (Bic3) results in reduced in vivo toxicity in mice. *Leuk Res* 29: 331–341.

van Horssen, P., Preijers, F., Voosterhout, Y., Witte, T.D. (1996) Highly potent CD22-recombinant ricin A results in complete cure of disseminated malignant B-cell xenografts in SCID mice but fails to cure solid xenografts in nude mice. *Int J Cancer* 68: 378–383.

van Horssen, P.J., van Oosterhout, Y.V.J.M., Evers, S., Backus, H.H.J., van Oijen, M.G.C.T., Bongaerts, R., deWitte, T. (1999) Influence of cytotoxicity enhancers in combination with human serum on the activity of CD22-recombinant ricin A against B cell lines, chronic and acute lymphocytic leukemia cells. *Leukemia* 13: 241–249.

Vasandani, V.M., Castelli, J.C., Hott, J.S., Saxena, S., Mikulski, S.M., Youle, R.J. (1999) Interferon enhances the activity of the anticancer ribonuclease, Onconase. *J Interferon Cytok Res* 19: 447–454.

Vitetta, E.S., Fulton, R.J., May, R.D., Till, M., Uhr, J.W. (1987) Redesigning nature's poisons to create anti-tumor reagents. *Science* 238: 1098–1104.

Vogelzang, N., Taub, R., Shin, D., Costanzi, J., Pass, H., Gutheil, J., Georgiadis, M., McAndrew, P., Kelly, K., Chun, H., Mittelman, A., McCachren, S., Shogen, K., Mikulski, S. (2000) Phase III randomized trial of Ranpirnase (Onc) vs doxorubicin (DOX) in patients with unresectable malignant mesothelioma: analysis of survival. *Proc Am Soc Clin Oncol* 19: 577a.

Vogelzang, N., Aklilu, M., Stadler, W.M., Dumas, M., Mikulski, S. (2001) A phase II trial of weekly intravenous ranpirnase (Onconase), a novel ribonuclease in patients with metastatic kidney cancer. *Investig New Drugs* 19: 255–260.

Wa, D. (2004) Tumor-activated prodrugs – a new approach to cancer therapy. *Cancer Invest* 22: 604–619.

Weaver, M., Laske, D.W. (2003) Transferrin receptor ligand-targeted toxin conjugate (TF-CRM107) for therapy of malignant gliomas. *J Neurooncol* 65: 3–13.

Whitlow, M., Bell, B.A., Feng, S.L., Filpula, D., Hardman, K.D., Hubert, S.L., Rollence, M.L., Wood, J.F., Schott, M.E., Milenic, D.E., Yokota, T., Schlom, J. (1993) An improved linker for single-chain FV with reduced aggregation and enhanced proteolytic stability. *Protein Eng* 6: 989–995.

Williams, D.P., Snider, C.E., Strom, T.B., Murphy, J.R. (1990) Structure/function analysis of interleukin-2-toxin (DAB486-IL-2. Fragment B sequences required for the delivery of fragment A to the cytosol of target cells. *J Biol Chem* 265: 11885–11889.

Woo, J.H., Liu, Y.Y., Neville, D.M. Jr. (2006) Minimization of aggregation of secreted bivalent anti-human T cell immunotoxin in *Pichia pastoris* bioreactor culture by optimizing culture conditions for protein secretion. *J Biotechnol* 121: 75–85.

Wu, A.M., Senter, P.D. (2005) Arming antibodies: Prospects and challenges for immunoconjugates. *Nat Biotechnol* 23: 1137–1146.

Wu, Y.N., Mikulski, S.M., Ardelt, W., Rybak, S.M., Youle, R.J. (1993) Cytotoxic ribonuclease: A study of the mechanism of Onconase cytotoxicity. *J Biol Chem* 268: 10686–10693.

Wu, Y.N., Saxena, S.K., Ardelt, W., Gadina, M., Mikulski, S.M., DeLorenzo, C., D'Alessio, G., Youle, R.J. (1995) A study of the intracellular routing of cytotoxic ribonucleases. *J Biol Chem* 270: 17476–17481.

Yokota, T., Milenic, D.E., Whitlow, M., Schlom, J. (1992) Rapid tumor penetration of a single-chain Fv and comparison with other immunoglobulin forms. *Cancer Res* 52: 3402–3408.

Yoon, J.M., Han, S.H., Kown, O.B., Kim, S.H., Park, M.H., Kim, B.K. (1999) Cloning and cytotoxicity of fusion proteins of EGF and angiogenin. *Life Sci* 64: 1435–1445.

Youle, R.J., Uckun, F.M., Vallera, D.A., Colombatti, M. (1986) Immunotoxins show rapid entry of diphtheria toxin but not ricin via the T3 antigen. *J Immunol* 136: 93–98.

Zewe, M., Rybak, S.M., Dubel, S., Coy, J.F., Welschof, M., Newton, D.L., Little, M. (1997) Cloning and cytotoxicity of a human pancreatic RNase immunofusion. *Immunotechnology* 3: 127–136.

**Part IV
Emerging Concepts**

4
Automation of Selection and Engineering

Zoltán Konthur

4.1
Introduction

The field of recombinant antibody technology first arrived with the development of two major milestones in molecular biology: the development of murine hybridoma technology (Köhler and Milstein 1975) and the discovery and ease of use of the polymerase chain reaction (PCR) to multiply DNA *in vitro* by Mullis and coworkers (Saiki et al. 1985; Mullis et al. 1986). Together these technologies allowed, in combination with antibody sequence information, the construction of many different recombinant antibody fragments with different specificities and a wide range of applications, revolutionizing diagnostic and therapeutic applications in medicine (Little et al. 2000). For therapeutic applications, however, the initial use of murine antibodies was greatly hampered due to difficulties in obtaining murine monoclonal antibodies for cross-species conserved antigens, a lack of biological function of murine effector domains in humans, and the immune intolerance provoked by mouse antibody sequences and glycosylation patterns during treatment – the so-called human anti-mouse antibody (HAMA) response (see Vol. I, Chap. 1 for more details). The generation of fully human antibodies was therefore desirable and technologies were developed, which meanwhile have become the standard for therapeutic antibodies (Weiner 2006).

4.1.1
Emergence of Antibody Phage Display

To overcome these limitations, in the early 1990s, the merger of recombinant antibody technology with bacteriophage surface display (Smith 1985) – yet another milestone development – resulted in a completely different strategy for the development of specific antibodies. For the first time, combinatorial human antibody libraries were generated (McCafferty et al. 1990; Barbas et al. 1991; Breitling et al. 1991; Clackson et al. 1991; Hoogenboom et al. 1991) allowing the isolation and production of functional antibody fragments based solely on the binding reaction

Handbook of Therapeutic Antibodies. Edited by Stefan Dübel
Copyright © 2007 WILEY-VCH Verlag GmbH & Co. KGaA, Weinheim
ISBN 978-3-527-31453-9

of the antibody, hence uncoupling the generation of antibodies from an immune response (Winter and Milstein 1991). The strategy for generating combinatorial phage display libraries as well as the general enrichment process called biopanning is covered in depth in Vol. I, Chap. 3.

4.1.2
Phage Display and Automation

In the last 15 years, phage display has become the most frequently and most successfully employed *in vitro* selection system for the generation of human antibodies geared towards therapeutic applications. Hence, phage display is exploited commercially by a number of companies (Table 4.1) with many antibodies being in clinical trials. Despite this technology's high-throughput potential being regarded as low (Li 2000), the growing interest and demand for therapeutic antibodies has initiated increased streamlining processes of all aspects of phage display and many technological improvements have been achieved within the last 5 years.

In this chapter I describe some thoughts and considerations during setting up a (semi-)automated selection pipeline for the generation, screening, and downstream evaluation of monoclonal antibodies. While the major focus of the chapter is on using phage display for the *in vitro* selection of antibodies rather than conventional immunization and hybridoma-based monoclonal antibody techniques,

Table 4.1 Non-exclusive collection of companies using phage display for generating human antibodies.

Company name	Antibody format for screening	Homepage
Affitech	scFv	http://www.affitech.com
Antibodies by Design (Div. of Morphosys)	scFv	http://www.antibodiesbydesign.com
BioInvent	scFv	http://www.bioinvent.com
Cambridge Antibody Technology	scFv	http://www.cambridgeantibody.com
Crucell	scFv	http://www.crucell.com
Domantis	single domain	http://www.domantis.com
Dyax	Fab	http://www.dyax.com
Genentech	scFv	http://www.gene.com
MorphoSys	scFv, Fab	http://www.morphosys.com
Symphogen	scFv	http://www.symphogen.com
Wyeth Pharmaceuticals[a]	scFv	http://www.wyeth.com
Xerion Pharmaceuticals[a]	scFv	http://www.xerion-pharma.com
Xoma[a]	scFv, Fab	http://www.xoma.com

a Companies licensing antibody libraries from others in this list.

many of the considerations in respect to screening applications – especially for downstream evaluation – will apply to both.

4.2
General Considerations for the Automation of Antibody Generation

At the beginning of a project in which a pipeline for the generation and screening of antibodies is to be established, there are many technologies and automation possibilities to investigate. One of the major decisions to take, however, is to consider whether the pipeline is going to follow the unit-automation design strategy or should rather be a fully automated system. Unit automation refers to systems in which human intervention is necessary and the individual steps in a process pipeline are partially automated independent of each other (Menke 2002), for example with individual workstations that pipette PCR reactions together. In unit automation, individual processes rather than complex assays are automated and many commercially available workstations and individual automation solutions are available "off the shelf" (Table 4.2).

Fully automated systems refer to pipelines in which all steps of the process or an assay are covered without human intervention, for example, where robotic arms and conveyer belt-like systems are used to move plates from integrated and automated equipment under the control of a scheduling software monitoring and synchronizing all processes (Cohen and Trinka 2002). The benefits and drawbacks of these automatic systems are clear. They can work 24 hours a day, through 7 days a week, and no mistakes due to human handling errors can happen. However, they are very expensive, take a much longer time to develop, and allow little variation once installed without the need for a complete redesign.

In principle, almost everything can be automated; it is primarily a matter of financial resources that limits the extent of automation. In practice, the degree of laboratory automation is dependent upon the scope and timeline of the project pursued (Hamilton 2002).

Before the start of the design process a number of decisions have to be considered and a "project inventory" has to be compiled, dealing non-exclusively with the following, rather generalized question catalogue:
- What is the scientific aim of the project?
- What is the time-scale of the project?
- What type of throughput is aimed for?
- What is the starting point?
- Are targets available or easily accessible?
- What types of applications are aimed for?
- How many individual steps and assays are needed?
- Does one need to have a fully automated pipeline or does one automate purely for the sake of automation?
- Does the pipeline justify the costs?

Table 4.2 A collection of manufacturers providing automation technology and equipment.

Company name	Equipment	Homepage
Abgene	Microplate handling and labeling	www.abgene.com
Affymetrix Inc.	Microarray and detection equipment	www.affymetrix.com
Agilent Technologies Inc.	Bioanalyzers, microfluidic devices	www.agilent.com
Applied Biosystems	Microplate automation, liquid handling, detection, workstations, LIMS	www.appliedbiosystems.com
The Automation Partnership Ltd.	Sample storage	www.autoprt.co.uk
Beckman Coulter	Liquid handling, detection systems, workstations, laboratory automation	www.beckman-coulter.com
BIAcore	Labor-free detection systems	www.biacore.com
Bioveris Corporated	Microplate detection systems, magnetic bead-based technology	www.bioveris.com
Brooks Automation Inc.	Laboratory automation, sample handling	www.automationonline.com
Caliper Life Sciences	Workstations, liquid handling, screening systems	www.caliperls.com
Corning Incorporated	Label-free microplate detection	www.corning.com
CyBio AG	Liquid handling, detection, software and system integration	www.cybio-ag.com
Dmetrix, Inc.	Microscopy imaging instrumentation	www.dmetrix.net
GE Healthcare	Chromatography, microplate detection systems	www.gehealthcare.com
Genetix Ltd	Arraying technology	www.genetx.com
Genomic Solutions Inc.	Liquid handling, arraying technology	www.genomicsolutions.com
GeSiM mbH	Noncontact arraying	www.gesim.de
Luminex	Bead-based assay system	www.luminexcorp.com
Molecular Devices	Microplate readers, fluorometric imaging, microarray analysis, liquid handling, cellular imaging	www.moleculardevices.com
PerkinElmer Life and Analytical Sciences, Inc.	Plate readers and imagers, LIMS, workstations, laboratory automation	www.perkinelmer.com
QIAGEN GmbH	Liquid handling, workstations	www.qiagen.com
Scienion AG	Noncontact arraying	www.scienion.com
SSI Robotics	Labeling, storage systems, microplate handling, scheduling software	www.ssirobotics.com
Tecan	Liquid handling, microarray and microplate detection	www.tecan.com
Thermo Electron Corporation	Microtiter plate equipment, washers, liquid handling, laboratory robotics, workstations, LIMS, magnetic bead automation	www.thermo.com
Titertek	Microplate handling and reading, liquid handling, integrated systems	www.titertek.com
Tomtec	Microplate washer and sealers, liquid handling, workstations	www.tomtec.com
Zinsser Analytic	Liquid handling, workstations	www.zinsser-analytic.com

LIMS, Laboratory Information Management System.

- What is the overall cost per individual selection or individual antibody?
- Is it possible to partially use high-throughput robotic technology already installed for other purposes?
- Survey of accessible equipment and resources.

4.3
Development of an Antibody Generation Pipeline

The remainder of this chapter deals mainly with the approach we have taken at the Max Planck Institute for Molecular Genetics within the "Antibody Factory" (http://www.antibody-factory.de), a German National Genome Research Network NGFN initiative (http://www.ngfn.de). Our major goal in this project is to develop a streamlined process, which will eventually allow antibody selections against up to 500 target molecules per year, preferably with multiple antibody libraries and multiple selections for each target.

The process follows the unit-automation design strategy, where all procedures involved are handled as individual modules, finally being placed in a "virtual conveyer belt" type pipeline (Fig. 4.1). This approach allows individual modules of the system to be modified easily as well as allowing the pipline to be completely changed or extra modules to be added at later stages.

4.3.1
Selection Targets

Depending on the type of project pursued, there are several different stages at which the antibody generation pipeline can be entered. While for many therapeutic applications the number of targets is fairly limited and target accessibility, production, and cost are not issues, in projects aiming at the establishment of large antibody resources for proteomic applications it is quite the opposite.

4.3.1.1 Target Availability
Target availability in suitable format, amounts, and the production costs for generating the targets is pivotal. Therefore, large-scale projects often rely primarily on earlier established clone resources, for example for structural genomics (Bussow et al. 2004). Such clone collections are also available at the German Resource Center for Genomic Research in Berlin, (http://www.rzpd.de), or from companies such as Invitrogen (http://www.invitrogen.com). The collections range from full-length ORF (open-reading frame) Gateway or Creator clones, through N-terminally tagged cDNA expressions clones representing a smaller proportion of full-length ORFs and mainly partial (C-terminal) gene constructs, to non-expressing full-length clones such as the clones from the I.M.A.G.E. consortium (http://image.dbnl.gov) which in turn can be used as a template for subcloning or PCR amplification of the whole protein or parts thereof, such as

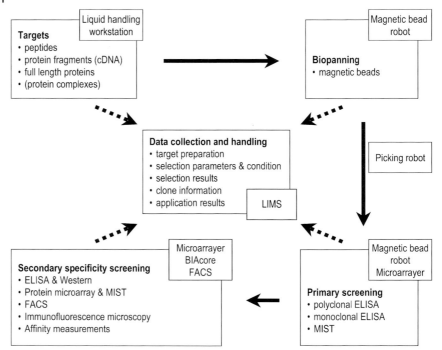

Fig. 4.1 Schematic representation of the antibody selection pipeline established at the Max Planck Institute for Molecular Genetics, Berlin. Unit automation preserves a highly open architecture allowing the individual modules and the pipeline to be easily modified and extended with novel applications.
*MIST, Multiple Spotting Technique.

specific domains of interest. In our laboratory, we are mainly using an arrayed human fetal brain cDNA expression library that was established in-house as a clone resource for protein expression (Bussow et al. 2000, 2004). In addition, targets of special interest are also cloned or shuttled into suitable expression vectors from Gateway Entry clones by recombination.

4.3.1.2 Expression Systems

Currently, the main expression system in use for low-cost protein generation of full-length or partial protein constructs is to use *Escherichia coli* as the expression host. However, target proteins can also be obtained in alternative expression systems, ranging from *in vitro* coupled transcription–translation systems (Kigawa et al. 2004), over *Saccharomyces cerevisiae* (Holz and Lang 2004) or *Pichia pastoris* (Boettner et al. 2002) to various cell culture formats, e.g. insect cells (Albala et al. 2000). In addition, peptides derived from the target proteins can be used for the generation of antibodies, similar to immunization strategies (see other chapter in the book).

Since the selection system we use in our pipeline can handle up to 96 targets at a time, we are currently using only proteins expressed in *E. coli* BL21 strains with helper plasmids providing rare tRNA codons for enhanced expression of human proteins. The proteins or fragments thereof are cloned in vectors that possess N-terminal affinity tags, such as the His6, biotin (AVI tag), or GST tag usable for protein purification and immobilization. For our purposes, the biotin tag is of major importance as it allows the directed immobilization of the selection target onto our preferred selection matrix, streptavidin-coated magnetic beads (see Section 4.3.2). The tag consists of a 14-amino-acid sequence representing an optimized recognition site for the biotin holoenzyme synthetase BirA, which catalyzes the transfer of biotin to a specific lysine residue in this sequence (Schatz 1993; Beckett et al. 1999). Hence, our vectors have an N-terminal His6 and biotin tag, for efficient purification and immobilization. Since the BirA enzyme is *E. coli* specific, this tag cannot be used to produce biotinylated proteins in eukaryotic expressions systems, unless the host is modified in order to also express the biotin ligase (Mechold et al. 2005; van Werven and Timmers 2006).

4.3.1.3 High-Throughput Target Protein Expression and Purification

Several high-throughput strategies concerning the expression of human proteins in *E. coli* have been reported. All have to rely on the arrayed organization of the protein expression clones, preferably in 96- or 384-well format. While the 384-well format is usually used for the storage of glycerol stocks of individual clone collections, the actual target protein expression is taking place in 96-well plates, ranging from normal over large capacity (0.5 mL per well) to deep-well microtiter plates with working volumes around 1 mL per well. Büssow and coworkers at the Protein Structure Factory (http://www.proteinstrukturfabrik.de) developed an expression protocol using a rich medium for culturing the cells up to an OD_{600} of around 1.5, before protein expression is induced by adding IPTG (isopropyl-beta-D-thiogalactopyranoside; final conc. 1 mmol L^{-1}) for 4 h (Bussow et al. 2000, 2004). In our laboratory, we have adapted the protocol to 200 µL culture volumes in 0.5 mL large-capacity microplates, obtaining similar quantities of recombinant protein by incubating the cells at 1200 rpm in microtiter plate incubators (iEMS, Thermo). High-throughput protein purification methods have been set up for parallel handling of up to 96 clones at a time using liquid handling robots, for instance a Zinsser Speedy (Scheich et al. 2003) or a Qiagen 8000 (Kersten et al. 2003; Lueking et al. 2003). Many more liquid-handling systems and workstations are available on the market that can do the same thing (Table 4.2).

Purification of the proteins is based on affinity chromatography using Ni-NTA agarose beads (Qiagen), Talon resin (BD Biosciences) or glutathione agarose (Sigma). In many cases, however, randomly chosen expression constructs originating from cDNA expression libraries will mainly produce insoluble protein (~60%) captured in inclusion bodies within the cytoplasm (Bussow et al. 2000, 2004). Hence, methods to systematically analyze expression constructs for solubility have been developed and partially unit automated. This is a very important issue, especially in the light of crystallography and *in vitro* antibody selection techniques. For instance, Stenvall and colleagues simply monitor the loss of protein concentration due to precipitation after dilution of the purified protein in presence or absence of a denaturing agent (urea) by measuring the protein concentration in the soluble fraction after centrifugation (Stenvall et al. 2005). Alternatively, methods exploit the property of green fluorescent protein (GFP) and its fusion proteins to fluoresce only in a properly folded state using a cell-free expression system (Coleman et al. 2004), or a flow cytometry-based approach is used, in which the whole cell fluores-cence is proportional to the amount of soluble protein in the *E. coli* cytoplasm (Hedhammar et al. 2005).

Currently, we are working on a two-step purification method, combining the benefits of metal ion-affinity chromatography with the selectivity of the streptavidin–biotin interaction. In the first purification step the recombinant protein is separated from the sole biotinylated *E. coli* protein BCCP (biotin carboxyl carrier protein), which does not bind to the chromatography support. In the second step, only the biotinylated recombinant human proteins are directly bound to the selection support matrix, separating the target protein from impurities as a result of unspecific binding of natural proteins to metal ions of the chromatography material.

4.3.2
Automating the Selection Procedure

The biopanning process in phage display runs over 3–5 rounds of selection and can be seen as an affinity-driven process during which binding molecules are continuously enriched out of an initially large pool of non-binders on an immobilized target until they finally represent the majority of all clones.

While variations in the selection process in itself are only relatively small, there are various strategies for target presentation. Proteins can be attached randomly or in a directed fashion to microtiter plate surfaces (Krebs et al. 2001), to magnetic beads (Walter et al. 2001), or to immunopins in the 96-well format (Lou et al. 2001). However, not all approaches are amenable for parallel selection and automation. For instance, selection efficacy on immunopins is limited by the rather small surface available for antigen presentation and lack of automation capability. Hence, I focus on approaches proven in respect of automation by published results and available equipment.

4.3.2.1 Panning in 96-well Microtiter Plates

Plastic 96-well microtiter plates were developed in 1952 in Hungary and have been used for countless immunological assays ever since the introduction of the enzyme-linked immunosorbent assay (ELISA) in the mid-1970s (Hamilton 2002). As a direct consequence, they have quickly become an indispensable tool in every routine diagnostic laboratory and hence, automated instrumentation for handling this format is widely available (Table 4.2).

For selection purposes, there are two ways of immobilizing the target molecules. The proteins are simply adsorbed to the polystyrene surface of the microtiter plate by applying 100 µL of antigen (10–100 µg mL^{-1}) in a bicarbonate buffer (pH 8) or phosphate-buffered saline (pH 7.4) and incubated overnight at 4°C. Alternatively, the target proteins are immobilized in a directed fashion using, for example, streptavidin-coated or antibody-coated plates directed against a tag sequence. The disadvantage of directed immobilization using antibodies is the much lower amount of bound target proteins, which can be as little as 5–10% compared with adsorption to plastic or directed immobilization of biotinylated proteins to streptavidin-coated surfaces.

For automation purposes, the general phage display biopanning can easily be adapted to the use of 96-well plates, since this process strongly resembles ELISA protocols with the difference that bound phage are amplified and not detected. Krebs and colleagues have successfully developed an automated protocol in which they use an ELISA washer for all washing cycles during the selection rounds and by simply modifying the phage amplification protocol during the individual biopanning cycles (Krebs et al. 2001). Surely, additional modifications to the selection process as well as a fully automated selection process utilizing a string of individual liquid handling robots, ELISA washers, incubators, and spectrophotometers under the control of a scheduling software will be seen in the near future.

4.3.2.2 Panning Using Magnetic Beads

As an alternative to microtiter plates, antigens can be coated onto magnetic particles. In our laboratory we use streptavidin-coated magnetic beads as the selection support and immobilize biotinylated target proteins (Walter et al. 2001). This has several advantages, such as directed immobilization, a uniform and dispersed presentation of the target, as well as a largely increased surface area in comparison to microtiter plates. Additionally, the use of magnetic beads easily allows specialized biopanning protocols, such as counterselections, to be performed. For example, adding a non-biotinylated isoform of the target protein can gear the selection towards antibodies recognizing only a certain desired conformation.

For unit automation of the biopanning procedure we use a pin-based magnetic particle processor (Kingfisher, Thermo), which enables the handling of 96 magnetic pins, corresponding to the positions of a 96-well microtiter plate. The processor has several positions for accommodating microtiter plates filled with individual buffers for washing and incubation and the individual steps of the biopanning procedure are performed by transferring the magnetic particles between wells by capture to and release from rod-shaped magnets covered with plastic caps (Rhyner et al. 2003). Controlling of the movements is software driven and parameters such as time, position, frequency, and strength of shaking movements can be adjusted, allowing reproducible control of each step of the phage display selection protocol for up to 96 parallel selections. Moving the beads from solution to solution instead of changing the solutions with liquid-handling robots is, in our eyes, an additional advantage of using magnetic bead selection over microtiter plate-based schemes. It results in reduced background binding, since liquid remaining in the system from dead volume in the container can be avoided and problems arising from leaky liquid-handling systems are eliminated (Konthur and Walter 2002).

In any case, we can conclude at this stage that automating the biopanning process on its own can largely increase the throughput of targets against which antibodies are selected, but it also shifts the bottleneck of the overall selection pipeline further towards the isolation and evaluation of monospecific binders.

4.3.3
Primary Screening

The screening of individual antibody selections and rounds thereof is the biggest task to handle and the methods applied can vary widely according to the downstream application of the antibodies. However, it is generally accepted that at this stage the numbers of individual clones and screens can dramatically expand in the range of 10^2–10^4 per selection target.

Within the antibody development process, the primary screening procedure is the first step in the determination of target binding. In our laboratory, we use this term to refer to screening the individual selection rounds (polyclonal pools) for binding to the target protein and when screening sets of monoclonals – that is single bacterial expression clones – for each target for the first time.

4.3.3.1 Screening Polyclonal Antibody Pools

To ensure the highest possible success rate, we decided to perform polyclonal screening in an assay that is as close as possible to the selection process. Since the selection is carried out using magnetic bead-bound antigens, we perform an ELISA with phage from the individual selection rounds for evaluating specific enrichment in the same assay format. As a negative control we generally use only blocked selection support material, that is streptavidin-coated magnetic beads (Invitrogen, Dynabeads M-280) blocked with 2% skimmed milk powder. The ELISA is carried out according to general protocols with 30- to 60-minute incubation times for the primary (phage antibody) and secondary reagents (anti-M13-HRP, GE Healthcare) in an automated fashion using the magnetic particle processor. As in the selection procedure, during the ELISA the magnetic beads are moved from microtiter plate to microtiter plate prefilled with all necessary solutions. After addition of the beads to the substrate (2′,2′-azino-bis(3-ethylbenzthiazoline-6-sulfonic acid) diammonium salt, Sigma), a first reading is taken after 30 min incubation at room temperature with a conventional spectrophotometer at OD_{405} and the values are exported and evaluated in Microsoft Excel (Fig. 4.2a).

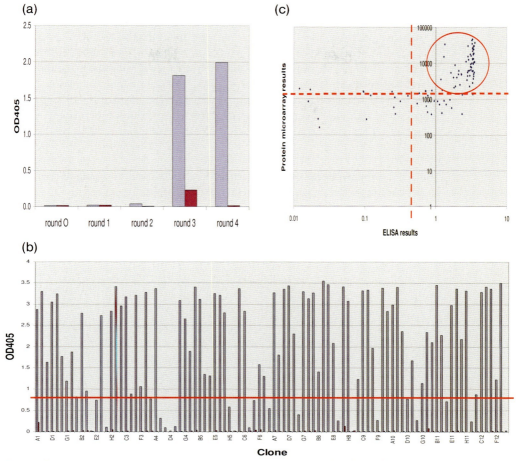

Fig. 4.2 Primary screening of an antibody phage display selection against a soluble human recombinant protein as a target. (a) An example of polyclonal enrichment of antibody fragments. (b) Conventional ELISA with 94 monoclonal soluble scFv fragments. blue: specific signal, red: background. (c) Correlation between the ELISA and the microarray binding assay MIST. Red dotted lines represent 50 times signal-to-background ratio; red circle represents clones positive in both assays.

Once a polyclonal enrichment for a specific selection target is successfully confirmed, we decide how many monoclonal entities we will further analyze from which selection rounds.

4.3.3.2 Generating Sets of Monoclonal Entities

To obtain single colonies for picking, we infect E. coli growing in the log-phase (OD_{600} ~0.4–0.6) with bacteriophage pools of the corresponding rounds and plate the infected cells onto 22 × 22 cm square dishes containing selective medium.

After overnight growth, individual colonies on the agar plates are visualized with a CCD camera mounted next to the picking gadget of a colony picker (Qbot, Genetix) and individual colonies are mapped and evaluated according to size and roundness with special image analysis software. Finally, up to 3000 selected colonies per hour are picked into bar-coded 384-well or 96-well plates prefilled with selective medium to generate master plates. After overnight incubation at 37°C without shaking, the master plates are replicated prior storage at −80°C to obtain two working copies each for further use. From the working copies, one is chosen for replicating into expression plates to obtain recombinant antibody fragments for characterization.

4.3.3.3 Primary Screening of Monoclonal Entities Without Changing Format

Depending on the phage display vector used, soluble antibody fragments can be obtained directly without subcloning simply by switching to a different *E. coli* host, for example HB2151. The underlying basis for this is the presence of an amber stop codon between the recombinant antibody fragment and the phage coat protein, which is read through only in amber-suppressor *E. coli* strains, such as TG1 and XL1-blue.

Conventional screening of specifically binding monoclonal entities involves the use of ELISA either in 384-well or 96-well plates and has been shown to be fully amenable to automation, reaching a throughput of 20.000 data points per day (Hallborn and Carlsson 2002). Another possibility to evaluate monoclonal binders at primary screening level is to make colony arrays consisting of around 20.000 *E. coli* clones, all harboring recombinant antibody molecules, which become accessible for analysis with directly labeled selection targets after lysis of the cells (de Wildt et al. 2000). However, this approach seems outdated ever since the introduction of protein microarray technology with multiplexing potential (Hultschig et al. 2006).

Currently there are two protein array methods used for the analysis of monoclonal antibody entities, which allow multiplexing. Sawyer and colleagues have set up a method to rapidly characterize primary cell fusions for the expression of mouse monoclonal antibodies obtained after a multiplexed immunization strategy (De Masi et al. 2005). For characterization, cell culture supernatants of cell fusions after only 12 days are spotted onto a glass microarray, which has previously been completely coated with 5 µg of an antigen used in the immunization process. Using a set of two different fluorescent-labeled secondary antibodies towards IgM and pan IgG of mouse, it was not only possible to find binding antibody molecules, but additionally to isotype them in a single multiplexing experiment.

In our pipeline, we are currently relying on classical ELISA as a primary screening tool to identify specific antibody fragments. However, we have developed a "Multiple Spotting Technique" (MIST), which allows the multiplexed analysis of hundreds of antibody fragments against a given set of target proteins on a single protein array (Angenendt et al. 2004). The technique is based on the simple but effective concept of addressing single positions on a chip multiple times

(Angenendt et al. 2003). After spotting multiple fields of antigens on the slide and blocking the remaining surface, a set of antibodies can be spotted onto the different antigens to not only find binders to each antigen but also to eliminate cross-specific binders at a very early stage of screening. Once we have thoroughly evaluated this technique and determined the maximum working capacity, we consider using MIST as the only primary screening tool in our antibody generation pipeline.

4.3.3.4 Primary Screening of Monoclonal Entities with Changing Format

In some laboratories the individual screening with monoclonal antibody fragments can only be conducted after the antibody fragments have been subcloned into different vectors (Krebs et al. 2001; Jostock et al. 2004). For example, McCafferty and coworkers subcloned the obtained scFv into an alkaline phosphatase fusion vector, which not only allowed the direct detection of the recombinant antibody without secondary detection reagents, but also increased the avidity of the antibody molecules, since alkaline phosphatase is a homodimer (Han et al. 2004).

Others are on the way to completely omit the screening of monoclonal entities produced in *E. coli* and primary screening of individual binders is directly performed in the desired application format. This can be especially useful when directly aiming at therapeutic application of the antibodies. For this purpose, the antibody heavy and light chain fragments are subcloned into specially designed acceptor vectors in batches before individual colonies are isolated and finally transfected to obtain fully human IgG1 antibodies (Jostock et al. 2004).

4.3.4
Secondary Screening of Monoclonal Entities

Re-evaluating identified binders by independent methods is referred to as secondary screening and normally results in additional information on the antibody, such as specificity, epitope binding, kinetic data, and the range of applications for which the antibody can be used.

4.3.4.1 Microtiter Plate and Bead-based Secondary Screening Assays

Next to ELISA, other microtiter plate-based assays are open to high-throughput screening such as the fluorometric microvolume assay technology (FMAT) (Hallborn and Carlsson 2002). Also common in use are homogeneous electrochemiluminescence assays on a magnetic bead basis (Schweitzer and Abriola 2002). Recently, a specially developed protocol for assessing all stages of monoclonal antibody production, including isotyping has been released (http://www.bioveris.com).

4.3.4.2 Cell-based and Tissue-based Secondary Screening

Additional information about antibodies is often desired, for instance when the antibodies are to target cell surface markers that are potential therapeutic targets in cancer. These assays frequently include fluorescence-activated cell sorting (FACS), but also complement-dependent cytotoxicity (CDC) and antibody-dependent cellular cytotoxicity (ADCC) and have been are covered in Chapter 1. Depending on the project pursued, the compatibility of the generated antibody with immunohistochemistry is of vital importance and multiple automated solutions for scoring data are available, which allow the investigation of protein expression patterns on a large scale using tissue microarrays (Warford et al. 2004; Uhlen and Ponten 2005).

4.3.4.3 Protein Microarray-based Secondary Screening Assays

Protein macro- and microarrays can be used at multiple steps within the pipeline. For example, we use protein microarrays containing a set of selection-relevant and -irrelevant recombinant human proteins for the monitoring of selection rounds with respect to increasing specificity. Templin and coworkers have used protein microarrays for the evaluation of specificity and affinity determination of individual phage display-derived Fab fragments (Poetz et al. 2005). Similarly, we have also used protein microarrays for the evaluation of oligoclonal mixtures and monoclonal antibody fragments and have also compared some of our antibody fragments with commercially available monoclonal antibodies in respect of specificity and sensitivity. Currently, we are also in the process of further advancing the use of protein microarray applications by combining the described features of the multiple spotting technique (Angenendt et al. 2003, 2004) with cell-free expression systems (Angenendt et al. 2006), which will allow the parallel testing of hundreds of single antibody-expressing clones at a time on a single chip in a multiplex assay format. In addition, the target protein can be represented by short oligopeptides (15-meres) directly synthesized onto a membrane support in a microarray format applying the SPOT synthesis and can be used for the mapping of the targeted epitope, provided it is linear (Frank 2002).

4.3.4.4 Secondary Screening Assays Determining Affinity

The most prominent technology for the determination of antibody affinities is the use of surface plasmon resonance (SPR). While classical BIAcore instrumentation is regarded as the gold standard, the product portfolio has been expanded recently by the Flexchip, a microarray-based SPR apparatus that allows the determination of up to 400 antibody affinity constants of in a parallel fashion (Wassaf et al. 2006). However, other alternative techniques have also been established, for example the Multiplexed Competitive Antibody Binning assay (MCAB), a bead-based assay applying the Luminex technology that allows fast affinity ranking of the antibodies under investigation (Jia et al. 2004). Another heterogeneous method to study antibody–antigen interactions is the Kinetic Exclusion Assay, a flow fluorimeter immunoassay based on the rapid separation and quantification of free and complexed antigen–antibody pairs monitored via a fluorescein-labeled secondary antibody in solution (Blake and Blake 2004).

4.3.5
Data Management

The selection of antibodies using phage display technology involves a great variety of methods and many individual experiments, which all need to be kept on record to guarantee good experimental practice and maintain high quality. Therefore, we have started to build a Laboratory Information Management System (LIMS) to store all antibody generation relevant data. The database is a Java-based web application built on the open source relational data base management system (RDBMS) PostgreSQL, Version 8.1 (http://www.postgresql.org). It covers areas as broad as "transcripts", "antibody_targets", "annotation", "expression_clones", "clones", "experiments", "experimentator" and "administrative data". In addition, we have linked our system to external databases to retrieve as much information to each target under way as possible with all our own data, for instance in which vector the target is expressed and whether it is a full-length product of a gene, or a mutant with amino acid substitution. The outline of the database was greatly inspired by the LIMS systems used for large-scale structural genomics projects, such as the Hex1 protein database of the Protein Structure Factory (http://www.proteinstructurfabrik.de) (Bussow et al. 2004, 2005).

4.4
Conclusion

Phage display has become a well-established and extremely powerful tool for the development of antibodies and, inevitably, plays a major role in research towards diagnostics and therapy.

During recent years, our general understanding of the phage display selection process as being difficult to automate has changed dramatically and many companies and organizations have started to develop (semi-)automated selection protocols similar to the one described in this chapter, or have successfully established downstream evaluation pipelines for the screening of thousands of monoclonal entities per day. With the increasing speed of novel technology developments in laboratory design and management, it is undoubtedly only a matter of time until fully automated selection systems with minimal user intervention are established, further raising the throughput of antibody development by phage display.

Acknowledgments

The author acknowledges financial support from the German Federal Ministry for Education and Research through the German National Genome Research Network (Grant No. 01GR0427), the EU Commission through the Proteome-Binder FP6 co-ordination action (Grant No. ZICA No. 026008) and the Max Planck Society.

References

Albala, J.S., Franke, K., McConnell, I.R., Pak, K.L., Folta, P.A., Rubinfeld, B., Davies, A.H., Lennon, G.G., Clark, R. (2000). From genes to proteins: high-throughput expression and purification of the human proteome. *J Cell Biochem* 80: 187–191.

Angenendt, P., Glokler, J., Konthur, Z., Lehrach, H., Cahill, D.J. (2003). 3D protein microarrays: Performing multiplex immunoassays on a single chip. *Anal Chem* 75: 4368–4372.

Angenendt, P., Wilde, J., Kijanka, G., Baars, S., Cahill, D.J., Kreutzberger, J., Lehrach, H., Konthur, Z., Glokler, J. (2004). Seeing better through a MIST: Evaluation of monoclonal recombinant antibody fragments on microarrays. *Anal Chem* 76: 2916–2921.

Angenendt, P., Kreutzberger, J., Glokler, J., Hoheisel, J.D. (2006). Generation of high-density protein microarrays by cell-free in situ expression of unpurified PCR products. *Mol Cell Proteomics* 5: 1658–1666.

Barbas, C.F., III, Kang, A.S., Lerner, R.A., Benkovic, S.J., (1991). Assembly of combinatorial antibody libraries on phage surfaces: the gene III site. *Proc Natl Acad Sci USA* 88: 7978–7982.

Beckett, D., Kovaleva, E., Schatz, P.J. (1999). A minimal peptide substrate in biotin holoenzyme synthetase-catalyzed biotinylation. *Protein Sci* 8: 921–929.

Blake, R.C., II and Blake, D.A. (2004). Kinetic exclusion assays to study high-affinity binding interactions in homogeneous solutions. *Methods Mol Biol* 248: 417–430.

Boettner, M., Prinz, B., Holz, C., Stahl, U., Lang, C. (2002). High-throughput screening for expression of heterologous proteins in the yeast *Pichia pastoris*. *J Biotechnol* 99: 51–62.

Breitling, F., Dubel, S., Seehaus, T., Klewinghaus, I., Little, M. (1991). A surface expression vector for antibody screening. *Gene* 104: 147–153.

Bussow, K., Nordhoff, E., Lubbert, C., Lehrach, H., Walter, G. (2000). A human cDNA library for high-throughput protein expression screening. *Genomics* 65: 1–8.

Bussow, K., Quedenau, C., Sievert, V., Tischer, C., Scheich, H., Seitz, B., Hieke, F., Niesen, H., Gotz, F., Harttig, U., Lehrach, H. (2004). A catalog of human cDNA expression clones and its application to structural genomics. *Genome Biol* 5: R71.

Bussow, K., Scheich, C., Sievert, V., Harttig, U., Schultz, J., Simon, B., Bork, P., Lehrach, H., Heinemann, U. (2005). Structural genomics of human proteins – target selection and generation of a public catalogue of expression clones. *Microb Cell Fact* 4: 21.

Clackson, T., Hoogenboom, H.R., Griffiths, A.D., Winter, G. (1991). Making antibody fragments using phage display libraries. *Nature* 352: 624–628.

Cohen, S., Trinka, R.F., (2002). Fully automated screening systems. *Methods Mol Biol* 190: 213–228.

Coleman, M.A., Lao, V.H., Segelke, B.W., Beernink, P.T. (2004). High-throughput, fluorescence-based screening for soluble protein expression. *J Proteome Res* 3: 1024–1032.

De Masi, F., Chiarella, P., Wilhelm, H., Massimi, M., Bullard, B., Ansorge, W., Sawyer, A. (2005). High throughput production of mouse monoclonal antibodies using antigen microarrays. *Proteomics* 5: 4070–4081.

de Wildt, R.M., Mundy, C.R., Gorick, B.D., Tomlinson, I.M. (2000). Antibody arrays for high-throughput screening of antibody–antigen interactions. *Nat Biotechnol* 18: 989–994.

Frank, R. (2002). The SPOT-synthesis technique. Synthetic peptide arrays on membrane supports–principles and applications. *J Immunol Methods* 267: 13–26.

Hallborn, J., Carlsson, R. (2002). Automated screening procedure for high-throughput generation of antibody fragments. *Biotechniques Suppl* 30–37.

Hamilton, S. (2002). Introduction to screening automation. *Methods Mol Biol* 190: 169–193.

Han, Z., Karatan, E., Scholle, M.D., McCafferty, J., Kay, B.K. (2004). Accelerated screening of phage-display output with alkaline phosphatase fusions. *Comb Chem High Throughput Screen* 7: 55–62.

Hedhammar, M., Stenvall, M., Lonneborg, R., Nord, O., Sjolin, O., Brismar, H., Uhlen, M., Ottosson, J., Hober, S. (2005). A novel flow cytometry-based method for analysis of expression levels in *Escherichia coli*, giving information about precipitated and soluble protein. *J Biotechnol* 119: 133–146.

Holz, C., Lang, C. (2004). High-throughput expression in microplate format in *Saccharomyces cerevisiae*. *Methods Mol Biol* 267: 267–276.

Hoogenboom, H.R., Griffiths, A.D., Johnson, K.S., Chiswell, D.J., Hudson, P., Winter, G. (1991). Multi-subunit proteins on the surface of filamentous phage: methodologies for displaying antibody (Fab) heavy and light chains. *Nucleic Acids Res* 19: 4133–4137.

Hultschig, C., Kreutzberger, J., Seitz, H., Konthur, Z., Bussow, K., Lehrach, H. (2006). Recent advances of protein microarrays. *Curr Opin Chem Biol* 10: 4–10.

Jia, X.C., Raya, R., Zhang, L., Foord, O., Walker, W.L., Gallo, M.L., Haak-Frendscho, M., Green, L.L., Davis, C.G. (2004). A novel method of Multiplexed Competitive Antibody Binning for the characterization of monoclonal antibodies. *J Immunol Methods* 288: 91–98.

Jostock, T., Vanhove, M., Brepoels, E., Van Gool, R., Daukandt, M., Wehnert, A., Van Hegelsom, R., Dransfield, D., Sexton, D., Devlin, M., Ley, A., Hoogenboom, H., Mullberg, J. (2004). Rapid generation of functional human IgG antibodies derived from Fab-on-phage display libraries. *J Immunol Methods* 289: 65–80.

Kersten, B., Feilner, T., Kramer, A., Wehrmeyer, S., Possling, A., Witt, I., Zanor, M.I., Stracke, R. Lueking, A., Kreutzberger, J., Lehrach H., Cahilll, D.J. (2003). Generation of *Arabidopsis* protein chips for antibody and serum screening. *Plant Mol Biol* 52: 999–1010.

Kigawa, T., Yabuki, T., Matsuda, N., Matsuda, T., Nakajima, R., Tanaka, A., Yokoyama, S. (2004). Preparation of *Escherichia coli* cell extract for highly productive cell-free protein expression. *J Struct Funct Genomics* 5: 63–68.

Köhler, G., Milstein, C. (1975). Continuous cultures of fused cells secreting antibody of predefined specificity. *Nature* 256: 495–497.

Konthur, Z., Walter, G. (2002). Automation of phage display for high-throughput antibody development. *TARGETS* 1: 30–36.

Krebs, B., Rauchenberger, R., Reiffert, S., Rothe, C., Tesar, M., Thomassen, E., Cao, M., Dreier, T., Fischer, D., Hoss, A., Inge, L., Knappik, A., Marget, M., Pack, P., Meng, X.Q., Schier, R., Sohlemann, P., Winter, J., Wolle, J., Kretzschmar, T. (2001). High-throughput generation and engineering of recombinant human antibodies. *J Immunol Methods* 254: 67–84.

Li, M. (2000). Applications of display technology in protein analysis. *Nat Biotechnol* 18: 1251–1256.

Little, M., Kipriyanov, S.M., Le Gall, F., Moldenhauer, G. (2000). Of mice and men: hybridoma and recombinant antibodies. *Immunol Today* 21: 364–370.

Lou, J., Marzari, R., Verzillo, V., Ferrero, F., Pak, D., Sheng, M., Yang, C., Sblattero, D., Bradbury, A. (2001). Antibodies in haystacks: how selection strategy influences the outcome of selection from molecular diversity libraries. *J Immunol Methods* 253: 233–242.

Lueking, A., Possling, A., Huber, O., Beveridge, A., Horn, M., Eickhoff, H., Schuchardt, J., Lehrach, H., Cahill, D. J. (2003). A nonredundant human protein chip for antibody screening and serum profiling. *Mol Cell Proteomics* 2: 1342–1349.

McCafferty, J., Griffiths, A.D., Winter, G., Chiswell, D.J. (1990). Phage antibodies: filamentous phage displaying antibody variable domains. *Nature* 348: 552–554.

Mechold, U., Gilbert, C., Ogryzko, V. (2005). Codon optimization of the BirA enzyme gene leads to higher expression and an improved efficiency of biotinylation of target proteins in mammalian cells. *J Biotechnol* 116: 245–249.

Menke, K.C. (2002). Unit automation in high throughput screening. *Methods Mol Biol* 190: 195–212.

Mullis, K., Faloona, F., Scharf, S., Saiki, R., Horn, G., Erlich, H. (1986). Specific enzymatic amplification of DNA *in vitro*: the polymerase chain reaction. *Cold Spring Harb Symp Quant Biol* 51 (Pt 1): 263–273.

Poetz, O., Ostendorp, R., Brocks, B., Schwenk, J.M., Stoll, D., Joos, T.O., Templin, M.F. (2005). Protein microarrays for antibody profiling: specificity and affinity determination on a chip. *Proteomics* 5: 2402–2411.

Rhyner, C., Konthur, Z., Blaser, K., Crameri, R. (2003). High-throughput isolation of recombinant antibodies against recombinant allergens. *Biotechniques* 35: 672–676.

Saiki, R.K., Scharf, S., Faloona, F., Mullis, K.B., Horn, G.T., Erlich, H.A., Arnheim, N. (1985). Enzymatic amplification of beta-globin genomic sequences and restriction site analysis for diagnosis of sickle cell anemia. *Science* 230: 1350–1354.

Schatz, P.J. (1993). Use of peptide libraries to map the substrate specificity of a peptide-modifying enzyme: a 13 residue consensus peptide specifies biotinylation in *Escherichia coli*. *Bio/Technology* 11: 1138–1143.

Scheich, C., Sievert, V., Bussow, K. (2003). An automated method for high-throughput protein purification applied to a comparison of His-tag and GST-tag affinity chromatography. *BMC Biotechnol* 3: 12.

Schweitzer, R.H., Abriola, L. (2002). Electrochemiluminescence. A technology evaluation and assay reformatting of the Stat6/P578 protein-peptide interaction. *Methods Mol Biol* 190: 87–106.

Smith, G.P. (1985). Filamentous fusion phage: novel expression vectors that display cloned antigens on the virion surface. *Science* 228: 1315–1317.

Stenvall, M., Steen, J., Uhlen, M., Hober, S., Ottosson, J. (2005). High-throughput solubility assay for purified recombinant protein immunogens. *Biochim Biophys Acta* 1752: 6–10.

Uhlen, M., Ponten, F. (2005). Antibody-based proteomics for human tissue profiling. *Mol Cell Proteomics* 4: 384–393.

van Werven, F.J., Timmers, H.T. (2006). The use of biotin tagging in Saccharomyces cerevisiae improves the sensitivity of chromatin immunoprecipitation. *Nucleic Acids Res* 34: e33.

Walter, G., Konthur, Z., Lehrach, H. (2001). High-throughput screening of surface displayed gene products. *Comb Chem High Throughput Screen* 4: 193–205.

Warford, A., Howat, W., McCafferty, J. (2004). Expression profiling by high-throughput immunohistochemistry. *J Immunol Methods* 290: 81–92.

Wassaf, D., Kuang, G., Kopacz, K., Wu, Q.L., Nguyen, Q., Toews, M., Cosic, J., Jacques, J., Wiltshire, S., Lambert, J. Pazmany, C.C., Hogan, S., Ladner, R.C., Nixon, A.E., Sexton, D.J. (2006). High-throughput affinity ranking of antibodies using surface plasmon resonance microarrays. *Anal Biochem* 351: 241–253.

Weiner, L.M. (2006). Fully human therapeutic monoclonal antibodies. *J Immunother* 29: 1–9.

Winter, G., Milstein, C. (1991). Man-made antibodies. *Nature* 349: 293–299.

5
Emerging Technologies for Antibody Selection

Mingyue He and Michael J. Taussig

5.1
Introduction

There is a continuing, intense demand for production of specific antibodies. As well as being the most widely used binding molecules in basic and medical research and their applications, antibodies are essential to the current rapid expansion of proteomic studies (Pandey and Mann 2000; Carter 2001; Nielsen et al. 2003; Holliger and Hudson 2005). For example, antibody microarrays (Mann and Jensen 2003), with their potential for rapid protein expression analysis and diagnostic biomarker detection, require thousands of antibodies to be immobilized on a solid surface. In order to meet the demands of such highly multiplexed systems, display technologies provide a means of producing recombinant monoclonal antibodies *in vitro* as a cost- and time-efficient alternative to conventional hybridomas. Through the physical linkage of genotype (DNA or RNA) with phenotype (antibody fragment), specific combining sites are selected from large combinatorial libraries (Hoogenboom 2005). Recombinant display technologies tap the exploitation potential of DNA diversity, creating and screening novel sequences that are inaccessible by *in vivo* processes. In addition, since selection takes place under defined *in vitro* conditions, the *in vivo* biological constraints on antibody production are exceeded, with affinities beyond the *in vivo* ceiling [Boder et al. 2000] as well as other required properties. Moreover, some of these technologies also provide a route to engineering of antibodies with enhanced properties through iterative cycles of mutation and selection.

A number of display methods have been established and are widely used (Hoogenboom 2005), among them phage display, cell surface display, and ribosome display. The cell-dependent methods (phage, cell surface display) have limitations which the fully cell-free system of ribosome display can overcome. In this chapter, we review the selection of antibodies by display *in vitro*, focusing particularly on ribosome display.

Handbook of Therapeutic Antibodies. Edited by Stefan Dübel
Copyright © 2007 WILEY-VCH Verlag GmbH & Co. KGaA, Weinheim
ISBN 978-3-527-31453-9

5.2
Display Technologies

In these methods, the physical association of phenotype and genotype ensures the simultaneous selection of a functional binding protein and the genetic information that encodes it. Figure 5.1 illustrates the cell-based and cell-free methods which have been developed through display of proteins on phage, cell surfaces, or ribosome complexes. Even though the selection steps in the procedure are carried out *in vitro*, phage display and cell surface display require that individual proteins are first expressed intracellularly, followed by assembly on the surface of the phage or transfer to the cell wall or membrane. In contrast, completely cell-free systems such as ribosome display and mRNA display do not require cell transformation, relying on polymerase chain reaction (PCR) and cell-free expression to produce libraries of stable protein–ribosome–mRNA or protein–mRNA complexes respectively; after co-selection of nascent protein and its encoding mRNA, the latter is subsequently reverse transcribed to DNA.

Currently, phage, yeast surface, and ribosome display are the most frequently used methods for selection and evolution of antibodies and ligand-binding scaffold proteins *in vitro* (Hoogneboom 2005). Each has its own particular advantages. Phage display is widely employed for antibodies in both Fab and single-chain (sc) Fv formats; it has been used to select high-affinity antibodies against free antigenic targets as well as antigens in their native location and conformation in cells

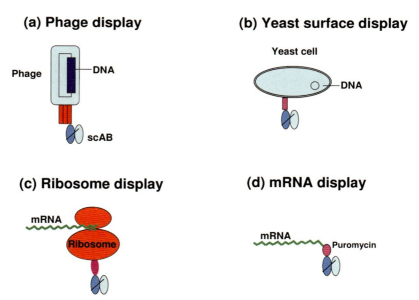

Fig. 5.1 Linkage of genotype and phenotype in four display technologies. scAB, single-chain antibody fragment carrying the antibody combining site.

Table 5.1 Comparison of phage display, yeast surface display and ribosome display.

	Phage display	Yeast surface display	Ribosome display
Maximum library size	10^{10}–10^{11}	10^7	10^{13}–10^{14}
Transformation required	Yes	Yes	No
Library form	Plasmid	Plasmid	PCR fragment or mRNA
Proteins to be displayed	Soluble, non-toxic, compatible with crossing membranes	Soluble, non-toxic, compatible with crossing membranes	Most proteins including cytotoxic, chemically modified and membrane proteins
Selection strategy	Panning	Sorting	Panning
Recovery	Elution, digestion or bacterial uptake. Strong binders may be lost	Cell sorting recovers strongest binders	RT-PCR potentially recovers all binders including strongest
Highest affinity antibody generated ($mol\,L^{-1}$)	10^{-11} [a]	10^{-14} [b]	10^{-12} [c]
Mutagenesis and protein evolution	DNA diversification followed by cloning	DNA diversification followed by cloning	DNA diversification without cloning

a Schier et al. 1996.
b Boder et al. 2000.
c Zahnd et al. 2004.

and tissues or through *in vivo* targeting. Yeast surface display allows direct screening by flow cytometry of individual antibody-displaying cells, providing a powerful tool for efficient sorting of antibody fragments with improved properties, and has produced affinities up to ~$10^{14}\,mol\,L^{-1}$ (Boder et al. 2000). However, these cell-dependent methods require cloning of DNA by cell transformation, restricting significantly the library size that can be displayed. Ribosome display (He and Khan 2005) overcomes this limitation through the use of PCR to create a DNA library encoding single-chain fragments which are then expressed in a cell-free system. Due to the ease with which very large PCR DNA libraries can be generated, ribosome display not only screens much larger populations but also allows continuous expansion of new diversity during the selection process; it is therefore suitable for rapid *in vitro* antibody evolution. Table 5.1 compares the properties of these three display systems.

5.3 Antibody Libraries

The diverse repertoire of antibody molecules is a result of the combination of six complementary determining regions (CDRs), three from the heavy (H) and three from light (L) chain variable domains (V_H, V_L), and contained within V domain frameworks. In B cells, the complete V regions result from DNA rearrangement and combinatorial assembly of different gene segments followed, after cell activation, by somatic mutation (Milstein 1993; Ohlin and Zouali 2003). Display technologies use recombinant DNA methods to generate combinatorial DNA libraries for selection. This allows expanded diversity and creation of novel sequences that are not available to the immune system *in vivo*. Library size has been shown to influence the properties of selected antibodies, with larger libraries providing increased probability of finding specific, high-affinity binders to particular epitopes (Ling 2003).

Different types of combinatorial antibody library have been constructed from animals and humans (Hoogenboom 2005; He and Khan 2005). Randomized assembly of V_H and V_L domains from different lymphoid sources (e.g. peripheral blood, bone marrow, spleen or tonsils) has been used to make "naive" single-chain antibody (scAb) libraries (Winter and Milstein 1991). The combinations create new diversity, though some pairs may not be favorably recombined (de Wildt et al. 1999). Through the introduction of randomized codons into CDRs, "synthetic" antibody libraries with new diversity can be obtained (Hoogenboom and Winter 1992). Naive and synthetic libraries are both generally antigen independent and particularly useful for unbiased selection of antibodies against any target antigen.

A novel design has been used to construct libraries with single V_H and V_L domain frameworks or consensus sequences. Diversity was built up by shuffling native CDR repertoires onto the same V_H (or V_L) frameworks (Knappik et al. 2000; Söderlind et al. 2000). This approach permits random recombination of all six CDRs, providing a large potential for generating novel variants. An advantage of this method is that the library should be functional, as CDRs derived from natural sequences usually form folded molecules (Söderlind et al. 2000). The use of an appropriate single framework facilitates downstream expression, since some framework sequences are not expressed in heterologous systems (Söderlind et al. 2000). A very large human antibody library (HuCAL) was generated by using a few modular consensus frameworks to display a full set of CDR repertoires (Knappik et al. 2000). Recently, a focused library has been constructed, which created biased diversity specific for binding to small molecule antigens such as haptens (Persson et al. 2006).

In contrast to naive libraries, those from immunized animals or humans provide enriched sequences for rapid selection of antibodies against the antigens used for the immunization (He et al. 1999; Felding-Habermann et al. 2005). The combination of B-cell activation *in vivo* with antibody selection *in vitro* offers a more rapid route to isolation of specific antibodies (He et al. 1999; Persson et al. 2006).

5.4
Antibody Selection and Maturation *In Vitro*

Antibodies are selected from display libraries under defined *in vitro* conditions, which can be adjusted in a controllable direction to enrich for desirable molecules. A number of selection strategies have been developed, including off-rate, specificity, stability, and antibody-guided selection.

Off-rate selection is designed to enrich for antibodies with slower dissociation rates and hence higher affinity. One method is to equilibrate the displayed library with a biotinylated antigen, followed by addition of an excess of unlabeled antigen; antibodies with slow off-rates remain bound to the biotinylated antigen for a longer time and are subsequently captured by immobilized streptavidin, while those with faster dissociation rates are released and their reassociation blocked by the unlabeled antigen. This strategy has led to isolation of molecules with affinities of up to $10^{-12}\,\text{mol}\,L^{-1}$ (Jermutus et al. 2001).

For improved specificity, epitope-blocked panning is a method in which irrelevant epitopes are blocked with antibodies prior to selection (Tsui et al. 2002). Alternating a particular hapten conjugated on different protein carriers in selection cycles can reduce cross-reactivity to the carrier proteins (Yau et al. 2003a). Direct selection on whole cells or frozen tissue sections has also produced specific antibodies recognizing specific epitopes *in situ* (Tanaka et al. 2002).

Antibodies have also been selected under conditions of thermal or chemical denaturation, leading to highly stable and aggregation-resistant binding fragments (Wörn and Plückthun 2001; Jespers et al. 2004). Inclusion of a reducing reagent such as dithiothreitol (DTT) during selection has isolated those with enhanced solubility and stability (Jermutus et al. 2001).

Guided selection (Osbourn et al. 2005) is a useful approach to generating a human combining site equivalent of an existing rodent antibody with similar antigen specificity, serving as an alternative to humanization. In this approach, the H- or L-chain of the rodent antibody was displayed as a capturing reagent to select human antibody H- and L-chain partners on the same antigen (Osbourn et al. 2005). Alternatively, humanization of rodent antibodies can be carried out by "reshaping" (Verhoeyen et al. 1988), which involves changing solvent-exposed residues in a murine framework to their human homologs, followed by screening of all variants that best preserve the original antibody properties. This approach overcomes the drawback of other humanization methods that require separate construction and analysis of individual antibody mutants (Rosok et al. 1996; Wang et al. 2004).

De novo design of antigens enables selection of antibodies with novel combining sites (e.g. catalytic antibodies and conformation-specific antibodies can be selected by using designed antigen analogs) (Cesaro-Tadic et al. 2003).

Combined with DNA mutagenesis, display technologies are an efficient means of developing antibody properties through evolutionary approaches. By repeated cycles of mutation and functional selection, antibody variants with improved properties can be isolated (Plückthun et al. 2000). For example, an *in vitro* antibody maturation strategy has selected mutants with affinities of up to $5 \times 10^{-14}\,\mathrm{mol\,L^{-1}}$ (Boder et al. 2000). Compared with *in vivo* somatic maturation processes, which usually provide antibodies of 10^9–$10^{10}\,\mathrm{mol\,L^{-1}}$, the possibility of a 10^3-fold affinity improvement makes *in vitro* maturation a relevant approach for the selection of potential therapeutic antibodies.

5.5
Linking Antibodies to mRNA: Ribosome and mRNA Display

Ribosome and mRNA display are cell-free methods which overcome some of the limitations of cell-based systems, particularly in regard to library size, by directly expressing large PCR libraries without the need for cloning. In ribosome display, phenotype and genotype are linked as protein–ribosome–mRNA complexes, which are achieved by stalling the ribosome at the end of translation, usually as a result of stop codon deletion. Since the presence of a stop codon is required to engage release factors, its absence means that the nascent protein does not dissociate from the ribosome, remaining associated with the encoding mRNA (Fig. 5.1). The generation of a library of ribosome complexes permits affinity selection (e.g. by immobilized ligand) of a nascent antibody and its encoding mRNA, which can be recovered as DNA by reverse transcriptase (RT)-PCR (Fig. 5.2). Cyclical reiteration of this process leads progressively to enrichment of antibodies originally present as rare species in a large population, by a factor or 10^3–10^5 per cycle (He and Taussig 1997). Both prokaryotic and eukaryotic ribosome display systems have been developed and applied successfully to antibody selection and evolution (He and Taussig 1997; Plückthun et al. 2000; He and Khan 2005).

In the method known as mRNA display, the protein and mRNA become covalently linked through a puromycin moiety attached to the 3′ end of the mRNA, which displaces the nascent protein at the end of translation (Roberts and Szostak 1997). The ribosome is no longer present in the complexes. mRNA display has been used for selection of alternative antibody mimics based on the fibronectin scaffold (Xu et al. 2002).

5.6
Advantages of Ribosome Display

Since there is no requirement for DNA cloning, ribosome display libraries of very large size can be generated with ease. For example, while a PCR library of 10^{12-14} members can be produced through a few reactions, up to 10^5 transformations would be required for cell-based display methods, as each transformation usually generates 10^{7-9} clones (Lamla and Erdmann 2003). The restriction for a ribosome display library is perhaps only the number of functional ribosomes in the reaction, which can be scaled up to 10^{14} per mL in rabbit reticulocyte lysate (He and Taussig 1997). The use of PCR DNA templates also provides a simple

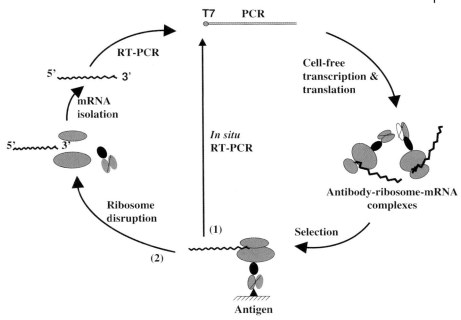

Fig. 5.2 Versions of the ribosome display cycle. (1) Rabbit reticulocyte lysate system with single-step recovery of DNA by RT-PCR. (2) *E. coli* S30 system with recovery by ribosome disruption and RT-PCR. T7, T7 promoter; RT-PCR, reverse transcriptase polymerase chain reaction.

tool for continuous introduction of additional diversity into the DNA pool for further selection cycles (Plückthun et al. 2000; He and Taussig 2002). Sequence changes can also be introduced at the mRNA level by inclusion of Qβ RNA-dependent RNA polymerases in the cell-free mixture (Irving et al. 2001). Thus, ribosome display provides an efficient system for antibody evolution *in vitro*, in contrast with phage display where such continuous "in-built" evolution is not possible.

5.7
Ribosome Display Systems

5.7.1
Prokaryotic: *E. coli* S30

The S30 cell-free lysate from *E. coli*, either with coupled or uncoupled transcription and translation, has been adapted for ribosome display (Mattheakis et al.

1994; Hanes and Plückthun 1997). In the coupled method, a DNA library is used to generate polyribosome complexes displaying proteins, which are captured with an immobilized ligand. To avoid any possible disruptive effect of DTT, prokaryotic ribosome display has more often been carried out in an uncoupled format (i.e the transcription and translation are performed separately) (Hanes and Plückthun 1997; Plückthun et al. 2000). In order to recover the associated mRNA, the polysome complexes are dissociated by chemical disruption and the released mRNA purified prior to RT-PCR (Fig. 5.2) (Mattheakis et al. 1994; Hanes and Plückthun 1997). The procedure has been modified to display folded antibody fragments by adding extra components such as protein disulfide isomerase, vanadyl ribonucleoside complexes, and anti-ssrA antisense oligonucleotide (Hanes and Plückthun 1997). Recent modification of *E. coli* ribosome display has included the generation of more stable ribosome complexes using a protein–mRNA interaction (Sawata et al. 2004).

5.7.2
Eukaryotic: Rabbit Reticulocyte

A eukaryotic system has also been developed for selection of antibody fragments using the coupled rabbit reticulocyte lysate. This technology was initially called ARM (antibody-ribosome-mRNA) display (He and Taussig 1997) (Fig. 5.2). In a modified version, oxidized/reduced glutathione and Qβ RNA-dependent RNA polymerase were included in the translation mixture to improve protein folding and introduce mutations (Irving et al. 2001). Wheatgerm cell-free lysate has also been adapted for ribosome display of folded proteins (Takahashi et al. 2002).

The main difference between the *E. coli* S30 and rabbit reticulocyte ribosome display systems lies in the DNA recovery step (Fig. 5.2). While the prokaryotic ribosome display method requires chemical disruption (e.g. EDTA chelation) to dissociate ribosome subunits and release mRNA prior to RT-PCR, rabbit reticulocyte lysate display employs an *in situ* recovery procedure in which RT-PCR is performed directly on the ribosome complexes without the need for prior dissociation. Successful *in situ* RT-PCR is achieved through the design of primers hybridizing slightly upstream of the 3′ end to avoid the region covered by the stalled ribosome [He and Taussig 1997]. *In situ* RT-PCR not only simplifies the recovery process but also avoids material losses incurred in disrupting complexes for mRNA isolation. It has been used to analyze the binding specificity of ribosome complexes through detection of the attached mRNA (He and Taussig 2005). *In situ* RT-PCR would also facilitate automation of the ribosome display process.

Interestingly, the prokaryotic ribosome disruption method seems to be a relatively poor procedure for releasing mRNA from rabbit reticulocyte ribosome complexes (Hanes et al. 1999; He and Taussig 2005; Douthwaite et al. 2006), possibly indicating a difference in stability between prokaryotic and eukaryotic complexes. A method has been described to disrupt rabbit reticulocyte complexes by heating above 70 °C (Bieberich et al. 2000).

5.7.3
Ribosome Display Constructs (Fig. 5.3)

DNA constructs for ribosome display should contain a T7 promoter and a translation initiation sequence such as Shine–Dalgarno for *E. coli* S30 or the Kozak sequence for eukaryotic systems. It is possible to generate a single prokaryotic/eukaryotic consensus sequence (Allen and Miller 1999). To enable the complete exit of the displayed portion of the nascent protein from the ribosome tunnel, a spacer domain of at least 23–30 amino acids is fused at the C-terminus (He and Taussig 1997; Plückthun et al. 2000). The DNA sequence of the spacer also provides a known region for designing an annealing primer for *in situ* RT-PCR recovery. Spacers reported for ribosome display of proteins include the constant region of Igκ L-chain (Cκ), the $C_H 3$ domain of human IgM (Plückthun et al. 2000), gene III of M13 phage, streptavidin, and GST (Zhou et al. 2002). Spacer length has been shown to affect display efficiency, with a longer spacer being more efficient (Schaffitzel et al. 1999). Constructs for *E. coli* display also require incorporation of sequences containing stem–loop structures at both the 5' and 3' ends of the DNA to prevent mRNA degradation by the high RNase activities in the *E. coli* S30 system (Hanes and Plückthun 1997). The diverse library sequences are placed in-frame between the initiation codon ATG and the spacer. To form stable ribosome complexes, the stop codon should be deleted from the construct by using a 3' primer without the stop codon.

5.7.4
Monosome versus Polysome Display

With the commonly used cell-free systems, both coupled and uncoupled methods have been developed for ribosome display of single-chain antibody fragments and

(a) *E.coli* **construct**

(b) Eukaryotic construct

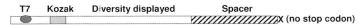

Fig. 5.3 Constructs used for prokaryotic and eukaryotic ribosome display. S/D, Shine–Dalgarno sequence for prokaryotic translation initiation; Kozak, eukaryotic translation initiation sequence.

other proteins (Plückthun et al. 2000; He and Taussig 2002). The choice of cell-free system depends mainly on the origins and properties of the proteins to be displayed and the downstream applications; some proteins may express better in one particular cell-free system than in another. One issue is whether the complexes are monoribosomal (monosome) or polyribosomal (polysome), the former having only one translatin ribosome per mRNA whereas the latter carry several. It has been shown that rabbit reticulocyte lysate produces mainly monosome complexes [Holet and Osbourn 2001], whereas *E. coli* S30 generates polysomes (Mattheakis et al. 1994). However, it is also possible to display monosomes in *E. coli* S30 by controlling the ratio of input mRNA to ribosome number (Fedorov and Baldwin 1998). The formation of polysome complexes raises the possibility of polyvalent display of incomplete nascent polypeptides and avidity effects which may lead to isolation of lower affinity combining sites (depending on the selection procedure used).

5.8
Antibody Generation by Ribosome Display

Ribosome display has been applied to selection, evolution, and humanization of antibodies *in vitro*, yielding molecules of high affinity and specificity (Schaffitzel et al. 1999; Plückthun et al. 2000; He and Taussig 2002). Single-domain mini-antibodies have also been successfully displayed (Yau et al. 2003). Through repeated rounds of mutation and *in vitro* selection, antibody variants with improved affinity (down to picomolar K_D), specificity, and stability have been isolated (Schaffitzel et al. 1999; He and Taussig 2002; Thom et al. 2006). Antibodies with novel binding sites recognizing conformation-specific epitopes or having catalytic activities have also been selected (Hanes et al. 2000; Amstutz et al. 2002). Recently, ribosome display has been utilized for antibody humanization by reshaping, which rapidly identifies humanized antibody variants from a shuffled DNA library (Wang et al. 2004), or through guided selection using the H- or L-chain of a rodent antibody as a capturing reagent to select human antibody partner chains with similar antigen specificity (Osbourn et al. 2005).

5.9
Summary

Display technologies provide powerful and versatile methods for selection and manipulation of recombinant antibodies. They are capable of generating binding molecules with optimized features or properties suitable for diagnostic and therapeutic applications, which may not be available through animal immunization. The operation of ribosome display is completely cell-free, overcoming the limitations of cell-based display methods by rapidly generating and screening very large libraries, which in turn increase the likelihood of finding high-affinity binding molecules. Potentially, the selection power of ribosome display could be further enhanced by automation or combination with protein arrays, leading to high-throughput selections of antibodies against genome-coded targets for proteomic studies.

Acknowledgments

Research at the Babraham Institute is supported by the BBSRC. Work at the Technology Research Group is also supported through the EC 6th Framework Programme Integrated Project "MolTools" (www.moltools.org).

References

Allen, S.V., Miller, E.S. (1999) RNA-binding properties of *in vitro* expressed histidine-tagged RB69 RegA translational repressor protein. *Anal Biochem* 269: 32–37.

Amstutz, P., Pelletier, J.N., Guggisberg, A. et al. (2002) In vitro selection for catalytic activity with ribosome display. *J Am Chem Soc* 124: 9396–9403.

Bieberich, E., Kapitonov, D., Tencomnao, T., Yu, R.K. (2000) Protein-ribosome-mRNA display: affinity isolation of enzyme-ribosome-mRNA complexes and cDNA cloning in a single-tube reaction. *Anal Biochem* 287: 294–298.

Boder, E.T., Midelfort, K.S., Wittrup, K.D. (2000) Directed evolution of antibody fragments with monovalent femtomolar antigen-binding affinity. *Proc Natl Acad Sci USA* 97: 10701–10705.

Carter, P. (2001) Improving the efficacy of antibody-based cancer therapies. *Nature Rev Cancer* 1: 118–129.

Cesaro-Tadic, S., Lagos, D., Honegger, A. et al. (2003) Turnover-based *in vitro* selection and evolution of biocatalysts from a fully synthetic antibody library *Nat Biotechnol* 21: 679–685.

de Wildt, R.M., Hoet, R.M., van Venrooij, W.J. et al. (1999) Analysis of heavy and light chain pairings indicates that receptor editing shapes the human antibody repertoire. *J Mol Biol* 285: 895–901.

Douthwaite, J.A., Groves, M.A., Dufner, P., Jermutus, L. (2006) An improved method for an efficient and easily accessible eukaryotic ribosome display technology. *Protein Eng Des Sel* 19: 85–90.

Fedorov, A.N., Baldwin, T.O. (1998) Protein folding and assembly in a cell-free expression system. *Methods Enzymol* 290: 1–17.

Felding-Habermann, B., Lerner, R.A., Lillo, A. et al. (2005) Combinatorial antibody libraries from cancer patients yield ligand-mimetic Arg-Gly-Asp-containing immunoglobulins that inhibit breast cancer metastasis *Proc Natl Acad Sci USA* 101: 17210–17215.

Hanes, J., Plückthun, A. (1997) *In vitro* selection and evolution of functional proteins by using ribosome display. *Proc Natl Acad Sci USA* 94: 4937–4942.

Hanes, J., Jermutus, L., Schaffitzel, C., Plückthun, A. (1999) Comparison of *E. coli* and rabbit reticulocyte ribosome display systems. *FEBS Lett* 450: 105–110.

Hanes, J., Schaffitzel, C., Knappik, A., Plückthun, A. (2000) Picomolar affinity antibodies from a fully synthetic naïve library selected and evolved by ribosome display. *Nat Biotechnol* 18: 1287–1292.

He, M., Khan, F. (2005) Ribosome display: next generation display technologies for production of antibodies *in vitro*. *Expert Rev Proteomics* 2: 421–430.

He, M., Taussig, M.J. (1997) ARM complexes as efficient selection particles for *in vitro* display and evolution of antibody combining sites. *Nucleic Acids Res* 25: 5132–5134.

He, M., Taussig, M.J. (2002) Ribosome display: Cell-free protein display technology. *Briefings Functional Genomics Proteomics* 1: 204–212.

He, M., Taussig, M.J. (2005) Ribosome display of antibodies: expression, specificity and recovery in a eukaryotic system. *J Immunol Methods* 297: 73–82.

He, M., Menges, M., Groves, M.A. et al. (1999) Selection of a human anti-progesterone antibody fragment from a transgenic mouse library by ARM ribosome display. *J Immunol Methods* 231: 105–117.

Holet, T., Osbourn, J. (2001) Improvements to ribosome display. International patent application (PCT/ WO 0175097).

Holliger, P., Hudson, P. (2005) Review: Engineered antibody fragments and the rise of single domains. *Nat Biotechnol* 23: 1126–1136.

Hoogenboom, H.R. (2005) Review: selection and screening recombinant antibody libraries. *Nat Biotechnol* 23: 1105–1116.

Hoogenboom, H.R., Winter, G. (1992) By-passing immunisation: Human antibodies from synthetic repertoires of germline VH gene segments rearranged *in vitro*. *J Mol Biol* 227: 381–388.

Irving, R.A., Coia, G., Roberts, A., Nuttall, S.D., Hudson, P.L. (2001) Ribosome display and affinity maturation: from antibodies to single V-domains and steps towards cancer therapeutics. *J Immunol Methods* 248: 31–45.

Jermutus, L., Honegger, A., Schwesinger, F., Hanes, J., Plückthun, A.(2001) Tailoring *in vitro* evolution for protein affinity or stability. *Proc Natl Acad Sci USA* 98: 75–80.

Jespers, L., Schon, O., Famm, F., Winter, G. (2004) Aggregation-resistant domain antibodies selected on phage by heat denaturation. *Nat Biotechnol* 22: 1161–1165.

Knappik, A., Ge, L., Honegger, A. et al. (2000) Fully synthetic human combinatorial antibody libraries (HuCAL) based on modular consensus frameworks and CDRs randomized with trinucleotides. *J Mol Biol* 296: 57–86.

Lamla, T., Erdmann, V.A. (2003) Searching sequence space for high affinity binding peptides using ribosome display. *J Mol Biol* 329: 381–388.

Ling, M.M. (2003) Large antibody display libraries for isolation of high affinity antibodies. *Comb Chem High Throughput Screen* 6: 421–432.

Mann, M., Jensen, O.N. (2003) Proteomic analysis of post-translational modifications *Nat Biotechnol* 21: 255–261.

Mattheakis, L.C., Bhatt, R.R., Dower, W.J. (1994) An *in vitro* polysome display system for identifying ligands from very large peptide libraries. *Proc Natl Acad Sci USA* 91: 9022–9026.

Milstein, C. (1993) From the structure of antibodies to the diversification of immune response. *Scand J Immunol* 37: 385–395.

Nielsen, U.B., Cardone, M.H., Sinskey, A.J. et al. (2003) Profiling receptor tyrosine kinase activation by using Ab microarrays. *Proc Natl Acad Sci USA* 100: 9330–9335.

Ohlin, M., Zouali, M. (2003) The human antibody repertoire to infectious agents: implications for disease pathogenesis. *Mol Immunol* 40: 1–11.

Osbourn, J., Groves, M., Vaughan, T. (2005) From rodent reagents to human therapeutics using antibody guided selection. *Methods* 36: 61–68.

Pandey, A., Mann, M. (2000) Proteomics to study genes and genomes. *Nature* 405: 837–846.

Persson, H., Lantto, J., Ohlin, M. (2006) A focused antibody library for improved hapten recognition. *J Mol Biol* 357: 607–620.

Plückthun, A., Schaffitzel, C., Hanes, J., Jermutus, L. (2000) *In vitro* selection and evolution of proteins. *Adv Protein Chem* 55: 367–403.

Roberts, R.W., Szostak, J.W. (1997) RNA-peptide fusion for the in vitro selection of peptides and proteins. *Proc Natl Acad Sci USA* 94: 12297–12302.

Rosok, M.J., Yelton, D.E., Harris, L.J. et al. (1996) A combinatorial library strategy for the rapid humanisation of anticarcinoma BR96 Fab. *J Biol Chem* 271: 22611–22618.

Sawata, S.Y., Suyama, E., Taira, K. (2004) A system based on specific protein-RNA interactions for analysis of target protein-protein interactions *in vitro*: successful selection of membrane-bound bak-Bcl-xL proteins *in vitro*. *Protein Eng Des Sel* 17: 501–508.

Schaffitzel, C., Hanes, J., Jermutus, L., Plückthun, A. (1999) Ribosome display: an *in vitro* method for selection and evolution of antibodies from libraries. *J Immunol Methods* 231: 119–135.

Schier, R., McCall, A., Adams, G.P. et al. (1996) Isolation of picomolar affinity anti-c-erbB-2 single-chain Fv by molecular evolution of the complementarity determining regions in the center of the antibody binding site. *J Mol Biol* 263: 551–567.

Söderlind, E., Strandberg, L., Jirholt, P. et al. (2000) Recombining germline-derived CDR sequences for creating diverse single-framework antibody libraries. *Nature Biotechnol* 18: 852–856.

Takahashi, F., Ebihara, T., Mie, M. et al. (2002) Ribosome display for selection of active dihydrofolate reductase mutants using immobilized methotrexate on agarose beads. *FEBS Lett* 514: 106–110.

Tanaka, T., Ito, T., Furuta, M. et al. (2002) *In situ* phage screening: A method for identification of subnanogram tissue components *in situ*. *J Biol Chem* 277: 30382–30387.

Thom, G., Cockroft, A.C., Buchanan, A.G. et al. (2006) Probing a protein-protein interaction by *in vitro* evolution. *Proc Natl Acad Sci USA* 103: 7619–7624.

Tsui, P., Tornetta, M.A , Ames, R.S. et al. (2002) Progressive epitope-blocked panning of a phage librray for isolation of human RSV antibodies. *J Immunol Methods* 263: 123–132.

Verhoeyen, M., Milstein, C., Winter, G. (1988) Reshaping human antibodies: grafting an anti-lysozyme activity. *Science* 239: 1534–1536.

Wang, X.B., Zhou, B., Yin, C.C., Lin, Q., Huang, H.L. (2004) A new approach for rapidly reshaping single-chain antibody *in vitro* by combining DNA shuffling with ribosome display. *J Biochem* 136: 19–28.

Winter, G., Milstein, C. (1991) Man-made antibodies. *Nature* 349: 293–299.

Wörn, A., Plückthun, A. (2001) Stability engineering of antibody single-chain Fv fragments. *J Mol Biol* 305: 989–1010.

Xu, L., Aha, P., Gu, K. et al. (2002) Directed evolution of high-affinity antibody mimics using mRNA display. *Chem Biol* 9: 933–942.

Yau, K.Y., Groves, M.A., Li, S., Sheedy, C., Lee, H., Tanha, J., MacKenzie, C.R., Jermutus, L., Hall, J.C. (2003) Selection of hapten-specific single-domain antibodies from a non-immunized llama ribosome display library. *J Immunol Methods* 281: 161–175.

Zahnd, C., Spinelli, S., Luginbuhl, B., Amstutz, P., Cambillau, C., Plückthun, A. (2004) Directed *in vitro* evolution and crystallographic analysis of a peptide-binding single chain antibody fragment (scFv) with low picomolar affinity. *J Biol Chem* 279: 18870–18877.

Zhou, J.M., Fujita, S., Warashina, M., Baba, T., Taira, K. (2002) A novel strategy by the action of ricin that connects phenotype and genotype without loss of the diversity of libraries. *J Am Chem Soc* 124: 538–543.

6
Emerging Alternative Production Systems
Thomas Jostock

6.1
Introduction

The market for recombinant protein pharmaceuticals is growing rapidly and reached a volume of about US$47 billion in 2004. Among those, recombinant antibodies are the fastest growing group and a large number of new antibody products are in clinical and preclinical development, which will lead to an increased need for production capacities and expression systems in the near future. Since conventional cost-effective microbial expression systems are not yet applicable for the expression of fully functional IgG antibodies, all currently marketed therapeutic recombinant antibody products are produced from animal cell culture processes that are associated with relative high operating expenses. The production costs of recombinant antibodies result in significantly higher expense for antibody-based therapies compared with most chemical drugs. Thus, to make a broad use of recombinant antibody therapeutics more competitive and affordable for the healthcare systems, alternative production systems with improved efficiency and reduced operating expenses are highly desirable.

IgG currently is the dominant antibody format for therapeutics. As a heterotetrameric molecule with numerous disulfide bonds and several glycosylation sites, IgG makes itself a challenging candidate for heterologous expression, especially in prokaryotes. Glycosylation of the Fc region has been shown to be essential for mediating effector functions such as antibody-dependent cell-mediated cytotoxicity (ADCC) and complement-dependent cytotoxicity (CDC) (Boyd et al. 1995). Furthermore, the glycosylation pattern strongly influences the efficiency of the induction of effector functions by antibodies (Umana et al. 1999; Niwa et al. 2005). For applications that depend on the immunological effector functions of the antibody therefore, eukaryotic production systems with suitable glycosylation capabilities are necessary.

Antibody fragments such as single-chain fragment variable (scFv) and fragment antigen binding (Fab) are less complex and the antigen-binding activity is independent of glycosylation, which makes them qualified for prokaryotic expression systems. Certainly, these fragments are incapable of mediating

Handbook of Therapeutic Antibodies. Edited by Stefan Dübel
Copyright © 2007 WILEY-VCH Verlag GmbH & Co. KGaA, Weinheim
ISBN 978-3-527-31453-9

immunological effector functions but for numerous applications this is not necessary or even unwanted.

Besides the two traditional expression systems – *E. coli* for antibody fragments and Chinese hamster ovary (CHO), baby hamster kidney (BHK), or myeloma cells for IgG antibodies – numerous alternative expression systems are currently under development. This includes the evaluation of new expression hosts such as filamentous fungi or transgenic plants as well as "tuning" of the more conventional expression hosts via genetic modifications. In the following section some emerging production systems for antibody expression are briefly described.

6.2
Production Systems

6.2.1
Prokaryotic Expression Systems

Fast growth, inexpensive culture media, and high resistance to shear forces usually are advantageous properties of prokaryotic organisms for recombinant protein production. The main disadvantages of prokaryotic expression systems, especially for antibody expression, are the limited glycosylation, folding, and secretion capabilities of the host cells. Currently the almost exclusive prokaryotic production system for antibodies is *E. coli*. Two members of the Gram-positive species *Bacillus* also have been reported to produce functional antibody fragments, *B. subtilis* (Inoue et al. 1997; Wu et al. 1998, 2002) and *B. brevis* (Bolhuis et al. 1999; Shiroza et al. 2001). The cell wall-less L-form of *Proteus mirabilis* was used to successfully produce scFv and mini-antibodies (Kujau et al. 1998; Rippmann et al. 1998). Using the prepro sequence of *S. hyicus* lipase, secretory production of a chimeric Fab was shown for *Staphylococcus carnosus* (Schnappinger et al. 1995). Several other prokaryotic expression systems are under development for the production of pharmaceutical proteins but to our knowledge have not been applied to produce antibodies yet, including *Ralstonia eutropha* and *Streptomyces* sp. with yields in the range of $100\,\text{mg}\,\text{L}^{-1}$ to $2\,\text{g}\,\text{L}^{-1}$ (Schmidt 2004). The following section describes the properties of some emerging prokaryotic systems.

6.2.1.1 Advanced and Alternative *E. coli* Expression Systems

The ease of handling, the well-characterized genetics, and the high transformation efficiency made the Gram-negative bacterium *E. coli* the most frequently used organism for recombinant protein expression. However, the lack of a eukaryotic-like folding apparatus and secretion machinery limits the applicability of *E. coli* as expression host for complex, multimeric, and glycosylated proteins like antibodies. Up to now there are only few examples of the successful production of entire IgG antibodies in *E. coli* (Simmons et al. 2002). In this report a careful balancing of the ratio of light to heavy chain expression and an adjustment of the translation levels to eliminate secretory blocks had to be performed to allow the formation and accumulation of tetrameric IgG molecules in the periplasm. As expected, the aglycosylated IgG antibodies bound to the Fcγn receptor but not to C1q or to the FcγI receptor. Thus, a long *in vivo* plasma half-life is achieved with this antibody product but no ability to mediate effector functions, which limits the spectrum of possible therapeutic applications to indications depending on neutralization or agonistic activities of the antibody.

Several genetic modifications of *E. coli* strains have been undertaken in order to improve the folding and secretion capabilities for recombinant protein production. This includes strains over- or coexpressing chaperones for improved folding (Bothmann and Plückthun 2000; Zhang et al. 2003) as well as strains overexpressing disulfide isomerases for more efficient disulfide bond formation (Humphreys et al. 1996; Kurokawa et al. 2001). Further, protease-deficient strains have been developed to reduce periplasmic degradation of the product (Wulfing and Rappuoli 1997; Kandilogiannaki et al. 2001; Chen et al. 2004). With a triple mutant *E. coli* host strain yields of up to $2\,g\,L^{-1}$ for a F(ab')$_2$ molecule have been reported (Chen et al. 2004).

As an alternative to the traditionally used Sec-secretion pathway for periplasmic expression of antibody fragments, the TAT (twin arginine translocation) pathway can be used for exporting recombinant proteins to the periplasm (DeLisa et al. 2003). In contrast to the Sec-secretion pathway, where unfolded polypeptides are transferred to the periplasm (van Wely et al. 2001), the TAT system exports proteins only in a folded state (Berks et al. 2000), which could increase the yield of correctly folded and biologically active material. Since disulfide bond formation is efficient only in an oxidative milieu, strains with an oxidizing cytoplasm are used for efficient antibody expression (DeLisa et al. 2003).

Purification of proteins accumulated in the periplasm of *E. coli* requires lysis of the outer membrane and the cell wall, which complicates the purification process. *E. coli* strains lacking the cell wall allow secretion of functional antibody product to the culture medium (Kujau et al. 1998).

6.2.1.2 *Proteus mirabilis*

The L-form of *Proteus mirabilis* is without a cell wall and is thus able to secrete signal peptide-containing proteins to the growth medium. Furthermore this *P. mirabilis* strain is easy to transform and can be cultured in fermenters. Mini-antibodies have been produced using *P. mirabilis* as production host with similar yields as with an L-form of *E. coli* (Kujau et al. 1998). In a comparative expression analysis of several scFv antibodies in *P. mirabilis* and the *E. coli* strain JM109, yields of 40–200 $mg\,L^{-1}$ active soluble antibody could be reached with *P. mirabilis*, which was about 60 times higher than with *E. coli* (Rippmann et al. 1998). An explanation could be the reduced toxic effects of the product which is released to the culture medium instead of accumulating in the periplasm.

6.2.1.3 *Bacillus subtilis* and *Bacillus brevis*

The lack of an outer membrane and endotoxins make Gram-positive bacteria attractive host cell candidates for the secretory expression of recombinant proteins such as antibodies. N-terminal fusion of appropriate signal peptides or secretion domains facilitates export of the recombinant protein to the culture media which is advantageous for purification. Problems related to endogenous proteases can be avoided using protease-deficient mutant strains (Wu et al. 1993; Takimura et al. 1997; Wu et al. 1998; Wu et al. 2002).

Bacillus subtilis and *Bacillus brevis* both have been successfully used to produce scFv or Fab antibody fragments (Wu et al. 1993; Inoue et al. 1997; Wu et al. 1998; Bolhuis et al. 1999; Shiroza et al. 2001; Wu et al. 2002).

In *B. subtilis* the formation of inclusion bodies can limit the secretory production of antibody fragments. Overexpression of intracellular and extracytoplasmic molecular chaperones could substantially reduce the problem of inclusion body formation of a scFv fragment. The percentage of intracellular insoluble scFv protein dropped from 60% to 6%, while the percentage of secreted scFv increased from 23% to 43%, giving yields of up to $12\,\text{mg}\,\text{L}^{-1}$ (Wu et al. 1998).

Bacillus brevis has the advantage of naturally secreting much fewer proteases than *B. subtilis* and of being able to secrete heterologous proteins with disulfide bonds, like human epidermal growth factor (EGF) (Yamagata et al. 1989). In *B. brevis* an scFv antibody with a yield of $10\,\text{mg}\,\text{L}^{-1}$ (Shiroza et al. 2001) as well as a chimeric Fab fragment with an estimated yield of $100\,\text{mg}\,\text{L}^{-1}$ (Inoue et al. 1997) were produced. For Fab production the light chain and the heavy chain fragment of the Fab were expressed from a dicistronic construct, secretion was achieved by N-terminal fusion of the signal peptide of the middle-wall protein. The polypeptide chains were correctly processed and formed disulfide-linked heterodimeric Fab fragments.

Possible limitations of using *Bacillus* as expression host are relative low transformation efficiencies and plasmid stability. Further, for *Bacillus subtilis* bottlenecks for efficient secretion of functional proteins can be processing by the signal peptidase, passage through the cell wall, degradation in the wall and growth medium and misfolding due to incorrect disulfide bond formation (Bolhuis et al. 1999).

6.2.2
Eukaryotic Expression Systems

Eukaryotic systems are the method of choice if the antibody product needs to be glycosylated for its biological functions. Protein secretion in eukaryotes occurs via the endoplasmic reticulum and Golgi apparatus and is aided by chaperones and cofactors. This is a complex process but the principles are the same for all eukaryotic organisms. However, the pattern of protein glycosylation differs between different species, even if they are very closely related. For therapeutic antibodies it is aimed to have a glycosylation pattern as close to the human one as possible to minimize immunogenicity and, for some applications, to maximize induction of immunological effector functions. Besides production levels and cost, the glycosylation pattern of eukaryotic expression hosts therefore is one of the most noted rating characteristics of the production system. The set of eukaryotic hosts that in the meantime have been developed for and applied to antibody expression ranges from simple organisms such as yeasts to transgenic mammals and plants.

6.2.2.1 Yeast

Yeasts share some advantageous properties with *E. coli*, such as rapid growth in cheap growth media, applicability for fermentation processes with high cell densities, and ease of genetic manipulation. In addition they provide a eukaryotic folding and secretion apparatus that allows secretion of functionally folded proteins to the culture media. This combination makes yeast an attractive expression system for large-scale production. Among the yeast species used for recombinant antibody expression are the well-characterized species *Saccaromyces cerevisiae*, as well as more unconventional hosts such as *Yarrowia lipolytica* and *Kluyveromyces lactis*, or the increasingly popular methylotrophic yeast *Pichia pastoris*, which has very promising secretion yields. The main issues for antibody expression in yeast are secretion rate, proteolytic activity, and glycosylation pattern.

Saccharomyces cerevisiae, which still is the most commonly used yeast species in pharmaceutical production (Schmidt 2004), has been used successfully for the expression of recombinant antibodies (Wood et al. 1985; Shusta et al. 1998). However, the secretion capabilities of *S. cerevisiae* for complex proteins seem to be limited (Rakestraw and Wittrup 2006) and hyperglycosylation of the recombinant proteins can lead to reduced pharmacological activity and further reduction of the secretion efficiency. Optimization of the gene expression and coexpression of chaperones or foldases lead to production yields of $20\,\mathrm{mg\,L^{-1}}$ for an scFv antibody (Shusta et al. 1998).

The two unconventional yeasts, *Y. lipolytica* and *K. lactis*, are regarded as safe (GRAS) organisms and are thought to have a good potential for the secretion of heterologous proteins. Swennen et al. (2002) report the production of an scFv antibody in *Y. lipolytica* and *K. lactis* with levels up to $10-20\,\mathrm{mg\,L^{-1}}$ showing the potential of both species for antibody production.

Within only a short period of time *Pichia pastoris* has become the most frequently used yeast species for recombinant protein production (Schmidt 2004; Gurkan and Ellar 2005; Macauley-Patrick et al. 2005). As a methylotrophic yeast *Pichia* is able to utilize methanol as sole carbon source. Since the alcohol oxidase enzyme of the methanol-assimilating pathway can reach up to 30% of the total protein of the cell, the endogenous alcohol oxidase promoter (AOX) provides a tightly regulated and powerful expression regulator. The very high homologous recombination rate of yeasts allows simple and efficient integration of the gene of interest into the AOX locus of the host cell genome, leading to production clones with high expression levels and long-term stability. By increasing the gene copy number an increasing expression of the recombinant protein can be achieved (Macauley-Patrick et al. 2005). However, in cases where folding and disulfide bond formation are the rate-limiting steps, a higher gene dose can also have no or even a negative effect (Hohenblum et al. 2004).

Pichia pastoris secretes only very low levels of endogenous proteins and can be cultured in well-defined minimal media with very low or no protein content (Hellwig et al. 2001), which predestines it for secretory expression and simplifies purification processes. Very high secretion levels for recombinant proteins have been reported using the S. cerevisiae alpha factor prepro peptide as secretion signal sequence for P. pastoris, reaching up to $15\,g\,L^{-1}$ for murine collagen (Werten et al. 1999). However, heterologous proteins secreted via the alpha factor prepro leader have been shown to commonly contain variable N-termini with different numbers of N-terminal amino acids added to the protein of interest (Brocca et al. 1998). This is assumed to be due to influences of the N-terminal protein sequence on the efficiency of the two-step processing of the leader peptide. Initially, a part of the leader peptide is cleaved by the Kex2 protease before the Ste13 protease removes the remaining amino acids. If the cleavage efficiency of Ste13 is affected by the protein sequence, a mixture of partially processed and completely processed material can be found in the culture media.

There is little information on P. pastoris proteases, but proteolytic degradation of the recombinant product in the culture medium can be a problem. Three different types of proteases have been identified in P. pastoris cultures during the optimization of the expression of an scFv antibody: aspartic, cysteine, and serine type proteases (Shi et al. 2003). Strategies for the reduction of proteolysis include adaptation of culture conditions (e.g. pH and temperature), addition of protease inhibitors, use of protease-deficient strains, and adaptation of the sequence of the recombinant protein (Macauley-Patrick et al. 2005).

The glycosylation pattern of P. pastoris differs from that of S. cerevisiae, although both mainly produce N-linked glycosylation of the high-mannose type. In P. pastoris, there is less hyperglycosylation found and the average length of the added oligosaccharide chains is much shorter (Tschopp et al. 1987; Grinna and Tschopp 1989). The terminal $\alpha1,3$ glycans found on the core oligosaccharides of S. cerevisiae are believed to be primarily responsible for the high antigenic nature of glycoproteins produced in S. cerevisiae. Since those $\alpha1,3$ glycans are missing in P. pastoris-derived glycoproteins they are thought to be more suitable for therapeutic applications (Cregg et al. 1993). Nevertheless, a human-like glycosylation pattern with complex N-glycans is highly desirable for therapeutic applications and developments towards the engineering of Pichia strains capable of generating human-like glycosylation are ongoing (Choi et al. 2003; Hamilton et al. 2003; Gerngross 2004). After inactivation of an initial yeast-type glycosylation step by deleting the OCH1 ($\alpha1,6$ mannosyltransferase) gene, a combinatorial approach of fusing different mannosidase catalytic domains with different yeast localization domains was used to select a strain that efficiently trims core glycans in vivo, a prerequisite for the conversion of high mannose-tye glycans to hybrid or complex N-glycans. To further humanize the glycosylation pattern a second gene library of GnTI ($\beta1,2$-N-acetylglucosaminyltransferase I) catalytic domains combined with localization signals was screened and led to the identification of a strain producing human-like hybrid N-glycosylation structures (Choi et al. 2003). However, further removal of mannose and addition of $\beta1,2$-GlcNac would be required to assemble complex N-glycans. Based on this work further adaptation of the glycosylation pattern was performed using a similar approach leading to the generation of a *Pichia* strain producing glycoproteins with uniform complex N-glycosylation (Hamilton et al. 2003). By analyzing Fc-receptor binding of antibodies produced in a set of *Pichia* strains with different glycosylation patterns, antibody glycoforms with improved ADCC efficiency could be identified (Li et al. 2006). *Pichia*-derived glycoforms with 10-fold improved binding affinity to the V158 allele of FcγRIIIa and 100-fold improved binding to the F158 allele of FCγRIIIa compared with commercial CHO-derived IgG (rituximab) were found. They were also shown to be more efficient than the mammalian material in a B-cell depletion assay.

The antibodies that so far have been produced in *Pichia* are mainly in the scFv format (Ridder et al. 1995; Luo et al. 1996; Eldin et al. 1997; Hellwig et al. 2001; Shi et al. 2003; Damasceno et al. 2004; Gurkan et al. 2004; Trentmann et al. 2004; Emberson et al. 2005), but production of Fab fragments (Lange et al. 2001; Ning et al. 2003), an scFv-Fc fusion (Powers et al. 2001), ds-diabodies (FitzGerald et al. 1997), and a bispecific tetravalent antibody fusion protein (Biburger et al. 2005) have also been reported. Expression of entire functional IgG antibodies in *Pichia* has been demonstrated but only a portion of the secreted antibody chains were assembled to intact tetrameric antibody (Ogunjimi et al. 1999). Recently, production of different glycoforms of IgG in *Pichia* has been reported, although there was no information on production yields (Li et al. 2006).

Using an online methanol control in fed-batch fermentation expression of an scFv antibody, yields of about 4 g L^{-1} could be reached, which to our knowledge is the highest yield of a yeast-expressed antibody reported so far (Damasceno et al. 2004). This clearly demonstrates the potential of *Pichia* for high-level production of antibodies in fermentation processes. However, with proteins getting more complex, the expression levels of functional protein are decreasing. For a Fab fragment the portion of correctly assembled heteromeric protein in the culture medium was found to be 30%, which is primarily attributed to the failure of one of the chains to find a partner (Lange et al. 2001). Yeast-expressed scFv-Fc fusion proteins might be of special therapeutic value since they share some important features with intact IgG antibodies. Disulfide-linked dimeric scFv-Fc fusions could be produced in *Pichia* with yields of 2 mg L^{-1} and were shown to efficiently mediate ADCC and to have a 12- to 30-fold prolonged *in vivo* plasma half-life compared with scFv alone (Powers et al. 2001).

6.2.2.2 Filamenteous Fungi

Filamenteous fungi such as *Aspergillus* or *Trichoderma* are known to be capable of secreting very high levels of homologous proteins ($30–40 \text{ g L}^{-1}$) and are well established in the industrial production of enzymes, antibiotics, etc. Further, fungi have a posttranslational modification apparatus that is more similar to mammalian cells than that of yeast, and several species possess the GRAS status (Schmidt 2004). For recombinant heterologous products, the secretion levels have so far failed to reach the levels achieved by endogenous proteins. While heterologous fungal enzymes are produced with yields of up to $4\,g\,L^{-1}$, yields of mammalian proteins usually stay below the $g\,L^{-1}$ range (Schmidt 2004). Antibody fragments have been produced in *Trichoderma reesei* (Nyyssonen et al. 1993) and in *Aspergillus niger* (Frenken et al. 1998). Recently, Ward et al. reported the expression of functional humanized IgG and Fab antibodies in *Aspergillus niger* (Ward et al. 2004). Both antibody chains were expressed with N-terminal fusion of the *Aspergillus* enzyme glucoamylase to improve the secretion efficiency. A KexB cleavage site was inserted between antibody chain and glucoamylase to mediate cleavage of the fusion protein in the Golgi apparatus by the native *Aspergillus* protease KexB. The cleavage efficiency did not reach 100% and the cleavage position varied, leading to heterogeneity of the N-termini. By adding a stuffer of three glycines between the antibody N-terminus and cleavage site the cleavage efficiency could be increased, leading to homogeneous N-termini. However, the three glycine residues thereby remain added to the N-termini of the processed antibody. The additional N-terminal amino acids had no influence on the binding activity of the antibodies and the *Aspergillus*-derived material was found to have similar affinities to antibodies derived from mammalian cell cultures. Approximately half of the secreted heavy chains were N-glycosylated with glycans of the high mannose type. Analysis of the *in vivo* serum half-life suggests that there is no significant difference to antibodies from mammalian production and the antibodies from *Aspergillus* production were shown to efficiently mediate ADCC. Secretion levels of up to $0.9\,g\,L^{-1}$ were achieved in this study which clearly demonstrates that *Aspergillus niger* is a promising expression host for industrial full-length antibody production. For therapeutic applications improvement of the processing leading to native N-termini and a human-like glycosylation pattern to reduce imunogenecity are desirable.

6.2.2.3 Insect Cells

The baculovirus expression vector system for recombinant protein production in insect cells is a well-established system, especially in research and development. Gene transfer to the expression host in this system is based on the infection of insect cell lines that are usually derived from *Spodoptera frugiperda* (SF9 and SF-21) or *Trichoplusia ni* (High Five) with recombinant baculovirus. Advantages of the system are high transduction efficiencies, posttranslational modifications of higher eukaryotes, and high expression levels driven by strong baculoviral promoters (Hu 2005). There are numerous examples of recombinant antibody expression using baculovirus-infected insect cells including scFv (Kretzschmar et al. 1996; Brocks et al. 1997; Lemeulle et al. 1998; Ailor et al. 1999; Yoshida et al. 1999; Reavy et al. 2000; Demangel et al. 2005), Fab (Bes et al. 2001), scFv-Fc (Brocks et al. 1997), scFv-based immunotoxins (Choo et al. 2002), fluorescent scFv fusions (Peipp et al. 2004), full-length IgG (Hasemann and Capra 1990; Liang et al. 1997; Tan and Lam 1999; Liang et al. 2001), and IgA antibodies (Carayannopoulos et al. 1994). Yields of up to $200\,mg\,L^{-1}$ for an scFv (Reavy et al. 2000) and $70\,mg\,L^{-1}$ for entire IgG (Tan and Lam 1999) have been achieved.

One of the major disadvantages associated with the baculovirus expression system are cell death and lysis of the host insect cells within days after the infection. For secretory proteins, like antibodies, this can raise the problem of the release of unprocessed immature antibody polypeptides to the culture medium, especially since the highest expression levels driven by the strong polyhedrin promoter are noticed briefly prior to cell lysis. In that phase processing of proteins is already seriously affected by the damage of the secretory pathway (Jarvis and Summers 1989; Hu 2005). Another problem connected with a lytic expression system is protein degradation due to the release of proteases. To overcome these problems nonlytic transient and stable expression systems for insect cells have been developed that are based on cell transfection instead of baculovirus infection (Jarvis et al. 1990). Stably transformed insect cells have been used to produce full-length IgG antibodies, but the yields were considerably lower than those from transient expression after baculovirus infection (Guttieri et al. 2000, 2003).

The N-glycosylation pattern of insect cells differs from that of mammalian cells (Tomiya et al. 2004; Hu 2005; Kost et al. 2005). Insect cells can assemble N-glycans of the high mannose and paucimannose type but typically fail to produce N-glycans of the complex mammalian type with terminal galactose or sialic acid residues. This limits the suitability of the expression system for therapeutic approaches, which led to intensive research on the humanization of the glycosylation pattern of insect cells (Jarvis 2003). Transgenic insect cell lines were generated that produce N-glycan-processing enzymes of mammalian origin. Insect cells expressing bovine β1,4-galactosyltransferase have been shown to produce proteins with terminally galactosylated N-glycans (Hollister et al. 1998). The addition of a rat α2,6-sialyltransferase transgene to this cell line led to the generation of monoantennary terminally sialylated N-glycans (Hollister and Jarvis 2001). Further efforts led to the generation of a transgenic insect cell line that expresses five mammalian glycosyltransferases (Hollister et al. 2002) and finally a cell line that in addition contains transgenes for two murine enzymes to allow *de novo* glycoprotein syalization was created that produced a recombinant protein with highly homogeneous biantennary sialylated N-glycans (Aumiller et al. 2003).

6.2.2.4 Mammalian Cells

Despite rather high costs, currently 60–70% of all recombinant biopharmaceuticals and virtually all therapeutic monoclonal antibodies are produced in mammalian cell culture processes (Wurm 2004). During the last two decades a 100-fold improvement of production yields has been achieved for industrial large-scale processes. In 1986, when t-PA (tissue plaminogen activator) was the first recombinant pharmaceutical protein produced in CHO cells, the yield was about 50 mg L^{-1}. Nowadays, yields of up to 4.7 g L^{-1}, as reported for a recombinant antibody, are obtainable (Wurm 2004). This increase is mainly due to higher volumetric production yields based on much higher cell densities in the production process. Since the specific production rates of the cells during this period improved only moderately, there might be still potential to further improve production yields by using alternative cell lines, genetic engineering of production cell lines, and optimization of expression vectors (Fussenegger et al. 1999; Wurm 2004; Dinnis and James 2005).

Currently, the most widely used cell line for antibody production is the CHO cell line. To a lesser extent the murine myeloma (NS0, SP2/0), BHK, and HEK293 (human embryonic kidney) cell lines are also used. Besides these more traditional host cell lines the human retinoblastoma cell line PER.C6 recently has been shown to efficiently produce recombinant IgG at high cell densities (Jones et al. 2003). Among the advantages of this cell line are the human origin, leading to a nonimmunogenic glycosylation pattern, and the ability to adapt quickly to serum-free suspension culture conditions.

Gene transfer in mammalian systems for stable expression is usually based on integration of the expression cassette into the host cell genome. Gene integration in principal can occur in three different ways: random integration, homologous recombination, and recombinase-mediated targeted integration. It is widely accepted that the site of integration in the host genome has a dominant impact on expression level and stability of the recombinant cell line (position effect). Furthermore, even after integration into expression-competent euchromatin, expression of the transgene can be silenced rapidly (Mutskov and Felsenfeld 2004). Gene silencing might be influenced by surrounding endogenous condensed chromatin and is correlated with histone hypoacetylation, loss of methylation at H3 lysine 4, increase of histone H3 lysine 9 methylation and CpG DNA methylation at the promoter. The histone modifications thereby seem to be the primary events in gene silencing (Mutskov and Felsenfeld 2004). Using random integration-based methods, the generation of stable transfectants with high expression levels and high stability of transgene expression therefore often requires screening of a high number of clones. Alternatively, homologous recombination can be used, which allows targeted integration of the transgene at genomic sites that are known to allow high and stable expression. However, in the mammalian cell lines that are usually used for recombinant protein expression, homologous recombination is a much less frequent event than in yeast cells, which makes this strategy quite inefficient for biotechnological processes. To overcome this limitation site-specific targeted integration methods with higher recombination frequencies are developed (Bode et al. 2000). Site-directed targeted integration is mediated by recombinases that catalyze DNA strand exchange reactions between short target sequences. The first described site-specific recombination systems are based on the recombinases Cre from the phage P1 and Flp from the yeast *Saccaromyces cerevisiae* with their recognition sites LoxP and FRT, respectively (O'Gorman et al. 1991; Fukushige and Sauer 1992). Both systems can be used to perform a targeted integration driven by recognition sites localized on a circular targeting vector and the genomic DNA. However, the efficiency of the integration reaction is limited by competition with the thermodynamically favored re-excision reaction (Bode et al. 2000). Pairs of mutant recognition sites were developed that allow advanced gene exchange procedures (RMCE, recombinase mediated cassette exchange) for Cre (Araki et al. 1997) and Flp (Seibler et al. 1998), like the recombinase-mediated exchange of a selection marker/ reporter gene for the gene of interest in the genome of an acceptor cell line. Acceptor cell lines can be generated using reporter or selection genes or a combination of both to find and label genomic integration sites with high expression activity and long-term stability. Another advantage of the engineered recognition sites is the possibility to only insert the gene of interest without cointegration of prokaryotic vector components that might promote gene silencing. However, so far high-efficiency application of this principle is mainly restricted to embryonic stem (ES) cells.

A different approach to overcome gene silencing and position effects is the use of scaffold- or matrix-attachment regions (S/MAR elements) of chromosomal DNA flanking the expression cassette, which can augment transgene expression and reduce negative influence by surrounding chromatin (Klehr et al. 1991; Fussenegger et al. 1999). Genetic insulator elements, which naturally mark boundaries between open chromatin and constitutively condensed chromatin, can also be used to protect transgenes against silencing and position effects (Burgess-Beusse et al. 2002). However, the enhancer-blocking function of the insulator elements at the same time might reduce the transcriptional activity of the promoter.

Since the generation of stable recombinant mammalian cell lines is a time-consuming procedure, transient expression systems are gaining importance, especially in antibody production for research and development. Mainly host cells that are capable of episomal replication of suitable expression vectors are used, like COS monkey cell lines or the derivatives of the human HEK293 cell line HEK293-EBNA and HEK293-T (Fussenegger et al. 1999; Meissner et al. 2001; Jostock et al. 2004). HEK293 cells are very popular, since they allow high transfection efficiencies and constitutively express the adenoviral E1A transactivator which enhances transcription (Fussenegger et al. 1999). Vectors containing an SV40 origin of replication (ori) are highly amplified in cells expressing the T antigen of simian virus 40 (SV40), like COS and HEK293-T, reaching up to 200000 copies per cell 48h after transfection. This leads to very high expression levels but also finally culminates in cell death, which limits the system to transient production processes (Van Craenenbroeck et al. 2000). Recombinant IgG expression levels of 15–20 mg L^{-1} have been reported for transient transfections of HEK293T cells (Jostock et al. 2004).

Episomal replication driven by the EBNA1 transactivator and oriP from Epstein–Barr virus (EBV) leads to much lower copy numbers (5–100) but very high retention time of the plasmids (Van Craenenbroeck et al. 2000). Retention rates of 92–98% per cell generation are typical of EBV vectors in the absence of selection. Under selective pressure transgene expression is stable over a long period of time with the EBV expression system. Transient expression in HEK293-EBNA cells yielded up to 40 mg L^{-1} of recombinant IgG (Meissner et al. 2001).

Besides these viral replication systems, vectors for epsiomal replication driven by chromosomal S/MAR elements are also in development (Piechaczek et al. 1999; Jenke et al. 2004).

Upscaling of transient transfection of suspension-adapted HEK293-EBNA cells in bioreactors has been reported up to the 100 L scale with a production yield of 0.5 g IgG antibody (Girard et al. 2002). Thus, transient large-scale transfection might become a very time-efficient alternative for therapeutic antibody production in an early phase of development.

Several genetic engineering strategies, based on the overexpression of individual chaperones and foldases, have been applied to improve specific antibody production levels by mammalian host cells. However, although it is generally presumed that antibody production is limited by the posttranslational folding and assembly reactions, so far no reliable increase of production levels has been achieved with these engineered cell lines (Dinnis and James 2005). It is proposed that a global expansion of the complete secretory machinery, as is found during the differentiation of B cells to high-level antibody-producing plasma cells, would be necessary to generally improve the production capabilities of host cell lines (Dinnis and James 2005).

6.2.2.5 Plants

The almost unlimited upscaling potential of transgenic plants as production hosts for biopharmaceuticals is clearly a conceptional advantage compared with bioreactor-based production in microbial and mammalian cells. In 1990, human serum albumin was the first recombinant protein with pharmaceutical potential to be produced in transgenic plants (Sijmons et al. 1990; Ma et al. 2005). Since then, substantial improvements of transgenic plant technology have been achieved and currently several plant-derived recombinant proteins are in clinical studies, including antibodies (Ma et al. 2005).

The initial step of protein production in plants is the transfer of the gene of interest into the host (transformation). Plant transformation using physical methods is rather inefficient and usually done by biolistic gene delivery (gene gun) in the case of monocots, such as wheat, rice, and corn. In the case of dicots, such as tobacco and pea, *Agrobacterium*-mediated gene transfer is the method of choice (Schillberg et al. 2005). Both stable and transient plant expression systems have been developed. The generation of stable transgenic plants is quite time consuming and depends on the genomic integration of the transgene in the nuclear or plastid DNA. Integration in the genome of plastids, like chloroplasts in tobacco, has the advantage of high copy numbers of the transgene resulting in high expression levels of the recombinant protein but also an increased risk of proteolysis of the product (Ma et al. 2005).

Transient expression is less time consuming but limited in scale. Commonly used transient expression systems are agroinfiltration using recombinant *Agrobacterium tumefaciens* and viral vectors based on plant viruses like the tobacco mosaic virus (Schillberg et al. 2005). ScFv as well as full-size antibodies have been transiently produced in plants (Verch et al. 1998; Galeffi et al. 2005).

Tobacco is among the most commonly used species for molecular farming. Several other species, such as maize, potato, rice, wheat, pea, tomato, and banana, are also amenable to recombinant protein production. Comparison of different production hosts with regards to production levels per kilogram of biomass of an scFv antibody revealed that the yields varied among different species and expression systems but were broadly comparable. In general, for all species tested, yields could be improved by accumulating the scFv in the endoplasmic reticulum via a KDEL ER-retention signal (Schillberg et al. 2005).

As higher eukaryotes, plants are able to synthesize and process even complex multi-subunit proteins such as full-size antibodies. Plant chaperones are homologous to those of mammalian cells, which ensures a high folding capacity, and the secretion apparatus of plants performs posttranslational modifications including N-linked glycosylation (Ma et al. 2005). However, plant and mammalian complex N-glycans differ in their structures, which might result in impaired effector functions of plant antibodies. Furthermore, antibodies produced in tobacco show a highly heterogeneous N-glycosylation pattern and plant N-glycans are potentially allergenic and immunogenic (Gomord et al. 2005). Several strategies are currently employed to eliminate unwanted sugar residues and to humanize the glycosylation pattern of host plants via genetically engineering (Gomord and Faye 2004).

For an IgA antibody a yield of $500\,mg\,kg^{-1}$ biomass from tobacco has been reported (Ma et al. 1998). Besides production rates for molecular farming several other key features, such as biomass yield, length of a production cycle and cost of processing of the host species have to be considered to evaluate the market potential (Schillberg et al. 2005). The cost for the production of recombinant proteins in transgenic plants is believed to be much lower than in mammalian cell culture (Hood et al. 2002). However, in the case of therapeutic antibodies, which are often administered intravenously, processing of the plant biomass to extract the antibody product in suitable quality and purity is likely to be cost intensive.

6.2.2.6 Transgenic Animals

Along with transgenic plants, transgenic animals might also become an alternative to bioreactor systems for the production of biopharmaceuticals. Mammary gland-specific expression of the transgene to target the recombinant protein to the milk of farm animals allows continuous recovery of the product during the lifespan of adult female animals. To achieve this, chimeric transgenes consisting of the regulatory elements of milk-specific genes and the coding regions of the gene of interest are used. Examples for such milk specific genes are the ovine β-lactoglobulin, rodent whey acid protein (WAP) and bovine α-s1-casein genes (Pollock et al. 1999). The chimeric transgenes are usually delivered to the host animals by pronuclei microinjection. Although well established, the efficiency of transgene integration into the host genome using this method is rather low and the time from initiating microinjection to full lactation is long (16–18 months for goats; Pollock et al. 1999).

The most commonly used animal species for transgenic protein expression are mouse, rabbit, pig, sheep, goat, and cattle. To date, recombinant antibodies have been produced in transgenic milk of mice and goats. The yields that are reported for IgG are between 0.4 and $5\,g\,L^{-1}$ in mice and up to $14\,g\,L^{-1}$ in goats (Limonta et al. 1995; Castilla et al. 1998; Pollock et al. 1999). Small animals such as mice and rabbits have the advantage of good reproductive characteristics but the disadvantage of low milk yields, which makes them most suitable for the testing of expression systems. Goats have an average milk output of 600–800 L per 300-day lactation, thus, herds of transgenic goats should be able to produce kilogram amounts of product per year (Pollock et al. 1999).

It is expected that antibodies derived from transgenic mammals have a glycan pattern similar to material from mammalian cell culture. Since the antigen-binding properties have been shown to be similar or equivalent (Pollock et al. 1999), they may have a comparable clinical efficacy. Among the main advantages of transgenic animals for antibody production are the flexible scalability and cost-efficient maintenance of the production facilities. Disadvantages are time- and labor-intensive generation of founder animals as well as safety issues regarding the animal-derived material if intravenous application of the product is planned. Milk from transgenic animals also contains endogenous antibodies which might be difficult to separate from the product and the animals are potential hosts for pathogenic microorganisms, viruses, and prions.

6.3
Outlook

Mammalian cell lines are currently the dominant system for production processes of full-sized antibodies, and *E. coli* for antibody fragments. Both systems still may be further optimized by genetic engineering of the host cell lines for improved folding, secretion, and growth characteristics. Besides those, *P. pastoris* and insect cells are well established, especially in research. Due to high growth rates, high production levels, ease of handling and rapid generation of stable transformants, *P. pastoris* has good potential to reduce cost and time-lines of production processes. The ongoing humanization of the glycosylation pattern of *P. pastoris* might allow the production of antibody products with clinical efficacies and pharmacokinetics comparable to material from mammalian cell culture processes. However, high expression levels are mainly reported for antibody fragments rather than for entire IgG molecules. For insect cell lines developments towards the humanization of the glycolsylation pattern are also ongoing but there is still a demand for the development of high cell density large-scale production processes and efficient technologies to generate stable production cell lines. Because of their intrinsic high-performance secretion machineries, Gram-positive bacteria and filamentous fungi are well suited in general to production processes and several species are established production hosts in the food industry, but the application of these hosts for recombinant heterologous protein production is still in the development stage. Gram-positive bacteria are unable to perform eukaryote-like posttranslational modifications and thus might be suitable for the production of aglycosylated antibodies and antibody fragments only. The glycosylation pattern of filamentous fungi would have to be humanized to obtain high-quality therapeutic products. Due to the virtually unlimited scalability and comparably low maintenance of the production facilities, transgenic plants and animals probably have the highest potential to reduce the costs of antibody production for applications with a high product demand. Long time-lines for the generation of the producer strains, complex and expensive downstream processes and finally unclear safety issues for the regulatory approval of the products and production facilities are the main hurdles that need to be cleared to make this approach state of the art.

In summary, substantial effort is currently underway to develop new alternative production systems for the growing market of recombinant antibody therapeutics. Some of the systems are close to market maturity while others are pretty much in an early phase of development. An overview of examples of the production systems presented in this chapter is given in Table 6.1. With the arrival of

Table 6.1 Examples of production levels for different expression systems.

Host	Format	Yield	Citation
Echerichia coli	F(ab')2	$2 g L^{-1}$	Chen et al. 2004
	IgG	$150 mg L^{-1}$	Simmons et al. 2002
Proteus mirabilis	scFv	$200 mg L^{-1}$	Rippmann et al. 1998
	Mini-antibody	$18 mg L^{-1}$	Kujau et al. 1998
Bacillus brevis	scFv	$10 mg L^{-1}$	Shirazoa et al. 2001
	Fab	$100 mg L^{-1}$	Inoue et al. 1997
Bacillus subtilis	scFV	$15 mg L^{-1}$	Wu et al. 2002
Pichia pastoris	scFv	$4 g L^{-1}$	Damasceno et al. 2004
	Fab	$40 mg/L$	Lange et al. 2001
	ds-diabody	$1 mg L^{-1}$	FitzGerald et al. 1997
	scFv-Fc	$2 mg L^{-1}$	Powers et al. 2001
	IgG	n.s.	Ogunjimi et al. 1999
			Li et al. 2006
Saccharomyces cerevisiae	scFv	$20 mg L^{-1}$	Shusta et al. 1998
	IgM	n.s.	Wood et al. 1985
Yarrowia lipolytica	scFv	$20 mg L^{-1}$	Swennen et al. 2002
Kluyveromyces lactis	scFv	$10 mg L^{-1}$	Swennen et al. 2002
Aspergillus niger	IgG	$0.9 g L^{-1}$	Ward et al. 2004
Insect cells (SF9)	IgG	$18 mg L^{-1}$	Liang et al. 2001
Insect cells (SF21)	IgG	$40 mg L^{-1}$	Tan et al. 1999
Insect cells (High Five)	IgG	$70 mg L^{-1}$	Tan et al. 1999
Mammalian cells (CHO)	IgG	$4.7 g L^{-1}$	Wurm F.M. 2004
Mammalian cells (PER.C6)	IgG	$0.5–1 g L^{-1}$	Jones et al. 2003
Tobacco	IgA	$80 mg kg^{-1}$	Ma et al. 1998
Transgenic mice	IgG	$5 g L^{-1}$	Castilla et al. 1998
Transgenic goat	IgG	$14 g L^{-1}$	Pollock et al. 1999

Volumetric productivity values for different expression hosts and antibody formats are shown. It should be noted that the comparability of the values to each other is very limited since no equalization of scale, conditions, and runtime of the expression or quality of the product was made.

biosimilars for therapeutics with expired patents, the pressure to reduce production costs will rise even further, since direct competition between different producers for the same biotherapeutic product will lead to significant market price reductions.

References

Ailor, E., Pathmanathan, J., Jongbloed, J.D., Betenbaugh, M.J. (1999) A bacterial signal peptidase enhances processing of a recombinant single chain antibody fragment in insect cells. *Biochem Biophys Res Commun* 255: 444–450.

Araki, K., Araki, M., Yamamura, K. (1997) Targeted integration of DNA using mutant lox sites in embryonic stem cells. *Nucleic Acids Res* 25: 868–872.

Aumiller, J.J., Hollister, J.R., Jarvis, D.L. (2003) A transgenic insect cell line engineered to produce CMP-sialic acid and sialylated glycoproteins. *Glycobiology* 13: 497–507.

Berks, B.C., Sargent, F., Palmer, T. (2000) The Tat protein export pathway. *Mol Microbiol* 35: 260–274.

Bes, C., Cerutti, M., Briant-Longuet, L., Bresson, D., Peraldi-Roux, S., Pugniere, M., Mani, J.C., Pau, B., Devaux, C., Granier, C., Devauchelle, G., Chardes, T. (2001) The chimeric mouse-human anti-CD4 Fab 13B8.2 expressed in baculovirus inhibits both antigen presentation and HIV-1 promoter activation. *Hum Antibodies* 10: 67–76.

Biburger, M., Weth, R., Wels, W.S. (2005) A novel bispecific tetravalent antibody fusion protein to target costimulatory activity for T-cell activation to tumor cells overexpressing ErbB2/HER2. *J Mol Biol* 346: 1299–311.

Bode, J., Schlake, T., Iber, M., Schubeler, D., Seibler, J., Snezhkov, E., Nikolaev, L. (2000) The transgeneticist's toolbox: novel methods for the targeted modification of eukaryotic genomes. *Biol Chem* 381: 801–813.

Bolhuis, A., Tjalsma, H., Smith, H.E., de Jong, A., Meima, R., Venema, G., Bron, S., van Dijl, J.M. (1999) Evaluation of bottlenecks in the late stages of protein secretion in *Bacillus subtilis*. *Appl Environ Microbiol* 65: 2934–2941.

Bothmann, H., Plückthun, A. (2000) The periplasmic *Escherichia coli* peptidylprolyl cis,trans-isomerase FkpA. I. Increased functional expression of antibody fragments with and without cis-prolines. *J Biol Chem* 275: 17100–17105.

Boyd, P.N., Lines, A.C., Patel, A.K. (1995) The effect of the removal of sialic acid, galactose and total carbohydrate on the functional activity of Campath-1H. *Mol Immunol* 32: 1311–1318.

Brocca, S., Schmidt-Dannert, C., Lotti, M., Alberghina, L., Schmid, R.D. (1998) Design, total synthesis, and functional overexpression of the *Candida rugosa* lip1 gene coding for a major industrial lipase. *Protein Sci* 7: 1415–1422.

Brocks, B., Rode, H.J., Klein, M., Gerlach, E., Dübel, S., Little, M., Pfizenmaier, K., Moosmayer, D. (1997) A TNF receptor antagonistic scFv, which is not secreted in mammalian cells, is expressed as a soluble mono- and bivalent scFv derivative in insect cells. *Immunotechnology* 3: 173–184.

Burgess-Beusse, B., Farrell, C., Gaszner, M., Litt, M., Mutskov, V., Recillas-Targa, F., Simpson, M., West, A., Felsenfeld, G. (2002) The insulation of genes from external enhancers and silencing chromatin. *Proc Natl Acad Sci USA* 99(Suppl 4): 16433–16437.

Carayannopoulos, L., Max, E.E., Capra, J.D. (1994) Recombinant human IgA expressed in insect cells. *Proc Natl Acad Sci USA* 91: 8348–8352.

Castilla, J., Pintado, B., Sola, I., Sanchez-Morgado, J.M., Enjuanes, L. (1998) Engineering passive immunity in transgenic mice secreting virus-neutralizing antibodies in milk. *Nat Biotechnol* 16: 349–354.

Chen, C., Snedecor, B., Nishihara, J.C., Joly, J.C., McFarland, N., Andersen, D.C., Battersby, J.E., Champion, K.M. (2004) High-level accumulation of a recombinant antibody fragment in the periplasm of *Escherichia coli* requires a triple-mutant (degP prc spr) host strain. *Biotechnol Bioeng* 85: 463–474.

Choi, B.K., Bobrowicz, P., Davidson, R.C., Hamilton, S.R., Kung, D.H., Li, H., Miele, R.G., Nett, J.H., Wildt, S., Gerngross, T.U. (2003) Use of combinatorial genetic libraries to humanize N-linked glycosylation in the yeast Pichia pastoris. *Proc Natl Acad Sci USA* 100: 5022–5027.

Choo, A.B., Dunn, R.D., Broady, K.W., Raison, R.L. (2002) Soluble expression of a functional recombinant cytolytic immunotoxin in insect cells. *Protein Expr Purif* 24: 338–347.

Cregg, J.M., Vedvick, T.S., Raschke, W.C. (1993) Recent advances in the expression of foreign genes in *Pichia pastoris*. *Biotechnology (NY)* 11: 905–910.

Damasceno, L.M., Pla, I., Chang, H.J., Cohen, L., Ritter, G., Old, L.J., Batt, C.A. (2004) An optimized fermentation process for high-level production of a single-chain Fv antibody fragment in *Pichia pastoris*. *Protein Expr Purif* 37: 18–26.

DeLisa, M.P., Tullman, D., Georgiou, G. (2003) Folding quality control in the export of proteins by the bacterial twin-arginine translocation pathway. *Proc Natl Acad Sci USA* 100: 6115–6120.

Demangel, C., Zhou, J., Choo, A.B., Shoebridge, G., Halliday, G.M., Britton, W.J. (2005) Single chain antibody fragments for the selective targeting of antigens to dendritic cells. *Mol Immunol* 42: 979–985.

Dinnis, D.M., James, D.C. (2005) Engineering mammalian cell factories for improved recombinant monoclonal antibody production: lessons from nature? *Biotechnol Bioeng* 91: 180–189.

Eldin, P., Pauza, M.E., Hieda, Y., Lin, G., Murtaugh, M.P., Pentel, P.R., Pennell, C.A. (1997) High-level secretion of two antibody single chain Fv fragments by *Pichia pastoris*. *J Immunol Methods* 201: 67–75.

Emberson, L.M., Trivett, A.J., Blower, P.J., Nicholls, P.J. (2005) Expression of an anti-CD33 single-chain antibody by *Pichia pastoris*. *J Immunol Methods* 305: 135–151.

FitzGerald, K., Holliger, P., Winter, G. (1997) Improved tumour targeting by disulphide stabilized diabodies expressed in *Pichia pastoris*. *Protein Eng* 10: 1221–1225.

Frenken, L.G., Hessing, J.G., Van den Hondel, C.A., Verrips, C.T. (1998) Recent advances in the large-scale production of antibody fragments using lower eukaryotic microorganisms. *Res Immunol* 149: 589–599.

Fukushige, S., Sauer, B. (1992) Genomic targeting with a positive-selection lox integration vector allows highly reproducible gene expression in mammalian cells. *Proc Natl Acad Sci USA* 89: 7905–7909.

Fussenegger, M., Bailey, J.E., Hauser, H., Mueller, P.P. (1999) Genetic optimization of recombinant glycoprotein production by mammalian cells. *Trends Biotechnol* 17: 35–42.

Galeffi, P., Lombardi, A., Donato, M.D., Latini, A., Sperandei, M., Cantale, C., Giacomini, P. (2005) Expression of single-chain antibodies in transgenic plants. *Vaccine* 23: 1823–1827.

Gerngross, T.U. (2004) Advances in the production of human therapeutic proteins in yeasts and filamentous fungi. *Nat Biotechnol* 22: 1409–1414.

Girard, P., Derouazi, M., Baumgartner, G., Bourgeois, M., Jordan, M., Jacko, B., Wurm, F.M. (2002) 100-liter transient transfection. *Cytotechnology* 38: 15–21.

Gomord, V., Faye, L. (2004) Posttranslational modification of therapeutic proteins in plants. *Curr Opin Plant Biol* 7: 171–181.

Gomord, V., Chamberlain, P., Jefferis, R., Faye, L. (2005) Biopharmaceutical production in plants: problems, solutions and opportunities. *Trends Biotechnol* 23: 559–565.

Grinna, L.S., Tschopp, J.F. (1989) Size distribution and general structural features of N-linked oligosaccharides from the methylotrophic yeast, *Pichia pastoris*. *Yeast* 5: 107–115.

Gurkan, C., Ellar, D.J. (2005) Recombinant production of bacterial toxins and their derivatives in the methylotrophic yeast *Pichia pastoris*. *Microb Cell Fact* 4: 33.

Gurkan, C., Symeonides, S.N., Ellar, D.J. (2004) High-level production in Pichia pastoris of an anti-p185HER-2 single-chain antibody fragment using an alternative secretion expression vector. *Biotechnol Appl Biochem* 39: 115–122.

Guttieri, M.C., Bookwalter, C., Schmaljohn, C. (2000) Expression of a human, neutralizing monoclonal antibody specific to puumala virus G2-protein in stably-transformed insect cells. *J Immunol Methods* 246: 97–108.

Guttieri, M.C., Sinha, T., Bookwalter, C., Liang, M., Schmaljohn, C.S. (2003) Cassette vectors for conversion of Fab fragments into full-length human IgG1 monoclonal antibodies by expression in stably transformed insect cells. *Hybrid Hybridomics* 22: 135–145.

Hamilton, S.R., Bobrowicz, P., Bobrowicz, B., Davidson, R.C., Li, H., Mitchell, T., Nett, J.H., Rausch, S., Stadheim, T.A., Wischnewski, H., Wildt, S., Gerngross, T.U. (2003) Production of complex human glycoproteins in yeast. *Science* 301: 1244–1246.

Hasemann, C.A., Capra, J.D. (1990) High-level production of a functional immunoglobulin heterodimer in a baculovirus expression system. *Proc Natl Acad Sci USA* 87: 3942–3946.

Hellwig, S., Emde, F., Raven, N.P., Henke, M., van Der Logt, P., Fischer, R. (2001) Analysis of single-chain antibody production in *Pichia pastoris* using on-line methanol control in fed-batch and mixed-feed fermentations. *Biotechnol Bioeng* 74: 344–352.

Hohenblum, H., Gasser, B., Maurer, M., Borth, N., Mattanovich, D. (2004) Effects of gene dosage, promoters, and substrates on unfolded protein stress of recombinant *Pichia pastoris*. *Biotechnol Bioeng* 85: 367–375.

Hollister, J.R., Jarvis, D.L. (2001) Engineering lepidopteran insect cells for sialoglycoprotein production by genetic transformation with mammalian beta 1,4-galactosyltransferase and alpha 2,6-sialyltransferase genes. *Glycobiology* 11: 1–9.

Hollister, J.R., Shaper, J.H., Jarvis, D.L. (1998) Stable expression of mammalian beta 1,4-galactosyltransferase extends the N-glycosylation pathway in insect cells. *Glycobiology* 8: 473–480.

Hollister, J., Grabenhorst, E., Nimtz, M., Conradt, H., Jarvis, D.L. (2002) Engineering the protein N-glycosylation pathway in insect cells for production of biantennary, complex N-glycans. *Biochemistry* 41: 15093–15104.

Hood, E.E., Woodard, S.L., Horn, M.E. (2002) Monoclonal antibody manufacturing in transgenic plants—myths and realities. *Curr Opin Biotechnol* 13: 630–635.

Hu, Y.C. (2005) Baculovirus as a highly efficient expression vector in insect and mammalian cells. *Acta Pharmacol Sin* 26: 405–416.

Humphreys, D.P., Weir, N., Lawson, A., Mountain, A., Lund, P.A. (1996) Co-expression of human protein disulphide isomerase (PDI) can increase the yield of an antibody Fab' fragment expressed in *Escherichia coli*. *FEBS Lett* 380: 194–197.

Inoue, Y., Ohta, T., Tada, H., Iwasa, S., Udaka, S., Yamagata, H. (1997) Efficient production of a functional mouse/human chimeric Fab' against human urokinase-type plasminogen activator by *Bacillus brevis*. *Appl Microbiol Biotechnol* 48: 487–492.

Jarvis, D.L. (2003) Developing baculovirus-insect cell expression systems for humanized recombinant glycoprotein production. *Virology* 310: 1–7.

Jarvis, D.L., Summers, M.D. (1989) Glycosylation and secretion of human tissue plasminogen activator in recombinant baculovirus-infected insect cells. *Mol Cell Biol* 9: 214–223.

Jarvis, D.L., Oker-Blom, C., Summers, M.D. (1990) Role of glycosylation in the transport of recombinant glycoproteins through the secretory pathway of lepidopteran insect cells. *J Cell Biochem* 42: 181–191.

Jenke, A.C., Stehle, I.M., Herrmann, F., Eisenberger, T., Baiker, A., Bode, J., Fackelmayer, F.O., Lipps, H.J. (2004) Nuclear scaffold/matrix attached region modules linked to a transcription unit are sufficient for replication and maintenance of a mammalian episome. *Proc Natl Acad Sci USA* 101: 11322–11327.

Jones, D., Kroos, N., Anema, R., van Montfort, B., Vooys, A., van der Kraats, S., van der Helm, E., Smits, S., Schouten, J., Brouwer, K., Lagerwerf, F., van Berkel, P., Opstelten, D.J., Logtenberg, T., Bout, A. (2003) High-level expression of recombinant IgG in the human cell line PER.C6. *Biotechnol Prog* 19: 163–168.

Jostock, T., Vanhove, M., Brepoels, E., Van Gool, R., Daukandt, M., Wehnert, A., Van Hegelsom, R., Dransfield, D., Sexton, D., Devlin, M., Ley, A., Hoogenboom, H., Müllberg, J. (2004) Rapid generation of functional human IgG antibodies derived from Fab-on-phage display libraries. *J Immunol Methods* 289: 65–80.

Kandilogiannaki, M., Koutsoudakis, G., Zafiropoulos, A., Krambovitis, E. (2001) Expression of a recombinant human anti-MUC1 scFv fragment in protease-deficient *Escherichia coli* mutants. *Int J Mol Med* 7: 659–664.

Klehr, D., Maass, K., Bode, J. (1991) Scaffold-attached regions from the human interferon beta domain can be used to enhance the stable expression of genes under the control of various promoters. *Biochemistry* 30: 1264–1270.

Kost, T. A., Condreay, J.P., Jarvis, D.L. (2005) Baculovirus as versatile vectors for protein expression in insect and mammalian cells. *Nat Biotechnol* 23: 567–575.

Kretzschmar, T., Aoustin, L., Zingel, O., Marangi, M., Vonach, B., Towbin, H., Geiser, M. (1996) High-level expression in insect cells and purification of secreted monomeric single-chain Fv antibodies. *J Immunol Methods* 195: 93–101.

Kujau, M.J., Hoischen, C., Riesenberg, D., Gumpert, J. (1998) Expression and secretion of functional miniantibodies McPC603scFvDhlx in cell-wall-less L-form strains of *Proteus mirabilis* and *Escherichia coli*: a comparison of the synthesis capacities of L-form strains with an *E. coli* producer strain. *Appl Microbiol Biotechnol* 49: 51–58.

Kurokawa, Y., Yanagi, H., Yura, T. (2001) Overproduction of bacterial protein disulfide isomerase (DsbC) and its modulator (DsbD) markedly enhances periplasmic production of human nerve growth factor in *Escherichia coli*. *J Biol Chem* 276: 14393–14399.

Lange, S., Schmitt, J., Schmid, R.D. (2001) High-yield expression of the recombinant, atrazine-specific Fab fragment K411B by the methylotrophic yeast *Pichia pastoris*. *J Immunol Methods* 255: 103–114.

Lemeulle, C., Chardes, T., Montavon, C., Chaabihi, H., Mani, J.C., Pugniere, M., Cerutti, M., Devauchelle, G., Pau, B., Biard-Piechaczyk, M. (1998) Anti-digoxin scFv fragments expressed in bacteria and in insect cells have different antigen binding properties. *FEBS Lett* 423: 159–166.

Li, H., Sethuraman, N., Stadheim, T.A., Zha, D., Prinz, B., Ballew, N., Bobrowicz, P., Choi, B.K., Cook, W.J., Cukan, M., Houston-Cummings, N.R., Davidson, R., Gong, B., Hamilton, S.R., Hoopes, J.P., Jiang, Y., Kim, N., Mansfield, R., Nett, J. H., Rios, S., Strawbridge, R., Wildt, S., Gerngross, T.U. (2006) Optimization of humanized IgGs in glycoengineered *Pichia pastoris*. *Nat Biotechnol* 24: 210–215.

Liang, M., Guttieri, M., Lundkvist, A., Schmaljohn, C. (1997) Baculovirus expression of a human G2-specific, neutralizing IgG monoclonal antibody to Puumala virus. *Virology* 235: 252–260.

Liang, M., Dübel, S., Li, D., Queitsch, I., Li, W., Bautz, E.K. (2001) Baculovirus expression cassette vectors for rapid production of complete human IgG from phage display selected antibody fragments. *J Immunol Methods* 247: 119–130.

Limonta, J., Pedraza, A., Rodriguez, A., Freyre, F.M., Barral, A.M., Castro, F.O., Lleonart, R., Gracia, C.A., Gavilondo, J.V., de la Fuente, J. (1995) Production of active anti-CD6 mouse/human chimeric antibodies in the milk of transgenic mice. *Immunotechnology* 1: 107–113.

Luo, D., Mah, N., Wishart, D., Zhang, Y., Jacobs, F., Martin, L. (1996) Construction and expression of bi-functional proteins of single-chain Fv with effector domains. *J Biochem (Tokyo)* 120: 229–232.

Ma, J.K., Hikmat, B.Y., Wycoff, K., Vine, N.D., Chargelegue, D., Yu, L., Hein, M.B., Lehner, T. (1998) Characterization of a recombinant plant monoclonal secretory antibody and preventive immunotherapy in humans. *Nat Med* 4: 601–606.

Ma, J.K., Barros, E., Bock, R., Christou, P., Dale, P.J., Dix, P.J., Fischer, R., Irwin, J., Mahoney, R., Pezzotti, M., Schillberg, S., Sparrow, P., Stoger, E., Twyman, R.M. (2005) Molecular farming for new drugs and vaccines. Current perspectives on the production of pharmaceuticals in transgenic plants. *EMBO Rep* 6: 593–599.

Macauley-Patrick, S., Fazenda, M.L., McNeil, B., Harvey, L.M. (2005) Heterologous protein production using the *Pichia pastoris* expression system. *Yeast* 22: 249–270.

Meissner, P., Pick, H., Kulangara, A., Chatellard, P., Friedrich, K., Wurm, F.M. (2001) Transient gene expression: recombinant protein production with suspension-adapted HEK293-EBNA cells. *Biotechnol Bioeng* 75: 197–203.

Mutskov, V., Felsenfeld, G. (2004) Silencing of transgene transcription precedes methylation of promoter DNA and histone H3 lysine 9. *EMBO J* 23: 138–149.

Ning, D., Junjian, X., Xunzhang, W., Wenyin, C., Qing, Z., Kuanyuan, S., Guirong, R., Xiangrong, R., Qingxin, L., Zhouyao, Y. (2003) Expression, purification, and characterization of humanized anti-HBs Fab fragment. *J Biochem (Tokyo)* 134: 813–817.

Niwa, R., Natsume, A., Uehara, A., Wakitani, M., Iida, S., Uchida, K., Satoh, M., Shitara, K. (2005) IgG subclass-independent improvement of antibody-dependent cellular cytotoxicity by fucose removal from Asn297-linked oligosaccharides. *J Immunol Methods* 306: 151–160.

Nyyssonen, E., Penttila, M., Harkki, A., Saloheimo, A., Knowles, J.K., Keranen, S. (1993) Efficient production of antibody fragments by the filamentous fungus *Trichoderma reesei*. *Biotechnology (NY)* 11: 591–595.

O'Gorman, S., Fox, D.T., Wahl, G.M. (1991) Recombinase-mediated gene activation and site-specific integration in mammalian cells. *Science* 251: 1351–1355.

Ogunjimi, A.A., Chandler, J.M., Gooding, C.M., Recinos III, A., Choudary, P.V. (1999) High-level secretory expression of immunologically active intact antibody from the yeast *Pichia pastoris*. *Biotechnol Lett* 21: 561–567.

Peipp, M., Saul, D., Barbin, K., Bruenke, J., Zunino, S.J., Niederweis, M., Fey, G.H. (2004) Efficient eukaryotic expression of fluorescent scFv fusion proteins directed against CD antigens for FACS applications. *J Immunol Methods* 285: 265–280.

Piechaczek, C., Fetzer, C., Baiker, A., Bode, J., Lipps, H.J. (1999) A vector based on the SV40 origin of replication and chromosomal S/MARs replicates episomally in CHO cells. *Nucleic Acids Res* 27: 426–428.

Pollock, D.P., Kutzko, J.P., Birck-Wilson, E., Williams, J.L., Echelard, Y., Meade, H.M. (1999) Transgenic milk as a method for the production of recombinant antibodies. *J Immunol Methods* 231: 147–157.

Powers, D.B., Amersdorfer, P., Poul, M., Nielsen, U.B., Shalaby, M.R., Adams, G.P., Weiner, L.M., Marks, J.D. (2001) Expression of single-chain Fv-Fc fusions in *Pichia pastoris*. *J Immunol Methods* 251: 123–135.

Rakestraw, A., Wittrup, K.D. (2006) Contrasting secretory processing of simultaneously expressed heterologous proteins in *Saccharomyces cerevisiae*. *Biotechnol Bioeng* 93: 896–905.

Reavy, B., Ziegler, A., Diplexcito, J., Macintosh, S.M., Torrance, L., Mayo, M. (2000) Expression of functional recombinant antibody molecules in insect cell expression systems. *Protein Expr Purif* 18: 221–228.

Ridder, R., Schmitz, R., Legay, F., Gram, H. (1995) Generation of rabbit monoclonal antibody fragments from a combinatorial phage display library and their production in the yeast *Pichia pastoris*. *Biotechnology (NY)* 13: 255–260.

Rippmann, J.F., Klein, M., Hoischen, C., Brocks, B., Rettig, W.J., Gumpert, J., Pfizenmaier, K., Mattes, R., Moosmayer, D. (1998) Procaryotic expression of single-chain variable-fragment (scFv) antibodies: secretion in L-form cells of *Proteus mirabilis* leads to active product and overcomes the limitations of periplasmic expression in *Escherichia coli*. *Appl Environ Microbiol* 64: 4862–4869.

Schillberg, S., Twyman, R.M., Fischer, R. (2005) Opportunities for recombinant antigen and antibody expression in transgenic plants – technology assessment. *Vaccine* 23: 1764–1769.

Schmidt, F.R. (2004) Recombinant expression systems in the pharmaceutical industry. *Appl Microbiol Biotechnol* 65: 363–372.

Schnappinger, D., Geissdorfer, W., Sizemore, C., Hillen, W. (1995) Extracellular expression of native human anti-lysozyme fragments in *Staphylococcus carnosus*. *FEMS Microbiol Lett* 129: 121–127.

Seibler, J., Schubeler, D., Fiering, S., Groudine, M., Bode, J. (1998) DNA cassette exchange in ES cells mediated by Flp recombinase: an efficient strategy for repeated modification of tagged loci by marker-free constructs. *Biochemistry* 37: 6229–6234.

Shi, X., Karkut, T., Chamankhah, M., Alting-Mees, M., Hemmingsen, S.M., Hegedus, D. (2003) Optimal conditions for the expression of a single-chain antibody (scFv) gene in *Pichia pastoris*. *Protein Expr Purif* 28: 321–330.

Shiroza, T., Shibata, Y., Hayakawa, M., Shinozaki, N., Fukushima, K., Udaka, S., Abiko, Y. (2001) Construction of a chimeric shuttle plasmid via a heterodimer system: secretion of an scFv protein from *Bacillus brevis* cells capable of inhibiting hemagglutination. *Biosci Biotechnol Biochem* 65: 389–395.

Shusta, E.V., Raines, R.T., Plückthun, A., Wittrup, K.D. (1998) Increasing the secretory capacity of Saccharomyces cerevisiae for production of single-chain antibody fragments. *Nat Biotechnol* 16: 773–777.

Sijmons, P.C., Dekker, B.M., Schrammeijer, B., Verwoerd, T.C., van den Elzen, P.J., Hoekema, A. (1990) Production of correctly processed human serum albumin in transgenic plants. *Biotechnology (NY)* 8: 217–221.

Simmons, L.C., Reilly, D., Klimowski, L., Raju, T.S., Meng, G., Sims, P., Hong, K., Shields, R.L., Damico, L.A., Rancatore, P., Yansura, D.G. (2002) Expression of full-length immunoglobulins in *Escherichia coli*: rapid and efficient production of aglycosylated antibodies. *J Immunol Methods* 263: 133–147.

Swennen, D., Paul, M.F., Vernis, L., Beckerich, J.M., Fournier, A., Gaillardin, C. (2002) Secretion of active anti-Ras single-chain Fv antibody by the yeasts *Yarrowia lipolytica* and *Kluyveromyces lactis*. *Microbiology* 148: 41–50.

Takimura, Y., Kato, M., Ohta, T., Yamagata, H., Udaka, S. (1997) Secretion of human interleukin-2 in biologically active form by *Bacillus brevis* directly into culture medium. *Biosci Biotechnol Biochem* 61: 1858–1861.

Tan, W., Lam, P.H. (1999) Expression and purification of a secreted functional mouse/human chimaeric antibody against bacterial endotoxin in baculovirus-infected insect cells. *Biotechnol Appl Biochem* 30: 59–64.

Tomiya, N., Narang, S., Lee, Y.C., Betenbaugh, M.J. (2004) Comparing N-glycan processing in mammalian cell lines to native and engineered lepidopteran insect cell lines. *Glycoconj J* 21: 343–360.

Trentmann, O., Khatri, N.K., Hoffmann, F. (2004) Reduced oxygen supply increases process stability and product yield with recombinant *Pichia pastoris*. *Biotechnol Prog* 20: 1766–1775.

Tschopp, J.F., Brust, P.F., Cregg, J.M., Stillman, C.A., Gingeras, T.R. (1987) Expression of the lacZ gene from two methanol-regulated promoters in *Pichia pastoris*. *Nucleic Acids Res* 15: 3859–3876.

Umana, P., Jean-Mairet, J., Moudry, R., Amstutz, H., Bailey, J.E. (1999) Engineered glycoforms of an antineuroblastoma IgG1 with optimized antibody-dependent cellular cytotoxic activity. *Nat Biotechnol* 17: 176–180.

Van Craenenbroeck, K., Vanhoenacker, P., Haegeman, G. (2000) Episomal vectors for gene expression in mammalian cells. *Eur J Biochem* 267: 5665–5678.

van Wely, K.H., Swaving, J., Freudl, R., Driessen, A.J. (2001) Translocation of proteins across the cell envelope of Gram-positive bacteria. *FEMS Microbiol Rev* 25: 437–454.

Verch, T., Yusibov, V., Koprowski, H. (1998) Expression and assembly of a full-length monoclonal antibody in plants using a plant virus vector. *J Immunol Methods* 220: 69–75.

Ward, M., Lin, C., Victoria, D.C., Fox, B.P., Fox, J.A., Wong, D.L., Meerman, H.J., Pucci, J.P., Fong, R.B., Heng, M.H., Tsurushita, N., Gieswein, C., Park, M., Wang, H. (2004) Characterization of humanized antibodies secreted by *Aspergillus niger*. *Appl Environ Microbiol* 70: 2567–2576.

Werten, M.W., van den Bosch, T.J., Wind, R.D., Mooibroek, H., de Wolf, F.A. (1999) High-yield secretion of recombinant gelatins by *Pichia pastoris*. *Yeast* 15: 1087–1096.

Wood, C.R., Boss, M.A., Kenten, J.H., Calvert, J.E., Roberts, N.A., Emtage, J.S. (1985) The synthesis and in vivo assembly of functional antibodies in yeast. *Nature* 314: 446–449.

Wu, S.C., Ye, R., Wu, X.C., Ng, S.C., Wong, S.L. (1998) Enhanced secretory production of a single-chain antibody fragment from *Bacillus subtilis* by coproduction of molecular chaperones. *J Bacteriol* 180: 2830–2835.

Wu, X.C., Ng, S.C., Near, R.I., Wong, S.L. (1993) Efficient production of a functional single-chain antidigoxin antibody via an engineered *Bacillus subtilis* expression-secretion system. *Biotechnology (NY)* 11: 71–76.

Wu, S.C., Yeung, J.C., Duan, Y., Ye, R., Szarka, S.J., Habibi, H.R., Wong, S.L. (2002) Functional production and characterization of a fibrin-specific single-chain antibody fragment from *Bacillus subtilis*: effects of molecular chaperones and a wall-bound protease on antibody fragment production. *Appl Environ Microbiol* 68: 3261–3269.

Wulfing, C., Rappuoli, R. (1997) Efficient production of heat-labile enterotoxin mutant proteins by overexpression of dsbA in a degP-deficient *Escherichia coli* strain. *Arch Microbiol* 167: 280–283.

Wurm, F.M. (2004) Production of recombinant protein therapeutics in cultivated mammalian cells. *Nat Biotechnol* 22: 1393–1398.

Yamagata, H., Nakahama, K., Suzuki, Y., Kakinuma, A., Tsukagoshi, N., Udaka, S. (1989) Use of *Bacillus brevis* for efficient synthesis and secretion of human epidermal growth factor. *Proc Natl Acad Sci USA* 86: 3589–3593.

Yoshida, S., Matsuoka, H., Luo, E., Iwai, K., Arai, M., Sinden, R.E., Ishii, A. (1999) A single-chain antibody fragment specific for the *Plasmodium berghei* ookinete protein Pbs21 confers transmission blockade in the mosquito midgut. *Mol Biochem Parasitol* 104: 195–204.

Zhang, Z., Song, L.P., Fang, M., Wang, F., He, D., Zhao, R., Liu, J., Zhou, Z.Y., Yin, C.C., Lin, Q., Huang, H.L. (2003) Production of soluble and functional engineered antibodies in *Escherichia coli* improved by FkpA. *Biotechniques* 35: 1032–1038, 1041–1042.

7
Non-Antibody Scaffolds
Markus Fiedler and Arne Skerra

7.1
Introduction

The use of so-called protein scaffolds with engineered ligand specificities as antibody surrogates both in biomedical research and in medical therapy has gained increasing attention in recent years. This development started with the notion that immunoglobulins owe their biochemical function to the combination of a conserved framework region with a spatially well-defined antigen-binding site, whereby the latter is composed of peptide segments that are hypervariable in amino acid sequence and conformation. Based on the current advanced methods for antibody engineering, together with biomolecular library techniques which permit the selection of functional antibody fragments on a routine level, several other protein classes have been employed for the construction of practically useful ligand-binding reagents. Properties such as small size of the engineered receptor protein, stability, and ease of production were initially in the focus. Once it had been demonstrated that protein scaffolds can be engineered to yield novel biomolecules with ligand affinities and specificities comparable to antibodies, efforts started to make these novel proteins amenable to application for human therapy as well. To develop such innovative biopharmaceutical compounds, aspects such as bioavailability, serum half-life and stability, tissue penetration, and immunogenicity have to be considered. Data from preclinical research carried out to date suggest that engineered protein scaffolds have potential to yield superior drugs with beneficial functions for the molecular recognition and targeting of tissues, cells or pathogens in the future.

7.2
Motivation for Therapeutic Use of Alternative Binding Proteins

Since the development of monoclonal antibody technology three decades ago (Köhler and Milstein 1975) and the invention of bacterial expression systems for

Handbook of Therapeutic Antibodies. Edited by Stefan Dübel
Copyright © 2007 WILEY-VCH Verlag GmbH & Co. KGaA, Weinheim
ISBN 978-3-527-31453-9

engineered antibody fragments in the late 1980s (Skerra and Plückthun 1988) recombinant immunoglobulins have made a tremendous impact on the field of modern biopharmaceuticals, with about 20 currently approved drugs for human therapy, as described in this monograph. Without doubt, this trend will accelerate even more in the near future as there are several hundred further antibody candidates in late preclinical and clinical development. However, despite this remarkable success there is also an increasing awareness of the limitations of antibody technology, both regarding the intrinsic molecular properties of immunoglobulins and their technological or even commercial aspects. Consequently, there is a growing demand for alternative reagents that can provide molecular recognition functions similar to those currently considered characteristic for antibodies, the products of the natural immune response (Skerra 2003).

The most problematic aspect for the broader application of intact (human or "humanized") antibodies in medical therapy is the large biotechnological effort and the associated high production cost for properly glycosylated, full-sized immunoglobulins, whose manufacturing requires elaborate cell culture fermentation systems, at least at the moment. Most of the currently approved antibody drugs make use of this format and, as significant therapeutic doses of the active compound are needed for many indications, the worldwide availability of sufficient production capacity has become an issue (Carson 2005). The complicated production is largely caused by the glycosylated Fc effector region of immunoglobulins, which mediates complement activation and antibody-dependent cellular cytotoxicity (ADCC) (Carter 2001) and is also responsible for the long serum half-life of antibodies (Ghetie and Ward 2002) as well as for their numerous interactions with immunological cell surface receptors (Gessner et al. 1998). However, for several medical applications the presence of this moiety is actually not needed, for example when merely binding or antagonistic inhibition of a target is desired. In these cases, bacterially produced antibody fragments can, in principle, be used, although many of them are expressed at low yields, have poor thermodynamic stability, and a propensity to oligomerize or aggregate (Binz et al. 2005).

Furthermore, with respect to the development and commercial exploitation of antibodies in biotech companies and the pharmaceutical industry the extremely complex intellectual property situation in this field has become a major obstacle (Owens 2005; Editor 2005; Van Brunt 2005). First, today there is a high burden involved in the protection of new antibody-based inventions. In particular, the US patent statutes §§ 102, 103, and 112, which demand novelty, non-obvious subject matter, and specification, greatly affect the patentability of new antibodies. A recent case (Smithkline Beecham vs. Aotex) cleary illustrates that novelty may be questioned for any antibody when considering the aspect of "inherent anticipation." Under this theory "a product that existed in trace amounts, although unknown and undetected and unisolated, is 'inherently anticipated' and barred from the patent system after it is discovered." As antibodies themselves and the ability to generate such antibodies is inherent to the immune system of a given species it is entirely possible that such a doctrine may prohibit the claim of antibodies themselves or of the antigens with which they interact (Lu et al. 2005).

Second, the development of new antibody-based reagents for therapy, diagnosis, or even purification purposes is hampered by a multitude of existing patents covering methods of generation, optimization, production, formulation, delivery, etc., which leads to the obligation to reserve a significant portion of any potential revenue stream for paying other parties (Baker 2005). The dimension of this indirect cost can be visualized by the fact that in 2004 just 17 companies participated with direct sales in the antibody market (Evans and Das 2005). All the other over one hundred antibody-related companies grabbed a piece of the US$10.3 billion market by dealing with their intellectual property. These commercial considerations will increasingly hinder the development of therapeutic antibodies by dramatically cutting into the profit.

Using alternative binding proteins as a basis for biopharmaceutical drug development may help to avoid such restrictions at least to some degree (Jeong et al. 2005). Utilizing such innovative approaches might even be essential in enabling entry into the biologics market, especially for small and medium-sized companies. Their success is further facilitated by the fact that alternative binding proteins selected against validated target molecules will benefit from preclinical assays, animal models, and clinical study designs originally established for monoclonal antibodies. This synergy may help to save time in drug development, which tends to require a prolonged period for biologicals, typically 4 years in the last half of the 1980s but already more than 7 years in the first half of this decade (Lahteenmaki and Baker 2004).

Antibody surrogates derived from alternative protein scaffolds can circumvent many of these technological and strategic limitations (Table 7.1). Their concept is based on the notion that certain polypeptide folds are observed in nature to occur in different context and with varying biochemical function. Hence, such protein scaffolds may be utilized to selectively reshape their active sites via protein engineering in order to create novel ligand-binding functions without having to worry about the protein folding problem in general (Skerra 2000a). Candidates for protein scaffolds should possess intrinsic conformational stability and they should be able to present surface segments, preferably flexible loops of varying sequence and length, including exposed hydrophobic residues, without undergoing significant changes in the structural framework. Naturally, the ligand-binding properties of such artificial receptor proteins will depend on the number, spatial distribution, and diversity of the variable regions. According to practical demands they should further be based on monomeric polypeptides, which are small in size (Fig. 7.1), robust, easily engineered, and efficiently produced by inexpensive prokaryotic expression systems. Several examples for medically useful protein scaffolds have emerged in recent years and will be discussed in the following chapters. For a more complete overview the reader is referred to some recent review articles (Nygren and Skerra 2004; Binz and Plückthun 2005; Binz et al. 2005; Hey et al. 2005; Hosse et al. 2006).

Table 7.1 Properties of artificial binding proteins versus antibodies (or their conventional fragments) with relevance for therapeutic application.

Property	Monoclonal antibodies (mABs)	Antibody fragments (Fab/scFv)	Scaffold-based proteins
Size (kDa)	150	50/25	≤20[a]
Polypeptide chains	4	2/1	1
E. coli production	−	+/−	+
Ease of modification	+/−	+	+
Fusion proteins	−	−/(+)	+
High specificity	+	+	+
High affinity	+	+	+
High stability	+	−	+
Human origin	+/−	+/−	+/−
Ig effector function	+	−	−
IP situation	−	−	+
Neutralizing activity	+	+	+
Intracellular activity	−	+/−	+/−
Targeted delivery	+/−	+	+
Non-invasive delivery	−	−	(+)
Tissue penetration	−	+	+
Clearance	Slow	Fast	Fast

a Size of the binding protein depends on the scaffold chosen (see text).

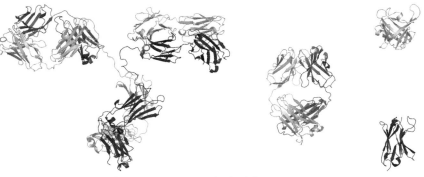

Fig. 7.1 Size comparison between an intact antibody (left), its Fab fragment (middle), a single immunoglobulin domain (right, bottom), and an anticalin as a typical scaffold protein (right, top).

In principle, engineered binding proteins offer a number of advantages, both when considered as drug candidates for therapy or for *in vivo* diagnostics and when applied for *in vitro* diagnostics or as laboratory research tools. First, the binding proteins are engineered entirely *in vitro*, without the need for animal immunization. Usually, a combinatorial library of the chosen scaffold protein is created by selective random mutagenesis of appropriate surface residues (Skerra 2003) and variants with the desired target or "antigen" specificity are subsequently selected by well-established techniques such as phage display (Lowman and Wells 1991; Pini et al. 2002), ribosome display (Groves and Osbourn 2005), ELISA or colony screening (Schlehuber et al. 2000). This *in vitro* selection and screening methodology also offers numerous strategies to improve and fine-tune, under suitable conditions, the biomolecular properties of an initially identified scaffold-derived binding protein. In this regard there is usually a desire for exquisite target affinity and specificity (especially if discrimination against closely related targets is needed), thermodynamic folding stability (mostly against thermal denaturation), shelf-life and solubility as well as chemical stability and protease resistance.

The binding proteins resulting from this process are usually amenable to microbial production at high yields and with simplified downstream processing. Normally, it is intended to use protein scaffolds that lack glycosylation and which can thus be manufactured using conventional *E. coli* expression systems. If the scaffold is devoid of disulfide bonds the binding protein may be produced in the bacterial cytoplasm and directly extracted in a soluble and folded state, which is a clear advantage over antibody fragments that usually cannot fold in a reducing environment (Wörn and Plückthun 2001). Furthermore, intracellular applications may be possible, for example as antagonists of cytosolic protein interaction networks. Even scaffolds that carry disulfide bonds may be produced according to the cytoplasmic route, albeit a secretion strategy that directly yields correctly folded protein is often preferred, at least at the laboratory scale. However, if necessary, subsequent refolding from cytoplasmic inclusion bodies is usually much more efficient than for antibody fragments, which are composed of several subunits or domains, because almost all currently exploited scaffold proteins comprise a monomeric polypeptide chain and possess a robust globular fold.

Aiming at therapeutic application, a series of pharmaceutical aspects has also to be considered in conjunction with the intended medical indication when choosing an appropriate scaffold protein: serum half-life and stability, bioavailability, tissue penetration, and immunogenicity. Unfortunately, in practice some of the desirable biomolecular parameters tend to be of opposing nature. For example, a small protein size usually leads to better tissue penetration but also to a shorter serum half-life (Holliger and Hudson 2005). Nevertheless, owing to their simple biomolecular architecture, their robustness, and their rather small dimensions, scaffold proteins usually can be further engineered for improved pharmacokinetic and functional properties. Serum half-life can be extended by site-specific PEGylation (Harris and Chess 2003), preferentially via introduction of a free Cys side-chain into a sterically well accessible position remote from the active site (Rosendahl et al. 2005). Alternatively, fusion with serum albumin (Osborn et al. 2002), with albumin-binding peptides (Dennis et al. 2002) or domains (Lejon et al. 2004) or even the generation of a novel albumin-binding site (Connelly 2005) can be employed to achieve prolonged circulation.

Most of the scaffold-based binding proteins are ideally suited to construct fusion proteins in order to adopt novel effector functions. In the simplest case one could produce a fusion protein with the Fc portion of an antibody (Chamow and Ashkenazi 1996). This would not only serve to recruit the natural immunologic effector functions, but also bivalent binding properties should arise, thus leading to an enhanced avidity. Furthermore, this could even mediate crosslinking of target cell surface receptors and thus trigger associated intracellular signaling events. Nevertheless, the construction of such a hybrid protein would not give much advantage over the use of a conventional antibody possessing the same antigen specificity.

Rather, the use of a single domain scaffold permits the preparation of smarter fusion proteins, which are also easier to produce because heterologous chain assembly is usually not required – in contrast with multidomain antibody fragments. Hence, dimerization of such binding proteins can be achieved, for example, by simple tandem fusion (Schlehuber and Skerra 2001) or via fusion with a small pair-forming module such as a helix bundle (Pack and Plückthun 1992), a single constant Ig domain (Li et al. 1997), or a homodimeric enzyme (Tudyka and Skerra 1997).

Fusion proteins with enzymes should not only be useful to generate practically applicable diagnostic tools such as in combination with alkaline phosphatase (Schlehuber et al. 2000), they may also offer interesting biopharmaceutical agents, especially for enzyme prodrug activation approaches in tumor therapy (Florent et al. 1998; Carter 2001). Fusion proteins with enzyme toxins from bacteria (e.g. *Pseudomonas* exotoxin) (Pastan 2003), or from plants, such as saporin (Palmisano et al. 2004), have demonstrated the principal applicability of this approach, but at the expense of significant side effects due to the considerable immunogenicity and some intrinsic unspecific toxicity of the fusion partners. Combinations of nonimmunogenic alternative scaffolds with less toxic effectors of human origin, for example RNase (Zewe et al. 1997) (see Chapter 3, Vol II), will allow the development and higher yield production of promising "immunotoxin" drugs of a next generation, thereby eliminating all of the major roadblocks of the current antibody–heterologous toxin constructs.

Finally, the intrinsically small size and robustness of scaffold-based binding proteins may open novel routes of parenteral administration, apart from injection or infusion. In the light of recent developments on inhalable insulin (Powell 2004), pulmonary delivery has attracted particular attention because bioavailability through the lungs should be much higher for certain scaffold proteins than for large antibodies (Agu et al. 2001). Transdermal delivery could also be of interest (Barry 2004), and even oral administration has been envisaged for some of the engineered scaffold proteins (Goldberg and Gomez-Orellana 2003).

During the last 15 years several protein scaffolds have been investigated, mainly in an academic environment, for the tailoring of non-antibody binding proteins. With the currently available repertoire, artificial binding proteins with high affinities and specificities for molecules differing remarkably in size and shape have been successfuly generated, thus addressing a variety of targets, ranging from low molecular weight compounds over carbohydrates and peptides to proteins. The different kinds of biomolecular architecture that underly the individual protein scaffolds can provide either cavities for small molecules, flexible loops which may enter substrate clefts in enzymes, for example, or extended interfaces for tight complex formation with larger proteins.

This review will focus on those examples that are already closest to medical application or that illustrate important structural and functional principles (Table 7.2, Fig. 7.2). Since this area has also seen increasing commercialization by small and medium-sized biotech companies in recent years, corresponding examples will be mentioned (Table 7.3). For a more complete compilation of the current scaffold protein approaches the reader should refer to recent review articles (Skerra 2000a; Nygren and Skerra 2004; Binz et al. 2005; Binz and Plückthun 2005; Hey et al. 2005; Hosse et al. 2006). Here, the various protein scaffold architectures will be grossly classified into four groups: single domain immunoglobulins, scaffold proteins presenting a contiguous hypervariable loop region, scaffold proteins for display of individual extended loops, and scaffold proteins providing a rigid secondary structure interface.

7.3
Single-Domain Immunoglobulins

Probably the best example of a protein scaffold that presents a set of structurally well-positioned hypervariable loops, giving rise to a contiguous interface for biomolecular interaction, is provided by the Ig class itself. While the immunological effector functions reside in the constant region of an antibody, its antigen-binding activity is exclusively provided by the pair of variable domains (V_H and V_L), which are located in close spatial neighborhood at the N-termini of each light and heavy chain (Padlan 1994). The fold of the mutually homologous variable domains is dominated by a sandwich of two antiparallel β-sheets that form a structurally conserved framework. Three loops at one end of the β-sandwich in each variable domain are thus brought together to form the so-called complementarity determining region (CDR) of an antibody. Due to the peculiar genetic mechanisms of the immune system each CDR loop is highly variable in its amino acid sequence such that altogether six peptide segments give rise to an extended antigen-combining site (Skerra 2003). The resulting structural diversity of the paratope explains the pronounced affinities and specificities that are typically observed for antibodies originating from a natural immune response.

Biotechnological methods for the generation of antibody fragments or even intact immunoglobulins with prescribed antigen specificities are well established and are described in accompanying chapters of this book. However, in spite of their still increasing use both as research tools and as protein therapeutics, the characteristic protein architecture of antibodies also causes some practical disadvantages. For example, immunoglobulins possess a rather large size, which

Table 7.2 Biomolecular characteristics of certain protein scaffolds that are being explored for medical purposes.

Name	Scaffold	Origin	Fold	No. of residues/ crosslinks/domains	Randomized structural elements	Selected references
Affibody	Three-helix bundle from Z-domain of Protein A	Bacterial	α_3	58/–/1	13 residues in 2 helices	Wikman et al. 2004
Affilin molecules	γ-Crystallin/ Ubiquitin	Human	β-Sandwich/α/β	74 and 174/–/1	8 residues in β-sheet	Hey et al. 2005
AdNectin	10th fibronectin type III domain	Human	β-Sandwich	94/–/1	2 to 3 loops	Xu et al. 2002
Anticalin	Lipocalins	Human and various species	β-Barrel	160–180/0,1,2 S-S/1	16 to 24 residues in 4 loops	Schlehuber and Skerra 2005
DARPin	Ankyrin repeat protein	Artificial consensus sequence	α_2/β_2 repeated	n × 33/–/combination of multiple domains	β-turn, 1 α-helix, 1 loop	Binz et al. 2003
Domain antibody (DAb)	Variable domain of antibody light or heavy chain	Human	β-Sandwich	ca. 120/1 S-S/1	3 loops (CDRs)	Holt et al. 2003
Evibody	Cytotoxic-associated antigen (CTLA-4)	Human	β-Sandwich	136/2 S-S/1	17.3 loops (CDR analogue)	Irving et al. 2001

Knottin	Different proteins from the knottin family	Multiple sources	β-Sheet or β-Sandwich	23–113/2 S-S/1	β-turn, β-sheet and/or loop	Craik et al. 2001
Kunitz-type domain	Trypsin inhibitor	Human and bovine	α/β	58/3 S-S/1	1 or 2 inserted loops	Dennis and Lazarus 1994
Maxibody/ Avimer	A-domain	Human	LDL receptor-like module, disulfide-rich calcium-binding fold	n × 35–45/3 S-S/ combination of multiple domains	mainly β-turns	Silverman et al. 2005
Nanobody	Variable domain of antibody heavy chain	Camelidae	β-Sandwich	ca. 120/1 S-S/1	3 loops (CDRs)	Conrath et al. 2005
Tetranectin	Monomeric or trimeric C-type lectin domain	Human	α/β	137/3 S-S/natural homo-trimer	2 loops	see Table 7.3
Trans-body	Serum transferrin	Human	α/β	329/8 S-S/2	1 or 2 inserted loops	see Table 7.3
V(NAR)	Variable domain of new antigen receptor	Shark	β-Sandwich	112/1 S-S/1	3 loops (CDRs)	Holliger and Hudson 2005
	β-Lactamase	Bacteria	α/β	263/1 S-S/1	1 or 2 inserted loops	Legendre et al. 2002

Table 7.3 List of companies and references to web sites for scaffold proteins that are in development for medical applications.

Company	Scaffold name	Status/Indication[a]
Ablynx (Ghent, Belgium) www.ablynx.com	Nanobody – single-domain antibodies from the camelid family	Preclinical[b]: rheumatoid arthritis (injectable), inflammatory bowel disease (oral administration), thrombosis associated with arterial stenosis (injectable)
Affibody (Bromma, Sweden) www.affibody.com	Affibody – Z-domain of Protein A from S. aureus	Preclinical[b]: breast cancer (*in vivo* diagnostics, radiotherapy)
Avidia (Mountain View, CA, USA) www.avidia.com	Maxibody – human A-domain	Preclinical[b]: inflammation
BioRexis (King of Prussia, PA, USA) www.biorexis.com	*Trans*-body – human transferrin	n.d.
Borean Pharma (Aarhus, Denmark) www.borean.com	Tetranectin monomeric or trimeric human C-type lectin domain	Preclinical[b]: inflammation
Compound Therapeutics (Waltham, MA, USA) www.compoundtherapeutics.com	AdNectin – human 10th fibronectin type III domain	Preclinical[b]: cancer, ophthalmology
Domantis (Cambridge, UK) www.domantis.com	Domain antibody-variable domain of human light or heavy chain	Preclinical[b]: inflammation, asthma, cancer
Dyax (Cambridge, MA, USA) www.dyax.com	Kunitz-type domain of human and bovine trypsin inhibitor	Clinical phase II: hereditary angioedema, on-pump heart surgery, cystic fibrosis
Evogenix (Sydney, Australia) www.evogenix.com	Evibody – CTLA-4	n.d.
Molecular Partners (Zürich, Switzerland) www.molecularpartners.com	DARPin – ankyrin repeat protein	n.d.
Pieris (Freising, Germany) www.pieris-ag.com	Anticalin – various human and insect lipocalins	Preclinical[b]: cancer, cardiovascular inflammatory diseases
Scil Proteins (Halle, Germany) www.scilproteins.com	Affilin molecules – human γ-crystallin/human ubiquitin	Preclinical[b]: cancer, ophthalmology, inflammation

a As disclosed on the company home page (in February 2006).
b Preclinical studies with respect to biological activity in cell culture, serum half-life, (serum) stability, *in vivo* efficacy, toxicity in animal models, etc.

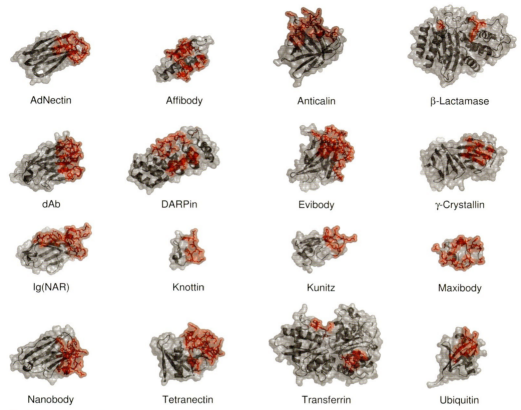

Fig. 7.2 Various scaffold proteins that are in development or considered for therapeutic application. Some of the examples shown are not discussed in the text, due to insufficient published data; their properties (as far as known) are listed in Tables 7.2 and 7.3.

results in poor production economy, not only with respect to their "specific antigen-binding activity" but also because the efficient production of intact antibodies usually requires eukaryotic cell culture. In addition, the voluminous molecular size results in poor tissue penetration. Even the smallest stable antibody fragment that carries an intact combining site, the scFv fragment, is composed of more than 250 residues. This is a disadvantage, especially when serving as an antigen recognition module, for example in a fusion protein with a reporter enzyme or a toxin.

Furthermore, the fact that the antigen-binding site is formed together by the light and heavy Ig chains turns out to be a problem in the biotechnological use of antibodies and, in particular, of their smaller antigen-binding fragments. On the one hand, two different coding regions have to be cloned and handled in parallel and, on the other, extra measures must be taken to stabilize the heterodimerization of the two chains at the level of the biosynthetic protein. Especially in the Fv fragment, which merely comprises the two variable domains (Skerra and Plückthun 1988), the noncovalent association between both chains often leads to limited stability. In order to rigidify the fragment for practical applications, the two domains can be connected by a flexible peptide linker, resulting in so-called single-chain (sc) Fv fragments (Bird and Walker 1991). Yet, the format of the scFv fragment frequently results in other undesired properties, like low folding efficiency upon expression in *E. coli*, enhanced aggregation, and a pronounced tendency to form oligomers (Atwell et al. 1999).

Consequently, there is a generally recognized need for alternative scaffolds that offer the possibility of displaying structurally variable loops on a monomeric protein architecture, thus avoiding problems with chain pairing and insufficient folding stability. One example in this respect is provided by nature itself with a special Ig class whose antigen-binding region is formed by a single polypeptide chain. Normally, an unpaired variable domain exposes a significant area of hydrophobic surface to solvent, which is otherwise shielded by association either with the second variable domain or, as often seen for light chains, via formation of so-called Bence Jones dimers (Stevens et al. 1991). Especially in the case of isolated heavy chain variable domains (V_H) this fundamental property usually causes aggregation (Glockshuber et al. 1990; Davies and Riechmann 1994) and nonspecific adsorption.

Interestingly, a family of peculiar antibodies that is devoid of light chains has been identified as a natural subclass of the IgG pool in camelids (Hamers-Casterman et al. 1993; De Genst et al. 2006). In such "heavy-chain" antibodies, which are composed of a pair of heavy chains lacking the C_H1 domain, the V_H domains have apparently evolved to remain soluble without heterodimerization. Sequence analysis of the natural heavy-chain antibodies from camels in conjunction with X-ray structural analyses of corresponding V_H antibody fragments (dubbed VHH) from camels, dromedaries, and also from llamas revealed the reasons for the observed high solubility of this type of Ig domain (Muyldermans et al. 2001; Conrath et al. 2005). In essence, the camelid V_H domains have an increased surface hydrophilicity in the region that faces the V_L domain in ordinary Igs. In addition, they have a much longer CDR-H3, which often participates in a second disulfide bridge within the V_H domain, thus partially shielding the interface region from solvent.

Up to now, camel VHH domains were mainly generated by classical immunization of animals. Due to the simpler nature of their antigen-binding site compared with conventional antibodies, with just three CDR loops that protrude from one edge of the protein, they can no longer form deep pockets. Hence, the binding of haptens has only been established in special cases (e.g. with a rather large azo dye compound) (Spinelli et al. 2001), otherwise it is accompanied by rather poor affinity (Yau et al. 2003). However, their arrangement of three hypervariable loops, among which the CDR3 often adopts an extended conformation, seems to be particularly well suited for creating inhibitor proteins that can bind to the active sites of enzymes (Lauwereys et al. 1998; Muyldermans et al. 2001; Conrath et al. 2001).

As a consequence of the loss of their hydrophobic interface VHH domains can often be expressed at high yields in *E. coli* – via secretion into the bacterial periplasm to ensure disulfide bond formation – and they possess favorable conformational stability with the feature of reversible denaturation, a property that distinguishes them from conventional V_H domains that usually show irreversible aggregation behavior (Dumoulin et al. 2002; Ewert et al. 2002). Currently, so-called "nanobodies" (cf. Tables 7.2 and 7.3) are in development against a number of medical targets, including enzymes, transcription factors, cytokines, and tumor markers, either for diagnostic purposes (Saerens et al. 2005) or for therapy in indications such as infectious diseases, rheumatoid arthritis, asthma, or solid tumors (Cortez-Retamozo et al. 2004).

Notably, there also exist some human or murine variable antibody domains whose natural functions seem not to depend on the association with a second variable domain – either V_L or V_H – so that they might be considered as human scaffolds for the engineering of single Ig domains. For example, a melanoma-specific "V_H antibody" was isolated from an scFv phage library derived from a patient that had been immunized with genetically modified autologous tumor cells (Cai and Garen 1996). In contrast, the monoclonal antibody NEMO, which was obtained by immunization of a mouse, consists solely of a κ light chain (Masat et al. 1994). Only the monomeric form of this Ig fragment was shown to be active and to recognize an antigen expressed by human cells of the melanocytic lineage. Furthermore, a natural V_H domain from a mouse hybridoma clone that carries a mutation at the interface with its V_L domain was successfully used as scaffold for the preparation of a phage display library after randomization of nine residues within CDR-H3, thus yielding soluble V_H domains with specific binding activities towards tumor necrosis factor (TNF) or immunoglobulins as targets (Reiter et al. 1999).

Thus, there seem to be structural solutions other than those originally found in camel VHH Igs which can confer solubility and specific binding properties onto isolated variable domains of human or murine antibodies. This notion is strengthened by the recent description of a conventional llama V_H domain that lacks the characteristic mutations of VHH antibodies within the interface region and is still highly soluble (Vranken et al. 2002). Moreover, the concept of selecting so-called Domain Antibodies (DAbs) from libraries of cloned human V_H or V_L domains has emerged (Holt et al. 2003; Holliger and Hudson 2005). DAbs (cf. Tables 7.2 and 7.3) have been successfully selected against targets such as TNF-α or human serum albumin, in the latter case resulting in so-called AlbuDabs, which can be fused to other therapeutic proteins to prolong their serum half-life (Connelly 2005). Pulmonary and oral routes of delivery are currently being explored for VHH domains of both human and cameloid origin.

Another source of single Ig domains that appear to be useful as scaffolds for the generation of artificial receptor proteins are cartilaginous fish, in particular wobbegong and nurse sharks (Holliger and Hudson 2005). The new antigen receptor (NAR) of the nurse shark, *Ginglymostoma cirratum*, consists of two protein chains, each with one variable and five constant domains, and seems to be the representative of the Ig superfamily for adaptive immune response in this animal (Dooley et al. 2003). Based on this perception, a number of NAR variable domains from the spotted wobbegong shark, *Orectolobus maculatus*, were cloned and used for the construction of protein libraries in which the CDR3 loop was randomized (Nuttall et al. 2001). From this library mutant Ig domains with specificity for the protease gingipain K were identified via phage display.

Another phage display library was generated by incorporating synthetic CDR3 regions with 15–18 residues into the NAR domain. Thus, a variant with high affinity for the 60 kDa cytosolic domain of the outer membrane translocase receptor of human mitochondria (Tom70) was selected (Nuttall et al. 2003). In an attempt to generate IgNAR antibodies with anti-infective activity, V(NAR) domains directed against the malarial apical membrane antigen-1 (AMA1) from *Plasmodium falciparum* were selected, whose initial affinity of about 0.2 micromolar was approximately 10-fold enhanced by *in vitro* affinity maturation (Nuttall et al. 2004). More recently, a synthetic library of V(NAR) domains was successfully constructed on the basis of an anti-lysozyme IgNAR domain with known three-dimensional structure, which had been obtained from an immunized nurse shark, as a single well-defined scaffold (Shao et al. 2006).

Taken together, single domain Igs offer a somewhat conservative approach in so far as the methods for their generation as well as their biophysical properties are strongly related to conventionally engineered antibody fragments, except that they do not rely on the light/heavy chain association. Hence, intellectual property issues in this field also largely apply to the resulting binding proteins. In addition, it is not quite clear whether all single-domain Ig fragments behave as truly monomeric proteins. At least in some cases X-ray structural analysis has indicated a mode of dimerization in the crystal packing (Jespers et al. 2004; Streltsov et al. 2004) that resembles the well-known V_H/V_L pairing of antibodies (Padlan 1994) or the structurally analogous light-chain homo-association that is typically observed for Bence Jones proteins (Stevens et al. 1991; Pokkuluri et al. 1998).

7.4
Scaffold Proteins Presenting a Contiguous Hypervariable Loop Region

An obvious extension of the single-domain antibody concept is the use of more remote members of the Ig superfamily as protein scaffolds to generate novel binding proteins. This strategy can no longer benefit from the diversification mechanisms of the natural immune system – as, for example, in the case of cameloid VHH domains – instead, techniques of targeted randomization and subsequent *in vitro* selection of variants with the desired specificity must be applied.

For example, the human cytotoxic T lymphocyte-associated antigen 4 (CTLA-4), a functionally important T-cell surface coreceptor (also described further below), was explored as a potential protein scaffold. Its extracellular domain exhibits a V-like Ig fold, albeit with two intramolecular disulfide bonds instead of one. Normally, CTLA-4 is expressed as a membrane protein on activated T cells and exists as a homodimer that is crosslinked by a disulfide bridge within a stalk region just outside the transmembrane segment. Even though monomeric CTLA-4 V-domains were successfully synthesized in eukaryotic Chinese hamster ovary (CHO) or *Pichia pastoris* expression systems, attempts to produce active soluble protein in *E. coli* remained unsuccessful. However, when the loops corresponding either to CDR1 or to CDR3 of antibodies were substituted with somatostatin – a 14-residue intra-disulfide-linked peptide hormone – CTLA-4 variants with superior solubility were obtained and successfully produced in the periplasm of *E. coli* (Nuttall et al. 1999).

In a subsequent study, the nine-amino-acid CDR3-like loop was replaced by a random sequence surrounding a fixed central RGD motif and several CTLA-4-based variants capable of binding to the human $\alpha_v\beta_3$ integrin were selected using phage display (Hufton et al. 2000). A similar but nonbiased CTLA-4 library was later used to select variants with affinity towards lysozyme via ribosome display (Irving et al. 2001). Hence, at least in principle, the CTLA-4 extracellular domain may be utilized to generate globular binding proteins toward protein targets, despite the fact that a specificity with therapeutic relevance has so far not been described.

Tendamistat, a 74-amino-acid inhibitor of α-amylase with a β-sheet sandwich topology, which is stabilized by two rather than one disulfide bond as in Igs, was employed as a more distantly related Ig-type scaffold in order to present conformationally constrained random peptides (McConnell and Hoess 1995). In this case two loops, comprising residues 38–40 and 60–65, were randomly mutagenized and a phage display library was prepared and subjected to selection against the monoclonal antibody A8, which recognizes the 21-residue peptide hormone endothelin. As result, tendamistat variants exhibiting a variety of sequences in both of the loops, none of which matched a linear sequence from endothelin, were isolated. In another study loop no. 1 of tendamistat was used for the preparation of specialized random libraries containing the RGD motif and specific integrin-binding variants were identified via phage display (Li et al. 2003). Hence, the tendamistat scaffold seems to tolerate variegation of its CDR-like loops and it will be interesting to see whether any binding specificities of medical importance may be generated.

The 10th domain of the 15 repeating units in human fibronectin type III (FN3) provides another small, monomeric β-sandwich protein which shows structural resemblance to a trimmed Ig V_H domain. It consists of 94 amino acids and possesses seven β-strands (instead of nine for a normal Ig variable domain) with three loops connecting the strands at one end of the β-sandwich. In contrast with conventional members of the Ig superfamily, FN3 is devoid of disulfide bonds. The loop that carries the integrin-binding RGD sequence in the natural fibronectin domain is topologically equivalent to the CDR3 of a V_H domain. FN3-type domains are ubiquitous and occur in cell adhesion molecules, cell surface hormone and cytokine receptors, chaperonins as well as in carbohydrate-binding domains, all of which are involved in molecular recognition.

Based on this FN3 domain a phagemid display library was prepared with 10 randomized residues, five in the BC loop (residues 26–30) and five in the FG loop (residues 77–81), and used for panning against ubiquitin. Indeed, one variant (initially called "monobody," later dubbed "Trinectin" or "AdNectin") with specific target-binding activity and a dissociation constant in the low micromolar range, was identified (Koide et al. 1998). Although this variant could be readily produced as a soluble protein in *E. coli*, its solubility was significantly lower than that of the recombinant wildtype FN3. In a subsequent study all three CDR-like loops were randomized at once and the mRNA display technique was applied for selection, leading to FN3 variants with high affinity towards TNF-α (Xu et al. 2002), a medically relevant target for the treatment of rheumatoid arthritis (Bang and Keating 2004). Recently, FN3-based binding proteins for the vascular endothelial growth factor receptor 2 (VEGF-R2), an important cell surface receptor in angiogenesis, have been described as well (Parker et al. 2005).

A different type of protein scaffold that presents a structurally hypervariable loop region on top of a rigid β-sheet secondary structure is derived from the lipocalin protein family (Skerra 2000b). The lipocalins represent a class of small, robust proteins that share a rigid β-barrel of eight antiparallel strands winding around a central axis as their central folding motif. Lipocalins are functionally diverse polypeptides of 160–180 residues, with rather weak sequence homology but high similarity at the tertiary structural level (Flower 1996). The family comprises several hundred members, which are found in almost all vertebrates, including humans (Breustedt et al. 2006), but also in insects and in bacteria. In most cases, their physiological role lies in the storage or transport of hydrophobic and/or chemically sensitive organic compounds. At the open end of the conical structure the β-strands are connected in a pair-wise fashion by four loops, which form the entrance to the ligand-binding pocket (Fig. 7.2). In contrast to the highly conserved β-barrel topology, this loop region differs considerably among individual lipocalins, both in conformation and length of the corresponding polypeptide segments. Hence, there appears to be a functional resemblance with the antigen-binding region of immunoglobulins (Skerra 2003).

Initially, the 174-residue bilin-binding protein (BBP) from *Pieris brassicae* with its rather wide and shallow ligand pocket – where biliverdin IX$_\gamma$ is complexed as a natural ligand – served as scaffold for the generation of novel binding proteins towards several low molecular weight molecules (Beste et al. 1999). Sixteen residues distributed across all four loop segments and commonly located at the center of the binding site were identified by molecular modeling and subjected to concerted random mutagenesis, followed by phagemid display selection. In the case of the ligand fluorescein, which was chosen as a well-known immunological hapten, several variants with high specificity for this compound and dissociation constants as low as 35.2 nanomolar were identified. Following X-ray analysis of the complex between the corresponding artificial binding protein and its cognate ligand (Korndörfer et al. 2003a), improved variants with K_D values for fluorescein around 1 nanomolar were constructed just by optimizing two side-chains in the binding pocket (Vopel et al. 2005). Thus, it has been demonstrated that engineered lipocalins with novel specificities – so-called "anticalins" – can provide hapten-binding proteins with affinities as they are typical for antibodies. Notably, the BBP variants recognize fluorescein or other small molecule targets as true haptens, without measurable context dependence concerning the carrier protein that was employed during selection. With their capability to provide deep and highly complementary ligand pockets anticalins distinguish themselves from most other scaffolds currently under investigation.

From the BBP mutant library an anticalin with specificity for the cardiac steroid digoxigenin was also selected (Schlehuber et al. 2000). Its initially rather moderate affinity was subsequently raised by selective random mutagenesis of the first hypervariable loop, followed by phagemid display and colony screening, resulting in a 10-fold lower K_D value of 30.2 nanomolar. Attempts to improve the affinity for digoxigenin even further were made with a combinatorial approach using the previously employed "loop-walking" randomization strategy (Schlehuber and Skerra 2002), and also by rational protein design based on the crystal structure of the anticalin (Korndörfer et al. 2003b). These approaches allowed the identification of several point mutations that led to K_D values as low as 800 picomolar for digoxin (i.e. the natural glycosylated derivative of digoxigenin) (Schlehuber and Skerra 2005).

The resulting anticalin, called Digical, may be directly suitable as a therapeutic agent for the treatment of digitalis intoxication. Although digitalis is widely applied in conjunction with heart insufficiency and arrhythmias (Hauptman and Kelly 1999), this drug has a very narrow therapeutic window and precise adjustment of digoxin plasma levels is essential to prevent intoxication with fatal outcome. When Digical was used in studies with a guinea-pig animal model of digitalis intoxication the anticalin appeared to be effective in reversing the digoxin-induced toxicity after administering just a moderate stoichiometric excess (Schlehuber and Skerra 2005). The neutralizing effect of the anticalin was likewise confirmed in preclinical studies with farmyard pigs, which more closely reflect the physiological situation in humans, thus demonstrating the acute protective effect of this anticalin on the cardiovascular system and its suitability as an antidote against digoxin.

Recently, the anticalin concept was extended in two directions. First, some human lipocalins, in particular tear lipocalin, siderocalin (also known as NGAL), and apolipoprotein D (Breustedt et al. 2006), were recruited as protein scaffolds for the generation of anticalins in order to reduce the risk of immunogenic side effects upon chronic medical treatment of patients. Second, specialized random libraries were constructed by mutagenizing residues at more exposed positions within the four hypervariable loops to yield anticalins with specificities for larger proteins (i.e. the more relevant class of targets in human therapy) instead of haptens (Schlehuber and Skerra 2005). In an initial proof of concept, an anticalin based on the ApoD scaffold was selected against hemoglobin (Vogt and Skerra 2004).

Another "human" anticalin with specificity for a protein target, based on the NGAL scaffold, was recently selected against CTLA-4 (cf. above). CTLA-4 (CD152) is known as an "immune brake" because it reverses CD28-dependent costimulation of T cells after initial activation. Hence, CTLA-4 has emerged as an attractive target for immunomodulatory drugs that can block its inhibitory function and concomitantly enhance T-cell activity, hence offering potential for cancer immunotherapy (Leach et al. 1996). Neutralizing antibodies against the extracellular domain of human CTLA-4 have been shown to be effective for cancer treatment in several preclinical and clinical studies (Keler et al. 2003). However, an Fc region is not required for the antagonistic function, and in fact it may even lead to undesired side effects. Therefore, an anticalin with high affinity, in the single-digit nanomolar range, was successfully selected against the extracellular region of CTLA-4. The CTLA-4-specific anticalin recognizes the intact target receptor protein on cells, both in immunohistochemistry and in fluorocytometry, and it was shown to bind in an antagonistic manner with the natural counter-receptors B7.1 and B7.2 (Schlehuber and Skerra 2005). Most importantly, it effectively blocks CTLA-4 in cell culture assays, thus providing a promising biopharmaceutical drug candidate for cancer therapy.

7.5
Scaffold Proteins for Display of Individual Extended Loops

Probably the simplest approach for the generation of alternative binding proteins is the modification of a single pre-existing exposed loop that is presented by a stable scaffold. Thus, the peptide loop acquires new binding properties and maintains conformationally fixed on the protein surface. Because of their typically small size and robust nature protease inhibitors provide attractive candidates in this respect. Their natural protease-binding site is mostly formed by a peptide loop of varying length and sequence, which predestines them to insert there novel peptide segments and thus create artificial activities.

Kunitz-type protease inhibitors are small α/β-proteins of about 60 amino acids with few secondary structure elements but stabilized by three disulfide bonds. They naturally act as slow but tight-binding, reversible inhibitors of serine proteases. Bovine pancreatic trypsin inhibitor (BPTI) (Roberts et al. 1992), Alzheimer's amyloid β-protein precursor inhibitor (APPI) (Dennis and Lazarus 1994), tendamistat (McConnell and Hoess 1995), human pancreatic secretory trypsin inhibitor (PSTI) (Röttgen and Collins 1995), and human lipoprotein-associated coagulation inhibitor D1 (LACI-D1) (Markland et al. 1996b) are examples of proteins that have been employed as scaffolds for the presentation of binding peptides.

The selection of novel binding proteins based on the first Kunitz domain of LACI-D1 from a corresponding phage display library yielded particularly promising results. Using procedures of iterative variegation it was possible to select potent inhibitors of human plasma kallikrein (pKAL), a serine protease that is an important mediator in the pathophysiology of hereditary angioedema (HAE), in two steps: randomization of five positions in the P1 loop of LACI-D1, followed by randomization of another four residues in a neighboring second loop, yielded highly pKAL-specific variants with inhibitory constants (K_i) in the picomolar range, which also exhibit excellent stability (Markland et al. 1996a).

Meanwhile, this approach has led to a first drug candidate (Williams and Baird 2003), DX-88, with potent and selective inhibitory activity towards human plasma kallikrein. In a phase II clinical trial DX-88 provided substantial therapeutic benefit in HAE patients who had experienced acute attacks, including life-threatening laryngeal attacks, irrespective of whether administered through the intravenous or subcutaneous route. Clinical response, defined as the beginning of improvement of HAE symptoms within four hours of dosing with DX-88, was observed at all dose levels. For the intravenous dosing ($5\,\mathrm{mg\,m^{-2}}$, $10\,\mathrm{mg\,m^{-2}}$, $20\,\mathrm{mg\,m^{-2}}$), the response rates ranged from 86 to 100%, while the subcutaneous outcome (30 mg fixed dose) showed a 100% response rate. A corresponding clinical phase III trial is ongoing (http://www.dyax.com).

A second drug candidate derived from a Kunitz-type domain scaffold, DX-890 or EPI-hNE4, is an engineered inhibitor for human neutrophil elastase (hNE). EPI-hNE4 was derived from the second Kunitz domain of inter-α-inhibitor protein (ITI-D2) and is a highly specific and potent inhibitor ($K_i = 4 \times 10^{-12}\,\mathrm{mol\,L^{-1}}$) of hNE. The protein of 56 amino acids is produced by fermentation in *Pichia pastoris* and is resistant to oxidative and proteolytic inactivation. EPI-hNE4 has completed two phase IIa trials and is currently subject to a phase IIb trial in Europe for the treatment of cystic fibrosis (http://www.debio.com). Both compounds have obtained orphan drug designation in the US and EU.

Another straightforward strategy for the construction of proteins with new binding properties is the fusion or insertion of a peptide which *per se* shows intrinsic target recognition into a permissible site. Thus, the peptide becomes incorporated into a larger carrier protein that can provide beneficial properties, for example with respect to prolonged serum half-life, and it ideally retains its pre-existing affinity and specificity for the target. This strategy can be extended such that instead of a predefined peptide being inserted, a library of sequences with varying amino acid composition and possibly length is generated in the context of the scaffold protein, followed by selection of variants with the desired target specificity.

Thioredoxin (TrxA) was employed as a first scaffold for the display of single conformationally constrained peptide sequences which were inserted into its active site loop. Naturally, TrxA is a small enzyme involved in the cytosolic thiol/disulfide equilibrium of *E. coli*. It is highly soluble, structurally rigid, can be overexpressed at elevated levels and has been exploited as a generic fusion partner for the bacterial production of recombinant proteins (LaVallie et al. 1993). In the oxidized state of the enzyme its short active site sequence Cys-Gly-Pro-Cys forms a tight, disulfide-constrained and solvent-accessible loop. This segment permits the insertion of diverse peptide sequences, whereby the enzymatic activity gets lost.

When a 20-residue random sequence was inserted into the active site loop, TrxA variants – dubbed "aptamers" – with affinity towards human cyclin-dependent kinase 2 (Cdk2) could be selected by means of the yeast two-hybrid system (Colas et al. 1996). The same selection strategy was applied towards the DNA-binding and dimerization domains of the E2F transcription factors as a target (Fabbrizio et al. 1999). In this case, the variegated segment alone appeared to remain active as a synthetic peptide with respect to blocking cell proliferation in the G_1 phase, hence suggesting that this peptide aptamer is sufficiently structured even in the absence of the thioredoxin framework. Since then several specific peptide aptamers have been selected as inhibitors of individual signaling components that are essential in cancer development and progression, thus providing potential new lead structures for drug development (Borghouts et al. 2005).

Members of the so-called "knottin family" represent small, 25- to 35-residue proteins, some of which are protease inhibitors while others function as sugar- or lipid-binding molecules (Craik et al. 2001). Even though naturally occurring knottins mutually share little sequence homology, they typically contain a small, triple-stranded antiparallel β-sheet and a cysteine-knot motif that arises from three interlocking disulfide bridges. Their small size and their high stability make them attractive as scaffold proteins for presenting an inserted peptide loop that may be highly variable both in sequence and in length. The general suitability of some members of the knottin family for the generation of specific binding proteins was demonstrated with the C-terminal cellulose-binding domain (CBD) of cellobiohydrolase I from the fungus *Trichoderma reesei* (Smith et al. 1998; Lehtio et al. 2000) and with the *Ecballium elaterium* trypsin inhibitor II (EETI-II) (Le Nguyen et al. 1990; Christmann et al. 1999; Hilpert et al. 2003; Souriau et al. 2005). CBD variants with novel binding activities were isolated from a library constructed by randomization of seven residues clustered either close to the N-terminus or close to the C-terminus of the protein. Selection via phagemid display led to the isolation of variants with specificity for alkaline phosphatase, revealing moderate K_D values in the micromolar range (Smith et al. 1998).

Certain classes of neurotoxins possessing specificities for voltage-gated ion channels represent another type of small scaffold that provides exposed loops which may be engineered to achieve novel ligand-binding activities. The structural motif of charybdotoxin, a scorpion toxin that also belongs to the knottin protein family (Norton and Pallaghy 1998), was used in order to transfer other functional sites onto its β-sheet, for instance the CDR2-like loop of human CD4, which conformationally resembles the corresponding β-hairpin in natural charybdotoxin (Vita et al. 1998). The chemically synthesized toxin variant was able to inhibit binding between soluble recombinant gp120 and a HSA-CD4 hybrid protein with an apparent IC_{50} of 20 micromolar (compared with 0.8 nanomolar for HSA-CD4 itself). More recently, the structurally related scyllatoxin, which gives higher yields and purity after chemical peptide synthesis, was employed as an alternative scaffold for CD4-mimetic peptides, which were even able to inhibit HIV-1 infection in cell culture (Dowd et al. 2002; Martin et al. 2003; Huang et al. 2005).

Another principle, originally devised for the biological delivery of peptide drugs, relies on a transport protein from blood, human serum transferrin (HST). This monomeric glycoprotein with a molecular mass of about 80 kDa was used for the functional insertion of a peptide sequence cleavable by HIV-1 protease (VSQNYPIVL) into a permissible surface loop, without altering its overall structure and physiological function (Ali et al. 1999b). HST possesses favorable intrinsic properties such as a long circulatory half-life, which makes it suitable as a therapeutic peptide drug carrier. Interestingly, the insertion of the otherwise highly immunogenic HIV-1 protease substrate peptide into the HST scaffold resulted in an attenuated antipeptide immune response in rabbits (Ali et al. 1999a), which might be an advantage for clinical applications.

An interesting extension to this approach is the use of a scaffold protein with intrinsic enzymatic activity to insert a peptide motif that confers targeting function. The class of bacterial TEM-1 β-lactamases provides extremely efficient enzymes, with high turnover numbers when catalyzing the nucleophilic cleavage of the amide bond in β-lactam antibiotics (Christensen et al. 1990). Notably, β-lactamase has no mammalian counterpart, which makes it of interest for selective prodrug activation in tumor chemotherapy (McDonagh et al. 2003; Cortez-Retamozo et al. 2004). Due to the uniqueness of the catalyzed reaction, a variety of specific substrates have been synthesized (Hakimelahi et al. 2002a,b).

Different combinatorial libraries of TEM-1 β-lactamase were constructed introducing one or two insertions of six random residues into permissible surface loops that surround the catalytic site of the enzyme. The libraries were first selected on ampicillin to remove inactive clones and then used for isolating variants with novel binding activities via phage display (Legendre et al. 1999). Variants were successfully selected for binding to monoclonal antibodies against the prostate-specific antigen (PSA) or to streptavidin. These engineered enzymes with their integrated target-binding activities were applied in homogeneous immunoassays as binding of the antigen markedly interfered with β-lactamase activity. In a following study β-lactamase variants with specificities towards ferritin and β-galactosidase could be isolated and were shown to retain nearly native enzyme activity (Legendre et al. 2002). Bifunctional proteins of this kind might prove useful to replace enzyme-conjugated antibodies in antibody-directed enzyme prodrug therapy (ADEPT), especially when further engineered for reduced immunogenicity (Harding et al. 2005).

Taken together, although the molecular diversity that may be obtained with a single peptide segment is more restricted than in the case of randomizing several loops at once, novel binding proteins with advantageous properties can be identified by this strategy. This is especially the case if the target protein offers a narrow cleft which provides access to a slim peptide motif for molecular interaction, as exemplified with the successful selection of novel inhbitors of serine proteases. In general, the presentation of a peptide by a protein scaffold offers an advantage with respect to the achievable affinity compared with the isolated peptide. While the latter usually assumes a flexible conformation in solution, fixation of the peptide backbone by the scaffold leads to a lower entropic cost upon complex formation. On the other hand, a peptide with pre-existing target affinity may lose its binding activity when presented by a protein scaffold as a consequence of sterical restriction. Indeed, the context dependence of such a peptide was recently shown when attempts were made to transfer peptide aptamers that were initially selected on the TrxA scaffold onto a different protein (e.g. to green fluorescent protein (GFP) or staphylococcal nuclease), resulting in loss of binding activity (Klevenz et al. 2002).

7.6
Scaffold Proteins Providing a Rigid Secondary Structure Interface

A different class of protein scaffolds does not make use of the loop-mediated binding mechanisms described in the previous sections. Rather, target recognition is accomplished through amino acids that are mostly situated within rigid secondary structure elements. In this case, solvent-exposed side-chains on the surface of an α-helix bundle or of a β-sheet are randomized in order to modify a pre-existed binding site or to generate an entirely novel interface for molecular recognition.

One of the first scaffold proteins investigated in this context was derived from protein Z and later dubbed "affibody." Protein Z represents the engineered domain B of the IgG-binding "Protein A" on the cell surface of *Staphylococcus aureus*, a small three-helix bundle of 58 amino acids (Uhlen et al. 1992; Nygren and Skerra 2004). Affibody libraries were generated by randomization of up to 13 exposed amino acids on the surface of the two α-helices that are naturally involved in binding to the Fc part of antibodies (Nord et al. 1995). Interestingly, this extensive randomization of secondary structure elements was possible without affecting the overall structure of the parental scaffold, even though some of the resulting variants may adopt molten globule structures in the absence of their target proteins (Wahlberg et al. 2003).

According to this strategy, protein-binding affibodies displaying micromolar affinities for *Taq* DNA polymerase, human insulin, and human apolipoprotein A-1, were selected by phage display (Nord et al. 1997) and, in one case, subsequently improved to achieve a dissociation constant in the nanomolar range (Gunneriusson et al. 1999). Affibodies were also selected against several therapeutically relevant targets, including human CD28, whereby the affibody was shown to block the interaction with the immunological counter-receptor CD80 that is involved in T-cell stimulation (Sandstrom et al. 2003). Similarly, an affibody was raised against the breast cancer target HER2/neu, exhibiting nanomolar affinity and, after radiolabeling, showing specific binding to the native receptor on HER2-expressing cells (Wikman et al. 2004). To take advantage of an avidity effect a dimeric version of this affibody was constructed, which seems to be a promising candidate for radionuclide-based detection of HER-2 expression in tumors *in vivo* (Steffen et al. 2005, 2006).

The ankyrin repeat proteins (ARPs) offer another rigid and also modular architecture that has been employed for the generation of artificial binding proteins. The ARP family comprises a variety of natural receptor proteins that are directed against other proteins, mostly in the cellular cytoplasm but also in the extracellular environment (Mosavi et al. 2004). Each of their characteristic repeats of 33 amino acids exhibits a β-turn and two antiparallel α-helices. This fold provided the basis for the generation of so-called "designed ankyrin repeat proteins" (DARPins) which contain up to three of such repeats flanked by N- and C-terminal capping modules (Binz et al. 2003).

Libraries of DARPins were generated starting from a consensus module as building block and randomizing six positions per repeat, whereby in principle a varying number of units may be employed (Binz et al. 2003; Forrer et al. 2004). Substitutions were mainly allowed within the β-turn and the first α-helix of each repeat. Using ribosome display, DARPins with nanomolar affinities and specificities for the *E. coli* maltose-binding protein as well as for the eukaryotic mitogen-activated kinases JNK2 and p38 were isolated (Binz et al. 2004). The beneficial biophysical properties of the parental ankyrin scaffold (high-level expression, solubility, and stability) were largely retained in these DARPins.

Recently, high-affinity inhibitors of aminoglycoside phosphotransferase (3')-IIIa (APH) were also isolated from a DARPin library. *In vitro* and *in vivo* assays showed complete enzyme inhibition, thus underlining the potential of DARPins for modulation of intracellular protein function (Amstutz et al. 2005). One of these inhibitors was co-crystallized with the target protein and its allosteric inhibition mechanism was elucidated (Kohl et al. 2005). Interestingly, so-called leucine-rich repeat (LRP) proteins, which share a certain structural resemblance with ARPs, were found to mediate the adaptive immune response of the sea lamprey (Pancer et al. 2004), a jawless vertebrate, indicating that protein architectures other than the immunoglobulin fold may be utilized in higher organisms for diversification and selection under natural conditions.

"Avimers" are artificial multidomain proteins derived from the so-called A-domain scaffold. This concept makes use of multiple interactions with the target protein, thus involving an avidity effect (Silverman et al. 2005). A native A-domain comprises just 35 amino acids and folds efficiently into a defined conformation that is stabilized by disulfide bond formation and complexation of calcium ions. A conserved sequence motif of merely 12 residues seems to be required to adopt this structure. Random libraries of individual A-domains were generated by taking advantage of their natural diversity, whereas positions with 90% sequence identity in an alignment of 197 homologous sequences were kept fixed. Degenerate codons at the approx. 28 variable positions were chosen to allow only amino acids that occur in natural family members.

Selection of multimers with novel binding activities was achieved by means of phage display, using selection of different binding domains which each recognize another epitope on the target protein and subsequent combination in a single fusion protein. According to this strategy avimers with subnanomolar affinities were finally obtained for IL-6, CD28, and CD40L (Silverman et al. 2005). Trimeric avimers directed against IL-6 were subjected to preclinical studies. To prolong their circulation, a domain with IgG-binding activity was added to the N-terminus, resulting in an expected half-life of around 178 h in human serum as predicted from pharmacokinetic studies in cynomolgus monkeys. A subpicomolar IC_{50} value was demonstrated for the best of these avimers in cell-based proliferation assays. Furthermore, this avimer completely abrogated acute phase protein induction by human IL-6 (but not by IL-1) in a dose-dependent manner in mice (Silverman et al. 2005).

Further to these scaffold examples, which were based on protein domains with certain natural binding functions, some proteins with rigid secondary structure that do not possess a binding activity of their own were engineered to implement molecular recognition. In this strategy, which has led to so-called "affilin" proteins, the choice of suitable frameworks was primarily driven by biotechnological aspects such as protein stability, solubility, and ease of recombinant production. For example, human γ-crystallin, a protein of 176 amino acids, folds into an overall β-sheet structure with extraordinary stability (Jaenicke and Slingsby 2001). Once deposited at high protein concentration in the eye lens during embryogenesis it fulfils its function (i.e. providing the refractory power necessary for vision) during the entire lifetime without any turnover. By randomization of eight surface-exposed amino acids on two adjacent β-strands it was possible to create a *de novo* binding site for predefined targets, for example steroid hormones (Fiedler and Rudolph 2001). Also human ubiquitin, a rather small 76-amino-acid protein that exhibits an exposed β-sheet, possesses remarkable stability against chemical and physical denaturation and has served to generate affilins in a similar approach (Fiedler et al. 2004). Affilins with binding activities in the nanomolar K_D range towards steroids and towards disease-relevant proteins were successfully generated using both protein scaffolds.

Whereas the binding sites of antibodies as well as of their surrogates that utilize flexible loops, as described in the sections above, can undergo substantial changes upon binding of a molecular target (called induced fit), artificial binding proteins that are based on secondary structure elements are likely to act more as rigid bodies. Nevertheless, many examples of such a classical lock-and-key mechanism of protein–protein interaction are known from nature and can indeed lead to very specific and tight complex formation. In fact, if both partners already present a geometry with mutual complementary, no structural adaptation is required. It is speculated (Binz et al. 2004) that such a rigid body interaction might be advantageous both for affinity, because of low entropic costs upon binding, and specificity, due to conformational restriction. However, in cases where no geometric complementarity exists in the first place the antigen itself may be forced to respond with an induced fit to complex formation, in particular if the binding protein exhibits a stiff fold. A nice example for such an inverse mechanism of conformational adaptation is seen in the crystal structure of a complex between a cognate DARPin with aminoglycoside phosphotransferase, where a whole loop is pulled out of the target protein and eventually leads to loss of enzymatic function via a conformational mechanism (Kohl et al. 2005).

7.7
Conclusions and Outlook: Therapeutic Potential and Ongoing Developments

Up to now the concept of engineering novel scaffold-based binding proteins with high medical target specificity and affinity was proven in various approaches, leading in some cases to convincing preclinical data packages or even to investigational drug candidates for clinical trials. Concerning their newly acquired binding functions, these engineered proteins resemble classical antibodies with their exquisite abilities to recognize antigens, but lack their immunological effector functions. Nevertheless, in order to gain wider acceptance as putative therapeutic agents in the future artificial binding proteins will have to fulfil not only the criteria in terms of specificity and high affinity but they also have to offer further advantages.

In this respect the worldwide limited manufacturing capacity for full-sized (humanized) antibodies and their high production cost (Werner 2004) probably provides one of the strongest arguments for the development and clinical application of simpler surrogates in order to serve the public with an innovative and beneficial class of biotherapeutics without overstraining current healthcare systems. In contrast to antibodies, most of the alternative scaffolds described so far are single-domain proteins and do not require posttranslational modification, which facilitates their production in microbial organisms at the industrial scale. When produced in *E. coli* the yields of some scaffold proteins can be in the gram per liter range at the fermenter scale. Furthermore, they may even fold properly under reducing conditions, thus permitting cytoplasmic expression in a soluble form, which might also open new therapeutic approaches by using intracellular antagonistic or inhibitory activities as an alternative to gene-based knockout techniques (Amstutz et al. 2005).

Another important point that has to be considered concerns pharmacokinetic and pharmacodynamic parameters such as clearance rate, serum stability, bioavailability, and tissue penetration. Because of their small size, most of the alternative scaffolds exhibit rapid clearance from the bloodstream, but for the same reason they show improved tissue penetration, making alternative binding proteins ideally suited for drug targeting applications. In this respect their amenability to genetic fusion with protein toxins or to the deliberate introduction of specific coupling sites for payloads such as radioisotopes or photosensitizers provides an additional benefit. However, a short serum half-life is disadvantageous if high drug levels have to be maintained for a longer period without using constant infusion. In this case, the rate of renal clearance can be slowed down by enlarging the size of the binding protein (e.g. by PEGylation) or by enabling complex formation with abundant serum proteins, such as HSA or Ig, which possess a much longer half-life.

As the appearance of neutralizing host antibodies to therapeutic proteins is an issue of general concern (Koren et al. 2002; Schellekens 2002), the immunogenicity of scaffold-based biopharmaceuticals deserves careful examination. To reduce the risk of eliciting an immune response in patients, efforts are taken to make non-human therapeutic proteins as similar to their human counterparts as possible. For antibodies, numerous examples of "humanization" by grafting rodent CDR residues onto human acceptor scaffolds, deimmunization by removing T-cell epitopes, and production of human antibodies in transgenic mice have been described (see Chapter 4, 6, 11, Vol I). On the other hand, an endogenous human scaffold protein that is employed for library construction should be less immunogenic right from the start even though novel T-cell epitopes might arise during the generation of the target-binding site. However, these may be identified and removed using bioinformatics methods (Schirle et al. 2001; Flower 2003; Bian and Hammer 2004). Nevertheless, even administration of entirely human proteins, for example insulin, can elicit antibodies without preventing clinical use. In the end, only clinical trials can give a reliable picture of a protein's behavior in humans and the next few years will certainly provide a lot of insight into the medical benefits, but possibly also the caveats, of currently emerging therapeutic antibody surrogates.

References

Agu, R.U., Ugwoke, M.I., Armand, M., Kinget, R., Verbeke, N. (2001) The lung as a route for systemic delivery of therapeutic proteins and peptides. *Respir Res* 2: 198–209.

Ali, S.A., Joao, H.C., Hammerschmid, F., Eder, J., Steinkasserer, A. (1999a) An antigenic HIV-1 peptide sequence engineered into the surface structure of transferrin does not elicit an antibody response. *FEBS Lett* 459: 230–232.

Ali, S.A., Joao, H.C., Hammerschmid, F., Eder, J., Steinkasserer, A. (1999b) Transferrin trojan horses as a rational approach for the biological delivery of therapeutic peptide domains. *J Biol Chem* 274: 24066–24073.

Amstutz, P., Binz, H.K., Parizek, P., Stumpp, M.T., Kohl, A., Grutter, M.G., Forrer, P., Plückthun, A. (2005) Intracellular kinase inhibitors selected from combinatorial libraries of designed ankyrin repeat proteins. *J Biol Chem* 280: 24715–24722.

Atwell, J.L., Breheney, K.A., Lawrence, L.J., McCoy, A.J., Kortt, A.A., Hudson, P.J. (1999) scFv multimers of the anti-neuraminidase antibody NC10: length of the linker between VH and VL domains dictates precisely the transition between diabodies and triabodies. *Protein Eng* 12: 597–604.

Baker, M. (2005) Upping the ante on antibodies. *Nat Biotechnol* 23: 1065–1072.

Bang, L.M., Keating, G.M. (2004) Adalimumab: a review of its use in rheumatoid arthritis. *BioDrugs* 18: 121–139.

Barry, B.W. (2004) Breaching the skin's barrier to drugs. *Nat Biotechnol* 22: 165–167.

Beste, G., Schmidt, F.S., Stibora, T., Skerra, A. (1999) Small antibody-like proteins with prescribed ligand specificities derived from the lipocalin fold. *Proc Natl Acad Sci USA* 96: 1898–1903.

Bian, H., Hammer, J. (2004) Discovery of promiscuous HLA-II-restricted T cell epitopes with TEPITOPE. *Methods* 34: 468–475.

Binz, H.K., Plückthun, A. (2005) Engineered proteins as specific binding reagents. *Curr Opin Biotechnol* 16: 459–469.

Binz, H.K., Stumpp, M.T., Forrer, P., Amstutz, P., Plückthun, A. (2003) Designing repeat proteins: well-expressed, soluble and stable proteins from combinatorial libraries of consensus ankyrin repeat proteins. *J Mol Biol* 332: 489–503.

Binz, H.K., Amstutz, P., Kohl, A., Stumpp, M.T., Briand, C., Forrer, P., Grutter, M.G., Plückthun, A. (2004) High-affinity binders selected from designed ankyrin repeat protein libraries. *Nat Biotechnol* 22: 575–582.

Binz, H.K., Amstutz, P., Plückthun, A. (2005) Engineering novel binding proteins from nonimmunoglobulin domains. *Nat Biotechnol* 23: 1257–1268.

Bird, R.E. and Walker, B.W. (1991) Single chain antibody variable regions. *Trends Biotechnol* 9: 132–137.

Borghouts, C., Kunz, C., Groner, B. (2005) Peptide aptamers: recent developments for cancer therapy. *Expert Opin Biol Ther* 5: 783–797.

Breustedt, D.A., Schönfeld, D.L., Skerra, A. (2006) Comparative ligand-binding analysis of ten human lipocalins. *Biochim Biophys Acta* 1764: 161–173.

Cai, X., Garen, A. (1996) A melanoma-specific VH antibody cloned from a fusion phage library of a vaccinated melanoma patient. *Proc Natl Acad Sci USA* 93: 6280–6285.

Carson, K.L. (2005) Flexibility – the guiding principle for antibody manufacturing. *Nat Biotechnol* 23: 1054–1058.

Carter, P. (2001) Improving the efficacy of antibody-based cancer therapies. *Nat Rev Cancer* 1: 118–129.

Chamow, S.M., Ashkenazi, A. (1996) Immunoadhesins: principles and applications. *Trends Biotechnol* 14: 52–60.

Christensen, H., Martin, M.T., Waley, S.G. (1990) Beta-lactamases as fully efficient enzymes. Determination of all the rate constants in the acyl-enzyme mechanism. *Biochem J* 266: 853–861.

Christmann, A., Walter, K., Wentzel, A., Kratzner, R., Kolmar, H. (1999) The cystine knot of a squash-type protease inhibitor as a structural scaffold for *Escherichia coli* cell surface display of conformationally constrained peptides. *Protein Eng* 12: 797–806.

Colas, P., Cohen, B., Jessen, T., Grishina, I., McCoy, J., Brent, R. (1996) Genetic selection of peptide aptamers that recognize and inhibit cyclin-dependent kinase 2. *Nature* 380: 548–550.

Connelly, R. (2005) Fully human domain antibody therapeutics: the best of both worlds. *Innovations Pharmac Technol* 42–45.

Conrath, K.E., Lauwereys, M., Galleni, M., Matagne, A., Frere, J.M., Kinne, J., Wyns, L., Muyldermans, S. (2001) β-Lactamase inhibitors derived from single-domain antibody fragments elicited in the Camelidae. *Antimicrob Agents Chemother* 45: 2807–2812.

Conrath, K., Vincke, C., Stijlemans, B., Schymkowitz, J., Decanniere, K., Wyns, L., Muyldermans, S., Loris, R. (2005) Antigen binding and solubility effects upon the veneering of a camel VHH in framework-2 to mimic a VH. *J Mol Biol* 350: 112–125.

Cortez-Retamozo, V., Backmann, N., Senter, P.D., Wernery, U., De Baetselier, P., Muyldermans, S., Revets, H. (2004) Efficient cancer therapy with a nanobody-based conjugate. *Cancer Res* 64: 2853–2857.

Craik, D. J., Daly, N.L., Waine, C. (2001) The cystine knot motif in toxins and implications for drug design. *Toxicon* 39: 43–60.

Davies, J., Riechmann, L. (1994) "Camelising" human antibody fragments: NMR studies on VH domains. *FEBS Lett* 339: 285–290.

De Genst, E., Saerens, D., Muyldermans, S., Conrath, K. (2006) Antibody repertoire development in camelids. *Dev Comp Immunol* 30: 187–198.

Dennis, M.S., Lazarus, R.A. (1994) Kunitz domain inhibitors of tissue factor-factor VIIa. I. Potent inhibitors selected from libraries by phage display. *J Biol Chem* 269: 22129–22136.

Dennis, M.S., Zhang, M., Meng, Y.G., Kadkhodayan, M., Kirchhofer, D., Combs, D., Damico, L.A. (2002) Albumin binding as a general strategy for improving the pharmacokinetics of proteins. *J Biol Chem* 277: 35035–35043.

Dooley, H., Flajnik, M.F., Porter, A.J. (2003) Selection and characterization of naturally occurring single-domain (IgNAR) antibody fragments from immunized sharks by phage display. *Mol Immunol* 40: 25–33.

Dowd, C.S., Leavitt, S., Babcock, G., Godillot, A.P., Van Ryk, D., Canziani, G. A., Sodroski, J., Freire, E., Chaiken, I.M. (2002) Beta-turn Phe in HIV-1 Env binding site of CD4 and CD4 mimetic miniprotein enhances Env binding affinity but is not required for activation of co-receptor/17b site. *Biochemistry* 41: 7038–7046.

Dumoulin, M., Conrath, K., Van Meirhaeghe, A., Meersman, F., Heremans, K., Frenken, L.G., Muyldermans, S., Wyns, L., Matagne, A. (2002) Single-domain antibody fragments with high conformational stability. *Protein Sci* 11: 500–515.

Editor (2005) King in the kingdom of uncertainty. *Nat Biotechnol* 23: 1025.

Evans, D., Das, R. (2005) *Monoclonal Antibodies: Evolving into a $30 Billion Market*. www.datamonitor.com.

Ewert, S., Cambillau, C., Conrath, K., Plückthun, A. (2002) Biophysical properties of camelid V(HH) domains compared to those of human V(H)3 domains. *Biochemistry* 41: 3628–3636.

Fabbrizio, E., Le Cam, L., Polanowska, J., Kaczorek, M., Lamb, N., Brent, R., Sardet, C. (1999) Inhibition of mammalian cell proliferation by genetically selected peptide aptamers that functionally antagonize E2F activity. *Oncogene* 18: 4357–4363.

Fiedler, U., Rudolph, R. (2001) Fabrication of beta-pleated sheet proteins with specific binding properties. International patent publication W00104144 (A3).

Fiedler, M., Fiedler, U., Rudolph, R. (2004) Generation of artificial binding proteins based on ubiquitin proteins. International patent publication W02004106368 (A1).

Florent, J.C., Dong, X., Gaudel, G., Mitaku, S., Monneret, C., Gesson, J.P., Jacquesy, J.C., Mondon, M., Renoux, B., Andrianomenjanahary, S., Michel, S., Koch, M., Tillequin, F., Gerken, M., Czech, J., Straub, R., Bosslet, K. (1998) Prodrugs of anthracyclines for use in antibody-directed enzyme prodrug therapy. *J Med Chem* 41: 3572–3581.

Flower, D.R. (1996) The lipocalin protein family: structure and function. *Biochem J* 318: 1–14.

Flower, D.R. (2003) Towards in silico prediction of immunogenic epitopes. *Trends Immunol* 24: 667–674.

Forrer, P., Binz, H.K., Stumpp, M.T., Plückthun, A. (2004) Consensus design of repeat proteins. *Chembiochem* 5: 183–189.

Gessner, J.E., Heiken, H., Tamm, A., Schmidt, R.E. (1998) The IgG Fc receptor family. *Ann Hematol* 76: 231–248.

Ghetie, V., Ward, E.S. (2002) Transcytosis and catabolism of antibody. *Immunol Res* 25: 97–113.

Glockshuber, R., Malia, M., Pfitzinger, I., Plückthun, A. (1990) A comparison of strategies to stabilize immunoglobulin F_v-fragments. *Biochemistry* 29: 1362–1367.

Goldberg, M., Gomez-Orellana, I. (2003) Challenges for the oral delivery of macromolecules. *Nat Rev Drug Discov* 2: 289–295.

Groves, M.A., Osbourn, J.K. (2005) Applications of ribosome display to antibody drug discovery. *Expert Opin Biol Ther* 5: 125–135.

Gunneriusson, E., Nord, K., Uhlen, M., Nygren, P. (1999) Affinity maturation of a Taq DNA polymerase specific affibody by helix shuffling. *Protein Eng* 12: 873–878.

Hakimelahi, G.H., Shia, K.S., Pasdar, M., Hakimelahi, S., Khalafi-Nezhad, A., Soltani, M.N., Mei, N.W., Mei, H.C., Saboury, A.A., Rezaei-Tavirani, M., Moosavi-Movahedi, A.A. (2002a) Design, synthesis, and biological evaluation of a cephalosporin-monohydroguaiaretic acid prodrug activated by a monoclonal antibody-beta-lactamase conjugate. *Bioorg Med Chem* 10: 2927–2932.

Hakimelahi, G.H., Shia, K.S., Xue, C., Hakimelahi, S., Moosavi-Movahedi, A.A., Saboury, A.A., Khalafi-Nezhad, A., Soltani-Rad, M.N., Osyetrov, V., Wang, K., Liao, J.H., Luo, F.T. (2002b) Design, synthesis, and biological evaluation of a series of beta-lactam-based prodrugs. *Bioorg Med Chem* 10: 3489–3498.

Hamers-Casterman, C., Atarhouch, T., Muyldermans, S., Robinson, G., Hamers, C., Songa, E.B., Bendahman, N., Hamers, R. (1993) Naturally occurring antibodies devoid of light chains. *Nature* 363: 446–448.

Harding, F.A., Liu, A.D., Stickler, M., Razo, O.J., Chin, R., Faravashi, N., Viola, W., Graycar, T., Yeung, V.P., Aehle, W., Meijer, D., Wong, S., Rashid, M.H., Valdes, A.M., Schellenberger, V. (2005) A beta-lactamase with reduced immunogenicity for the targeted delivery of chemotherapeutics using antibody-directed enzyme prodrug therapy. *Mol Cancer Ther* 4: 1791–1800.

Harris, J.M., Chess, R.B. (2003) Effect of pegylation on pharmaceuticals. *Nat Rev Drug Discov* 2: 214–221.

Hauptman, P.J., Kelly, R.A. (1999) Digitalis. *Circulation* 99: 1265–1270.

Hey, T., Fiedler, E., Rudolph, R., Fiedler, M. (2005) Artificial, non-antibody binding proteins for pharmaceutical and industrial applications. *Trends Biotechnol* 23: 514–522.

Hilpert, K., Wessner, H., Schneider-Mergener, J., Welfle, K., Misselwitz, R., Welfle, H., Hocke, A.C., Hippenstiel, S., Höhne, W. (2003) Design and characterization of a hybrid miniprotein that specifically inhibits porcine pancreatic elastase. *J Biol Chem* 278: 24986–24993.

Holliger, P., Hudson, P.J. (2005) Engineered antibody fragments and the rise of single domains. *Nat Biotechnol* 23: 1126–1136.

Holt, L.J., Herring, C., Jespers, L.S., Woolven, B.P., Tomlinson, I.M. (2003) Domain antibodies: proteins for therapy. *Trends Biotechnol* 21: 484–490.

Hosse, R.J., Rothe, A., Power, B.E. (2006) A new generation of protein display scaffolds for molecular recognition. *Protein Sci* 15: 14–27.

Huang, C.C., Stricher, F., Martin, L., Decker, J.M., Majeed, S., Barthe, P., Hendrickson, W.A., Robinson, J., Roumestand, C., Sodroski, J., Wyatt, R., Shaw, G.M., Vita, C., Kwong, P.D. (2005) Scorpion-toxin mimics of CD4 in complex with human immunodeficiency virus gp120 crystal structures, molecular mimicry, and neutralization breadth. *Structure* 13: 755–768.

Hufton, S.E., van Neer, N., van den Beuken, T., Desmet, J., Sablon, E., Hoogenboom, H.R. (2000) Development and application of cytotoxic T lymphocyte-associated antigen 4 as a protein scaffold for the generation of novel binding ligands. *FEBS Lett* 475: 225–231.

Irving, R.A., Coia, G., Roberts, A., Nuttall, S.D., Hudson, P.J. (2001) Ribosome display and affinity maturation: from antibodies to single V-domains and steps towards cancer therapeutics. *J Immunol Methods* 248: 31–45.

Jaenicke, R., Slingsby, C. (2001) Lens crystallins and their microbial homologs: structure, stability, and function. *Crit Rev Biochem Mol Biol* 36: 435–499.

Jeong, K.J., Mabry, R., Georgiou, G. (2005) Avimers hold their own. *Nat Biotechnol* 23: 1493–1494.

Jespers, L., Schon, O., James, L.C., Veprintsev, D., Winter, G. (2004) Crystal structure of HEL4: a soluble, refoldable human V(H) single domain with a germ-line scaffold. *J Mol Biol* 337: 893–903.

Keler, T., Halk, E., Vitale, L., O'Neill, T., Blanset, D., Lee, S., Srinivasan, M., Graziano, R.F., Davis, T., Lonberg, N., Korman, A. (2003) Activity and safety of CTLA-4 blockade combined with vaccines in cynomolgus macaques. *J Immunol* 171: 6251–6259.

Klevenz, B., Butz, K., Hoppe-Seyler, F. (2002) Peptide aptamers: exchange of the thioredoxin-A scaffold by alternative platform proteins and its influence on target protein binding. *Cell Mol Life Sci* 59 1993–1998.

Kohl, A., Amstutz, P., Parizek, P., Binz, H. K., Briand, C., Capitani, G., Forrer, P., Pluckthun, A., Grutter, M.G. (2005) Allosteric inhibition of aminoglycoside phosphotransferase by a designed ankyrin repeat protein. *Structure (Camb)* 13: 1131–1141.

Köhler, G., Milstein, C. (1975) Continuous cultures of fused cells secreting antibody of predefined specificity. *Nature* 256: 495–497.

Koide, A., Bailey, C.W., Huang, X., Koide, S. (1998) The fibronectin type III domain as a scaffold for novel binding proteins. *J Mol Biol* 284: 1141–1151.

Koren, E., Zuckerman, L.A., Mire-Sluis, A.R. (2002) Immune responses to therapeutic proteins in humans – clinical significance, assessment and prediction. *Curr Pharm Biotechnol* 3: 349–360.

Korndörfer, I.P., Beste, G., Skerra, A. (2003a) Crystallographic analysis of an "anticalin" with tailored specificity for fluorescein reveals high structural plasticity of the lipocalin loop region. *Proteins Struct Funct Genet* 53: 121–129.

Korndörfer, I.P., Schlehuber, S., Skerra, A. (2003b) Structural mechanism of specific ligand recognition by a lipocalin tailored for the complexation of digoxigenin. *J Mol Biol* 330: 385–396.

Lahteenmaki, R., Baker, M. (2004) Public biotechnology 2003 – the numbers. *Nat Biotechnol* 22: 665–670.

Lauwereys, M., Ghahroudi, M.A., Desmyter, A., Kinne, J., Hölzer, W., De Genst, E., Wyns, L., Muyldermans, S. (1998) Potent enzyme inhibitors derived from dromedary heavy-chain antibodies. *EMBO J* 17: 3512–3520.

LaVallie, E.R., DiBlasio, E.A., Kovacic, S., Grant, K.L., Schendel, P.F., McCoy, J.M. (1993) A thioredoxin gene fusion expression system that circumvents inclusion body formation in the E. coli cytoplasm. *Biotechnology (NY)* 11: 187–193.

Le Nguyen, D., Heitz, A., Chiche, L., Castro, B., Boigegrain, R.A., Favel, A., Coletti-Previero, M.A. (1990) Molecular recognition between serine proteases and new bioactive microproteins with a knotted structure. *Biochimie* 72: 431–435.

Leach, D.R., Krummel, M.F.and Allison, J.P. (1996) Enhancement of antitumor immunity by CTLA-4 blockade. *Science* 271: 1734–1736.

Legendre, D., Soumillion, P., Fastrez, J. (1999) Engineering a regulatable enzyme for homogeneous immunoassays. *Nat Biotechnol* 17: 67–72.

Legendre, D., Vucic, B., Hougardy, V., Girboux, A.L., Henrioul, C., Van Haute, J., Soumillion, P., Fastrez, J. (2002) TEM-1 beta-lactamase as a scaffold for protein recognition and assay. *Protein Sci* 11: 1506–1518.

Lehtio, J., Teeri, T.T., Nygren, P.A. (2000) Alpha-amylase inhibitors selected from a combinatorial library of a cellulose binding domain scaffold. *Proteins* 41: 316–322.

Lejon, S., Frick, I.M., Bjorck, L., Wikstrom, M., Svensson, S. (2004) Crystal structure and biological implications of a bacterial albumin binding module in complex with human serum albumin. *J Biol Chem* 279: 42924–42928.

Li, E., Pedraza, A., Bestagno, M., Mancardi, S., Sanchez, R., Burrone, O. (1997) Mammalian cell expression of dimeric small immune proteins (SIP). *Protein Eng* 10: 731–736.

Li, R., Hoess, R.H., Bennett, J.S., DeGrado, W.F. (2003) Use of phage display to probe the evolution of binding specificity and affinity in integrins. *Protein Eng* 16: 65–72.

Lowman, H.B., Wells, J.A. (1991) Monovalent phage display: a method for selecting variant proteins from random libraries. *Methods* 3 205–216.

Lu, D.L., Collison, A.M., Kowalski, T.J. (2005) The patentability of antibodies in the United States. *Nat Biotechnol* 23: 1079–1080.

Markland, W., Ley, A.C., Ladner, R.C. (1996a) Iterative optimization of high-affinity protease inhibitors using phage display. 2. Plasma kallikrein and thrombin. *Biochemistry* 35: 8058–8067.

Markland, W., Ley, A.C., Lee, S.W., Ladner, R.C. (1996b) Iterative optimization of high-affinity proteases inhibitors using phage display. 1. Plasmin. *Biochemistry* 35: 8045–8057.

Martin, L., Stricher, F., Misse, D., Sironi, F., Pugniere, M., Barthe, P., Prado-Gotor, R., Freulon, I., Magne, X., Roumestand, C., Menez, A., Lusso, P., Veas, F., Vita, C. (2003) Rational design of a CD4 mimic that inhibits HIV-1 entry and exposes cryptic neutralization epitopes. *Nat Biotechnol* 21: 71–76.

Masat, L., Wabl, M., Johnson, J.P. (1994) A simpler sort of antibody. *Proc Natl Acad Sci USA* 91: 893–896.

McConnell, S.J., Hoess, R.H. (1995) Tendamistat as a scaffold for conformationally constrained phage peptide libraries. *J Mol Biol* 250: 460–470.

McDonagh, C.F., Beam, K.S., Wu, G.J., Chen, J.H., Chace, D.F., Senter, P.D., Francisco, J.A. (2003) Improved yield and stability of L49-sFv-beta-lactamase, a single-chain antibody fusion protein for anticancer prodrug activation, by protein engineering. *Bioconjug Chem* 14: 860–869.

Mosavi, L.K., Cammett, T.J., Desrosiers, D.C., Peng, Z.Y. (2004) The ankyrin repeat as molecular architecture for protein recognition. *Protein Sci* 13: 1435–1448.

Muyldermans, S., Cambillau, C., Wyns, L. (2001) Recognition of antigens by single-domain antibody fragments: the superfluous luxury of paired domains. *Trends Biochem Sci* 26: 230–235.

Nord, K., Nilsson, J., Nilsson, B., Uhlen, M., Nygren, P.A. (1995) A combinatorial library of an alpha-helical bacterial receptor domain. *Protein Eng* 8: 601–608.

Nord, K., Gunneriusson, E., Ringdahl, J., Stahl, S., Uhlen, M., Nygren, P.A. (1997) Binding proteins selected from combinatorial libraries of an alpha-helical bacterial receptor domain. *Nat Biotechnol* 15: 772–777.

Norton, R.S., Pallaghy, P.K. (1998) The cystine knot structure of ion channel toxins and related polypeptides. *Toxicon* 36: 1573–1583.

Nuttall, S.D., Rousch, M.J., Irving, R.A., Hufton, S.E., Hoogenboom, H.R., Hudson, P.J. (1999) Design and expression of soluble CTLA-4 variable domain as a scaffold for the display of functional polypeptides. *Proteins* 36: 217–227.

Nuttall, S.D., Krishnan, U.V., Hattarki, M., De Gori, R., Irving, R.A., Hudson, P.J. (2001) Isolation of the new antigen receptor from wobbegong sharks, and use as a scaffold for the display of protein loop libraries. *Mol Immunol* 38: 313–326.

Nuttall, S.D., Krishnan, U.V., Doughty, L., Pearson, K., Ryan, M.T., Hoogenraad, N.J., Hattarki, M., Carmichael, J.A., Irving, R. A., Hudson, P.J. (2003) Isolation and characterization of an IgNAR variable domain specific for the human mitochondrial translocase receptor Tom70. *Eur J Biochem* 270: 3543–3554.

Nuttall, S.D., Humberstone, K.S., Krishnan, U.V., Carmichael, J.A., Doughty, L., Hattarki, M., Coley, A.M., Casey, J.L., Anders, R.F., Foley, M., Irving, R.A., Hudson, P.J. (2004) Selection and affinity maturation of IgNAR variable domains targeting Plasmodium falciparum AMA1. *Proteins* 55: 187–197.

Nygren, P.A., Skerra, A. (2004) Binding proteins from alternative scaffolds. *J Immunol Methods* 290: 3–28.

Osborn, B.L., Olsen, H.S., Nardelli, B., Murray, J.H., Zhou, J.X., Garcia, A., Moody, G., Zaritskaya, L.S., Sung, C. (2002) Pharmacokinetic and pharmacodynamic studies of a human serum albumin-interferon-alpha fusion protein in cynomolgus monkeys. *J Pharmacol Exp Ther* 303: 540–548.

Owens, J. (2005) Genentech calm Cabilly storm. *Nat Rev Drug Discov* 4: 876.

Pack, P., Plückthun, A. (1992) Miniantibodies: use of amphipathic helices to produce functional, flexibly linked dimeric F_V fragments with high avidity in *Escherichia coli*. *Biochemistry* 31: 1579–1584.

Padlan, E.A. (1994) Anatomy of the antibody molecule. *Mol Immunol* 31: 169–217.

Palmisano, G.L., Tazzari, P.L., Cozzi, E., Bolognesi, A., Polito, L., Seveso, M., Ancona, E., Ricci, F., Conte, R., Stirpe, F., Ferrara, G.B., Pistillo, M.P. (2004) Expression of CTLA-4 in nonhuman primate lymphocytes and its use as a potential target for specific immunotoxin-mediated apoptosis: results of in vitro studies. *Clin Exp Immunol* 135: 259–266.

Pancer, Z., Amemiya, C.T., Ehrhardt, G.R., Ceitlin, J., Gartland, G.L., Cooper, M.D. (2004) Somatic diversification of variable lymphocyte receptors in the agnathan sea lamprey. *Nature* 430: 174–180.

Parker, M.H., Chen, Y., Danehy, F., Dufu, K., Ekstrom, J., Getmanova, E., Gokemeijer, J., Xu, L., Lipovsek, D. (2005) Antibody mimics based on human fibronectin type three domain engineered for thermostability and high-affinity binding to vascular endothelial growth factor receptor two. *Protein Eng Des Sel* 18: 435–444.

Pastan, I. (2003) Immunotoxins containing Pseudomonas exotoxin A: a short history. *Cancer Immunol Immunother* 52: 338–341.

Pini, A., Ricci, C., Bracci, L. (2002) Phage display and colony filter screening for high-throughput selection of antibody libraries. *Comb Chem High Throughput Screen* 5: 503–510.

Pokkuluri, P.R., Huang, D.B., Raffen, R., Cai, X., Johnson, G., Stevens, P.W., Stevens, F.J., Schiffer, M. (1998) A domain flip as a result of a single amino-acid substitution. *Structure* 6: 1067–1073.

Powell, K. (2004) Inhaled insulin products puff along. *Nat Biotechnol* 22: 1195–1196.

Reiter, Y., Schuck, P., Boyd, L.F., Plaksin, D. (1999) An antibody single-domain phage display library of a native heavy chain variable region: isolation of functional single-domain VH molecules with a unique interface. *J Mol Biol* 290: 685–698.

Roberts, B.L., Markland, W., Ley, A.C., Kent, R.B., White, D.W., Guterman, S.K., Ladner, R.C. (1992) Directed evolution of a protein: selection of potent neutrophil elastase inhibitors displayed on M13 fusion phage. *Proc Natl Acad Sci USA* 89: 2429–2433.

Rosendahl, M.S., Doherty, D.H., Smith, D.J., Bendele, A.M., Cox, G.N. (2005) Site-specific protein PEGylation – Application to cysteine analogs of recombinant human granulocyte colony-stimulating factor. *BioProcess Int* 52–60.

Röttgen, P., Collins, J. (1995) A human pancreatic secretory trypsin inhibitor presenting a hypervariable highly constrained epitope via monovalent phagemid display. *Gene* 164: 243–250.

Saerens, D., Frederix, F., Reekmans, G., Conrath, K., Jans, K., Brys, L., Huang, L., Bosmans, E., Maes, G., Borghs, G., Muyldermans, S. (2005) Engineering camel single-domain antibodies and immobilization chemistry for human prostate-specific antigen sensing. *Anal Chem* 77: 7547–7555.

Sandstrom, K., Xu, Z., Forsberg, G., Nygren, P.A. (2003) Inhibition of the CD28-CD80 co-stimulation signal by a CD28-binding affibody ligand developed by combinatorial protein engineering. *Protein Eng* 16: 691–697.

Schellekens, H. (2002) Immunogenicity of therapeutic proteins: clinical implications and future prospects. *Clin Ther* 24: 1720–1740; discussion 1719.

Schirle, M., Weinschenk, T., Stevanovic, S. (2001) Combining computer algorithms with experimental approaches permits the rapid and accurate identification of T cell epitopes from defined antigens. *J Immunol Methods* 257: 1–16.

Schlehuber, S., Skerra, A. (2001) Duocalins: engineered ligand-binding proteins with dual specificity derived from the lipocalin fold. *Biol Chem* 382: 1335–1342.

Schlehuber, S., Skerra, A. (2002) Tuning ligand affinity, specificity, and folding stability of an engineered lipocalin variant – a so-called "anticalin" – using a molecular random approach. *Biophys Chem* 96: 213–228.

Schlehuber, S., Skerra, A. (2005) Lipocalins in drug discovery: from natural ligand-binding proteins to "anticalins". *Drug Discov Today* 10: 23–33.

Schlehuber, S., Beste, G., Skerra, A. (2000) A novel type of receptor protein, based on the lipocalin scaffold, with specificity for digoxigenin. *J Mol Biol* 297: 1105–1120.

Shao, C.Y., Secombes, C.J., Porter, A.J. (2006) Rapid isolation of IgNAR variable single-domain antibody fragments from a shark synthetic library. *Mol Immunol* 44: 656–665.

Silverman, J., Lu, Q., Bakker, A., To, W., Duguay, A., Alba, B.M., Smith, R., Rivas, A., Li, P., Le, H., Whitehorn, E., Moore, K.W., Swimmer, C., Perlroth, V., Vogt, M., Kolkman, J., Stemmer, W.P. (2005) Multivalent avimer proteins evolved by exon shuffling of a family of human receptor domains. *Nat Biotechnol* 23: 1556–1561.

Skerra, A. (2000a) Engineered protein scaffolds for molecular recognition. *J Mol Recognit* 13: 167–187.

Skerra, A. (2000b) Lipocalins as a scaffold. *Biochim Biophys Acta* 1482: 337–350.

Skerra, A. (2003) Imitating the humoral immune response. *Curr Opin Chem Biol* 7: 683–693.

Skerra, A., Plückthun, A. (1988) Assembly of a functional immunoglobulin Fv fragment in Escherichia coli. *Science* 240: 1038–1041.

Smith, G.P., Patel, S.U., Windass, J.D., Thornton, J.M., Winter, G., Griffiths, A.D. (1998) Small binding proteins selected from a combinatorial repertoire of knottins displayed on phage. *J Mol Biol* 277: 317–332.

Souriau, C., Chiche, L., Irving, R., Hudson, P. (2005) New binding specificities derived from Min-23: a small cystine-stabilized peptidic scaffold. *Biochemistry* 44: 7143–7155.

Spinelli, S., Tegoni, M., Frenken, L., van Vliet, C., Cambillau, C. (2001) Lateral recognition of a dye hapten by a llama VHH domain. *J Mol Biol* 311: 123–129.

Steffen, A.C., Orlova, A., Wikman, M., Nilsson, F.Y., Stahl, S., Adams, G.P., Tolmachev, V., Carlsson, J. (2006) Affibody-mediated tumour targeting of HER-2 expressing xenografts in mice. *Eur J Nucl Med Mol Imaging* 33: 631–638.

Steffen, A.C., Wikman, M., Tolmachev, V., Adams, G.P., Nilsson, F.Y., Stahl, S., Carlsson, J. (2005) In vitro characterization of a bivalent anti-HER-2 affibody with potential for radionuclide-based diagnostics. *Cancer Biother Radiopharm* 20: 239–248.

Stevens, F.J., Solomon, A., Schiffer, M. (1991) Bence Jones proteins: a powerful tool for the fundamental study of protein chemistry and pathophysiology. *Biochemistry* 30: 6803–6805.

Streltsov, V.A., Varghese, J.N., Carmichael, J.A., Irving, R.A., Hudson, P.J., Nuttall, S.D. (2004) Structural evidence for evolution of shark Ig new antigen receptor variable domain antibodies from a cell-surface receptor. *Proc Natl Acad Sci USA* 101: 12444–12449.

Tudyka, T., Skerra, A. (1997) Glutathione S-transferase can be used as a C-terminal, enzymatically active dimerization module for a recombinant protease inhibitor, and functionally secreted into the periplasm of *Escherichia coli*. *Protein Sci* 6: 2180–2187.

Uhlen, M., Forsberg, G., Moks, T., Hartmanis, M., Nilsson, B. (1992) Fusion proteins in biotechnology. *Curr Opin Biotechnol* 3: 363–369.

Van Brunt, J. (2005) The monoclonal maze. *SIGNALS* www.signalsmag.com.

Vita, C., Vizzavona, J., Drakopoulou, E., Zinn-Justin, S., Gilquin, B., Menez, A. (1998) Novel miniproteins engineered by the transfer of active sites to small natural scaffolds. *Biopolymers* 47: 93–100.

Vogt, M., Skerra, A. (2004) Construction of an artificial receptor protein ("anticalin") based on the human apolipoprotein D. *Chembiochem* 5: 191–199.

Vopel, S., Mühlbach, H., Skerra, A. (2005) Rational engineering of a fluorescein-binding anticalin for improved ligand affinity. *Biol Chem* 386: 1097–1104.

Vranken, W., Tolkatchev, D., Xu, P., Tanha, J., Chen, Z., Narang, S., Ni, F. (2002) Solution structure of a llama single-domain antibody with hydrophobic residues typical of the VH/VL interface. *Biochemistry* 41: 8570–8579.

Wahlberg, E., Lendel, C., Helgstrand, M., Allard, P., Dincbas-Renqvist, V., Hedqvist, A., Berglund, H., Nygren, P.A., Hard, T. (2003) An affibody in complex with a target protein: structure and coupled folding. *Proc Natl Acad Sci USA* 100: 3185–3190.

Werner, R.G. (2004) Economic aspects of commercial manufacture of biopharmaceuticals. *J Biotechnol* 113: 171–182.

Wikman, M., Steffen, A.C., Gunneriusson, E., Tolmachev, V., Adams, G.P., Carlsson, J., Stahl, S. (2004) Selection and characterization of HER2/neu-binding affibody ligands. *Protein Eng Des Sel* 17: 455–462.

Williams, A., Baird, L.G. (2003) DX-88 and HAE: a developmental perspective. *Transfus Apher Sci* 29: 255–258.

Wörn, A., Plückthun, A. (2001) Stability engineering of antibody single-chain Fv fragments. *J Mol Biol* 305: 989–1010.

Xu, L., Aha, P., Gu, K., Kuimelis, R.G., Kurz, M., Lam, T., Lim, A.C., Liu, H., Lohse, P.A., Sun, L., Weng, S., Wagner, R.W., Lipovsek, D. (2002) Directed evolution of high-affinity antibody mimics using mRNA display. *Chem Biol* 9: 933–942.

Yau, K.Y., Groves, M.A., Li, S., Sheedy, C., Lee, H., Tanha, J., MacKenzie, C.R., Jermutus, L., Hall, J.C. (2003) Selection of hapten-specific single-domain antibodies from a non-immunized llama ribosome display library. *J Immunol Methods* 281: 161–175.

Zewe, M., Rybak, S.M., Dubel, S., Coy, J.F., Welschof, M., Newton, D.L., Little, M. (1997) Cloning and cytotoxicity of a human pancreatic RNase immunofusion. *Immunotechnology* 3: 127–136.

8
Emerging Therapeutic Concepts I: ADEPT

Surinder K. Sharma, Kerry A. Chester, and Kenneth D. Bagshawe

8.1
Introduction and Basic Principles of ADEPT

Although the production of antisera in various species was well underway in the 1930s it was not until the late 1950s that the idea of using these as a diagnostic tool took root. Following the discovery of carcinoembryonic antigen (CEA) (Gold and Freedman 1965), pathologists began to use antisera to identify antigens on tissue sections. If antibodies could bind to antigens on tissue sections and to antigens in solution the question arose whether, when injected intravenously, they would localize for instance on CEA expressed on colorectal cancers. That this could be achieved was demonstrated by Mach et al. (1974) and Goldenberg et al. (1974). The advent of hybridoma technology followed by the introduction of monoclonal antibodies (Köhler and Milstein 1975) focused attention on their enormous potential in diagnosis and therapy.

So could antibodies deliver cytotoxic agents with less toxicity than by conventional means? Several approaches had been developed to target cytotoxic agents (Levy et al. 1975; Pimm et al. 1982) and toxins (Thorpe et al. 1978; Gilliard et al. 1980; Pastan and Kreitman 2002) by linking to monoclonal antibodies. One of the limitations of this antibody–drug conjugate approach was that an intact antibody could only carry about 10–15 molecules of a cytotoxic drug without losing its antigen-binding ability. Moreover large molecules could gain entry to tumor sites but diffused poorly (Pedley et al. 1993) so that they failed to deliver their pay load where it was wanted. This failure was compounded by the fact that in carcinomas, as opposed to the lymphomas, there is marked heterogeneity in the distribution of the known antigenic targets so that many viable tumor cells fail to express the target antigen (Edwards et al. 1985).

It was recognized that if cytotoxic agents could be generated and restricted to cancer sites this would have the potential to deliver more drug to tumors and avoid normal tissue toxicity. The idea of relatively nontoxic prodrugs that would be converted by tumor-located enzymes into potent cell killing agents had been explored for several years but the failure to find enzymes exclusively located in

Handbook of Therapeutic Antibodies. Edited by Stefan Dübel
Copyright © 2007 WILEY-VCH Verlag GmbH & Co. KGaA, Weinheim
ISBN 978-3-527-31453-9

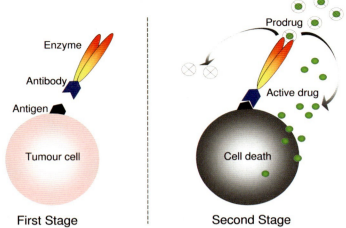

Fig. 8.1 Basic principle of ADEPT. An antibody–enzyme construct (AEC) is injected and allowed to localize to tumors. After blood clearance of enzyme activity, a nontoxic prodrug is given. The enzyme cleaves the protective moiety to release the cytotoxic active drug. The drug is generated extracellularly and can diffuse throughout the tumor killing both antigen-positive as well as antigen-negative cells, thus providing a "bystander" effect.

cancer cells had proved the stumbling block. So it was suggested that antibodies directed at tumor antigens could be used to vector enzymes not found in human tissues to tumor sites where they could activate appropriate prodrugs (Bagshawe 1987). This formed the basis of what is described as ADEPT, or Antibody Directed Enzyme Prodrug Therapy (Bagshawe et al. 1988; Senter et al. 1988) (Fig. 8.1).

It was clear from the outset that there would be several obstacles to overcome. It was recognized that if foreign enzymes were used, and those of bacterial origin seemed the most likely, there would be an immunogenicity issue that would have to be addressed. The same consideration also applied to the murine monoclonals that were then the norm (Reilly et al. 1995), although there were good reasons to believe that human or humanized antibodies would resolve that issue and so it has proved (Jones et al. 2004; Winter and Harris 1993). Second, it was known from work with radiolabeled intact IgG antibodies in humans that they remain in the blood for many days and it would not be useful to give prodrug whilst antibody–enzyme conjugate (AEC) was still in the blood since to do so would simply mimic conventional therapy. Similarly, it would be necessary for the drug generated at tumor sites to have a short half-life so that it did not leak out of tumors on a scale big enough to produce systemic toxicity.

Although it was evident that not all these requirements could be achieved quickly, the approach had big potential advantages. Each enzyme molecule located within a tumor mass would be able to activate a large number of prodrug molecules, thereby providing an amplification factor. The drugs generated would be small molecules that would diffuse through the tumor mass more readily than the AEC. The drug molecules would be able to attack not only cells expressing the target antigen but also those that failed to express the antigen, the so-called bystander effect.

8.2 Preclinical Studies

8.2.1 CPG2 and Benzoic Mustard Prodrugs

It seemed important to test the system as soon as possible. In our studies, we have utilized a bacterial enzyme carboxypeptidase G2 (CPG2), isolated from a *Pseudomonas* species (Sherwood et al. 1985). At the Trophoblastic Disease Centre at Charing Cross Hospital, London, one of the tumor cell lines under regular study was a choriocarcinoma xenograft that was resistant to the cytotoxic agents (Bagshawe 1989) available when used singly or in combination. The serum of nude mice bearing these xenografts contained quite high levels of human chorionic gonadotrophin (hCG). This model was selected for testing the ADEPT approach *in vivo* (Bagshawe et al. 1988). Nude mice with established choriocarcinoma xenografts (CC3) were given an AEC comprising a murine monoclonal IgG antibody directed at hCG chemically conjugated (Searle et al. 1986) to the bacterial enzyme carboxypeptidase (CPG2). It was known from previous studies (Begent et al. 1986) that antibodies could localize in the tumors despite the high level of hCG in blood. In the CC3 xenograft model, a prodrug was given at 48 h after the AEC when there was no detectable enzyme in the blood. This resulted in the elimination of 9 out of 12 tumors without regrowth of these tumors followed up to 1 year (Springer et al. 1991).

This first custom-made prodrug was glutamated benzoic acid mustard (Springer et al. 1990), which was activated by cleavage of the glutamate by the CPG2. Its most immediate limitation was poor aqueous solubility so that it could only be administered in dimethyl sulfoxide (DMSO).

The next xenograft target was a poorly differentiated colon cancer model LS174T. Although it expressed CEA at the cellular level there was no detectable CEA in the blood of xenograft-bearing mice. Mice bearing established xenografts received antibody directed at CEA conjugated to CPG2 followed at 72 h by the prodrug. All the mice died. It was evident that in the CC3 experiment the AEC had cleared quickly from the blood through immune complex formation with the hCG in the blood and that in the LS174T model there was still enzyme in blood at 72 h, which activated the prodrug and caused fatal toxicity. The experiment was repeated with a time interval of 7 days between giving the AEC and the prodrug. This time there was no toxicity but little tumor response.

These results had been anticipated and the use of second antibodies directed at another antibody had been used in radioimmunoassay for many years. One of the antienzyme antibodies produced (SB43) inactivated the enzyme (Sharma et al. 1990) and this was used to clear enzyme activity from blood in the next experiment and marked growth delay was then obtained with the LS174T xenografts (Sharma et al. 1991). To prevent this antibody inactivating enzyme at tumor sites it was galactosylated to ensure its rapid clearance. Hence a three-component system was developed: (1) the AEC, (2) the galactosylated enzyme-inactivating antibody, and (3) the prodrug (Sharma et al. 1991, 1994a).

Another approach to accelerate clearance of CPG2 from blood was also explored. This involved galactosylation of the AEC so that it cleared from blood and tissues via the carbohydrate receptors in the liver. Localization in tumors was achieved by blocking carbohydrate receptors in the liver for a period of time using an asialo-bovine submaxillary gland mucin. As this blocking agent cleared from the liver, AEC rapidly cleared from blood and other tissues but was retained in the tumor, allowing prodrug to be given safely in the LS174T xenograft model and resulting in a growth delay that was comparable with that achieved in the three-component ADEPT system (Sharma et al. 1994a). However, this approach was not used clinically as the blocking agent was likely to be immunogenic and its safety and side effects were not known. The three-component system subsequently studied in the clinic utilized the galactosylated SB43 as it had the advantage of inactivating CPG2 in blood within minutes without affecting tumor enzyme levels and without immune complex toxicity. ADEPT studies with the same antibody and enzyme/prodrug system resulted in growth delay of a drug resistant ovarian xenograft model (Sharma et al. 1994b) and using a different antibody produced regressions and cures in a human breast xenograft model (Eccles et al. 1994).

8.2.2
Other Enzyme/Prodrug Systems

A number of preclinical studies of ADEPT have been reported since the ADEPT approach was first proposed (Bagshawe et al. 1988; Senter et al. 1988). These include enzymes of mammalian and nonmammalian origin in combination with prodrugs of current chemotherapy agents (reviewed in Senter and Springer 2001; Bagshawe et al. 2004). The mammalian enzymes, including human, studied were alkaline phosphatase (Senter et al. 1988), carboxypeptidase A (Smith et al. 1997), and β-glucuronidase (Bosslet et al. 1994). The nonmammalian enzymes with mammalian homolog include bacterial nitroreductase (Mauger et al. 1994) and β-glucuronidase (Bignami et al. 1992), which has a substrate specificity different from that of its human counterpart (Houba et al. 1996). Both these are potentially immunogenic in humans. The nonmammalian enzymes with no human analog include CPG2 (described earlier), β-lactamase (Alexander et al. 1991; Meyer et al. 1993, 1995; Cortez-Retamozo et al. 2004; Roberge et al. 2006), cytosine deaminase (Wallace et al. 1994), and penicillin G-amidase (Bignami et al. 1992). Virtually all of these studies have used prodrug versions of existing, approved cytotoxic agents. Many of these studies give no data on the half-life of the drugs and the importance of this characteristic is not widely appreciated.

Various subsequent studies have reported the use of human enzymes but none have yet been reported in clinical trials.

8.2.3
Catalytic Antibodies

The first catalytic antibodies (or abzymes) were made by immunizing rabbits, and later mice, with transition state analogs (Nevinsky and Buneva 2003; Xu et al. 2004). The resulting antibodies acted as catalysts by interacting to stabilize the transition state in a similar manner to enzymes. The potential of using these antibodies to catalyze a prodrug had been anticipated in the early days of ADEPT (Bagshawe 1989). Although progress in this field has been initially relatively slow and the catalytic antibodies generated remained less efficient than naturally occurring enzymes (Wentworth et al. 1996; Shabat et al. 1999). However, recent advances in the development of efficient catalytic antibodies (Rader et al. 2003; Shamis et al. 2004; Sinha et al. 2004) are more encouraging and catalytic antibodies are now being employed in enzyme–prodrug therapy of cancer (Kakinuma et al. 2002), for example in neuroblastoma where *in vivo* tumor growth was delayed and there did not appear to be conversion of prodrug by endogenous enzymes (Shabat et al. 2001).

Indeed, the potential of current antibody technology to create human catalytic antibodies is enormous. Active human catalytic antibodies can be selected from synthetic antibody libraries (Cesaro-Tadic et al. 2003) and directed evolution of enzymes can be achieved using phage display (Fernandez-Gacio et al. 2003). Thus, it is possible that bispecific antitumor/abzyme molecules could eventually provide an entirely nonimmunogenic approach for targeted cancer chemotherapy.

8.3
Clinical Studies

8.3.1
F(ab)$_2$ Fragments Conjugated to CPG2

Following our extensive preclinical studies in xenograft models, a pilot clinical study in patients with advanced metastatic colorectal cancer followed. It was necessary in this exploratory study to incorporate prodrug dose escalation and exploratory dose levels of the AEC and the enzyme-inactivating antibody. A total of 17 patients entered the study and of the eight that received the highest doses of prodrug four achieved partial responses and one a mixed response (Bagshawe et al. 1991, 1995; Bagshawe and Begent 1996).

A second, small-scale clinical study with the same three-component system followed but with a lower dose of the AEC. This was an important mechanistic study (Napier et al. 2000) which resulted in one partial response and most of the other patients had stable disease for several months. An important result of this study was that biopsies of liver metastases, after the inactivating antibody had been given, showed tumor-to-blood ratios far in excess of 10000:1, demonstrating the efficiency of the inactivating antibody and confirming that the myelosuppressive effects seen in both clinical studies resulted not from enzyme in blood but from the long half-life of the drug generated at tumor sites (Martin et al. 1997).

In response to these findings a bis-iodo phenol mustard prodrug was developed (Springer et al. 1995). This prodrug is also an alkylating agent and is activated by CPG2 to generate a drug with very short half-life. In combination with AEC this was shown to be highly potent *in vivo* (Blakey et al. 1996) and the mechanism of cell death to be mainly apoptosis (Monks et al. 2001). This bis-iodo phenol mustard prodrug in combination with AEC was used in the next clinical study but the enzyme inactivation step was omitted in the cause of simplicity. There were no responses in any of the 28 patients entered (Francis et al. 2002).

8.3.2
Recombinant scFv-CPG2 Fusion Protein

The clinical studies highlighted a need for better enzyme delivery systems and at this time advances in molecular manipulation offered the opportunity to end the limitations of chemical conjugation and to be able to build fusion proteins of antibody and enzyme with characteristics to order. A new anti-CEA single-chain Fv (scFv) antibody (MFE23) with higher binding affinity than the antibody used previously had been developed (Chester et al. 2000). The gene for MFE was fused with the gene for CPG2 to result in a fusion protein MFE23-CPG2 (Michael et al. 1996). Expressed in *E. coli* this showed improved pharmacokinetics *in vivo* as compared with the AEC (Bhatia et al. 2000) but the yield was low. The subsequent expression of this fusion protein (MFECP) in *Pichia pastoris* resulted in high yield of a stable protein which was also glycosylated (Medzihradszky et al. 2004). MFECP localized effectively in colon carcinoma xenografts with rapid clearance of enzyme activity in blood, resulting in tumor-to-plasma ratios of 1400:1 within 6h after injection. Therapy studies in combination with the bis-iodo phenol mustard prodrug showed growth delay of the LS174T and regressions in the SW1222 colon carcinoma xenografts without apparent toxicity (Sharma et al. 2005).

A phase I clinical trial with a single cycle of ADEPT comprising MFECP1 and the bis-iodo phenol prodrug has been completed successfully (Mayer et al. 2004a; Mayer and Begent, unpublished data) and a repeated cycle ADEPT study is in progress.

8.4
Immunogenicity

Successful cancer therapy requires multiple cycles of treatment and one of the limitations of the ADEPT approach using a murine monoclonal antibody conjugated to a bacterial enzyme was that of immunogenicity and as such was anticipated from the first (Bagshawe 1989). All patients receiving AEC showed the presence of human anti-mouse antibodies (HAMA) and human anti-CPG2 antibodies (HACA) (Sharma et al. 1992). However, using an immunosuppressive agent, up to three ADEPT cycles could be given within a 21-day period (Bagshawe et al. 1995; Bagshawe and Sharma 1996).

The clinical studies with MFECP1 have shown a lower frequency of HACA than with AEC (Mayer et al. 2004b). This may be due to the addition of a hexa-histidine tag (His-tag) to the C-terminus of CPG2 where an immunodominant B-cell epitope had been identified (Spencer et al. 2002).

The B-cell modification approach to reduce immunogenicity may be limited in that removing antigenic epitopes may not necessarily reduce overall immunogenicity because repeated administration with the modified protein can elicit an antibody response to a different set of epitopes on the same molecule. Another approach, modification of T-cell epitopes, may be more successful in creating nonimmunogenic enzymes for ADEPT, because T cell help is required to mount a long-lived, isotype-switched, and high-affinity antibody response (see review, Chester et al. 2005).

Enzymes present a challenge for the T-cell epitope modification strategy, because changes in amino acid sequence can readily lead to loss of catalytic activity. However, the potential effectiveness of this approach makes it very attractive and it has already met with (*in vitro*) success with beta-lactamase from *Enterobacter cloacae* (Harding et al. 2005). Lactamases are useful enzymes for ADEPT as they can activate a wide variety of prodrugs, but they are immunogenic due to their bacterial origin. Harding et al. mutated T-cell epitopes in beta-lactamase to create a variant which retained enzyme activity but which induced significantly less T-cell stimulation in human and mouse cell assays. The results are promising although it remains to be seen to what extent its immunogenicity has been overcome in humans.

The possibility of removing T-cell epitopes from CPG2 is also being explored. Eight potential immunogenic regions have been identified using T-cell proliferation assays and *in silico* analysis (Chester et al. 2005). It is proposed that suitable amino acid substitutions in these regions will lead to a CPG2 molecule with reduced immunogenicity *in vivo*.

The immunological response may also be addressed by using human, rather than foreign, enzymes for ADEPT. However, this increases the risk of unwanted activation of prodrug by endogenous human enzyme in nontarget organs. One way of preventing this is to mutate human enzymes so they will activate prodrugs not recognized by their wildtype human equivalent. This has been shown to be possible in principle with a mutant of human carboxypeptidase A1 which has been modified to activate prodrugs of methotrexate *in vitro* (Wolfe et al. 1999), although as yet this system does not appear to be effective *in vivo*. Another approach is to use a human enzyme which has little or no activity in human blood, for example human beta-glucuronidase (de Graaf et al. 2002; Biela et al. 2003). An ADEPT system using a recombinant thermostable human prolyl endopeptidase is also promising (Heinis et al. 2004).

New, nonimmunogenic enzymatic activities may also be obtained from combinatorial libraries of designed amino acid sequences (Wei and Hecht 2004) or by novel screening and selection technologies from enzyme-encoding gene repertoires (Aharoni et al. 2005). These approches all indicate that progress is underway in the development of less immunogenic ADEPT enzymes. However, further work is clearly necessary to overcome the immunogenicity hurdle, whether by further elimination of troublesome epitopes, mutated human enzymes, abzymes, or new recombinant enzymes.

8.5
Important Considerations/Outlook

It has been recognized that the drug generated in an ADEPT system should have a very short half-life to avoid leak back into blood where it can result in toxicity. It has also been recognized that it is necessary to continue to eliminate immunogenic epitopes on the bacterial enzymes used in ADEPT.

It is evident from numerous studies that the concentration of enzyme, as AEC, in tumors tends to equilibrate with that in blood. If the AEC clears quickly from blood the concentration in tumors never attains the peak concentration in blood and also clears quickly. It is therefore necessary to maintain a high concentration of AEC in blood for a prolonged period. Studies are needed to determine the optimum duration of that period.

Prolonged residence of AEC in tumors also has the benefit of delayed escape of AEC from tumors when the blood level falls. The fall in concentration in tumors is then slower than that in blood and provides a basis for a time window in which there is adequate enzyme in tumors and zero enzyme in blood. It is now known how to construct AECs with the necessary characteristics.

For an ADEPT system to be a practical procedure in the clinic it is useful to minimize the effects of individual variation on the clearance of antibody–enzyme constructs from blood. Early studies with a galactosylated enzyme inactivating antibody proved safe and precisely defined a time window for prodrug administration and this may be a way forward for future ADEPT systems.

Understanding of ADEPT has increased with the successful development of preclinical therapies and knowledge of essential requirements from clinical trials. Tools to address the limitations of ADEPT are available and it now remains for ADEPT to fulfil its potential and promise in the clinic.

Acknowledgments

The authors would like to thank Professor Richard Begent for critical reading of this review, Cancer Research UK, Royal Free Cancer Research Trust and NTRAC for financial support.

References

Aharoni, A., Amitai, G., Bernath, K., Magdassi, S., Tawfik, D.S. (2005) High-throughput screening of enzyme libraries: thiolactonases evolved by fluorescence-activated sorting of single cells in emulsion compartments. *Chem Biol* 12: 1281–1289.

Alexander, R.P., Beeley, N.R.A., Odriscoll, M. et al. (1991) Cephalosporin nitrogen mustard carbamate prodrugs for ADEPT. *Tetrahedron Lett* 32: 3269–3272.

Bagshawe, K.D. (1987) Antibody directed enzymes revive anti-cancer prodrugs concept. *Br J Cancer* 56: 531–532.

Bagshawe, K.D. (1989) The First Bagshawe lecture. Towards generating cytotoxic agents at cancer sites. *Br J Cancer* 60: 275–281.

Bagshawe, K.D., Begent, R.H.J. (1996) First clinical experience with ADEPT. *Adv Drug Deliv Rev* 22: 365–367.

Bagshawe, K.D., Springer, C.J., Searle, F., Antoniw, P., Sharma, S.K., Melton, R.G., Sherwood, R.F. (1988) A cytotoxic agent can be generated selectively at cancer sites. *Br J Cancer* 58: 700–703.

Bagshawe, K.D., Sharma, S.K., Springer, C.J., Antoniw, P., Boden, J.A., Rogers, G.T., Burke, P.J., Melton, R.G., Sherwood, R.F. (1991) Antibody directed enzyme prodrug therapy (ADEPT): clinical report. *Dis Markers* 9: 233–238.

Bagshawe, K.D., Sharma, S.K., Springer, C.J., Antoniw, P. (1995) Antibody-directed enzyme prodrug therapy: a pilot scale clinical trial. *Tumor Targeting* 1: 17–29.

Bagshawe, K.D., Sharma, S.K. (1996) Cyclosporine delays host immune response to antibody enzyme conjugate in ADEPT. *Transplant Proc* 28: 3156–3158.

Bagshawe, K.D., Sharma, S.K., Begent, R.H. (2004) Antibody-directed enzyme prodrug therapy (ADEPT) for cancer. *Expert Opin Biol Ther* 4: 1777–1789.

Begent, R.H., Keep, P.A., Searle, F., Green, A.J., Mitchell, H.D., Jones, B.E., Dent, J., Pendower, J.E., Parkins, R.A., Reynolds, K.W. (1986) Radioimmunolocalization and selection for surgery in recurrent colorectal cancer. *Br J Surg* 73: 64–67.

Bhatia, J., Sharma, S.K., Chester, K.A., Pedley, R.B., Boden, R.W., Read, D.A., Boxer, G.M., Michael, N.P., Begent, R.H. (2000) Catalytic activity of an in vivo tumor targeted anti-CEA scFv: carboxypeptidase G2 fusion protein. *Int J Cancer* 85: 571–577.

Biela, B.H., Khawli, L.A., Hu, P., Epstein, A.L. (2003) Chimeric TNT-3/human beta-glucuronidase fusion proteins for antibody-directed enzyme prodrug therapy (ADEPT). *Cancer Biother Radiopharm* 18: 339–353.

Bignami, G.S., Senter, P.D., Grothaus, P.G., Fischer, K.J., Humphreys, T., Wallace, P.M. (1992) N-(4'-hydroxyphenylacetyl)palytoxin: a palytoxin prodrug that can be activated by a monoclonal antibody-penicillin G amidase conjugate. *Cancer Res* 52: 5759–5764.

Blakey, D.C., Burke, P.J., Davies, D.H., Dowell, R.I., East, S.J., Eckersley, K.P., Fitton, J.E., McDaid, J., Melton, R.G., Niculescu-Duvaz, I.A., Pinder, P.E., Sharma, S.K., Wright, A.F., Springer, C.J. (1996) ZD2767: an improved system for antibody-directed enzyme prodrug therapy that results in tumor regressions in colorectal tumor xenografts. *Cancer Res* 56: 3287–3292.

Bosslet, K., Czech, J., Hoffmann, D. (1994) Tumor-selective prodrug activation by fusion protein-mediated catalysis. *Cancer Res* 54: 2151–2159.

Cesaro-Tadic, S., Lagos, D., Honegger, A., Rickard, J.H., Partridge, L.J., Blackburn, G.M., Plückthun, A. (2003) Turnover-based in vitro selection and evolution of biocatalysts from a fully synthetic antibody library. *Nat Biotechnol* 21: 679–685.

Chester, K.A., Bhatia, J., Boxer, G., Cooke, S.P., Flynn, A.A., Huhalov, A., Mayer, A., Pedley, R.B., Robson, L., Sharma, S.K., Spencer, D.I., Begent, R.H. (2000) Clinical applications of phage-derived sFvs and sFv fusion proteins. *Dis Markers* 16: 53–62.

Chester, K.A., Baker, M., Mayer, A. (2005) Overcoming the immonological response to foreign enzymes in cancer therapy. *Expert Rev Clin Immunol* 1: 549–559.

Cortez-Retamozo, V., Backmann, N., Senter, P.D., Wernery, U., De Baetselier, P., Muyldermans, S., Revets, H. (2004) Efficient cancer therapy with a nanobody-based conjugate. *Cancer Res* 64: 2853–2857.

de Graaf, M., Boven, E., Oosterhoff, D., van der Meulen-Muileman, I.H., Huls, G.A., Gerritsen, W.R., Haisma, H.J., Pinedo, H.M. (2002) A fully human anti-Ep-CAM scFv-beta-glucuronidase fusion protein for selective chemotherapy with a glucuronide prodrug. *Br J Cancer* 86: 811–818.

Eccles, S.A., Court, W.J., Box, G.A., Dean, C.J., Melton, R.G., Springer, C.J. (1994) Regression of established breast carcinoma xenografts with antibody-directed enzyme prodrug therapy against c-erbB2 p185. *Cancer Res* 54: 5171–5177.

Edwards, P.A., Skilton, R.A., Payne, A.W., Ormerod, M.G. (1985) Antigenic heterogeneity of breast cell lines detected by monoclonal antibodies and its relationship with the *Cell Cycle J Cell Sci* 73: 321–333.

Fernandez-Gacio, A., Uguen, M., Fastrez, J. (2003) Phage display as a tool for the directed evolution of enzymes. *Trends Biotechnol* 21: 408–414.

Francis, R.J., Sharma, S.K., Springer, C., Green, A.J., Hope-Stone, L.D., Sena, L., Martin, J., Adamson, K.L., Robbins, A., Gumbrell, L., O'Malley, D., Tsiompanou, E., Shahbakhti, H., Webley, S., Hochhauser, D., Hilson, A.J., Blakey, D., Begent, R.H. (2002) A phase I trial of antibody directed enzyme prodrug therapy (ADEPT) in patients with advanced colorectal carcinoma or other CEA producing tumours. *Br J Cancer* 87: 600–607.

Gilliland, D.G., Steplewski, Z., Collier, R.J., Mitchell, K.F., Chang, T.H., Koprowski, H. (1980) Antibody-directed cytotoxic agents: use of monoclonal antibody to direct the action of toxin A chains to colorectal carcinoma cells. *Proc Natl Acad Sci USA* 77: 4539–4543.

Gold, P., Freedman, S.O. (1965) Specific carcinoembryonic antigens of the human digestive system. *J Exp Med* 122: 467–481.

Goldenberg, D.M., Preston, D.F., Primus, F.J., Hansen, H.J. (1974) Photoscan localization of GW-39 tumors in hamsters using radiolabeled anticarcinoembryonic antigen immunoglobulin G. *Cancer Res* 34: 1–9.

Harding, F.A., Liu, A.D., Stickler, M., Razo, O.J., Chin, R., Faravashi, N., Viola, W., Graycar, T., Yeung, V.P., Aehle, W., Meijer, D., Wong, S., Rashid, M.H., Valdes, A.M., Schellenberger, V. (2005) A beta-lactamase with reduced immunogenicity for the targeted delivery of chemotherapeutics using antibody-directed enzyme prodrug therapy. *Mol Cancer Ther* 4: 1791–1800.

Heinis, C., Alessi, P., Neri, D. (2004) Engineering a thermostable human prolyl endopeptidase for antibody-directed enzyme prodrug therapy. *Biochemistry* 43: 6293–6303.

Houba, P.H., Boven, E., Haisma, H.J. (1996) Improved characteristics of a human beta-glucuronidase-antibody conjugate after deglycosylation for use in antibody-directed enzyme prodrug therapy. *Bioconjug Chem* 7: 606–611.

Jones, R.L., Smith, I.E. (2004) Efficacy and safety of trastuzumab. *Expert Opin Drug Saf* 3: 317–327.

Kakinuma, H., Fujii, I., Nishi, Y. (2002) Selective chemotherapeutic strategies using catalytic antibodies: a common pro-moiety for antibody-directed abzyme prodrug therapy. *J Immunol Methods* 269: 269–281.

Köhler, G., and Milstein, C. (1975) Continuous cultures of fused cells secreting antibody of predefined specificity. *Nature* 256: 495–497.

Levy, R., Hurwitz, E., Maron, R., Arnon, R., Sela, M. (1975) The specific cytotoxic effects of daunomycin conjugated to antitumor antibodies. *Cancer Res* 35: 1182–1186.

Mach, J.P., Carrel, S., Merenda, C., Sordat, B., Cerottini, J.C. (1974) In vivo localisation of radiolabelled antibodies to carcinoembryonic antigen in human colon carcinoma grafted into nude mice. *Nature* 248: 704–706.

Martin, J., Stribbling, S.M., Poon, G.K. et al. (1997) Antibody-directed enzyme prodrug therapy: Pharmacokinetics and plasma levels of prodrug and drug in a phase I clinical trial. *Cancer Chemother Pharmacol* 40: 189–201.

Mauger, A.B., Burke, P.J., Somani, H.H., Friedlos, F., Knox, R.J. (1994) Self-immolative prodrugs: candidates for antibody-directed enzyme prodrug therapy in conjunction with a nitroreductase enzyme. *J Med Chem* 37: 3452–3458.

Mayer, A., Francis, R., Sharma, S.K. et al. (2004a) A phase I/II trial of antibody directed enzyme prodrug therapy (ADEPT) with MFECP1 and ZD2767P. *Br J Cancer* 91(Suppl 1): S8.

Mayer, A., Sharma, S.K., Tolner, B. et al. (2004b) Modifying an immunogenic epitope on a therapeutic protein: a step towards an improved system for antibody-directed enzyme prodrug therapy (ADEPT). *Br J Cancer* 90: 2402–2410.

Medzihradszky, K.F., Spencer, D.I., Sharma, S.K., Bhatia, J., Pedley, R.B., Read, D.A., Begent, R.H., Chester, K.A. (2004) Glycoforms obtained by expression in Pichia pastoris improve cancer targeting potential of a recombinant antibody-enzyme fusion protein. *Glycobiology* 14: 27–37.

Meyer, D.L., Jungheim, L.N., Law, K.L., Mikolajczyk, S.D., Shepherd, T.A., Mackensen, D.G., Briggs, S.L., Starling, J.J. (1993) Site-specific prodrug activation by antibody-beta-lactamase conjugates: regression and long-term growth inhibition of human colon carcinoma xenograft models. *Cancer Res* 53: 3956–3963.

Meyer, D.L., Law, K.L., Payne, J.K., Mikolajczyk, S.D., Zarrinmayeh, H., Jungheim, L.N., Kling, J.K., Shepherd, T.A., Starling, J.J. (1995) Site-specific prodrug activation by antibody-beta-lactamase conjugates: preclinical investigation of the efficacy and toxicity of doxorubicin delivered by antibody directed catalysis. *Bioconjug Chem* 6: 440–446.

Michael, N.P., Chester, K.A., Melton, R.G., Robson, L., Nicholas, W., Boden, J.A., Pedley, R.B., Begent, R.H., Sherwood, R.F., Minton, N.P. (1996) In vitro and in vivo characterisation of a recombinant carboxypeptidase G2::anti-CEA scFv fusion protein. *Immunotechnology* 2: 47–57.

Monks, N.R., Blakey, D.C., Curtin, N.J., East, S.J., Heuze, A., Newellm, D.R. (2001) Induction of apoptosis by the ADEPT agent ZD2767: comparison with the classical nitrogen mustard chlorambucil and a monofunctional ZD2767 analogue. *Br J Cancer* 85: 764–771.

Napier, M.P., Sharma, S.K., Springer, C.J., Bagshawe, K.D., Green, A.J., Martin, J., Stribbling, S.M., Cushen, N., O'Malley, D., Begent, R.H. (2000) Antibody-directed enzyme prodrug therapy: efficacy and mechanism of action in colorectal carcinoma. *Clin Cancer Res* 6: 765–772.

Nevinsky, G.A., Buneva, V.N. (2003) Catalytic antibodies in healthy humans and patients with autoimmune and viral diseases. *J Cell Mol Med* 7: 265–276.

Pastan, I., Kreitman, R.J. (2002) Immunotoxins in cancer therapy. *Curr Opin Investig Drugs* 3: 1089–1091.

Pedley, R.B., Boden, J.A., Boden, R., Dale, R., Begent, R.H. (1993) Comparative radioimmunotherapy using intact or F(ab')2 fragments of 131I anti-CEA antibody in a colonic xenograft model. *Br J Cancer* 68: 69–73.

Pimm, M.V., Embleton, M.J., Perkins, A.C., Price, M.R., Robins, R.A., Robinson, G.R., Baldwin, R.W. (1982) In vivo localization of anti-osteogenic sarcoma 791T monoclonal antibody in osteogenic sarcoma xenografts. *Int J Cancer* 30: 75–85.

Rader, C., Turner, J.M., Heine, A., Shabat, D., Sinha, S.C., Wilson, I.A., Lerner, R.A., Barbas, C.F. (2003) A humanized aldolase antibody for selective chemotherapy and adaptor immunotherapy. *J Mol Biol* 332: 889–899.

Reilly, R.M., Sandhu, J., Alvarez-Diez, T.M., Gallinger, S., Kirsh, J., Stern, H. (1995) Problems of delivery of monoclonal antibodies. Pharmaceutical and pharmacokinetic solutions. *Clin Pharmacokinet* 28: 126–142.

Roberge, M., Estabrook, M., Basler, J., Chin, R., Gualfetti, P., Liu, A., Wong, S.B., Rashid, M.H., Graycar, T., Babe, L., Schellenberger, V. (2006) Construction and optimization of a CC49-based scFv-beta-lactamase fusion protein for ADEPT. *Protein Eng Des Sel* 19: 141–145.

Searle, F., Bier, C., Buckley, R.G., Newman, S., Pedley, R.B., Bagshawe, K.D., Melton, R.G., Alwan, S.M., Sherwood, R.F. (1986) The potential of carboxypeptidase G2-antibody conjugates as anti-tumour agents.

I. Preparation of antihuman chorionic gonadotrophin-carboxypeptidase G2 and cytotoxicity of the conjugate against JAR choriocarcinoma cells in vitro. *Br J Cancer* 53: 377–384.

Senter, P.D., Springer, C.J. (2001) Selective activation of anticancer prodrugs by monoclonal antibody-enzyme conjugates. *Adv Drug Deliv Rev* 53: 247–264.

Senter, P.D., Saulnier, M.G., Schreiber, G.J., Hirschberg, D.L., Brown, J.P., Hellstrom, I., Hellstrom, K.E. (1988) Anti-tumor effects of antibody-alkaline phosphatase conjugates in combination with etoposide phosphate. *Proc Natl Acad Sci USA* 85: 4842–4846.

Shabat, D., Rader, C., List, B., Lerner, R.A., Barbas, C.F., III (1999) Multiple event activation of a generic prodrug trigger by antibody catalysis. *Proc Natl Acad Sci USA* 96: 6925–6930.

Shabat, D., Lode, H.N., Pertl, U., Reisfeld, R.A., Rader, C., Lerner, R.A., Barbas, C.F., III (2001) In vivo activity in a catalytic antibody-prodrug system: Antibody catalyzed etoposide prodrug activation for selective chemotherapy. *Proc Natl Acad Sci USA* 98: 7528–7533.

Shamis, M., Lode, H.N., Shabat, D. (2004) Bioactivation of self-immolative dendritic prodrugs by catalytic antibody 38C2. *J Am Chem Soc* 126: 1726–1731.

Sharma, S.K., Bagshawe, K.D., Burke, P.J., Boden, R.W., Rogers, G.T. (1990) Inactivation and clearance of an anti-CEA carboxypeptidase G2 conjugate in *Blood* after localisation in a xenograft model. *Br J Cancer* 61: 659–662.

Sharma, S.K., Bagshawe, K.D., Springer, C.J., Burke, P.J., Rogers, G.T., Boden, J.A., Antoniw, P., Melton, R.G., Sherwood, R.F. (1991) Antibody directed enzyme prodrug therapy (ADEPT): a three phase system. *Dis Markers* 9: 225–231.

Sharma, S.K., Bagshawe, K.D., Melton, R.G., Sherwood, R.F. (1992) Human immune response to monoclonal antibody-enzyme conjugates in ADEPT pilot clinical trial. *Cell Biophys* 21: 109–120.

Sharma, S.K., Bagshawe, K.D., Burke, P.J., Boden, J.A., Rogers, G.T., Springer, C.J., Melton, R.G., Sherwood, R.F. (1994a) Galactosylated antibodies and antibody-enzyme conjugates in antibody-directed enzyme prodrug therapy. *Cancer* 73: 1114–1120.

Sharma, S.K., Boden, J.A., Springer, C.J., Burke, P.J., Bagshawe, K.D. (1994b) Antibody-directed enzyme prodrug therapy (ADEPT). A three-phase study in ovarian tumor xenografts. *Cell Biophys* 24–25: 219–228.

Sharma, S.K., Pedley, R.B., Bhatia, J., Boxer, G.M., El Emir, E., Qureshi, U., Tolner, B., Lowe, H., Michael, N.P., Minton, N., Begent, R.H., Chester, K.A. (2005) Sustained tumor regression of human colorectal cancer xenografts using a multifunctional mannosylated fusion protein in antibody-directed enzyme prodrug therapy. *Clin Cancer Res* 11: 814–825.

Sherwood, R.F., Melton, R.G., Alwan, S.M., Hughes, P. (1985) Purification and properties of carboxypeptidase G2 from Pseudomonas sp. strain RS-16. Use of a novel triazine dye affinity method. *Eur J Biochem* 148: 447–453.

Sinha, S.C., Li, L.S., Miller, G.P., Dutta, S., Rader, C., Lerner, R.A. (2004) Prodrugs of dynemicin analogs for selective chemotherapy mediated by an aldolase catalytic Ab. *Proc Natl Acad Sci USA* 101: 3095–3099.

Smith, G.K., Banks, S., Blumenkopf, T.A., Cory, M., Humphreys, J., Laethem, R.M., Miller, J., Moxham, C.P., Mullin, R., Ray, P.H., Walton, L.M., Wolfe, L.A., III (1997) Toward antibody-directed enzyme prodrug therapy with the T268G mutant of human carboxypeptidase A1 and novel in vivo stable prodrugs of methotrexate. *J Biol Chem* 272: 15804–15816.

Spencer, D.I., Robson, L., Purdy, D., Whitelegg, N.R., Michael, N.P., Bhatia, J., Sharma, S.K., Rees, A.R., Minton, N.P., Begent, R.H., Chester, K.A. (2002) A strategy for mapping and neutralizing conformational immunogenic sites on protein therapeutics. *Proteomics* 2: 271–279.

Springer, C.J., Antoniw, P., Bagshawe, K.D., Searle, F., Bisset, G.M., Jarman, M. (1990) Novel prodrugs which are activated to cytotoxic alkylating agents by carboxypeptidase G2. *J Med Chem* 33: 677–681.

Springer, C.J., Bagshawe, K.D., Sharma, S.K., Searle, F., Boden, J.A., Antoniw, P., Burke, P.J., Rogers, G.T., Sherwood, R.F., Melton, R.G. (1991) Ablation of human choriocarcinoma xenografts in nude mice by antibody-directed enzyme prodrug therapy (ADEPT) with three novel compounds. *Eur J Cancer* 27: 1361–1366.

Springer, C.J., Dowell, R., Burke, P.J., Hadley, E., Davis, D.H., Blakey, D.C., Melton, R.G., Niculescu-Duvaz, I. (1995) Optimization of alkylating agent prodrugs derived from phenol and aniline mustards: a new clinical candidate prodrug (ZD2767) for antibody-directed enzyme prodrug therapy (ADEPT). *J Med Chem* 38: 5051–5065.

Thorpe, P.E., Ross, W.C., Cumber, A.J., Hinson, C.A., Edwards, D.C., Davies, A.J. (1978) Toxicity of diphtheria toxin for lymphoblastoid cells is increased by conjugation to antilymphocytic globulin. *Nature* 271: 752–755.

Wallace, P.M., MacMaster, J.F., Smith, V.F., Kerr, D.E., Senter, P.D., Cosand, W.L. (1994) Intratumoral generation of 5-fluorouracil mediated by an antibody-cytosine deaminase conjugate in combination with 5-fluorocytosine. *Cancer Res* 54: 2719–2723.

Wei, Y., Hecht, M.H. (2004) Enzyme-like proteins from an unselected library of designed amino acid sequences. *Protein Eng Des Sel* 17: 67–75.

Wentworth, P., Datta, A., Blakey, D., Boyle, T., Partridge, L.J., Blackburn, G.M. (1996) Toward antibody-directed "abzyme" prodrug therapy, ADAPT: carbamate prodrug activation by a catalytic antibody and its in vitro application to human tumor cell killing. *Proc Natl Acad Sci USA* 93: 799–803.

Winter, G., Harris, W.J. (1993) Humanized antibodies. *Trends Pharmacol Sci* 14: 139–143.

Wolfe, L.A., Mullin, R.J., Laethem, R., Blumenkopf, T.A., Cory, M., Miller, J.F., Keith, B.R., Humphreys, J., Smith, G.K. (1999) Antibody-directed enzyme prodrug therapy with the T268G mutant of human carboxypeptidase A1: in vitro and in vivo studies with prodrugs of methotrexate and the thymidylate synthase inhibitors GW1031 and GW1843. *Bioconjug Chem* 10: 38–48.

Xu, Y., Yamamoto, N., Janda, K.D. (2004) Catalytic antibodies: hapten design strategies and screening methods. *Bioorg Med Chem* 12: 5247–5268.

9
Emerging Therapeutic Concepts II: Nanotechnology

Dimiter S. Dimitrov, Igor A. Sidorov, Yang Feng, Ponraj Prabakaran, Michaela A.E. Arndt, Jürgen Krauss, and Susanna M. Rybak

9.1
Introduction

It has been known for decades that in colloid systems solid particles, liquid droplets, gas bubbles, and liquid films between them possess different properties from those of the same material in bulk when their size becomes smaller than 100–1000 nm (reviewed in Dimitrov 1983). Recent advances in physics, chemistry, materials sciences, engineering, and molecular biology have allowed the development of nanoparticles (size 1–100 nm and rarely up to 1000 nm) by combining atoms or molecules one at a time, and in arrangements that do not occur in nature (reviewed in Ferrari 2005). Such particles have attracted much attention because of their unique mechanical, electrical, and optical properties. This has resulted in a renewed interest in various nanoparticles already known about for many years (e.g. liposomes) and the development of new ones (e.g. quantum dots and gold nanoshells).

For biomedical applications, any particle with a size typically in the range from 1 to 100 nm can be referred to as a nanoparticle. A multifunctional nanoparticle has important additional properties and contains a synthetic component. Thus biological molecules of this size or their assemblies alone are excluded from this definition. To distinguish from generic liposomes (vesicles from a bilayer lipid membrane), such as cell-size liposomes, we will call a liposome in the nanosize range a nanoliposome, which is approximately equivalent to a small unilamellar vesicle (SUV) (Fig. 9.1). If this nanoliposome contains a drug encapsulated by its membrane, and directing molecules, for example a targeting antibody, on the surface, it becomes a multifunctional nanoparticle because it: (1) increases the half-life and decreases the systemic toxicity and potential immunogenicity of the drug, and (2) specifically binds to molecules of choice, in particular, cell surface receptors, that can ensure directed delivery of the drug to targeted cancer cells. An ideal multifunctional nanoparticle for use against cancer would also have a signaling component so it can diagnose cancer and assess the therapeutic

Handbook of Therapeutic Antibodies. Edited by Stefan Dübel
Copyright © 2007 WILEY-VCH Verlag GmbH & Co. KGaA, Weinheim
ISBN 978-3-527-31453-9

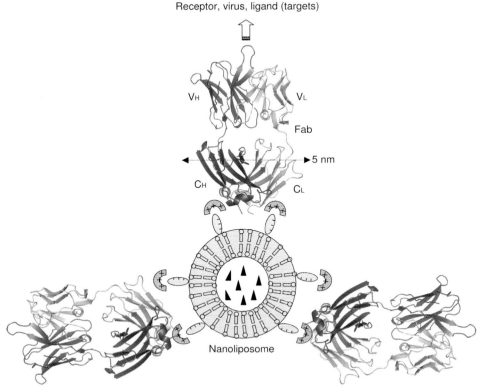

Fig. 9.1 Sketch of a nanoliposome–antibody conjugate. The nanoliposome with a diameter typically in the range from 20 to 100 nm is not in scale. Three Fab molecules are shown with positive charges interacting with patches of negative charges on the liposome surface. Dark triangles indicate drug molecules.

effect. In addition, it could have a triggering property so it can be made to release the drug only after reaching the target.

9.2
Nanoliposomes, Gold Nanoshells, Quantum Dots, and Other Nanoparticles with Biomedical Potential

Many nanoparticles, including carbon nanotubes (Sinha and Yeow 2005), nanodroplets (Garcia-Fuentes et al. 2005; Prego et al. 2005) including lipid emulsions (Hashida et al. 2005), solid lipid nanoparticles (Manjunath et al. 2005), and polymer nanoparticles (Soppimath et al. 2001; Ravi et al. 2004; Csaba et al. 2005) have potential biomedical applications. Although nanoliposomes are perhaps best investigated as vehicles for drug delivery, recently quantum dots have shown potential for imaging, and gold nanoshells for triggered drug delivery and imaging. These nanoparticles are briefly reviewed below.

Liposomes were discovered by Alec D. Bangham 40 years ago (Bangham et al. 1965a,b,c) and have been developed subsequently as vehicles for the sustained delivery of various chemotherapeutic agents, genes, and vaccines and their enhanced targeting (Szoka and Papahadjopoulos 1980; Szoka 1990; Gregoriadis 1995; Gregoriadis et al. 2002; Park et al. 2004; Hashida et al. 2005; Torchilin 2005). The basic structure of a liposome is either a small or large unilamellar, multilamellar, or multivesicular membrane consisting of alternating aqueous and lipid bilayers. This structural design consequently permits the incorporation of either water-soluble or lipophilic drugs. Amphipathic drugs are actually encapsulated within the aqueous and/or the lipid bilayers; the weakly lipophilic components are partially distributed within the aqueous compartment. The lipid bilayer typically consists of naturally occurring biocompatible materials such as phospholipids (derived from soy or egg sources) and cholesterol. Liposomes therefore resemble biological membranes, which should confer them with low immunogenicity and toxicity.

Another system with low, if any, immunogenicity and toxicity that allows separation of compounds from the environment, and provides a skeleton for attachment of functional groups hence does not occur in nature or resemble any biological entity, is the nanoshell. Nanoshells are either hollow or consist of a dielectric core surrounded by a thin metal (typically gold) shell and offer significant advantages over conventional imaging probes including continuous and broad wavelength tunability, far greater scattering and absorption coefficients, increased chemical stability, and improved biocompatibility (Loo et al. 2005a,b). Based on the relative dimensions of the shell thickness and core radius, nanoshells may be designed to scatter and/or absorb light over a broad spectral range, including the near-infrared (NIR), a wavelength region that provides maximal penetration of light through tissue. The ability to control both wavelength-dependent scattering and absorption of nanoshells offers the opportunity to design nanoshells which provide, in a single nanoparticle, both diagnostic and therapeutic capabilities. Nanoshells can be engineered to both scatter light in the NIR, enabling optical molecular cancer imaging, and to absorb light, allowing destruction of targeted cells through photothermal therapy (Loo et al. 2005b).

Hollow nanoshells can be also made, allowing encapsulation of compounds, as in nanoliposomes. For example, hollow nanoshells of gold (diameter below 100 nm) entrapping an enzyme, horseradish peroxidase (HRP), in the cavity have been prepared in reverse micelles by leaching out silver chloride (AgCl) from Au(shell)AgCl(core) nanoparticles with diluted ammonia solution (Kumar et al. 2005). Small substrate molecules such as o-dianisidine can easily enter through the pores of the nanoshell and can undergo enzymatic oxidation by H_2O_2. When the substrate is chemically conjugated with dextran molecule (10 kDa), the enzymatic reaction is practically completely prevented perhaps by the inability of dextran–o-dianisidine conjugate to penetrate the pores of the nanoshells; HRP does not show any activity when trapped inside solid gold nanoparticles.

Quantum dots (qdots) cannot encapsulate compounds but offer unique imaging possibilities that have generated a great deal of interest recently. They are fluorescent semiconductor nanocrystals combining high brilliance, photostability, broad excitation but very narrow emission spectra, size-tunable emission (from the UV to the IR), narrow spectral linewidths, continuous absorption profiles, and stability against photobleaching. Their large surface area-to-volume ratio also makes them appealing for the design of more complex nanosystems and their surface chemistry is compatible with biomolecular conjugation (Han et al. 2001; Lidke and Arndt-Jovin 2004; Lidke et al. 2004; Ozkan 2004; Arya et al. 2005; Gao et al. 2005; Hotz 2005; Michalet et al. 2005). Research on qdots has evolved over the past two decades from electronic materials science to biological applications. Recent examples of their experimental use include the observation of diffusion of individual glycine receptors in living neurons and the identification of lymph nodes in live animals by NIR emission during surgery (Michalet et al. 2005).

New generations of qdots have far-reaching potential for the study of intracellular processes at the single-molecule level, high-resolution cellular imaging, long-term *in vivo* observation of cell trafficking, and as tumor targeting agents, and in diagnostics.

9.3
Conjugation of Antibodies to Nanoparticles

We briefly review here several approaches for conjugation of antibodies to liposomes, including nanoliposomes, that could be also used for other nanoparticles although with major modifications. Conjugation methodologies can be divided into two major groups: through noncovalent interactions or through covalent chemical bonds. A major noncovalent approach for conjugation is simply mixing liposomes and antibodies which can bind to each other if appropriate charge complementarity, polar interactions, and hydrophobic interactions are available. We have developed a model which may help to design liposomes and antibodies that can spontaneously associate (Prabakaran and Dimitrov, in preparation). The noncovalent interactions have not been utilized much due to the nature of the weak interactions between liposome and other ligands, and uncontrolled distribution of the ligands on the liposome surface (Nobs et al. 2004). In the case of antibodies, we considered the possibility of protein engineering for the structurally well-studied antibody templates to provide a charged surface at the constant regions of heavy and light chains of antibody. Antibodies are in the typical nanodimension with a width of 5 nm for a Fab (Fig. 9.1).

The liposome surface can be readily altered with various modifications to introduce surface charges. To achieve noncovalent interactions between liposomes and antibodies based on electrostatic properties, we have to incorporate a negative charge through chemical modification such as carboxylate, phosphate, or sulfate moieties on the liposome surface, or by modulating the liposome membrane composition to form rafts of negatively charged lipids. On the antibody surface, amino acid residues of constant domains were selected based on the knowledge of crystal structures; residues that are not involved in the structural stability and are highly exposed were mutated to basic residues such as lysine, arginine, or histidine. This methodology offers the possibility of conjugation similar to that observed with other covalent linkage methods and also gives control over the orientation of the antibodies. A simple model of this approach is given in Fig. 9.1, in which the liposome particle with a diameter of about 20 nm holds three Fabs. The Fabs shown are from an anti-HIV antibody (m18) in which the heavy and light chains are shown by dark and light gray colors, respectively. Each of the constant domains (C_H and C_L) has a cluster of several residues bearing significant positive charges as indicated by plus signs. The negative charges residing on the liposome are shown by the minus signs. Depending on the size and charge of the liposome molecule, the bulkiness of the mutated residues in the antibody, and charge complementarities, the noncovalent interactions between liposome and antibody are stabilized, leading to the formation of an antibody–nanoliposome conjugate.

Another noncovalent antibody–liposome conjugation approach is based on the use of hydrophobic anchors, including long-chain fatty acids such as palmitic acid and phospholipids such as phosphatidylethanolamine (PE) and phosphatidylinositol (PI). These anchors provide long spacer arms to prevent steric hindrance. Typically these anchors are precoupled with the antibodies and then mixed with the other liposome components to form liposomes. In this case antibodies can be also trapped inside the liposomes. To avoid such problems anchors can be incorporated in the liposomes and used for conjugation with the antibodies.

The covalent conjugation group of approaches are based on the use of five types of covalent bonds: thiolether, disulfide, amide, hydrazone, and amine-amine (Nobs et al. 2004). The thiolether bond is formed by antibody thiol groups and maleimide groups on liposomes. Native thiol groups are readily available in antibody molecules, and can be also engineered by placing cysteines far from the antigen-binding site. The partner on the liposomes for this covalent bond is often N-[4-(p-maleimidophenyl)butyryl]phosphatidylethanolamine (or MPB-PE). Coupling can be achieved by adding antibodies with available thiol groups to preformed liposomes with incorporated maleimide. No crosslinker is required. This method has also been extended to couple antibodies to stealth (PEGylated) liposomes at the distal end of PEG (Nobs et al. 2004). Studies have been reported to use reversed function groups: a thiol function on liposomes with maleimide-containing antibodies. In this case, antibodies should be treated with a bifunctional crosslinker such as succinimidyl 4-(p-maleimidopheny)butyrate (or SPMB).

The coupling chemistry through disulfide bonds is efficient, rapid, and easy to carry out. The thiol group on the antibody molecule can react with a pyridyldithio group on anchor molecules. A drawback of this method is the relative instability of disulfide bonds in circulation.

Coupling through amide bonds is based on the use of anchoring molecules with a free carboxy group, which can form an amide bond with primary amines of antibodies. The reaction is mediated by 1-ethyl-3-(3-dimethylamino-propyl) carbodiimide (EDAC) and N-hydroxysulfosuccinimide (NHS). No pretreatment is required.

The hydrazone bond is formed between carbohydrate group and hydrazide groups incorporated in liposomes. The carbohydrate on the constant region of the antibody heavy chain is oxidated by galactose oxidase or sodium periodate into aldehyde groups. Lauric acid hydrazide is often used as anchor on liposomes for this coupling. The obvious advantadge of this technique is that antibodies are favorably orientated, leaving the antigen-binding region free. The shortfall of this method is the relatively low coupling efficiency.

Coupling though primary amine–amine bonds is based on the use of primary amine groups on antibodies that are crosslinked with anchors such as glutaldehyde or suberimidate. The only pretreatment required is the activation of the primary amine group on liposomes; no manipulation of antibodies is needed. This otherwise efficient method has not been widely used because of the difficulty of controlling the process of homopolymerization of antibodies and liposomes.

9.4
Nanoparticle–Antibody Conjugates for the Treatment of Cancer and Other Diseases

Liposomes conjugated with antibodies, also termed immunoliposomes, have been evaluated for their potential as drug delivery systems mostly for the treatment of cancer and gene delivery (Felnerova et al. 2004; Noble et al. 2004; Park et al. 2004; Schnyder and Huwyler 2005; Torchilin 2005). Several liposomal preparations, including PEGylated liposomal doxorubicin, are in clinical use (Tsatalas et al. 2003). PEGylated liposomes exhibit increased plasma half-life (several days) compared with pure lipid systems (several hours). They can extravasate in tumor tissue (Fig. 9.2) in accordance with an enhanced permeability and retention effect. Due to their relatively long half-life, the PEGylated (sterically stabilized) liposomes can gradually accumulate in tumors but do not bind and enter tumor cells directly. Because these molecules rather stay within tumor stroma and release drug for diffusion into the tumor cells, relative specificity is provided by enhanced permeability and retention of tumor tissue. However, truly targeted immunoliposomes are usually developed by conjugating them with antibody fragments.

Immunoliposomes have been used successfully to target cancer cells. For example, anti-HER2 immunoliposomes were developed for drug delivery to cancer cells overexpressing the oncogenic receptor tyrosine kinase HER2 (Park et al. 2004). These immunoliposomes were optimized for clinical trials and such trials are ongoing. A similar approach has been used for other targets, including EGFR (Mamot et al. 2003). Although immunoliposomes specifically target cells expressing the respective receptor they do not fuse with the plasma or endosomal membrane but gradually release drugs when becoming degraded. The fusion of

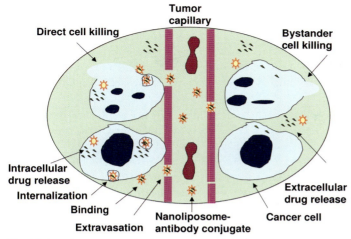

Fig. 9.2 Nanoliposome–antibody conjugates for the treatment of cancer. Liposomes extravasate in tumors and kill cancer cells either by entering the cells and releasing the drug intracellularly (direct killing) or by releasing the drug in the extracellular space but close to cancer cells, which then diffuses into the cancer cells (bystander cell killing). Targeting cancer cells by antibodies specific for cancer-related antigens enhances direct cell killing by an enhanced intracellular delivery of the toxic drug.

such liposomes could increase their efficiency. In this respect reconstitution of viral fusogenic proteins in liposomes (virosomes) (Sarkar et al. 2002) can lead to liposome–cell fusion. However, viral proteins are immunogenic and target only their own receptor. In an attempt to combine specificity of targeting and fusion ability, we have recently constructed a fusion protein of an antibody targeting the IGF-IR and a nonspecific fusogenic protein, and are currently evaluating its properties (Dimitrov et al., in preparation). Such nanoliposome systems could have an increased efficiency of delivery. However, they could fuse nonspecifically with any cell to which they bind. A more sophisticated system should include a trigger that would allow fusion only after interaction of the targeting antibody with the targeted receptor. Such systems are under development.

Gold nanoshells, gold nanocapes, and gold nanoparticles in general, conjugated to antibodies have been recently developed for various purposes, including imaging and therapy of cancer (Chen et al. 2005; El Sayed et al. 2005; Loo et al. 2005a,b), and as a substrate for an immunoassay that is capable of detecting subnanogram levels of analyte within whole blood on the order of minutes (Hirsch et al. 2005). Recently, qdots have also been used as conjugates with antibodies for imaging and detection of cancer (Kaul et al. 2003; Arya et al. 2005; Nida et al. 2005). In general, conjugation of nanoparticles to antibodies allows their targeting to appropriate receptors. In the next section an example of such targeting of macromolecules (protein toxins), not nanoparticles, is discussed in detail to allow a comparison between protein–antibody fusions and nanoparticle–antibody conjugates.

9.5
Comparison Between Nanoparticle–Antibody Conjugates and Fusion Proteins for Cancer Diagnosis and Treatment

9.5.1
Fusion Proteins

Various proteins have been fused to antibodies for improvement of their therapeutic activity by targeting to appropriate receptors. Members of the ribonuclease (RNase) A superfamily have been linked to antibodies against tumor-associated antigens to selectively kill the tumor cell (Rybak and Newton 1999) without the associated toxicities of current strategies employing plant and bacterial toxins (Rybak and Youle 1991; Frankel et al. 2000). Angiogenin (ANG) is an RNase that is a normal component of human plasma (Shapiro 1987), is not itself cytotoxic to cultured tumor cells yet is a potent inhibitor of protein synthesis in both cell-free systems (St Clair 1988) and when injected into a living cell (Saxena et al. 1992). These studies prompted the development of methods to specifically internalize ANG into tumor cells.

Recombinant derivatives comprising robust, small-sized monovalent and bivalent scFvs (Arndt et al. 2003, 2004; Krauss et al. 2003) derived from the anti-CD22 monoclonal antibodies LL2 (Pawlak-Byczkowska 1989) and RFB4 (Amlot et al. 1993) were fused to ANG (Fig. 9.3). Two monomeric fusion proteins carrying ANG either at the N- or C-terminal end were created (Fig. 9.3b,d), expressed as soluble proteins in *E. coli* and purified to homogeneity (Arndt et al. 2005). The orientation of the RNase relative to the antibody moiety played a significant role towards susceptibility of the fusion protein to degradation from bacterial proteases. Only the ANG-spacer-scFv aligned fusion protein (Fig. 9.3b) was cleaved. This phenomenon was also observed for other bacterially produced scFv-RNase fusion proteins when generated in the same orientation (Krauss, unpublished observations). This problem could be solved by producing these fusion proteins in mammalian cells (Fig. 9.3c) (Krauss et al. 2005). To create a dimeric fusion protein carrying two ribonuclease effector domains, ANG was fused via a $(G_4S)_3$ spacer peptide to the C-terminal end of the stable dimeric anti-CD22 V_L-V_H zero-linker scFv MLT-7 (Arndt et al. 2004, 2005) (Fig. 9.3e). Comparative studies with the homogenously purified dimeric fusion protein (Fig. 9.3e) and its monovalent counterpart (Fig. 9.3d) revealed that both constructs specifically bound to the target antigen and retained ribonucleolytic activity. However, they exhibited a markedly different capability for killing CD22-positive tumor cells. The monomeric construct (ANG-scFv$_{LL2}$) inhibited protein synthesis of target cells in a dose-dependent manner but 50% inhibition (IC$_{50}$) could only be achieved at concentrations of >360 nmol L^{-1}. In contrast, the dimeric fusion protein efficiently killed CD22-positive Raji and Daudi tumor cell lines with IC$_{50}$ values of 74 nmol L^{-1} and 118 nmol L^{-1}, respectively (Table 9.1). These results show that the therapeutic potential of scFv-ANG fusion proteins can be markedly enhanced by engineering dimeric derivatives.

Monomeric fusion proteins

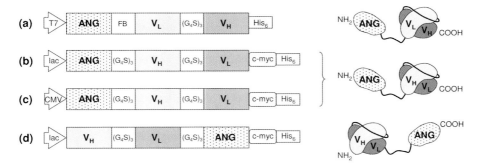

Dimeric fusion protein

Fig. 9.3 Single-chain Fv–angiogenin fusion proteins. (a–d) Monovalent scFv-ANG fusion proteins with angiogenin expressed either at the N-terminal or the C-terminal end of the antibody. (a) Antitransferrin receptor scFv-ANG fusion protein refolded from inclusion bodies in E. coli; (b–d) anti-CD22 scFv-ANG fusion proteins expressed (b,d) as soluble protein into the periplasm of E. coli; (c) from Chinese hamster ovary cells. (e) Non-covalent dimers forming fusion protein with ANG at the C-terminal of the monoclonal antibody LL2-derived anti-CD22 zero-linker scFv MJ7. C-myc and His$_6$-tags for purification are indicated. ANG, human RNase angiogenin; FB (staphylococcal protein A residues 48–60, AKKLNDAQAPKSD) and (G$_4$S)$_3$ spacer peptides are separating ANG from the scFv.

9.5.2
Nanoparticle-conjugated Antibodies

Compared with protein–antibody fusions such as the previously described examples the conjugation of antibodies to nanoparticles provides a number of advantages (Table 9.2). Hollow nanoparticles have the key property of being able to isolate the internal content from the environment. On the one hand the encapsulated material is protected from a hostile environment and on the other hand the environment is protected from the adverse effects of the enclosed component. The modular organization of the nanoparticle technology also enables a combinatorial approach in which a repertoire of monoclonal antibody fragments can be used in conjunction with a series of encapsulated drugs to yield a new generation of molecularly targeted agents. Possible disadvantages in some cases are the larger size of the nanoparticles, which can decrease their penetration into solid tumors and their increased cost of production.

Table 9.1 In vitro potency of angogenin–scFv fusion proteins.

Fusion protein	Target antigen	Cell line[a]	IC$_{50}$ (nmol L^{-1})[b]	Reference
F(ab')$_2$CH2ANG	Transferrin receptor	K562	0.05	Rybak et al. 1992
ANG-FB-scFv	Transferrin receptor	MDA-MB-231^{mdr1}	10	Newton et al. 1996
		HT-29^{mdr1}	7	
		HS578T	>100	
		ACHN	4	
		SF539	17	
		Malme	>100	
scFv$_{LL2}$-(G$_4$S)$_3$-ANG (monomer)	CD22	Raji	360	Arndt et al. 2005
		Daudi	>380	
scFv$_{LL2}$-(G$_4$S)$_3$-ANG (dimer)	CD22	Raji	74	Arndt et al. 2005
		Daudi	118	
ANG-(G$_4$S)$_3$-scFv$_{RFB4}$	CD22	Daudi	56	Krauss et al. 2005

a All cell lines are of human origin and are as follows: K562, erythroleukemia; HT-29^{par}, colon carcinoma; HT-29^{mdr1}, multidrug-resistant colon carcinoma; MDA-MB-231^{mdr1}, multidrug-resistant breast carcinoma; HS578T, breast carcinoma; ACHN, renal carcinoma; SF539, CNS carcinoma; MALME, melanoma; Daudi and Raji, B cell lymphoma.
b IC$_{50}$, concentration of fusion protein required to inhibit protein synthesis or cause cell death by 50%.

To compare the efficacy of targeted liposome-encapsulated doxorubicin (DOX) versus DOX alone, a simple model was recently developed that assumes that the cytotoxic potency of a drug is a function of the intracellular drug level in a critical compartment (Eliaz et al. 2004). Upon exposure to drug, cell death commences after a lag time, and the cell killing rate is dependent on the amount of drug in the critical intracellular compartment. The calculated number of cells in the culture at any time after exposure to the drug takes into account the cell proliferation rate, the cell killing rate, the average intracellular drug concentration, and a lag time for cell killing. This model was applied to DOX encapsulated in liposomes targeted to CD44 on B16F10 melanoma cells in culture. CD44 is the surface receptor that binds to hyaluronan and is overexpressed on various cancer cells, including B16F10. The results demonstrated that the enhanced potency of the encapsulated drug could stem from its enhanced uptake. However, in certain cases, where larger amounts of the free drug were added, such that the intracellular amounts of drug exceeded those obtained from the encapsulated drug, the numbers of viable cells were still significantly smaller for the encapsulated drug.

This finding demonstrated that for given amounts of intracellular DOX, the encapsulated form was more efficient in killing B16F10 cells than the free drug. The outcome was expressed in the kinetic model as a 5- to 6-fold larger rate constant of cell killing potency for the encapsulated drug versus the free drug. The model provided a quantitative framework for comparing the cytotoxic effect

9.5 Comparison Between Nanoparticle

Table 9.2 Comparison between nanoliposome–antibody conjugates and antibody conjugates with small molecules and proteins used as drugs and imaging agents currently under development for diagnosis and treatment of cancer.

Property	Small molecule–antibody	Protein–antibody fusion	Nanoliposome–antibody
Protection from environment	None	None	Yes
Carrying capacity	Several molecules	One	High
Toxicity	High	High	Low
Specific penetration of solid tumors	Moderate	Moderate to low	low
Bystander killing of cancer cells	No	No	Yes
Antibody format	Usually IgG but also others	Usually IgG but also others	scFv, Fab or IgG
Valency of binding	1 or 2	1 or 2	Much higher
Requirement for high-affinity antibody	Yes	Yes	No
Conjugation	Chemical	Fusion protein or chemical	Direct conjugation not required
Stability	Variable	Variable	High
Immunogenicity	Variable	High	None
Half-life in circulation	Long if stable	Long if not immunogenic	Long
Design	Direct	Direct	Modular
Production	Depends on the small molecule	Depends on the protein	Standard for liposomes, antibody

in cultured cells when applying the drug in the free form or in a delivery system.

We have developed a somewhat similar model to compare the toxicity of encapsulated versus free drugs to normal cells and to targeted cancer cells. As we discussed above, toxins and their derivatives as antibody–toxin fusion proteins have shown antitumor activity in humans. However, their efficacy is limited by nonspecific toxicity caused by binding to and killing of normal cells. Encapsulation of toxins in liposomes could decrease the extent of nonspecific toxicity. Major differences between binding of liposome encapsulated and free toxins relate to half-life, penetration, avidity, and capacity (ratio of toxin to targeting molecules – for free single-chain antibody–toxin fusion protein this ratio equals 1). In addition, intracellular delivery of toxin by liposomes may not depend on the use of internalizing antibodies. In an initial attempt to quantify the contribution of each of these factors we developed a model based on the assumption that the targeting molecule (e.g. scFv) is the same, and the distribution in the body is the same for both formulations – toxin-loaded liposome-conjugated antibody and toxin-conjugated antibody (see Appendix, Sidorov, Blumenthal and Dimitrov, in preparation). The results, which are being validated in experimental systems, could be useful in the design of toxin-encapsulating antibody-conjugated nanoliposomes with reduced toxicity compared with free toxin–antibody fusion proteins.

9.6
Conclusions

In recent years the interest in antibody-guided nanoparticles for cancer imaging and treatment has increased significantly due to successes in both nanotechnology and engineered therapeutic antibodies. The combination of these two promising technologies in antibody–nanoparticle conjugates offers new possibilities for early diagnosis, prevention, and treatment of cancer. The major direction of research appears to continue with the development of various improved systems based on drug-bearing nanoliposomes being conjugated with antibodies that target various surface-associated molecules, including growth factor receptors. In addition, imaging promises to be another major area of research where the marriage of nanotechnology and antibodies could produce entirely new possibilities.

9.7
Summary

Solid particles, liquid droplets, gas bubbles, and liquid films between them in colloid systems have been known for decades to possess size-dependent properties that differ from those of the same material in bulk when size is smaller than about 100 nm. Recent advances in physics, chemistry, materials sciences, engineering, and molecular biology have allowed the development of nanoparticles (size 1–100 nm) by combining atoms or molecules one at a time in arrangements that do not occur in nature. Such particles have attracted much attention because of their unique mechanical, electrical, and optical properties. This resulted in a renewed interest in various nanoparticles already known about for many years (e.g. liposomes) and the development of new ones (e.g. quantum dots and gold nanoshells). Such particles conjugated to antibodies can improve their binding or/and effector functions or confer new functions (e.g. cytotoxicity, size-dependent fluorescence, and light scattering).

Compared with engineered antibodies based on fusion proteins or chemical conjugates made with other compounds, nanoparticle–antibody conjugates have the fundamental capacity to separate compounds loaded inside the particle (e.g. inside liposomes) from the environment, thus avoiding possible toxicity and immunogenicity. In addition, high local concentrations of loaded active substances (e.g. imaging agents) can be achieved. Nanoparticles can also serve as skeletons for the construction of multifunctional nanoparticle–antibody conju-

References

Amlot, P.L., Stone, M.J., Cunningham, D., Fay, J., Newman, J., Collins, R., May, R., McCarthy, M., Richardson, J., Ghetie, V. et al. (1993) A phase I study of an anti-CD22-deglycosylated ricin A chain immunotoxin in the treatment of B-cell lymphomas resistant to conventional therapy. *Blood* 82: 2624–2633.

Arndt, M.A.E., Krauss, J., Schwarzenbacher, R., Vu, B.K., Green, S., Rybak, S.M. (2003) Generation of a highly stable internalizing anti-CD22 single-chain Fv fragment for targeting non-Hodgkin's lymphoma. *Int J Cancer* 107: 822–829.

Arndt, M.A.E., Krauss, J., Rybak, S.M. (2004) Antigen binding and stability properties of non-covalently linked anti-CD22 single-chain Fv dimers. *FEBS Lett* 578: 257–261.

Arndt, M.A., Krauss, J., Vu, B.K., Newton, D.L., Rybak, S.M. (2005) A dimeric angiogenin immunofusion protein mediates selective toxicity toward CD22+ tumor cells. *J Immunother* 28: 245–251.

Arya, H., Kaul, Z., Wadhwa, R., Taira, K., Hirano, T., Kaul, S.C. (2005) Quantum dots in bio-imaging: Revolution by the small. *Biochem Biophys Res Commun* 329: 1173–1177.

Bangham, A.D., Standish, M.M., Miller, N. (1965a) Cation permeability of phospholipid model membranes: effect of narcotics. *Nature* 208: 1295–1297.

Bangham, A.D., Standish, M.M., Watkins, J.C. (1965b) Diffusion of univalent ions across the lamellae of swollen phospholipids. *J Mol Biol* 13: 238–252.

Bangham, A.D., Standish, M.M., Weissmann, G. (1965c) The action of steroids and streptolysin S on the permeability of phospholipid structures to cations. *J Mol Biol* 13: 253–259.

Chen, J., Saeki, F., Wiley, B.J., Cang, H., Cobb, M.J., Li, Z.Y., Au, L., Zhang, H., Kimmey, M.B., Li, X., Xia, Y. (2005) Gold nanocages: bioconjugation and their potential use as optical imaging contrast agents. *Nano Lett* 5: 473–477.

Csaba, N., Caamano, P., Sanchez, A., Dominguez, F., Alonso, M.J. (2005) PLGA: poloxamer and PLGA: poloxamine blend nanoparticles: new carriers for gene delivery. *Biomacromolecules* 6: 271–278.

Dimitrov, D.S. (1983) Dynamic interactions between approaching surfaces of biological interest. *Progr Surface Sci* 14: 295–424.

El Sayed, I.H., Huang, X., El Sayed, M.A. (2005) Surface plasmon resonance scattering and absorption of anti-EGFR antibody conjugated gold nanoparticles in cancer diagnostics: applications in oral cancer. *Nano Lett* 5: 829–834.

Eliaz, R.E., Nir, S., Marty, C., Szoka, F.C., Jr. (2004) Determination and modeling of kinetics of cancer cell killing by doxorubicin and doxorubicin encapsulated in targeted liposomes. *Cancer Res* 64: 711–718.

Felnerova, D., Viret, J.F., Gluck, R., Moser, C. (2004) Liposomes and virosomes as delivery systems for antigens, nucleic acids and drugs. *Curr Opin Biotechnol* 15: 518–529.

Ferrari, M. (2005) Cancer nanotechnology: opportunities and challenges. *Nat Rev Cancer* 5: 161–171.

Frankel, A.E., Kreitman, R.J., Sausville, E.A. (2000) Targeted toxins. *Clin Can Res* 6: 326–334.

Gao, X., Yang, L., Petros, J.A., Marshall, F.F., Simons, J.W., Nie, S. (2005) In vivo molecular and cellular imaging with quantum dots. *Curr Opin Biotechnol* 16: 63–72.

Garcia-Fuentes, M., Alonso, M.J., Torres, D. (2005) Design and characterization of a new drug nanocarrier made from solid-liquid lipid mixtures. *J Colloid Interface Sci* 285: 590–598.

Gregoriadis, G. (1995) Engineering liposomes for drug delivery: progress and problems. *Trends Biotechnol* 13: 527–537.

Gregoriadis, G., Bacon, A., Caparros-Wanderley, W., McCormack, B. (2002) A role for liposomes in genetic vaccination. *Vaccine* 20(Suppl 5): B1–B9.

Han, M., Gao, X., Su, J.Z., Nie, S. (2001) Quantum-dot-tagged microbeads for multiplexed optical coding of biomolecules. *Nat Biotechnol* 19: 631–635.

Hashida, M., Kawakami, S., Yamashita, F. (2005) Lipid carrier systems for targeted drug and gene delivery. *Chem Pharm Bull (Tokyo)* 53: 871–880.

Hirsch, L.R., Halas, N.J., West, J.L. (2005) Whole-blood immunoassay facilitated by gold nanoshell-conjugate antibodies. *Methods Mol Biol* 303: 101–111.

Hotz, C.Z. (2005) Applications of quantum dots in biology: an overview. *Methods Mol Biol* 303: 1–17.

Kaul, Z., Yaguchi, T., Kaul, S.C., Hirano, T., Wadhwa, R., Taira, K. (2003) Mortalin imaging in normal and cancer cells with quantum dot immuno-conjugates. *Cell Res* 13: 503–507.

Krauss, J., Arndt, M.A., Martin, A.C., Liu, H., Rybak, S.M. (2003) Specificity grafting of human antibody frameworks selected from a phage display library: generation of a highly stable humanized anti-CD22 single-chain Fv fragment. *Protein Eng* 16: 753–759.

Krauss, J., Arndt, M.A., Vu, B.K., Newton, D.L., Rybak, S.M. (2005) Targeting malignant B-cell lymphoma with a humanized anti-CD22 scFv-angiogenin immunoenzyme. *Br J Haematol* 128: 602–609.

Kumar, R., Maitra, A.N., Patanjali, P.K., Sharma, P. (2005) Hollow gold nanoparticles encapsulating horseradish peroxidase. *Biomaterials* 26: 6743–6753.

Lidke, D.S., Arndt-Jovin, D.J. (2004) Imaging takes a quantum leap. *Physiology (Bethesda)* 19: 322–325.

Lidke, D.S., Nagy, P., Heintzmann, R., Arndt-Jovin, D.J., Post, J.N., Grecco, H.E., Jares-Erijman, E.A., Jovin, T.M. (2004) Quantum dot ligands provide new insights into erbB/HER receptor-mediated signal transduction. *Nat Biotechnol* 22: 198–203.

Loo, C., Hirsch, L., Lee, M.H., Chang, E., West, J., Halas, N., Drezek, R. (2005a) Gold nanoshell bioconjugates for molecular imaging in living cells. *Opt Lett* 30: 1012–1014.

Loo, C., Lowery, A., Halas, N., West, J., Drezek, R. (2005b) Immunotargeted nanoshells for integrated cancer imaging and therapy. *Nano Lett* 5: 709–711.

Mamot, C., Drummond, D.C., Greiser, U., Hong, K., Kirpotin, D.B., Marks, J.D., Park, J.W. (2003) Epidermal growth factor receptor (EGFR)-targeted immunoliposomes mediate specific and efficient drug delivery to. *Cancer Res* 63: 3154–3161.

Manjunath, K., Reddy, J.S., Venkateswarlu, V. (2005) Solid lipid nanoparticles as drug delivery systems. *Methods Find Exp Clin Pharmacol* 27: 127–144.

Michalet, X. and others (2005) Quantum dots for live cells, in vivo imaging, and diagnostics. *Science* 307: 538–544.

Newton, D.L., Xue, Y., Olson, K.A., Fett, J.W., Rybak, S.M. (1996) Angiogenin single-chain immunofusions: influence of peptide linkers and spacers between fusion protein domains. *Biochemistry* 35: 545–553.

Nida, D.L., Rahman, M.S., Carlson, K.D., Richards-Kortum, R., Follen, M. (2005) Fluorescent nanocrystals for use in early cervical cancer detection. *Gynecol Oncol* 99(3 Suppl 1): S89–94.

Noble, C.O., Kirpotin, D.B., Hayes, M.E., Mamot, C., Hong, K., Park, J.W., Benz, C.C., Marks, J.D., Drummond, D.C. (2004) Development of ligand-targeted liposomes for cancer therapy. *Expert Opin Ther Targets* 8: 335–353.

Nobs, L., Buchegger, F., Gurny, R., Allemann, E. (2004) Current methods for attaching targeting ligands to liposomes and nanoparticles. *J Pharm Sci* 93: 1980–1992.

Ozkan, M. (2004) Quantum dots and other nanoparticles: what can they offer to drug discovery? *Drug Discov Today* 9: 1065–1071.

Park, J.W., Benz, C.C., Martin, F.J. (2004) Future directions of liposome- and immunoliposome-based cancer therapeutics. *Semin Oncol* 31: 196–205.

Pawlak-Byczkowska, E.J. (1989) Two new monoclonal antibodies, EPB-1 and EPB-2, reactive with human lymphoma. *Cancer Res* 49: 4568–4577.

Prego, C., Garcia, M., Torres, D., Alonso, M.J. (2005) Transmucosal macromolecular drug delivery. *J Control Release* 101: 151–162.

Ravi, K.M., Hellermann, G., Lockey, R.F., Mohapatra, S.S. (2004) Nanoparticle-mediated gene delivery: state of the art. *Expert Opin Biol Ther* 4: 1213–1224.

Rybak, S.M., Newton, D.L. (1999) Natural and engineered cytotoxic ribonucleases: therapeutic potential. *Exp Cell Res* 253: 325–335.

Rybak, S.M., Youle, R.J. (1991) Clinical use of immunotoxins: monoclonal antibodies conjugated to protein. *Immunol Allergy Clin N Am* 11: 359–380.

Rybak, S.M., Hoogenboom, H.R., Meade, H.M., Raus, J.C.M., Schwartz, D., Youle, R.J. (1992) Humanization of immunotoxins. *Proc Natl Acad Sci USA* 89: 3165–3169.

Sarkar, D.P., Ramani, K., Tyagi, S.K. (2002) Targeted gene delivery by virosomes. *Methods Mol Biol* 199: 163–173.

Saxena, S.K., Rybak, S.M., Davey, R.T., Jr., Youle, R.J., Ackerman, E.J. (1992) Angiogenin is a cytotoxic, tRNA-specific ribonuclease in the RNase A superfamily. *J Biol Chem* 267: 21982–21986.

Schnyder, A., Huwyler, J. (2005) Drug transport to brain with targeted liposomes. *Neurorx* 2: 99–107.

Shapiro, R. (1987) Isolation of angiogenin from normal human plasma. *Biochemistry* 26: 5141–5146.

Sinha, N., Yeow, J.T. (2005) Carbon nanotubes for biomedical applications. *IEEE Trans NanobioSci* 4: 180–195.

Soppimath, K.S., Aminabhavi, T.M., Kulkarni, A.R., Rudzinski, W.E. (2001) Biodegradable polymeric nanoparticles as drug delivery devices. *J Control Release* 70: 1–20.

St Clair, D.K. (1988) Angiogenin abolishes cell-free protein synthesis by specific ribonucleolytic inactivation of 40S ribosomes. *Biochemistry* 27: 7263–7268.

Szoka, F.C., Jr. (1990) The future of liposomal drug delivery. *Biotechnol Appl Biochem* 12: 496–500.

Szoka, F., Jr., Papahadjopoulos, D. (1980) Comparative properties and methods of preparation of lipid vesicles (liposomes). *Annu Rev Biophys Bioeng* 9: 467–508.

Torchilin, V.P. (2005) Recent advances with liposomes as pharmaceutical carriers. *Nat Rev Drug Discov* 4: 145–160.

Tsatalas, C., Martinis, G., Margaritis, D., Spanoudakis, E., Kotsianidis, I., Karpouzis, A., Bourikas, G. (2003) Long-term remission of recalcitrant tumour-stage mycosis fungoides following chemotherapy with pegylated liposomal doxorubicin. *J Eur Acad Dermatol Venereol* 17: 80–82.

Appendix

A model of cancer cell killing by toxin-loaded liposomes conjugated with targeting antibodies.

Model equations To estimate quantitatively the efficacy of killing normal and tumor cells by toxin-encapsulating liposome and free drug molecules we have developed a dynamic model. The model consists of four equations and describes the dynamics of total intracellular toxin (I, μg) and concentrations of toxin encapsulated in liposomes (L, μg mL^{-1}), free toxin (T, μg mL^{-1}), and tumor or normal cells (C, cells mL^{-1}):

$$\frac{dL}{dt} = -k_L CL - a_L L$$

$$\frac{dT}{dt} = -k_T CT + a_L nL - a_T T$$

$$\frac{dI}{dt} = k_T T + k_L L - a_I I$$

$$\frac{dC}{dt} = gC - dC$$

where:

$$d = d_0 + d_1 \frac{I}{I+K}$$

k_L and k_T characterize the avidity of binding liposome and free toxin molecules to cells and transition into cell (mL per cell h^{-1}); n is the number of toxin molecules encapsulated in one liposome; a_L and a_T are the rate constants of liposome leaking and free toxin molecules destruction (h^{-1}), respectively; a_I is the rate constant of intracellular toxin destruction; d_0 is the death rate for normal or tumor cells (h^{-1}); d_1 is the maximum increase in value cell death rate for high intracellular toxin concentration (h^{-1}); K is the amount of intracellular toxin giving an increase of 50% of maximum in death rate (μg); and g is the rate of cell growth (h^{-1}). In this mode we assume that cell death rate d depends nonlinearly on total amount of intracellular toxin.

Parameter values Using the data for doxorubicin (DOX) uptake by the B16F10 cells (1.6×10^5 cells mL^{-1}) after 3 h of exposure to five different concentrations of free and liposome-encapsulated toxin (Eliaz et al. 2004) we can estimate parameters $k_L = 5.5 \times 10^{-8}$ mL cells^{-1} h^{-1} and $k_T = 1.8 \times 10^{-8}$ mL cells^{-1} day^{-1} as the result of data fitting. These values of these constants can provide uptake of femtomoles (from 0.6 to 32) of DOX per cell after 3 h by cells with concentration of 1.6×10^5 cells mL^{-1} in presence of micromoles (from 0.5 to 20) of DOX in media as reported by Eliaz et al. (2004). Rate constant of B16F10 cells proliferation was also estimated in this work: $g = 0.030\,h^{-1}$. Let us assume that $d_0 = 0.028\,h^{-1}$ and d_1 is 5 times larger (= $0.14\,h^{-1}$). A constant value of $K = 200\,\mu g$ can be estimated using the dose–response curves and the data about intracellular toxin amount. Here we assume that there is no liposome leakage and intracellular/extracellular toxin molecules are stable ($a_L = a_T = d_1 = 0$).

Simulations Figure 9.4 shows the results of simulation for the experiments represented in (Eliaz et al. 2004) with the parameter values described above. B16F10 cells were treated with HAL liposomes encapsulating DOX and free DOX molecules. Cytotoxic effect was determined either immediately at 3, 6, 12, 24, 48, 72, and 96 h after treatment or with a 96-h delay irrespective of treatment duration

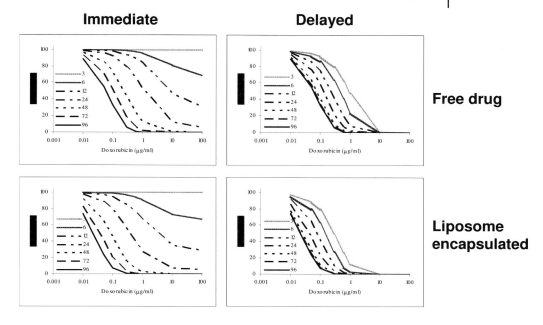

Fig. 9.4 Simulations of immediate and delayed cytotoxic effect of free and liposome-encapsulated DOX (experimental conditions as described in Eliaz et al. 2004) (see Appendix). B16F10 cell were treated with HAL liposome-encapsulated and free DOX molecules. Cytotoxic effect was determined either immediately at 3, 6, 12, 24, 48, 72, and 96 h after treatment or measured following a 96-h delay. The model parameters are described in the text.

(3, 6, 12, 24, 48, 72, and 96 h). It was noted that there is a lag-time in drug effect with no cell killing for about 3 h for very large drug concentration. We modeled this effect by holding the cell death rate at a value of d_0 for the first 3 h of treatment ($d = d_0$, for $0 \leq t < 3$ h). The results of the simulations represented in Fig. 9.3 describe the main features of immediate and delayed treatment: sigmoidal dose–response; increase in effect with prolonged treatment; increase in cell killing with the increase of free or liposome-encapsulated drug concentration; increased effect for delayed treatment compared with immediate one and for liposome-encapsulted drug compared with the free drug.

We used the model with the parameters described above to simulate an *in vitro* experiment: treatment of three different mixtures of normal and tumor cells (10%, 50%, and 80% of normal cells, respectively) by the liposome-encapsulated antibody-conjugated liposomes. It was assumed that normal and tumor cells (total concentration 10^7 cells mL^{-1}) have different avidity to the liposomes (different number of receptors specific to the liposome-conjugated antibodies): k_L for normal cells is 10 times less than for tumor cells (5.5×10^{-8} and 5.5×10^{-7} mL cells^{-1} h^{-1},

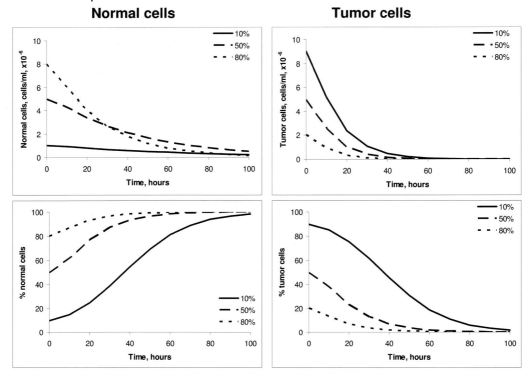

Fig. 9.5 Treatment of mixtures of tumor and normal cells (10%, 50%, and 80% of normal cells) with 10% µg mL^{-1} of liposome-encapsulated toxin. The model parameters are described in the Appendix.

respectively). All other parameters are the same as described above. Liposome-encapsulated toxin concentration was 10 µg/ml. Dynamics normal cells, tumor cells and percentage of cell in the population were calculated. As shown in Fig. 9.5 the toxic effect for normal cells was delayed comparing to the tumor cells for all initial ratios. Although concentrations of both normal and tumor cells decreased with time after treatment, it was about 100% of live normal cells and after 80 h of treatment for all tested ratios. These results show that the therapeutic effect of a liposome-encapsulated drug can be higher for normal cells than for tumor cells.

10
Emerging Therapeutic Concepts III: Chimeric Immunoglobulin T Cell Receptors, T-Bodies

Thomas Schirrmann and Gabriele Pecher

10.1
Introduction

The potential of monoclonal antibodies as targeted therapeutics has attracted a huge amount of interest over recent decades. However, there are still some obstacles to overcome. One is that antibodies mostly do not mediate effector functions themselves against virus-infected or tumor cells. Only in rare cases antibodies directly induce antiproliferatory, apoptotic, or cytotoxic effects against target cells, depending on the signaling properties of the antigen. Antibodies also need to recruit effector components or cells of the immune system. They are also rapidly removed from the tumor cell surface by processes such as capping, shedding, or endocytosis. To overcome the requirement of additional immunological effectors, therapeutic antibodies are usually fused to toxins, prodrug-activating enzymes, or radionuclides. Nevertheless, the inefficient access of large macromolecules such as antibodies or immunotoxins to poorly vascularized tumors as well as those protected by endothelia, surrounding tissue, or the blood–brain barrier limits their therapeutic efficacy.

In contrast to antibodies, immunological effector lymphocytes such as cytotoxic T lymphocytes (CTLs) or natural killer (NK) cells are capable of migrating into solid tissues or tumors and of mediating potent cytotoxic effector functions. However, the effectiveness and specificity of patient's lymphocytes is not efficient enough to completely destroy all tumor cells. Furthermore, the isolation and expansion of tumor-specific T cells from individual tumor patients has been proven to be problematic (Hoffman et al. 2000). Therefore, a strategy combining the antigen specificity of antibodies with the effector properties of cytotoxic lymphocytes is needed to achieve more efficient cancer immunotherapy.

Two major strategies have been directed towards this objective. First, bispecific antibodies binding to a tumor-associated antigen (TAA) and signal molecules such as CD3 on T cells or CD16 on NK cells have been successfully used to specifically redirect and activate effector lymphocytes against tumor cells (Perez

Handbook of Therapeutic Antibodies. Edited by Stefan Dübel
Copyright © 2007 WILEY-VCH Verlag GmbH & Co. KGaA, Weinheim
ISBN 978-3-527-31453-9

et al. 1985; Canevari et al. 1995). However, bispecific antibodies suffer from several limitations, such as the need for large amounts of recombinant protein, their inefficient migration into solid tumors (Jain 1990) or through endothelial barriers and their limited presence on the targeted effector lymphocytes due to dissociation. In addition, T cells stimulated by anti-(CD3 × TAA) bispecific antibodies can lose their signal transduction and effector properties following target cell recognition (Blank-Voorthuis et al. 1993) and require an additional costimulatory signal for full activation, for instance by ligation of CD28 using $CD80^+$ cells or certain anti-CD28 antibodies (Kipriyanov et al. 2002).

The second approach, which is discussed in this chapter, is based on adoptive cellular immunotherapy employing effector lymphocytes gene modified with tumor-specific chimeric receptor genes consisting of an antigen recognition domain of antibodies and a signal domain, triggering their cytolytic mechanisms. These chimeric immunoglobulin TCRs (cIgTCRs) are also termed "T-bodies" (Eshhar et al. 1996) because most of the studies employed T cells and signaling components of the T-cell receptor (TCR). In this review, we describe the development and design of cIgTCR constructs. Furthermore, we take a look at the genetic modification of effector lymphocytes and preclinical studies using cIgTCR gene-modified effector lymphocytes. Finally, we consider the therapeutic aspects of cIgTCR gene-modified effector lymphocytes in the context of other adoptive immunotherapy approaches and the first clinical studies.

10.2
Chimeric Immunoglobulin T-Cell Receptors – "T-Bodies"

10.2.1
Antigen Recognition of Antibodies and T-Cell Receptors

Adaptive immunity has evolved into two major recognition systems for non-self antigens which are expressed by T cells and B cells (Table 10.1). Both lymphocyte types express individual specific antigen recognition molecules that are not determined in the germline. T cells are responsible for the cellular arm of adaptive immunity and are the main protagonists of immune regulation. T cells expressing the α/β TCRs recognize short peptide antigen molecules displayed in noncovalent association of major histocompatibility complexes (MHC) on the surface of antigen-presenting cells and target cells. Antigenic peptides of 9 or 13–17 amino acids are generated by proteolytic digestion of proteins from either intracellular or extracellular origin presented in MHC classes I or II, respectively. TCR binding to the MHC–antigen complex is of low affinity and supported by the coreceptors CD8 or CD4. These coreceptors do not influence the antigen specificity of the TCR complex but enhance its signal over three magnitudes, that the recognition of about 100 antigen–MHC complexes is sufficient to activate the T cell.

B cells are the major players of the humoral arm of adaptive immunity. In contrast to TCRs, antibodies recognize a broad spectrum of antigens independent of cellular or MHC context and often with very high affinity.

Table 10.1 Comparison of antibody and TCR antigen recognition.

	Antibodies	α/β-TCR
Cell type	B lymphocytes, plasma cells	T lymphocytes
Expression	Membrane-bound (IgM, IgD), soluble (all isotypes)	Membrane-bound
Class switch	Yes	No
Antigen epitopes	Linear peptides, structural epitopes, haptens, soluble and cell-bound antigens	Linear peptides in MHC context, cell-bound[a]
MHC restriction	No[b]	Yes[a]
Antigen origin	Extracellular	Intracellular (MHC class I) Extracellular (MHC class II)[c]
Antigen on target cells	Usually 10 000–1 000 000 molecules per cell	Usually 10–1000 peptide antigen–MHC complexes per cell
Inhibition by soluble antigen	Yes	No[d]
Coreceptor	No	Th: CD4 for MHC class II CTL: CD8 for MHC class I
Antigen-driven affinity maturation	Yes	No
Affinity	IgG: very high (10^{-9} mol L^{-1}) IgM: usually median (10^{-7} mol L^{-1})	Low (10^{-5}–5×10^{-6} mol L^{-1})
Valency	IgG: bivalent; IgA: tetravalent IgM: 10 or 12 binding sites	Monovalent

a MHC-unrestricted recognition demonstrated for haptens.
b Antibody fragments with TCR-like MHC-restricted peptide antigen recognition obtained by phage display.
c Dendritic cells are able to present extracellular peptides also in MHC-I.
d Artificial antagonistic peptides can block TCR binding.

Despite their fundamentally different antigen recognition systems, antibodies and TCRs are very similar in molecular structure. They are formed by two different disulfide-linked chains, light (L) and heavy (H) chains for antibodies and α/β chains for TCRs. Each chain consists of immunoglobulin domains forming a constant (C) and a variable (V) region. The V region is responsible for the antigen recognition of antibodies as well as of TCRs. The C domains stabilize the antigen recognition domain, assemble the different chains, mediate effector or signal functions and thus transmembrane association. Immunoglobulin and TCR gene loci are similarly organized. During the early phases of T or B cell development both antibody and TCR chains are formed from separate genetic germline V(D)J elements organized in large and diverse clusters by somatic recombination. This mechanism provides the major part of the tremendous repertoire of antigen recognition of adaptive immunity. In contrast to TCRs, however, immunoglobulin genes undergo an additional process of somatic hypermutation and antigen-driven affinity maturation.

At the end of the 1980s, investigators made use of the structural similarity of immunoglobulins and TCRs by constructing chimeric TCR variants with antibody V regions to facilitate their studies of the TCR complex (Kuwana et al. 1987). They could also demonstrate that the antigen recognition of antibodies and the TCR signaling properties can be combined in chimeric receptor constructs. Based on these results efforts have been made to develop a technology towards the generation of T cells grafted with antibody specificity. Immunotherapy employing T cells recognizing antigens in an antibody-like fashion might overcome the MHC downregulation frequently seen in virus-infected and tumor cells.

10.2.2
General Design of Chimeric Immunoglobulin T-cell Receptors

Chimeric immunoglobulin TCRs generally consist of an appropriate N-terminal signal peptide followed by an extracellular antibody-derived antigen recognition domain, a transmembrane domain, and a cytoplasmatic signal domain (Fig. 10.1). These domains are required to enable the cell surface expression of the receptor and to couple the antibody-mediated antigen recognition with the effector function of the gene-modified immune cell. Additional domains, in particular extracellular spacer domains between antigen recognition domain and transmembrane domain, can dramatically improve the expression, antigen binding, and signal function of the cIgTCR. Novel cIgTCR designs contain additional motifs of costimulatory receptors or utilize downstream signal molecules to enhance the effector functions of the gene modified immune cells.

10.2.3
Antibody Fv and Fab Fragments in Double Chain T-Bodies

Initially, chimeric TCR variants were constructed by the substitution of both TCR V regions by those of antibodies to facilitate the study of the TCR in T cells. These studies did not reveal any structural advantage of the combination V_H with C_α and V_L with C_β or vice versa (Kuwana et al. 1987; Goverman et al. 1990). The chimeric TCR chains also interacted with endogenous TCR chains, which often resulted in nonfunctional TCR hybrid complexes. Interestingly, chimeric TCR chains with the V_H region of hapten-specific antibodies were often sufficient for hapten binding after association with V_α or V_β of the endogenous TCR chains (Gross et al. 1989a; Gorochov et al. 1992). Most antibodies recognize more complex epitopes and need the assembly of both antibody V regions to form a functional Fv fragment. Another problem with this approach of double chain T-bodies is that the chimeric TCR chains must assemble correctly with endogenous TCR signal molecules for signal transduction, thus limiting this approach to T cells.

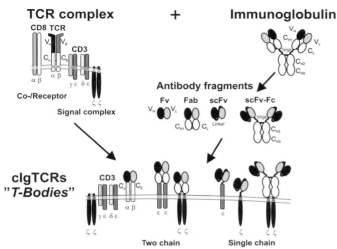

Fig. 10.1 Chimeric immunoglobulin T-cell receptors – "T-bodies." Effector lymphocytes can be grafted with antibody specificity by gene transfer of cIgTCRs. These chimeric receptors are also called "T-bodies," because they combine both the antigen recognition of antibodies with the signal properties of the T-cell receptor (TCR) complex. TCR V regions can be substituted by those of antibodies which require the assembly with the TCR signal complex, limiting this approach to T cells. To circumvent the assembly with the TCR complex, antibody Fv or Fab fragments were directly fused to TCR signal chains such as the CD3 ε or ζ chain (or the FcεRI γ chain, not shown). The employment of scFv antibody fragments allowed single-chain receptor constructs, which is the basis for most currently used cIgTCR designs.

The fusion of antibody V regions to one of the TCR signal chains like the CD3ε or ζ chain or the FcεRI γ chain allowed the extension of T-body approach to other immune cells than T cells. The fusion of CD4 to the cytoplasmatic domains of the ζ chain resulted in a chimeric single-chain receptor gene which could be expressed in T cells (Roberts et al. 1994) as well as in NK cells (Tran et al. 1995). Effector lymphocytes transfected with the chimeric CD4-ζ receptor showed a specific cytotoxicity against gp120$^+$ or HIV-infected target cells. Moreover, the transduction of stem cells with the CD4-ζ construct allowed the subsequent differentiation to NK cells and neutrophils expressing the chimeric receptor (Roberts et al. 1998). Accordingly, the introduction of an appropriate signal chain into cIgTCR constructs can extend the T-body approach to a broad set of immune cell types, including T cells, NK cells, neutrophils, mast cells (Bach et al. 1994) and monocytes (Biglari et al. 2006). In contrast to antibody Fv fragments, the introduction of Fab fragments fused to CD3ε, ζ or FcεRI γ chain enhanced the formation of functional antigen-binding domains stabilized by the disulfide bridge between C_L and C_H1 domains and reduced the interaction with endogenous TCR signal chains (Nolan et al. 1999; Yun et al. 2000). However, cIgTCR constructs employing Fab as well as Fv fragments still require the transfection of two genes.

10.2.4
Single-Chain Antibody Fragments in Single-Chain T-Bodies

First, the use of single-chain Fv (scFv) antibody fragments consisting of antibody V domains joined by a flexible peptide linker (Bird et al. 1988) allowed cIgTCR constructs consisting of one polypeptide chain. Single-chain Fv fragments can be directly obtained by phage display from human antibody gene libraries, whereas V regions derived from hybridoma antibodies must be genetically engineered and do not always result in functional scFv fragments (Asano et al. 2000). It should be noted that scFv fragments constructed of antibody V regions of mammalian origin cannot always be expressed in E. coli. To circumvent problems of bacterial expression scFv fragments can be fused with the human IgG1 Fc part (Schirrmann and Pecher 2002, 2005). To achieve the secretory expression of these scFv-Fc proteins in mammalian cell lines an N-terminal secretory signal peptide was introduced. The scFv-Fc fusion proteins are secreted as homodimers and are more stable than scFv fragments. Finally, the complete scFv-Fc fragment, including its secretory signal peptide, can be directly used in cIgTCR constructs with the Fc domain as spacer domain between antibody fragment and transmembrane moiety (see also Section 10.2.9). For this receptor design, the scFv-Fc protein represents the complete extracellular part of the cIgTCR, thus allowing more accurate analysis of the antigen-binding properties of the receptor in comparison to scFv fragments.

10.2.5
Signal Chains

CTLs and NK cells can mediate their cytotoxicity by two different mechanisms (Fig. 10.2). owever, enhanced target cell binding by receptor-grafted effector lymphocytes alone is not sufficient to obtain an efficient tumor targeting (Schirrmann and Pecher 2001). The effector cell requires a signal for the release of its cytotoxic factors upon the target cell binding. Therefore, an appropriate signal chain must be included in the cIgTCR construct or the receptor must assemble with cellular signal molecules triggering the cytotoxic mechanisms of the effector cell. The TCR on T cells and most of the activatory and cytotoxic immune receptors on T and NK cells (Moretta et al. 2001) are noncovalently associated with transmembrane signal molecules which can serve as signal domains for cIgTCR constructs. In most studies, the ζ chain of the TCR complex and the FcεRI γ chain have been used as signal domains. These are disulfide-linked homodimers with a very short extraplasmatic domain and their signal transduction is mainly mediated by cytoplasmic activatory sequence motifs containing two tyrosine phosphorylation sites (YxxL/I) separated by 6–8 amino acids. These immunoreceptor tyrosine activation motifs (ITAMs) recruit nonreceptor

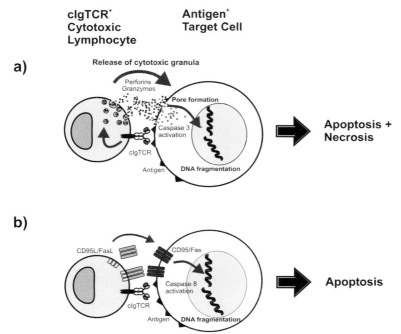

Fig. 10.2 Effector mechanisms of cIgTCR⁺ cytotoxic lymphocytes (CTLs). CTLs and NK cells are able to mediate the cytotoxicity by two different mechanisms, mediated by cIgTCR gene-modified effector lymphocytes. (a) Cytotoxic factors (perforins, granzymes) stored in specialized granules are released by exocytosis upon target cell binding. Released perforins form pores in the target cell membrane. Granzymes penetrate through these pores and activate the apoptosis cascade in the target cell. The membrane disintegration can also lead to a direct target cell lysis (necrosis). (b) The second mechanism is mediated by Fas ligand (FasL), which is expressed on most activated cytotoxic lymphocytes. FasL is a homotrimer secreted by proteolytic cleavage from the effector cell surface with a short half-life in solution. It binds to Fas⁺ (CD95, Apo-1) target cells and induces apoptosis.

protein tyrosine kinases (PTKs) of the Src family such as Syk and ZAP70 (ζ-associated protein 70) via their SH2 (Src homology 2) domains. The PTKs activate a multitude of signal pathways (Eshhar and Fitzer-Attas 1998). Figure 10.3 shows the activation of the phosphatidylinositol-3-kinase (PI3K) in the center of the cytotoxicity, which activates the mitogen activated protein kinase (MAPK) pathway. Finally, the activation of the extracellular signal regulatory kinase 2 (ERK-2) induces the mobilization of cytotoxic granules toward the bound target cell (Jiang et al. 2000).

The ζ chain contains three ITAMs whereas the FcεRI γ chain and the other signal molecules of the TCR complex contain only one ITAM. The high number of activatory sequence motifs in the ζ chain is not essential for signal transduction since CD3 ε, FcεRI γ chain, and artificial ζ chain variants with one ITAM also

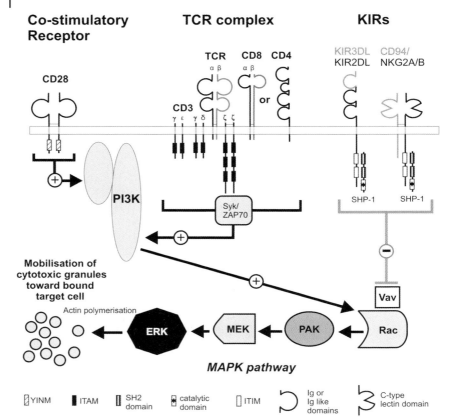

Fig. 10.3 Signal transduction in T cells. The α/β-TCR assembles with a signal complex consisting of CD3 molecules and the ζ chain. The coreceptors CD8 and CD4 enhance the sensitivity of the TCR complex. CD3 signal molecules and the ζ chain recruit nonreceptor protein tyrosine kinases (PTKs) of the Src family, such as Syk and ZAP70. These PTKs activate a multitude of signal pathways. The activation of phosphoinositide-3 kinase (PI3K) central to cytotoxicity. PI3K activates the MAPK pathway and finally ERK-2, which induces the mobilization of cytotoxic granules towards the bound target cell. The costimulatory receptor CD28 binds directly to the p85 subunit of PI3K and supports the T-cell activation. On the other hand, NK cells and T cells (at least subsets) express killer inhibitory receptors (KIRs) for HLA-C and nonclassical major histocompatibility complex (MHC) class Ib molecules. After ligand binding, KIRs recruit via their cytoplasmatic immunoreceptor tyrosine inhibitory motifs (ITIMs) the SH2-containing protein tyrosine phosphatase SHP-1, which dephosphorylates the guanosine nucleotide exchange factor Vav-1. The inactivation of Vav-1 inhibits the MAPK pathway and the granule-mediated cytotoxicity.

trigger the cytolytic function of T cells (Romeo et al. 1992; van Oers et al. 1998; Nolan et al. 1999). But the individual ITAMs of the ζ chain are recognized by different downstream signal molecules (Zenner et al. 1996) suggesting not only additional signaling events but also stronger regulation. The positive selection and T-cell maturation is more efficiently supported by the ζ chain than by the γ chain (Shores et al. 1997), whereas both signal chains seem to equally mediate T-cell development and function. T cells transfected with cIgTCR gene constructs containing the ζ chain instead of the FcεRI γ chain showed a greater capacity for tumor control (Haynes et al. 2001).

10.2.6
Signal Domains Employing Downstream Signal Molecules

Although sporadic human tumors show significant changes in their antigen expression which should activate adaptive immunity, they frequently develop mechanisms to evade or to actively suppress the tumor-directed immune response (Finke et al. 1999). T cells of tumor patients often have an altered TCR composition, blunted calcium responses, as well as functional deficiencies in the CD3 ζ chain, the PTKs p56^{lck} and p59^{fyn} and the transcription factor NFκB p65 (Salvadori et al. 1994; Correa et al. 1997; Bukowski et al. 1998; Whiteside 1999). The CD3 ζ chain seems to be proteolytically degraded by caspases 3 and 7, since no change in the mRNA level was observed. Chronic antigen stimulation without costimulation also leads to a downregulation of the ζ chain and to a substitution by the γ chain (Mizoguchi et al. 1992). To bypass receptor proximal signaling defects and to overcome the problems with tumor recognition frequently seen in cancer patients, novel approaches have been developed introducing downstream signal molecules of the TCR signal transduction pathway into cIgTCR constructs (Eshhar and Fitzer-Attas 1998). Since none of the TCR complex subunits possess their own intrinsic enzymatic activity, the activation of cytoplasmatic nonreceptor PTKs is one of the earliest signal transduction events of the TCR.

Receptor clustering is thought to stimulate PTKs of the Src family to phosphorylate the ITAMs of CD3 subunits and ζ chain, creating docking sites for the SH2 domains of ZAP70 or Syk. These Syk family PTKs are subsequently activated by phosphorylation. Together with Src family PTKs they phosphorylate downstream signal proteins involved in the PI3K activation and in the stimulation of the Ras pathway.

ZAP70 and Syk have been studied as signal domains of cIgTCR constructs (Eshhar and Fitzer-Attas 1998). Chimeras consisting of scFv-ZAP70 were totally ineffective in transducing signals for IL-2 secretion or target cell lysis (Fitzer Attas et al. 1998). In contrast, cIgTCRs utilizing Syk instead of ZAP70 were able trigger the IL-2 production and cytolysis in T cells (Fitzer Attas et al. 1997). The employment of Syk may overcome signaling defects and anergic states of the TCR as well as the need for costimulatory signals to activate T cells.

10.2.7
The Transmembrane Domain – More than Just a Membrane Anchor?

The transmembrane domain of the cIgTCR constructs is not just a membrane anchor but also plays an important role in the association with other signal components. Activating cytotoxic receptors on T or NK cells usually have a positive charge on their transmembrane domain, whereas signal chains such as the ζ or γ chain contain a corresponding negative charge. It has been noted that transmembrane moieties of CD3 components may negatively affect cIgTCR responses (Willemsen et al. 2003), but there are also results showing that the exchange of transmembrane domain did not affect the signal transduction capacity of the ζ chain (Romeo et al. 1992).

Independent studies have described different effects of transmembrane domains chosen from CD4 and CD8 in cIgTCR constructs (Roberts et al. 1994; Fitzer Attas et al. 1998; McGuinness et al. 1999). For example, the transmembrane domain of CD8α was crucial for cIgTCRs containing cytoplasmatic nonreceptor PTKs as signal domain, probably caused by different assembly and clustering of the chimeric receptor (Kolanus et al. 1993; Eshhar and Fitzer-Attas 1998). The choice of the optimal transmembrane domain may therefore depend on the respective cIgTCR construct.

10.2.8
Extracellular Spacer Domains Promote cIgTCR Expression and Function

Many cIgTCR constructs were gained with the introduction of extracellular domains between scFv fragment and transmembrane moiety. They served as spacers to increase the distance between the antigen recognition domain and the plasma membrane. Furthermore, hinge-like properties improved the accessibility of the antigen-binding region to the antigen (Moritz and Groner 1995).

Several groups have introduced the human IgG1 Fc fragment (hinge-CH_2-CH_3 domains) as a spacer domain into the cIgTCR constructs (Schirrmann and Pecher 2002, 2005). In addition to its hinge and spacer function, the Fc fragment also promotes and stabilizes cIgTCR homodimerization and suppresses an interaction with the endogenous ζ chain (Hombach et al. 1998, 1999). In contrast, cIgTCR constructs without a dimerization domain are often associated with the endogenous ζ chain. Heterodimeric complexes with the endogenous ζ chain are able to mediate cIgTCR functions but their antigen recognition and signaling may be reduced (Hombach et al. 2000a). Heterodimers formed by an scFv-γ receptor and the endogenous ζ chain reduced the antigen response of T-cell hybridomas (Annenkov et al. 1998). Other groups have employed the D3/D4 Ig-like domains of CD4, the hinge region of CD8α (Moritz and Groner 1995; Ren Heidenreich et al. 2000), and domains of CD28 (Eshhar et al. 2001). In some cIgTCR constructs the introduction of extracellular domains stabilized cell surface expression (Patel et al. 1999) and were necessary for an efficient antigen recognition and T-cell activation (Moritz and Groner 1995; Fitzer Attas et al. 1998).

In general, chimeric receptor constructs containing the ζ chain took more advantage from spacer domains than receptor constructs containing the γ chain (Patel et al. 1999). This is probably due to the stabilization of cIgTCR expression on the cell surface, preventing intracellular proteolytic breakdown of the ζ chain.

10.2.9
Accessory and Cosignaling Elements

In addition to the TCR-mediated signal, T cells require a costimulatory signal to become fully activated and to prevent anergy after TCR crosslinking. One major costimulatory receptor on T cells is CD28. The introduction of CD28 signal domains resulted in cIgTCR constructs with costimulatory properties. Double transfectants with scFv-CD28 and scFv-ζ showed synergistic signal effects (Alvarez-Vallina and Hawkins 1996). Novel strategies introduced signal domains of CD28 and ζ chain into one receptor construct. These cIgTCR constructs mediated both cytotoxic and costimulatory signals in T cells (Eshhar et al. 2001; Haynes et al. 2002a,b). Nevertheless, there are also controversial results regarding the requirement of CD28 costimulation in cIgTCR-transfected T cells (Hombach et al. 2001c; Ren-Heidenreich et al. 2002). The introduction of CD28 and CD134 (OX40) domains into one cIgTCR construct showed a more complex activation of T cells after antigen stimulation (Pule et al. 2005).

10.3
Preclinical Studies

10.3.1
Retroviral Gene Transfer into T Lymphocytes

Most studies used retroviral systems for the gene transfer of cIgTCRs into primary effector lymphocytes and effector cell lines. Optimized retroviral gene delivery systems allow the transduction of primary T cells with high efficiency (Finer et al. 1994; Weijtens et al. 1998a; Engels et al. 2003). Most of these systems are based on the Moloney murine leuke mia virus (MoMuLV), the amphotropic murine leukemia virus (A-MLV), or lentiviruses. The pseudotyping of the viral surface of MoMuLV vectors with the envelop of A-MLV or gibbon ape leukemia virus (GaLV) significantly improved the transfection efficiencies of human T cells (Gladow et al. 2000; Stitz et al. 2000; Uckert et al. 2000). In contrast, pseudotyping of HIV (human immunodeficiency virus)-derived lentiviral vectors did not improve the transfection of $CD4^+$ T lymphocytes (Muhlebach et al. 2003). Coculturing primary T lymphocytes with the retrovirus resulted in transduction efficiencies of over 75% (Altenschmidt et al. 1996, 1997). Retroviruses and derived vectors integrate into the chromosome of the host cell allowing stable, long-term transgene expression.

The tendency of retroviruses to integrate in transcription-active genomic loci, however, can also lead to interference or destruction of host genes. Studies have revealed a nonrandom preference of MoMuLV integration into fragile chromosomal sites (Bester et al. 2006). In addition, retroviral long terminal repeats (LTR) include enhancer elements that can upregulate cellular genes over long distances of more than 10 kb (Williams and Baum 2003).

Considering that there are over 100 proto-oncogenes in the human genome, oncogene dysregulation may occur in about 0.1–1% of all retroviral gene transfer events (Baum et al. 2003). The upregulation of oncogenes located near a retroviral chromosomal integration site may cause uncontrolled proliferation of the transduced cells, as was shown in a retroviral gene therapy study of X-SCID (X chromosomal linked severe cellular immunodeficiency) in young children with a defective gene of the common γ chain of the IL-2 receptor (Hacein-Bey-Abina et al. 2003a,b). However, extensive studies with retroviral transduced T cells did not reveal similar side effects after infusion into patients (Rosenberg et al. 1990; Rosenberg 1991). Therefore, the transfection of undifferentiated hematopoietic stem and precursor cells may have side effects different to those produced by gene modification of differentiated lymphocytes.

Another shortcoming of retroviral transduction is that only proliferating cells can be transduced. Accordingly, peripheral blood lymphocytes must be activated by IL-2 or mitogens like phytohemagglutinin (PHA) before retroviral transfection. This unspecific activation may lead to nonspecific activity or activation-induced cell death.

10.3.2
"Naked" DNA Gene Delivery Systems

In contrast to retroviral transduction, most plasmid vector-based gene delivery systems do not achieve high transfection efficiencies in primary lymphocytes and many lymphoid cell lines. Furthermore, they generally suffer from an ineffective chromosomal integration in human cells. The selection of transfected cells using antibiotics according to the selection marker gene does not ensure the cIgTCR expression in all transfected cells. In particular, cDNA-based transgenes are often downregulated or inactivated. Therefore, the use of genomic sequences instead of pure cDNA sequences should be considered to stabilize and increase transgene expression (McKnight et al. 1995). Alternatively, the insertion of intron sequences into the nontranslated 5' sequence or into the transgene can stabilize expression (Yew et al. 1997).

The chromosomal integration site, methylation, and chromatin accessibility of a transgene have a great influence on the level and stability of its expression (Garrick et al. 1996; Schubeler et al. 2000). Homologous genome sequences (Li and Baker 2000) or certain viral sequences (Kogure et al. 2001) flanking the transgene can enhance the probability and localization of chromosomal integration. The cIgTCR gene transfer of primary T cells using plasmid vectors has been demonstrated by electroporation (Jensen et al. 2000, 2003). Novel "naked" DNA transfection protocols such as the Nucleofector technique promise very high transfection rates, comparable to those of retroviral vectors for primary cells as well as effector cell lines (Gresch et al. 2004; Maasho et al. 2004).

10.3.3
Enrichment of cIgTCR-Transfected Effector Cells

The clinical employment of cIgTCR-grafted effector lymphocytes requires stable and high expression of the receptor. Therefore, enrichment of cIgTCR-expressing cells is often required. For example, 2 days after electroporation only 1–5% cells of human NK cell line YT showed cIgTCR expression, and this did not increase to more than 10–15% cIgTCR$^+$ YT cells after antibiotic selection. However, after enrichment using the magnetic activated cell sorting (MACS) system and antibodies against the Fc spacer domain, levels of over 90% cIgTCR$^+$ YT cells could be obtained (Schirrmann and Pecher 2002). Immunological enrichment was also described for cIgTCR-transfected T cells (Weijtens et al. 1998a; Nolan et al. 1999; Beecham et al. 2000; Hombach et al. 2001c). However, crosslinking of cIgTCRs by antibodies may lead to apoptosis or reciprocal lysis of cIgTCR$^+$ lymphocytes. Therefore, antibodies for cell sorting should be carefully chosen and tested. Importantly, T cells undergo activation-induced cell death if their TCR or CD3 complex is stimulated without costimulation (Li et al. 2000) as well as NK cells stimulated with IL-2 or IL-12 after CD16 crosslinking (Ortaldo et al. 1995; Jewett 2001).

10.3.4
Effector Functions of cIgTCR Gene-modified Effector Lymphocytes

The specific effector response of T and NK cells and cell lines transfected with different cIgTCR gene constructs has been demonstrated in many independent studies as summarized in Table 10.2. After cIgTCR gene transfer, T cell lines and hybridomas, peripheral blood T lymphocytes, tumor-infiltrating lymphocytes (TILs), and NK cell lines are capable of intracellular Ca^{2+} mobilization and the secretion of different cytokines such as IL-2 and IFNγ upon stimulation by antiidiotypic antibodies as well as antigen bound on plastic surfaces or expressed on target cells. Thus, cIgTCR crosslinking specifically induces signal transduction in different gene-modified effector cell types. Furthermore, cIgTCR$^+$ cytotoxic lymphocytes specifically lyse target cells expressing the corresponding antigen in an MHC-unrestricted antibody-like manner. The lysis is specifically mediated against target cells expressing the antigen without major bystander killing. However, cIgTCR-transfected peripheral T cells often show an unspecific lysis caused by T-cell activation with IL-2 or phytohemagglutinin (PHA) required for retroviral transduction (Beecham et al. 2000; Eshhar et al. 2001). Gene-modified T cells with high cIgTCR expression were able to lyse target cells with low antigen expression, in contrast to T cells with a low cIgTCR expression (Weijtens et al. 2000). Therefore, high cIgTCR expression on gene-modified effector cells can overcome problems caused by low antigen expression on target cells. In addition,

Table 10.2 Studies with chimeric immunoglobulin receptors.

Recognition domain[a] (specifcity)	Spacer	TM	Signal domain	Effector cell	Response and remarks	References
Two-chain Fv (PC)	Cα/Cβ	Cα/Cβ	Cα/Cβ	Murine EL4 T cell line	Intracellular calcium release	Kuwana et al. 1987
Two-chain Fv (TNP)	Cα/Cβ	Cα/Cβ	Cα/Cβ	Murine CTL hybridoma MD45	Lysis of hapten-modified target cells	Gross et al. 1989b
Only VH (PC)	Cα/Cβ	Cα/Cβ	Cα/Cβ	EL4	VHCα/VHCβ and VHCβ/VHCβ dimers	Goverman et al. 1990
Sp6 (TNP)	No	ζ; γ	ζ; γ	MD45	Lysis of hapten-modified target cells	Eshhar et al. 1993
N29 (ErbB-2)	No	ζ	ζ	MD45	IL-2 secretion; lysis of ErbB-2$^+$ cells	Stancovski et al. 1993
Sp6; N29; MOv18 (TNP; ErbB-2; FBP)	No	ζ or γ	ζ or γ	Rat basophilic leukemia mast cell line RBL-2H3	Ag-specific degranulation	Bach et al. 1994
84.1c (mouse IgE)	No	Cβ	Cβ	Human leukemia line Jurkat; MD45	Lysis of IgE-secreting BC; suppression of IgE production	Lustgarten and Eshhar 1995
Sp6 (TNP). MOV18 (FBP)	No	ζ or γ	ζ or γ	Human CD8$^+$ TIL	GM-CSF secretion; target cell lysis	Hwu et al. 1993a
TR66 (human CD3e)	CD8α hinge	ζ	ζ	Murine T cell hybridoma BW5147	Ag-specific production of IL-2 and IL-3	Brocker et al. 1993
TR66 (human CD3e)	No	Cβ	Cβ	BW5147	Assembly with endogenous TCR α chain	Brocker et al. 1996

10.3 Preclinical Studies | 547

Target	Spacer	Chain 1	Chain 2	Cell type	Result	Reference
CD4; 98.6 (gp120; gp41)	CD4; Fc	CD4; CD4	ζ; ζ	Jurkat; human CTL	IL-2 secretion; proliferation and target lysis (only PBL)	Roberts et al. 1994
FRP5 (ErbB-2)	CD8α	ζ	ζ	Murine CTL line Cl96 (C57/BL6 mice)	Target lysis; localisation at tumor sites; retarded ErbB-2+ tumor growth in nude mice	Moritz et al. 1994
CD4 (gp120, HIV)	CD4	CD4	ζ	Human NK cell clone NK3.3	Transfection of NK cells, lysis of gp120+ B cell line	Tran et al. 1995
MOv18 (FBP)	No	γ	γ	Murine CD8+ TILs	In vivo model	Hwu et al. 1995
FRP5 (ErbB-2)	CD4 or CD8α	ζ	ζ	Jurkat; Cl96	IFNγ production; target lysis; requirement of spacer domain	Moritz and Groner 1995
1.1ASML (CD44 exon 6 variant)	CD8α	ζ	ζ	Cl96	Target lysis; suppression of tumor cell xenografts	Hekele et al. 1996
C11; FRP5 (rat; human ErbB2)		ζ	ζ	Murine and rat TC; TC lines	Optimized retroviral gene transfer into TC	Altenschmidt et al. 1996
G250 (CAIX)	No	CD4; γ	γ	Human activated PBLs	Target lysis 4.5 months receptor expression; recycling of lytic activity	Weijtens et al. 1996
B1.8; NQ10, 12.5 (NIP;) Hawkins 1996	CD28: no	CD28: ζ	CD28: ζ	Jurkat	CD28 like costimulation, enhanced IL-2 production, synergy with scFvζ receptor	Alvarez-Vallina and phOx
Sp6 (TNP)	CD8α	CD8α; CD4	Syk; ZAP70	CTL hybridoma MD45.27]	IL-2 production; target lysis; Syk promotes CBL but not PLCγ phosphorylation	Fitzer Attas et al. 1997 1998
B72.3 (TAG72)	No	γ; ζ	γ; ζ	MD45	Only scFvγ expression	Hombach et al. 1997
CC49 (TAG72)	Fc	γ; ζ	γ; ζ		scFv-Fcγ/ζ expression	Hombach et al. 1998

Table 10.2 Continued

Recognition domain[a] (specificity)	Spacer	TM	Signal domain	Effector cell	Response and remarks	References
FRP5 (ErbB2)	No	ζ	ζ	Murine TC	Tumor regression in syn-geneic mice; no antibody response to modified TC	Altenschmidt et al. 1997
C2 (Kollagen-II)	No	γ	γ	Murine CTL hybridomas	Heterodimers with ζ chain reduce antigen response	Annenkov et al. 1998
CD4 (gp120)	CD4	CD4	ζ	Hematopoietic stem and precursor cells	Differentiation to transgeneic NK cells and neutrophils	Roberts et al. 1998
G250 (CAIX)	Diverse	Div.	γ; ζ; ε	Human CTL	STITCH retroviral vector	Weijtens et al. 1998a
G250 (CAIX)	Fc	γ	γ	Human TC	Lysis supported by high cIgTCR expression and adhesion molecules	Weijtens et al. 1998b, Weijtens et al. 2000
447D (gp120)	Fc; CD4; CD7; CD8α	CD4; CD8α	ζ	Human TC	Impact of extracellular domains	Patel et al. 1999
BW431 (CEA)	Fc	γ; ζ	γ; ζ	MD45; human TC	No inhibition by 25 µg mL^{-1} soluble CEA	Hombach et al. 1999, 2000b
763.74 (HMW-MAA)	No	γ	γ	MD45	Melanoma	Reinhold et al. 1999
MN14 as scFv and Fab' (CEA)	CD8α; no	ζ; ε	ζ; ε	Human TC	scFv- and Fab-ζ/ε	Nolan et al. 1999
ErbB2	No	γ	γ	MD45	FasL-mediated cytotoxicity	Haynes et al. 1999
CC49 (TAG72)	Fc:hinge -CH3	CD4	ζ	Human TC	No "bystander killing," xenograft mouse model	McGuinness et al. 1999

scFv (target)	Spacer	Signal	Cells	Comments	Reference
CO17.1A, GA733 (EGP40)	no	γ	Human PBL, murine 14.1 cell line	Only GA733-clgTCR with specific cytokine response	Daly et al. 2000
TR66 (human CD3ε)	No; no	ζ; ε	clgTCR-transgeneic mice	IL-2 stimulation required; syngeneic tumor model	Brocker 2000
Col1; CC49; 98.6, 447D, CD4 (CEA; TAG72; HIV)	CEA:a3b3; Fc; CD4	CD4	ζ	Human TC	Bispecific clgTCRs; dual specificity	Patel et al. 2000
GA733.2 (EGP-2)	No; CD8α	γ; γ or ζ	γ; γ or ζ	Human TC	Hinge domain improves properties, costimulation not necessary	Ren Heidenreich et al. 2000, 2002
MB3.6 (GD3)	CD8α; no	ζ; ε	ζ; ε	Human TC	Target lysis, no inhibition by 100μgmL⁻¹ soluble GD3	Yun et al. 2000
MN14 (CEA)	CD8α	ζ	ζ	Human TC	Target lysis; selection of CEA-tumor cell population	Beecham et al. 2000
HRS3 (CD30)	No; Fc	γ	γ	Human TC	Impact of Fc spacer domain	Hombach et al. 2000a,b
BW431; HRS3 (CEA; CD30)	Fc; no	γ; ζ	γ; ζ	Human CD4⁺ TC	Target lysis by clgTCR+ CD4⁺ TC	Hombach et al. 2001a
BW431; HRS3 (CEA; CD30)	Fc	γ; ζ	γ; ζ	Human CD4⁺ TC; PBL	ext. CD28 costimulation enhances IL-2 but not IFNγ, proliferation and cytotoxicity	Hombach et al. 2001c
A5B7?; (CEA) 14.G2a (GD2)	CD8α No	γ; ζ ζ	γ; ζ ζ	MD45 Human TC; EBV specific TC	ζ stronger signal than γ Lysis of neuroblastoma; dual specificity against EBV⁺ and GD2⁺ targets	Haynes et al. 2001 Rossig et al. 2001; Roessig et al. 2002
A5B7?; FRP5? (CEA; ErbB2)	CD8α	CD28	CD28⁺ζ	Murine TC	Enhanced IFNγ secretion; syn-/xenogeneic model	Haynes et al. 2002a,b

Table 10.2 Continued

Recognition domain[a] (specificity)	Spacer	TM	Signal domain	Effector cell	Response and remarks	References
FRP5 (ErbB2)	CD8α	ζ	ζ	Human NK cell line NK92	Lysis of Erb+ tumor cell lines	Uherek et al. 2002
BW431/26 (CEA)	Fc	ζ	ζ	Human NK cell line YT	Target lysis; tumor growth inhibition by irradiated clgTCR+ YT cells in vivo	Schirrmann and Pecher 2002
C2 (Collagen-II)	No; CD8α	γ; ζ	γ; ζ	Murine KLH-specific Th1 and Th2 cells	clgTCR stimulation optimal for IFNγ but not for IL-4	Annenkov et al. 2003
FRP5? (ErbB2)	CD8α	CD28	CD28+ ζ	Murine TC	Functional role of internal CD28 domain	Moeller et al. 2004
HuM195 (CD33)	Fc	ζ	ζ	YT	Lysis of AML cell line	Schirrmann and Pecher 2005
Hu3S193 (LeY)	?	CD28	CD28+ ζ	Human TC	Lysis of ovarian tumor	Westwood et al. 2005
14g2a (GD2)	IgG1-hinge	CD28	CD28+ CD134+ ζ	Human PBL	CD28+CD134 domains augment TC activation	Pule et al. 2005

a scFv fragments if not otherwise indicated.

Ag, antigen; AML, acute myeloid leukemia; BC, B cells; Cα/Cβ, constant TCR domains; CAIX, carboanhydrase IX on RCC; CD4, TCR coreceptor, binds to HIV gp120; CD7, antigen on TC; CD8α, TCR coreceptor, hinge (aa 105–165); CD19, B-cell lineage differentiation marker; CD28, costimulatory receptor, binds to CD80 (B7.1); CD30, Hodgkin lymphoma marker; CD33, myeloid leukemia marker; CD134 (OX40), costimulatory receptor; CEA, carcinoembryonic antigen; CTL, cytotoxic T lymphocyte; EGP, epithelial glycoprotein; ErbB-2 (Neu/HER2), human breast cancer-associated antigen; FasL, Fas or CD95 ligand; FBP, folate-binding protein; Fc, IgG1 Fc domains (hinge, CH2CH3); GD2, ganglioside on neuroblastoma; GD3, ganglioside on melanoma; gp41 and gp120, glycoprotein 41 and 120 of HIV; HIV, human immunodeficiency virus; HMW-MAA, high molecular weight melanoma-associated antigen; KLH, keyhole limpet hemocyanin; NIP, 4-hydroxy-5-iodo-3-nitrophenyl-acetyl; NK, natural killer; PBL, peripheral blood lymphocytes; PC, phosphorylcholin; phOx, 2-phenyl-2-oxazolin-5-one; PLCγ, phospholipase C γ; PTK, protein tyrosine kinase; RCC, renal cell carcinoma; Syk, non-receptor PTK; TAG72, mucin antigen on adenocarcinoma; TC, T cells; Th, T helper cells; TIL, tumor infiltrating lymphocyte; TM, transmembrane; TNP, 2,4,6-trinitrophenyl; ZAP70, ζ associated protein 70, nonreceptor PTK.

the cytotoxicity of cIgTCR$^+$ effector lymphocytes is also supported by adhesion molecules (Weijtens et al. 1998b).

10.3.5
Influence of Soluble Antigen

One major limitation of antibody-mediated antigen recognition is the interference with soluble antigen. This must also be considered for cIgTCR-transfected effector cells (Goverman et al. 1990; Eshhar et al. 1993; Daly et al. 2000), which is of particular interest if the tumor antigen occurs in soluble form in the serum. For example, patients with colorectal carcinomas frequently show serum levels of 1 µg mL^{-1} carcinoembryonic antigen (CEA) (Moertel et al. 1986). Higher concentrations at the tumor site must be expected. In studies using the NK cell line YT transfected with a CEA-specific cIgTCR construct, no inhibition of cytotoxicity against CEA$^+$ target cells was detected in the presence of 10 µg mL^{-1} soluble CEA (Schirrmann and Pecher 2002). Similar results for T cells transfected with CEA-specific cIgTCR constructs were demonstrated by other groups (Hombach et al. 1999; Beecham et al. 2000).

The cytotoxicity of T cells transfected with a GD3-specific cIgTCR was not blocked in the presence of 100 µg mL^{-1} soluble GD3 ganglioside (Yun et al. 2000). The large number of receptor–antigen interactions between cIgTCR$^+$ effector cells and antigen-expressing target cells seem to prevent in most cases efficient inhibition of target cell lysis by physiologically relevant concentrations of soluble antigen. It was approximated that over 80% of the cIgTCRs on the transfected effector lymphocytes can be blocked by soluble antigen without inhibiting cytotoxicity (Beecham et al. 2000). The cytotoxicity of effector cells transfected with a cIgTCR construct containing an Fc spacer domain was inhibited at about 5–10 times lower concentrations of soluble antigen than cIgTCR constructs without an Fc spacer domain (Hombach et al. 2000a). But in this study, a homodimeric Fc fusion protein of the antigen was used, which binds more efficiently to Fc spacer-containing homodimeric cIgTCR constructs than to Fc spacer-less monomeric cIgTCR variants. Thus the introduction of multimerization domains into cIgTCR constructs should be considered if the antigen occurs as multimeric form in the serum.

10.3.6
Animal Models

Several studies demonstrated that cIgTCR gene-modified effector lymphocytes can inhibit the development of tumors *in vivo* if injected together with tumor cells in mice. For example, the simultaneous injection of CEA$^+$ tumor cells and irradiated cIgTCR gene-modified cells of the NK cell line YT significantly inhibited tumor growth in NOD/SCID (nonobese diabetic/severe cellular immunodeficient) mice (Schirrmann and Pecher 2002). After subcutaneous coinjection of human T cells transduced with a TAG72-specific cIgTCR construct and TAG72$^+$

tumor cells into NOD/SCID mice three out of four mice survived and did not develop tumors. Furthermore, in four of four mice the intraperitoneal coinjection of these cIgTCR$^+$ T cells inhibited the development of TAG72$^+$ KLE B cell tumors (McGuinness et al. 1999).

However, the simultaneously injection of tumor cells and effector cells is an artificial approach that is far away from describing the situation for the therapy of tumor patients. Tumors are usually diagnosed when the tumor has already reached a certain size. Therefore, experimental approaches should be adapted to this situation. So far, only a few studies have been published investigating the treatment of established tumors with cIgTCR-grafted effector cells. Hanson et al. (2000) showed a kinetic dependency between tumor size and number of adoptively transferred murine CTLs expressing a tumor-specific TCR in a syngeneic mouse tumor model. They could even control bigger tumors. The dependency of tumor size and the number of effector cells required for adoptive immunotherapy was also shown for irradiated cells of the cytotoxic T cell line TALL-104. Multiple intratumoral injections of these MHC nonrestricted TALL-104 cells reduced smaller implanted human tumors (<150 mg) about 50–75% and suppressed the formation of lung metastasis (Cesano et al. 1998).

10.4
Therapeutic Considerations

10.4.1
Adoptive Cellular Immunotherapy

The basic concept of adoptive cellular immunotherapy can be dated back to the early 1960s when J.L. Gowans identified lymphocytes as most important mediators of immune response and as the carriers of immunological memory (Gowans and Uhr 1966). He also demonstrated that immunological competence can be transferred from one individual to another by the adoptive transfer of lymphocytes. In animal experiments, the adoptive immunotherapy of tumor diseases was successfully demonstrated (Rosenstein et al. 1984; Hanson et al. 2000). However, adoptive immunotherapy strongly depends on the availability of a large number of immunological effector cells and their tumor specificity. In contrast to animal models, it is difficult, time-consuming, and expensive to provide sufficient numbers of tumor-specific lymphocytes for the individual human patient. These effector cell resources could be improved by gene modification with tumor-specific cIgTCR gene constructs.

10.4.2
Autologous Approaches

Human autologous lymphocytes have been extensively studied for use in adoptive immunotherapy (Table 10.3). Peripheral blood lymphocytes can be activated to

Table 10.3 Characteristics of cytotoxic lymphocytes useful for adoptive immunotherapy.

Effector type	CTLs	TILs	NK cells	LAK cells	CIK cells
Source	Peripheral blood lymphocytes	Tumors, metastatic or tumor draining lymph nodes	Peripheral blood and bone marrow	NK cells and CTL activated by IL-2	Subset of T lymphocytes activated by cytokines
In vitro expansion					
Tumor stimulation	Yes	None	None	None	None
Requirement of IL-2 for response	++++	++++	++++ (CD56 dim) + (CD56 bright)	++++	++++[a]
Duration of culture	6 weeks	4 weeks	2–3 weeks	2–5 days	2–5 weeks
Target cells *in vitro*	Allogeneic cells	Autologous tumor cells	K562	Raji, Daudi	Autologous and allogeneic tumor cells
In vitro cytotoxicity					
Specificity	MHC restricted to allogeneic cells	MHC–TAA restricted to autologous tumor	Spontaneous lysis of virus infected and certain tumor cells	Lyse a wide spectrum of tumor cells (also NK resistant cells)	Cytotoxic activity superior to LAK; lyse autologous or allogeneic CML blasts
ADCC	CD16+ subsets	None	Yes	CD16+ subsets	Reverse ADCC
Effector phenotype	CD3+/4+,CD3+/8+,CD3+/8+/16+	CD3+/8+/56+	CD3-/16+/56+	CD56+/25+	CD3+/56+

a Additional stimulation by IFNγ, IL-12, and anti-CD3 monoclonal antibody.
b ADCC, Antibody dependent cellular cytotoxicity; CIK, Cytokine-induced killer; CML, Chronic myeloid leukemia; CTL, cytotoxic T lymphocytes; LAK, lymphokine-activated killer; NK, natural killer; TIL, tumor infiltrating lymphocytes.

produce lymphokin-activated killer (LAK) cells in the presence of high concentrations of IL-2. Although these lyse a broad range of tumor cell lines *in vitro*, therapies employing LAK cells showed only limited success in clinical trials due to their low tumor specificity and temporary activation (Bordignon et al. 1999). Peripheral blood lymphocytes were also activated with IL-2, IL-12, IFNγ, and anti-CD3 monoclonal antibodies to cytokine-induced killer (CIK) cells, mediating enhanced cytotoxic properties but still with limited specificity (Schmidt-Wolf et al. 1993; Kornacker et al. 2006). Tumor-infiltrating lymphocytes (TILs) isolated from solid tumors and tumor-draining lymph nodes frequently contain tumor-specific T cells that can be reactivated *ex vivo* with IL-2. The combination of TIL therapy and IL-2 treatment has shown some therapeutical effects against metastatic melanoma and renal cell carcinoma (Yannelli et al. 1996; Bordignon et al. 1999), but in contrast to LAK cells, their preparation is complicated and unreliable for individual patients (Hoffman et al. 2000).

Autologous T cells can also be sensitized *in vitro* (Lipshy et al. 1997). Human CTLs have been stimulated with Epstein–Barr virus (EBV)-immortalized autologous B cells *in vitro* and used for the prevention of EBV-associated lymphoproliferative diseases after bone marrow transplantation (Rooney et al. 1995, 1998).

NK cells are another type of effector lymphocytes. Unlike T cells, they do not express individual antigen receptors. They mediate an MHC-unrestricted cytotoxicity against a broad spectrum of virus-transformed cells or tumor cells without prior sensitization (Cervantes et al. 1996; Uharek et al. 1996). The ability of NK cells to eliminate tumor cells suggests that they are also involved in the control of cancer and that their presence and state of activation could be important for the outcome of the disease and finally in the treatment of tumors (Barlozzari et al. 1983). In the presence of high doses of IL-2, adherent NK cells can be generated with an advanced activity against tumor and metastatic cells (Vujanovic et al. 1995). Although there are many hopeful data from animal experiments, only few clinical studies using autologous NK cells have been published so far (Lister et al. 1995; Nalesnik et al. 1997).

10.4.3
Allogeneic Approaches

Autologous approaches generally suffer from the requirement to isolate, modify, and expand lymphocytes from each patient, which is unreliable in the context of a clinical routine application. Furthermore, the production of sufficient autologous tumor-specific lymphocytes is time-consuming and cannot be guaranteed because advanced tumor diseases are often accompanied by immune suppression (Finke et al. 1999). Contamination with the patient's tumor cells must also be excluded. In view of these limitations, the employment of allogeneic donor lymphocytes could be advantageous because their provision does not depend on a single patient. Allogeneic donor lymphocytes can mediate a graft-versus-tumor effect. However, allogeneic T lymphocytes are also responsible for graft-versus-host disease (GvHD), which may cause the death of the recipient (Bordignon et al. 1999). The control of GvHD is still a major problem in allogeneic stem cell transplantations and donor lymphocyte infusions. In recent years, allogeneic NK cells have become the focus of attention for allogeneic stem cell transplantation. Allogeneic NK cells are able to mediate an antitumor effect and reduce rejection of the transplant by eliminating residual immune cells of the recipient. But unlike allogeneic T cells, they do not cause GvHD (Ruggeri et al. 2001, 2002).

10.4.4
Gene Modification of Lymphocytes to Enhance Specificity

The specificity and effector properties of cytotoxic lymphocytes can be improved by gene modification. The development of recombinant TCR gene constructs (Eshhar 1997; Calogero et al. 2000), together with efficient retroviral gene transfer systems for activated human T cells and TILs (Finer et al. 1994; Uckert et al. 1998; Weijtens et al. 1998a; Engels et al. 2003), could overcome the limitations of the LAK and TIL therapies, such as the lack of specificity or complicated isolation and expansion of the effector lymphocytes, since these technologies allow a large number of tumor-specific T cells to be generated within a short time-scale. Human T cells transfected with recombinant TCR genes have been successfully targeted to virus-infected and tumor cells *in vitro* and *in vivo* (Table 10.2).

10.4.5
Effector Cell Lines – The Way to Cell-based Therapeutics?

Therapeutic strategies employing primary autologous or allogeneic lymphocytes can be performed only for individual patients or small patient groups. The "individualized" provision of tumor-specific lymphocytes is time-consuming, expensive, and success cannot be guaranteed. Therefore, allogeneic effector cell lines with antitumor properties could be an alternative source for adoptive immunotherapy. Allogeneic effector cell lines can be expanded without limitation. Furthermore, their employment is not limited to the individual patient. The production of effector cell lines is independent of patients or donors and can be performed in large-scale processes, allowing a high degree of standardization combined with a reduction of technical effort, time, and costs. Tumor-specific effector cell lines may open the way for the broad clinical use of adoptive immunotherapy comparable to immunopharmaceuticals such as antibodies or cytokines.

In clinical trials, however, the unlimited growth of effector cell line must be prevented, for example, by irradiation. Allogeneic cytotoxic cell lines with antitumor properties have been investigated in preclinical and clinical studies. The MHC nonrestricted cytotoxic T cell line TALL-104 was examined for adoptive immunotherapy of tumor diseases in mice models (Cesano et al. 1996b, 1997b), in dogs (Cesano et al. 1996a, Visonneau et al. 1999), and in a phase I trial in patients with advanced breast cancer (Visonneau et al. 2000). TALL-104 cells were well tolerated without any cellular immunization. Moreover, the human NK cell line NK92 irradiated with 1000 rad was used for the purging of bone marrow samples in leukemia patients (Maki et al. 2003) and has been tested in a phase I clinical trial in patients with advanced cancer (Tonn et al. 2001). The expansion and production of standardized lots of human effector cell lines TALL-104 (Visonneau et al. 2000) and NK92 (Tam et al. 2003) has been described.

Nevertheless, there are only a small number of human effector cell lines available. Therefore, it would be interesting to extend the spectrum of recognized tumor types of established effector cells by gene modification with tumor-specific cIgTCR genes. Schirrmann and Pecher (2002, 2005) were able to successfully redirect the IL-2-independent growing human NK cell line YT by cIgTCR gene transfer against CEA^+ and $CD33^+$ tumor cells without the need for retroviral gene delivery systems. After simultaneous transfer of the irradiated $cIgTCR^+$ YT cells and CEA^+ MC32A tumor cells tumor growth was significantly inhibited in NOD/SCID mice. Irradiation decreased the vitality and cytotoxicity of the $cIgTCR^+$ YT cells within days, but cytotoxicity was still sufficient. The gene modification of this effector cell line with suicide genes that convert a nontoxic prodrug into a cytotoxic compound could be an alternative approach to irradiation to prevent unlimited growth (Lal et al. 2000).

10.4.6
Clinical Studies with cIgTCR Gene-modified T Lymphocytes

Table 10.4 gives an overview of clinical trials with receptor gene-modified autologous peripheral T lymphocytes, although only the results of single studies have been published so far. Walker et al. (2000) studied the transfer of T cells obtained from a syngeneic twin and gene modified with a CD4-ζ receptor. Single and multiple infusions of 10^{10} gene-modified $CD8^+$ T cells resulted in a peak fraction of 10^4–10^5 gene-modified cells per 10^6 mononuclear cells after 24–48 h, which declined 100–1000 times within 8 weeks. In order to provide a longer, high-level persistence of the transferred gene-modified T cells, a second series of infusions containing gene-modified $CD4^+$ and $CD8^+$ T cells were used and costimulated *ex vivo* with anti-CD3- and anti-CD28-coated beads. Sustained fractions of about 10^3–10^4 gene-modified T cells per 10^6 total T cells persisted for at least a year. The cell infusions were well tolerated and were not associated with substantive immunological or virological changes. Subsequent studies with CD4-ζ gene-modified T cells achieved prolonged survival in HIV-infected adults (Mitsuyasu et al. 2000).

A phase I/II trial of adoptive immunotherapy against metastatic renal cell cancer (RCC) using autologous T lymphocytes transduced *ex vivo* with a gene encoding a cIgTCR containing an scFv fragment derived from the carbonic anhydrase IX (CAIX)-specific murine monoclonal antibody G250, the CD4 transmembrane domains and the FcεRI γ signal chain were initiated (Bolhuis et al. 1998). The results for the first three patients have been published (Lamers et al. 2006). After retroviral gene transfer 52–76% scFv(G250)-γ expression was achieved on the transduced T cells with a mean number of 2–7 copies per scFv(G250)-$γ^+$ T cell. All patients were treated with the transduced T cells in an inpatient dose-escalation scheme of intravenous doses of 2×10^7 cells at day 1, 2×10^8 cells at day 2, 2×10^9 cells at days 3–5 (cycle 1) and 2×10^9 cells at days 17–19 (cycle 2), in combination with IL-2 (5×10^5 U m^{-2} twice daily administered subcutaneously at days 1–10 and days 17–26). The time period during which the transduced cells could be detected in the circulation were up to 32 days by flow cytometry and up to 53 days by PCR, depending on the sensitivity of the method of 14 copies per 100 ng DNA or 0.008% and 0.06%, respectively (Lamers et al. 2005). Infusions

Table 10.4 Clinical trials with receptor gene-modified lymphocytes.

Phase	Disease	Antigen	Receptor construct	Effector cells	References and remarks
I[a]	Metastatic melanoma	MART-1 (HLA-A2)	TCR	Allogeneic CTL line C Cure 709	Gene-modified effector cell line; intratumoral injections (Duval et al. 2006)
I/II[b]	RCC	CAIX	scFv (G250)-TM (CD4)-γ	Autologous TC	(Lamers et al. 2005, 2006)
I[c]	CD20$^+$ B cell lymphoma	CD20	scFv-Fc-ζ	Autologous PBL	Naked plasmid DNA gene transfer (Jensen et al. 2000, 2003)
I/II[a]	HIV	gp120	CD4-ζ	Syngeneic/autologous PBL	(Mitsuyasu et al. 2000, Walker et al. 2000)
I[a]	CEA$^+$ tumors	CEA	scFv-ζ	Autologous PBL	(Referred in Ma et al. 2002)
I[c]	Ovarian cancer	FBP	scFv-γ	Autologous PBL	Trial number 96C0011
I[d]	Prostate cancer	PSMA	scFv-ζ	Autologous PBL	Preclinical data in (Gade et al. 2005)
I/II[d]	Neuroblastoma	GD2	scFv(14g2a)-CD8α/hinge-CD28-CD134-ζ	Autologous PBL	Texas Children's Hospital, Houston, USA (NCT00085930)

Abbreviations as in Table 10.2 except: PSMA, prostate-specific membrane antigen.
a Clinical study performed, published.
b Clinical study in progress, preliminary data published.
c Clinical study performed, no report available.
d Intended studies.

of these gene-modified T cells were initially well-tolerated. However, after four or five infusions, liver enzyme disturbances developed, reaching National Cancer Institute Common Toxicity Criteria grades 2 to 4, probably caused by a specific attack of the scFv(G250)-γ$^+$ T cells against CAIX$^+$ bile duct epithelial cells. Furthermore, all three patients developed low levels of anti-scFv(G250)-γ antibodies between 37 and 100 days after the start of T cell therapy, which were directed against the G250 idiotype. Remarkably, these responses were less frequent in RCC patients treated with weekly infusions of 50 mg chimeric G250 mAb (6–30% of patients), indicating that the expression of scFv(G250)-γ on the cell membrane of T cells elicits a relatively efficient immune response against the murine G250 idiotype of the cIgTCR. In order to prevent liver toxicity in future patients, the infusion of 5 mg chimeric G250 antibody 3 days before the first infusion of gene-modified T cells will be included into the clinical protocol.

Recently, the results of a phase I dose-escalation trial with C Cure 709 cells, an allogeneic CTL cell line transduced with an HLA-A2-specific TCR encoding gene recognizing MART-1⁺ tumor cells, have been published (Duval et al. 2006). Fifteen patients received a total of 24 intratumoral injection cycles of C Cure 709 cells. Toxicity was minor to moderate and most common injection site reactions were fever, fatigue, nausea/vomiting, and arthralgia/myalgia. Side effects disappeared in general within 24 h. Toxicity was not dose-dependent. One patient obtained a partial response, encompassing both metastases used and not used for intratumoral injections. The remaining patients did not achieve an overall response. In addition, local regression of metastases used for injection in two patients and of metastases not used for injection in one patient was observed. Intratumoral injections of C Cure 709 are feasible, safe, and capable of inducing tumor regression.

10.5
Perspectives

10.5.1
Multispecific Approaches

The treatment of sporadic human tumors by monospecific cIgTCR⁺ effector cells results in a selection of antigen-negative tumor cells because of their heterogenEous antigen expression (Beecham et al. 2000). The employment of multispecific effector cells could reduce the probability of selection of antigen-negative tumor cells. Allogeneic immunized T cells transfected with tumor-specific cIgTCR showed antigen specificity paired with alloreactivity and antitumor activity *in vivo* (Kershaw et al. 2002). The transfection of EBV-specific T cells with a CD19- or GD2-specific cIgTCR construct led to a dual specificity (Roessig et al. 2002; Rossig et al. 2002). Furthermore, only EBV-specific T cells but not primary T cells could be maintained long term in the presence of EBV-infected B cells. The gene transfer of bispecific cIgTCR constructs into T cells expanded their recognition properties to two different antigens (Patel et al. 2000). The transfection of two or more cIgTCR constructs into cytotoxic lymphocytes could also enhance their target recognition. However, the transfection and expression of several cIgTCRs in one cell may cause conflicts because of heterodimer formation and the competition for the same signal transduction pathways. Hence, it seems to be more practicable to use a mixture of different effector lymphocytes transfected with one cIgTCR gene construct.

The effector properties of effector lymphocytes could be enhanced by transfection with cytokine genes (Hwu et al. 1993b), but cytokines such as GM-CSF (granulocyte-macrophage colony-stimulating factor) and TNF-α (tumor necrosis factor α) can also promote the growth of tumors *in vivo* (Bordignon et al. 1999). In addition, TNF-α gene transfer induced apoptosis in transfected T cells (Ebert et al. 1997).

10.5.2
Tumor Taxis and Application of cIgTCR⁺ Effector Cells

Another important property of effector cells in adoptive immunotherapy is the ability to accumulate in the tumor, a prerequisite for the systemic application of cIgTCR⁺ effector cells. After injection into the tail vein of mice the murine CTL line CI96 transfected with an ErbB2-specific cIgTCR specifically accumulated in established ErbB2-transfected NIH3T3 tumors. The tumor growth was even slightly retarded (Moritz et al. 1994). Tumor-specific chemotaxis mediated by the ζ signal chain of the cIgTCR seems to be doubtful and other mechanisms of the CI96 cell have to be considered. Biodistribution experiments with radioactive labeled cells of the MHC nonrestricted CTL line TALL-104 injected intravenously into mice revealed primary accumulation in the lung and later in the liver, spleen, and kidney. There was also continuous accumulation of TALL-104 cells in the tumor and its metastasis (Cesano et al. 1999).

Effector cells can also be transfected with chemokine receptor genes. Chemokine receptors specifically influence the polarity and motility of cells depending on the concentration of their ligand (Mellado et al. 2001). Tumors secrete different chemokines, for example IL-8 by melanoma (Payne and Cornelius 2002). However, chemokines are not tumor specific. Fusion proteins consisting of an antibody fragment and a chemokine portion could combine the recognition of tumor specific antibodies with the chemotactic properties of chemokines (Challita-Eid et al. 1998).

10.5.3
Neovascularization of Solid Tumors – Barrier or Target?

After reaching a size of 1–2 mm solid tumors require their own blood vessels to ensure an adequate supply of oxygen and nutrients. Tumor-associated vessels are formed by the recruitment of endothelial cells and prevent the direct contact of immune cells or antibodies from blood and lymph with the tumor. In contrast to normal endothelium, endothelial cells from neovascular vessels express vascular endothelial growth factor (VEGF) receptor complexes (Brekken and Thorpe 2001). The targeting of effector lymphocytes to tumor-associated vessels could be accomplished by gene modification with recombinant receptors. Lymphocytes transfected with a chimeric receptor specific for the angiogenic endothelial receptor KDR achieved the lysis of KDR⁺ cells. Furthermore, the secretion of the chemokine IL-8 and the expression of vascular cell adhesion molecule (VCAM) and E-selectin was induced in human umbilical vein endothelial cells (HUVECs), playing an important role in the tumor immune response (Kershaw et al. 2000). Lymphocytes transfected with a chimeric receptor gene consisting of the disintegrin kistrin, which binds to the integrin $\alpha_v\beta_3$ on angiogenic endothelial cells, and the adhesion molecule CD31 (PECAM-1) accumulated in the tumor (Wiedle et al. 1999).

10.5.4
Rejection of Receptor Gene-modified Effector Lymphocytes

The clinical efficacy and safety of receptor gene-modified T cells still needs to be subject to comprehensive studies since only few long-term data are available regarding their *in vivo* behavior in patients (see Section 10.4.6). The adoptive transfer of gene-modified T cells did not induce any humoral response in a syngeneic mouse model (Altenschmidt et al. 1997). Against that, human autologous T cells transfected with an HIV-specific receptor were eliminated by receptor-specific CTLs (Riddell et al. 1996). The widely used retroviral vector pLXSN sensibilized transduced T cells against autologous NK cells and antigen-specific CTLs caused by the selection marker neo (Bordignon et al. 1999; Liberatore et al. 1999), which could be bypassed by selection marker-less vector systems. Murine antibody recognition domains of murine antibodies can induce a human anti-mouse antibody (HAMA) response, which can block but also activate cIgTCR$^+$ effector cells (Lamers et al. 2006). Therefore non-human domains should be avoided. However, anti-idiotypic responses against human antibody V regions are not excluded.

In particular for allogeneic effector cells and cell lines, rejection by the patient's immune system might be the major drawback during adoptive immunotherapy. Interestingly, studies with the CTL line TALL-104 demonstrated that allogeneic effector cell lines can be well-tolerated by different patients. The adoptive transfer of up to 10^8 TALL-104 cells per kg body weight into patients with advanced breast cancer was well tolerated and no cellular immunization against this cell line was observed, even after repeated administrations (Visonneau et al. 2000).

10.5.5
Combination of Conventional Tumor Therapies and Adoptive Immunotherapy

The synergistic effects of chemotherapy and the adoptive transfer of LAK cells (+ IL-2) have been demonstrated (Kawata et al. 1990, 1995). The surgical excision of human tumors subcutaneously implanted in mice improved the efficacy of adoptive transferred, irradiated TALL-104 cells (Cesano et al. 1998). Chemotherapy with adriamycin gave a better prognosis in combination with adoptive immunotherapy using TALL-104 cells (Cesano et al. 1997a). Treatment with cyclophosphamide reduced tumor-induced suppression of T cells (North 1982; Awwad and North 1988). T cells transfected with an ErbB2-specific cIgTCR achieved a dramatic increase in the survival of mice in an adjuvant setting of day 8 metastatic disease that was significantly greater than that afforded by either doxorubicin, 5-fluorouracil, or herceptin (Kershaw et al. 2004).

10.6
Conclusions

Despite past therapeutic failures, novel immunotherapy approaches such as adoptive immunotherapy employing receptor gene-modified effector lymphocytes grafted with the tumor specificity of antibodies are showing promise. There has been recent progress in the technological development of gene modification of primary T lymphocytes. However, individualized approaches are expensive, time-consuming, and still some way from routine application. Strategies to produce tumor-specific effector lymphocytes may need to be developed in new directions. The employment of effector cell lines could overcome some of the limitations, but they are also connected with problems regarding their allogenicity and unlimited growth. In the next few years more results of clinical studies employing receptor gene-modified effector lymphocytes should be expected.

References

Altenschmidt, U., Kahl, R., Moritz, D., Schnierle, B.S., Gerstmayer, B., Wels, W., Groner, B. (1996) Cytolysis of tumor cells expressing the Neu/erbB-2, erbB-3, and erbB-4 receptors by genetically targeted naive T lymphocytes. *Clin Cancer Res* 2: 1001–1008.

Altenschmidt, U., Klundt, E., Groner, B. (1997) Adoptive transfer of in vitro-targeted, activated T lymphocytes results in total tumor regression. *J Immunol* 159: 5509–5515.

Alvarez-Vallina, L., Hawkins, R.E. (1996) Antigen-specific targeting of CD28-mediated T cell co-stimulation using chimeric single-chain antibody variable fragment-CD28 receptors. *Eur J Immunol* 26: 2304–2309.

Annenkov, A.E., Moyes, S.P., Eshhar, Z., Mageed, R.A., Chernajovsky, Y. (1998) Loss of original antigenic specificity in T cell hybridomas transduced with a chimeric receptor containing single-chain Fv of an anti-collagen antibody and Fc epsilonRI-signaling gamma subunit. *J Immunol* 161: 6604–6613.

Annenkov, A.E., Daly, G.M., Brocker, T., Chernajovsky, Y. (2003) Clustering of immunoreceptor tyrosine-based activation motif-containing signalling subunits in CD4(+) T cells is an optimal signal for IFN-gamma production, but not for the production of IL-4. *Int Immunol* 15: 665–677.

Asano, R., Takemura, S., Tsumoto, K., Sakurai, N., Teramae, A., Ebara, S., Katayose, Y., Shinoda, M., Suzuki, M., Imai, K., Matsuno, S., Kudo, T., Kumagai, I. (2000) Functional construction of the anti-mucin core protein (MUC1) antibody MUSE11 variable regions in a bacterial expression system. *J Biochem (Tokyo)* 127: 673–679.

Awwad, M., North, R.J. (1988) Cyclophosphamide (Cy)-facilitated adoptive immunotherapy of a Cy-resistant tumour. Evidence that Cy permits the expression of adoptive T-cell mediated immunity by removing suppressor T cells rather than by reducing tumour burden. *Immunology* 65: 87–92.

Bach, N.L., Waks, T., Schindler, D.G., Eshhar, Z. (1994) Functional expression in mast cells of chimeric receptors with antibody specificity. *Cell Biophys* 24–25: 229–236.

Barlozzari, T., Reynolds, C.W., Herberman, R.B. (1983) In vivo role of natural killer cells: involvement of large granular lymphocytes in the clearance of tumor cells in anti-asialo GM1-treated rats. *J Immunol* 131: 1024–1027.

Baum, C., Dullmann, J., Li, Z., Fehse, B., Meyer, J., Williams, D.A., von Kalle, C. (2003) Side effects of retroviral gene transfer into hematopoietic stem cells. *Blood* 101 2099–2114.

Beecham, E.J., Ortiz Pujols, S., Junghans, R.P. (2000) Dynamics of tumor cell killing by human T lymphocytes armed with an anti-carcinoembryonic antigen chimeric immunoglobulin T-cell receptor. *J Immunother* 23: 332–343.

Bester, A.C., Schwartz, M., Schmidt, M., Garrigue, A., Hacein-Bey-Abina, S., Cavazzana-Calvo, M., Ben-Porat, N., Von Kalle, C., Fischer, A., Kerem, B. (2006) Fragile sites are preferential targets for integrations of MLV vectors in gene therapy. *Gene Ther* 3: 1057–1059.

Biglari, A., Southgate, T.D., Fairbairn, L.J., Gilham, D.E. (2006) Human monocytes expressing a CEA-specific chimeric CD64 receptor specifically target CEA-expressing tumour cells in vitro and in vivo. *Gene Ther* 13: 602–610.

Bird, R.E., Hardman, K.D., Jacobson, J.W., Johnson, S., Kaufman, B.M., Lee, S.M., Lee, T., Pope, S.H., Riordan, G.S., Whitlow, M. (1988) Single-chain antigen-binding proteins. *Science* 242: 423–426.

Blank-Voorthuis, C.J., Braakman, E., Ronteltap, C.P., Tilly, B.C., Sturm, E., Warnaar, S.O., Bolhuis, R.L. (1993) Clustered CD3/TCR complexes do not transduce activation signals after bispecific monoclonal antibody-triggered lysis by cytotoxic T lymphocytes via CD3. *J Immunol* 151: 2904–2914.

Bolhuis, R.L., Willemsen, R.A., Lamers, C.H., Stam, K., Gratama, J.W., Weijtens, M.E. (1998) Preparation for a phase I/II study using autologous gene modified T lymphocytes for treatment of metastatic renal cancer patients. *Adv Exp Med Biol* 451: 547–555.

Bordignon, C., Carlo-Stella, C., Colombo, M.P., De Vincentiis, A., Lanata, L., Lemoli, R.M., Locatelli, F., Olivieri, A., Rondelli, D., Zanon, P., Tura, S. (1999) Cell therapy: achievements and perspectives. *Haematologica* 84: 1110–1149.

Brekken, R.A., Thorpe, P.E. (2001) VEGF-VEGF receptor complexes as markers of tumor vascular endothelium. *J Control Release* 74: 173–181.

Brocker, T. (2000) Chimeric Fv-zeta or Fv-epsilon receptors are not sufficient to induce activation or cytokine production in peripheral T cells. *Blood* 96: 1999–2001.

Brocker, T., Peter, A., Traunecker, A., Karjalainen, K. (1993) New simplified molecular design for functional T cell receptor. *Eur J Immunol* 23: 1435–1439.

Brocker, T., Riedinger, M., Karjalainen, K. (1996) Redirecting the complete T cell receptor/CD3 signaling machinery towards native antigen via modified T cell receptor. *Eur J Immunol* 26: 1770–1774.

Bukowski, R.M., Rayman, P., Uzzo, R., Bloom, T., Sandstrom, K., Peereboom, D., Olencki, T., Budd, G.T., McLain, D., Elson, P., Novick, A., Finke, J.H. (1998) Signal transduction abnormalities in T lymphocytes from patients with advanced renal carcinoma: clinical relevance and effects of cytokine therapy. *Clin Cancer Res* 4: 2337–2347.

Calogero, A., de Leij, L.F., Mulder, N.H., Hospers, G.A. (2000) Recombinant T-cell receptors: an immunologic link to cancer therapy. *J Immunother* 23: 393–400.

Canevari, S., Stoter, G., Arienti, F., Bolis, G., Colnaghi, M.I., Di Re, E.M., Eggermont, A.M., Goey, S.H., Gratama, J.W., Lamers, C.H. et al. (1995) Regression of advanced ovarian carcinoma by intraperitoneal treatment with autologous T lymphocytes retargeted by a bispecific monoclonal antibody. *J Natl Cancer Inst* 87: 1463–1469.

Cervantes, F., Pierson, B.A., McGlave, P.B., Verfaillie, C.M., Miller, J.S. (1996) Autologous activated natural killer cells suppress primitive chronic myelogenous leukemia progenitors in long-term culture. *Blood* 87: 2476–2485.

Cesano, A., Visonneau, S., Jeglum, K.A., Owen, J., Wilkinson, K., Carner, K., Reese, L., Santoli, D. (1996a) Phase I clinical trial with a human major histocompatibility complex nonrestricted cytotoxic T-cell line (TALL-104) in dogs with advanced tumors. *Cancer Res* 56: 3021–3029.

Cesano, A., Visonneau, S., Pasquini, S., Rovera, G., Santoli, D. (1996b) Antitumor efficacy of a human major histocompatibility complex nonrestricted cytotoxic T-cell line (TALL-104) in immunocompetent mice bearing syngeneic leukemia. *Cancer Res* 56: 4444–4452.

Cesano, A., Visonneau, S., Rovera, G., Santoli, D. (1997a) Synergistic effects of adriamycin and TALL-104 cell therapy against a human gastric carcinoma in vivo. *AntiCancer Res* 17: 1887–1892.

Cesano, A., Visonneau, S., Wolfe, J.H., Jeglum, K.A., Fernandez, J., Gillio, A., O'Reilly, R.J., Santoli, D. (1997b) Toxicological and immunological evaluation of the MHC-non-restricted cytotoxic T cell line TALL-104. *Cancer Immunol Immunother* 44: 125–136.

Cesano, A., Visonneau, S., Santoli, D. (1998) TALL-104 cell therapy of human solid tumors implanted in immunodeficient (SCID) mice. *AntiCancer Res* 18: 2289–2295.

Cesano, A., Visonneau, S., Tran, T., Santoli, D. (1999) Biodistribution of human MHC non-restricted TALL-104 killer cells in healthy and tumor bearing mice. *Int J Oncol* 14: 245–251.

Challita-Eid, P.M., Abboud, C.N., Morrison, S.L., Penichet, M.L., Rosell, K.E., Poles, T., Hilchey, S.P., Planelles, V., Rosenblatt, J.D. (1998) A RANTES-antibody fusion protein retains antigen specificity and chemokine function. *J Immunol* 161: 3729–3736.

Correa, M.R., Ochoa, A.C., Ghosh, P., Mizoguchi, H., Harvey, L., Longo, D.L. (1997) Sequential development of structural and functional alterations in T cells from tumor-bearing mice. *J Immunol* 158: 5292–5296.

Daly, T., Royal, R.E., Kershaw, M.H., Treisman, J., Wang, G., Li, W., Herlyn, D., Eshhar, Z., Hwu, P. (2000) Recognition of human colon cancer by T cells transduced with a chimeric receptor gene. *Cancer Gene Ther* 7: 284–291.

Duval, L., Schmidt, H., Kaltoft, K., Fode, K., Jensen, J.J., Sorensen, S.M., Nishimura, M.I., von der Maase, H. (2006) Adoptive transfer of allogeneic cytotoxic T lymphocytes equipped with a HLA-A2 restricted MART-1 T-cell receptor: a phase I trial in metastatic melanoma. *Clin Cancer Res* 12: 1229–1236.

Ebert, O., Finke, S., Salahi, A., Herrmann, M., Trojaneck, B., Lefterova, P., Wagner, E., Kircheis, R., Huhn, D., Schriever, F., Schmidt-Wolf, I.G. (1997) Lymphocyte apoptosis: induction by gene transfer techniques. *Gene Ther* 4: 296–302.

Engels, B., Cam, H., Schuler, T., Indraccolo, S., Gladow, M., Baum, C., Blankenstein, T., Uckert, W. (2003) Retroviral vectors for high-level transgene expression in T lymphocytes. *Hum Gene Ther* 14: 1155–1168.

Eshhar, Z. (1997) Tumor-specific T-bodies: towards clinical application. *Cancer Immunol Immunother* 45: 131–136.

Eshhar, Z., Fitzer-Attas, C.J. (1998) Tyrosine kinase chimeras for antigen-selective T-body therapy. *Adv Drug Deliv Rev* 31: 171–182.

Eshhar, Z., Waks, T., Gross, G., Schindler. D.G. (1993) Specific activation and targeting of cytotoxic lymphocytes through chimeric single chains consisting of antibody-binding domains and the gamma or zeta subunits of the immunoglobulin and T-cell receptors. *Proc Natl Acad Sci USA* 90: 720–724.

Eshhar, Z., Bach, N., Fitzer-Attas, C.J., Gross, G., Lustgarten, J., Waks, T., Schindler, D.G. (1996) The T-body approach: potential for cancer immunotherapy. *Springer Semin Immunopathol* 18: 199–209.

Eshhar, Z., Waks, T., Bendavid, A., Schindler, D.G. (2001) Functional expression of chimeric receptor genes in human T cells. *J Immunol Methods* 248: 67–76.

Finer, M.H., Dull, T.J., Qin, L., Farson, D., Roberts, M.R. (1994) kat: a high-efficiency retroviral transduction system for primary human T lymphocytes. *Blood* 83: 43–50.

Finke, J., Ferrone, S., Frey, A., Mufson, A., Ochoa, A. (1999) Where have all the T cells gone? Mechanisms of immune evasion by tumors. *Immunol Today* 20: 158–160.

Fitzer Attas, C.J., Schindler, D.G., Waks, T., Eshhar, Z. (1997) Direct T cell activation by chimeric single chain Fv-Syk promotes Syk-Cbl association and Cbl phosphorylation. *J Biol Chem* 272: 8551–8557.

Fitzer Attas, C.J., Schindler, D.G., Waks, T., Eshhar, Z. (1998) Harnessing Syk family tyrosine kinases as signaling domains for chimeric single chain of the variable domain receptors: optimal design for T cell activation. *J Immunol* 160: 145–154.

Gade, T.P., Hassen, W., Santos, E., Gunset, G., Saudemont, A., Gong, M.C., Brentjens, R., Zhong, X.S., Stephan, M., Stefanski, J., Lyddane, C., Osborne, J.R., Buchanan, I.M., Hall, S.J., Heston, W.D., Riviere, I., Larson, S.M., Koutcher, J.A., Sadelain, M. (2005) Targeted elimination of prostate cancer by genetically directed human T lymphocytes. *Cancer Res* 65: 9080–9088.

Garrick, D., Sutherland, H., Robertson, G., Whitelaw, E. (1996) Variegated expression of a globin transgene correlates with chromatin accessibility but not methylation status. *Nucleic Acids Res* 24: 4902–4909.

Gladow, M., Becker, C., Blankenstein, T., Uckert, W. (2000) MLV-10A1 retrovirus pseudotype efficiently transduces primary human CD4+ T lymphocytes. *J Gene Med* 2: 409–415.

Gorochov, G., Lustgarten, J., Waks, T., Gross, G., Eshhar, Z. (1992) Functional assembly of chimeric T-cell receptor chains. *Int J Cancer* Suppl. 7: 53–57.

Goverman, J., Gomez, S.M., Segesman, K.D., Hunkapiller, T., Laug, W.E., Hood, L. (1990) Chimeric immunoglobulin-T cell receptor proteins form functional receptors: implications for T cell receptor complex formation and activation. *Cell* 60: 929–939.

Gowans, J.L., Uhr, J.W. (1966) The carriage of immunological memory by small lymphocytes in the rat. *J Exp Med* 124: 1017–1030.

Gresch, O., Engel, F.B., Nesic, D., Tran, T.T., England, H.M., Hickman, E.S., Korner, I., Gan, L., Chen, S., Castro-Obregon, S., Hammermann, R., Wolf, J., Muller-Hartmann, H., Nix, M., Siebenkotten, G., Kraus, G., Lun, K. (2004) New non-viral method for gene transfer into primary cells. *Methods* 33: 151–163.

Gross, G., Gorochov, G., Waks, T., Eshhar, Z. (1989a) Generation of effector T cells expressing chimeric T cell receptor with antibody type-specificity. *Transplant Proc* 21: 127–130.

Gross, G., Waks, T., Eshhar, Z. (1989b) Expression of immunoglobulin-T-cell receptor chimeric molecules as functional receptors with antibody-type specificity. *Proc Natl Acad Sci USA* 86: 10024–10028.

Hacein-Bey-Abina, S., von Kalle, C., Schmidt, M., Le Deist, F., Wulffraat, N., McIntyre, E., Radford, I., Villeval, J.L., Fraser, C.C., Cavazzana-Calvo, M., Fischer, A. (2003a) A serious adverse event after successful gene therapy for X-linked severe combined immunodeficiency. *N Engl J Med* 348: 255–256.

Hacein-Bey-Abina, S., Von Kalle, C., Schmidt, M., McCormack, M.P., Wulffraat, N., Leboulch, P., Lim, A., Osborne, C.S., Pawliuk, R., Morillon, E., Sorensen, R., Forster, A., Fraser, P., Cohen, J.I., de Saint Basile, G., Alexander, I., Wintergerst, U., Frebourg, T., Aurias, A., Stoppa-Lyonnet, D., Romana, S., Radford-Weiss, I., Gross, F., Valensi, F., Delabesse, E., Macintyre, E., Sigaux, F., Soulier, J., Leiva, L.E., Wissler, M., Prinz, C., Rabbitts, T.H., Le Deist, F., Fischer, A., Cavazzana-Calvo, M. (2003b) LMO2-associated clonal T cell proliferation in two patients after gene therapy for SCID-X1. *Science* 302: 415–419.

Hanson, H.L., Donermeyer, D.L., Ikeda, H., White, J.M., Shankaran, V., Old, L.J., Shiku, H., Schreiber, R.D., Allen, P.M. (2000) Eradication of established tumors by CD8+ T cell adoptive immunotherapy. *Immunity* 13: 265–276.

Haynes, N.M., Smyth, M.J., Kershaw, M.H., Trapani, J.A., Darcy, P.K. (1999) Fas-ligand-mediated lysis of erbB-2-expressing tumour cells by redirected cytotoxic T lymphocytes. *Cancer Immunol Immunother* 47: 278–286.

Haynes, N.M., Snook, M.B., Trapani, J.A., Cerruti, L., Jane, S.M., Smyth, M.J., Darcy, P.K. (2001) Redirecting mouse CTL against colon carcinoma: superior signaling efficacy of single-chain variable domain chimeras containing TCR-zeta vs Fc epsilon RI-gamma. *J Immunol* 166: 182–187.

Haynes, N.M., Trapani, J.A., Teng, M.W., Jackson, J.T., Cerruti, L., Jane, S.M., Kershaw, M.H., Smyth, M.J., Darcy, P.K. (2002a) Rejection of syngeneic colon carcinoma by CTLs expressing single-chain antibody receptors codelivering CD28 costimulation. *J Immunol* 169: 5780–5786.

Haynes, N.M., Trapani, J.A., Teng, M.W., Jackson, J.T., Cerruti, L., Jane, S.M., Kershaw, M.H., Smyth, M.J., Darcy, P.K. (2002b) Single-chain antigen recognition receptors that costimulate potent rejection of established experimental tumors. *Blood* 100: 3155–3163.

Hekele, A., Dall, P., Moritz, D., Wels, W., Groner, B., Herrlich, P., Ponta, H. (1996) Growth retardation of tumors by adoptive transfer of cytotoxic T lymphocytes reprogrammed by CD44v6-specific scFv: zeta-chimera. *Int J Cancer.* 68: 232–238.

Hoffman, D.M., Gitlitz, B.J., Belldegrun, A., Figlin, R.A. (2000) Adoptive cellular therapy. *Semin Oncol* 27: 221–233.

Hombach, A., Heuser, C., Sircar, R., Tillmann, T., Diehl, V., Kruis, W., Pohl, C., Abken, H. (1997) T cell targeting of TAG72+ tumor cells by a chimeric receptor with antibody-like specificity for a carbohydrate epitope. *Gastroenterology* 113: 1163–1170.

Hombach, A., Sircar, R., Heuser, C., Tillmann, T., Diehl, V., Kruis, W., Pohl, C., Abken, H. (1998) Chimeric anti-TAG72 receptors with immunoglobulin constant Fc domains and gamma or zeta signalling chains. *Int J Mol Med* 2: 99–103.

Hombach, A., Koch, D., Sircar, R., Heuser, C., Diehl, V., Kruis, W., Pohl, C., Abken, H. (1999) A chimeric receptor that selectively targets membrane-bound carcinoembryonic antigen (mCEA) in the presence of soluble CEA. *Gene Ther* 6: 300–304.

Hombach, A., Heuser, C., Gerken, M., Fischer, B., Lewalter, K., Diehl, V., Pohl, C., Abken, H. (2000a) T cell activation by recombinant FcepsilonRI gamma-chain immune receptors: an extracellular spacer domain impairs antigen-dependent T cell activation but not antigen recognition. *Gene Ther* 7: 1067–1075.

Hombach, A., Schneider, C., Sent, D., Koch, D., Willemsen, R.A., Diehl, V., Kruis, W., Bolhuis, R.L., Pohl, C., Abken, H. (2000b) An entirely humanized CD3 zeta-chain signaling receptor that directs peripheral blood t cells to specific lysis of carcinoembryonic antigen-positive tumor cells. *Int J Cancer* 88: 115–120.

Hombach, A., Heuser, C., Marquardt, T., Wieczarkowiecz, A., Groneck, V., Pohl, C., Abken, H. (2001a) Cd4(+) T cells engrafted with a recombinant immunoreceptor efficiently lyse target cells in a mhc antigen- and fas-independent fashion. *J Immunol* 167: 1090–1096.

Hombach, A., Muche, J.M., Gerken, M., Gellrich, S., Heuser, C., Pohl, C., Sterry, W., Abken, H. (2001b) T cells engrafted with a recombinant anti-CD30 receptor target autologous CD30(+) cutaneous lymphoma cells. *Gene Ther* 8: 891–895.

Hombach, A., Sent, D., Schneider, C., Heuser, C., Koch, D., Pohl, C., Seliger, B., Abken, H. (2001c) T-cell activation by recombinant receptors: CD28 costimulation is required for interleukin 2 secretion and receptor-mediated T-cell proliferation but does not affect receptor-mediated target cell lysis. *Cancer Res* 61: 1976–1982.

Hwu, P., Shafer, G.E., Treisman, J., Schindler, D.G., Gross, G., Cowherd, R., Rosenberg, S.A., Eshhar, Z. (1993a) Lysis of ovarian cancer cells by human lymphocytes redirected with a chimeric gene composed of an antibody variable region and the Fc receptor gamma chain. *J Exp Med* 178: 361–366.

Hwu, P., Yannelli, J., Kriegler, M., Anderson, W.F., Perez, C., Chiang, Y., Schwarz, S., Cowherd, R., Delgado, C., Mule, J. et al. (1993b) Functional and molecular characterization of tumor-infiltrating lymphocytes transduced with tumor necrosis factor-alpha cDNA for the gene therapy of cancer in humans. *J Immunol* 150: 4104–4115.

Hwu, P., Yang, J.C., Cowherd, R., Treisman, J., Shafer, G.E., Eshhar, Z., Rosenberg, S.A. (1995) In vivo antitumor activity of T cells redirected with chimeric antibody/T-cell receptor genes. *Cancer Res* 55: 3369–3373.

Jain, R.K. (1990) Physiological barriers to delivery of monoclonal antibodies and other macromolecules in tumors. *Cancer Res* 50: 814s–819s.

Jensen, M.C., Clarke, P., Tan, G., Wright, C., Chung-Chang, W., Clark, T.N., Zhang, F., Slovak, M.L., Wu, A.M., Forman, S.J., Raubitschek, A. (2000) Human T lymphocyte genetic modification with naked DNA. *Mol Ther* 1: 49–55.

Jensen, M.C., Cooper, L.J., Wu, A.M., Forman, S.J., Raubitschek, A.(2003) Engineered CD20-specific primary human cytotoxic T lymphocytes for targeting B-cell malignancy. *Cytotherapy* 5: 131–138.

Jewett, A. (2001) Activation of c-Jun N-terminal kinase in the absence of NFkappaB function prior to induction of NK cell death triggered by a combination of anti-class I and anti-CD16 antibodies. Hum Immunol 62: 320–331.

Jiang, K., Zhong, B., Gilvary, D.L., Corliss, B.C., Hong-Geller, E., Wei, S., Djeu, J.Y. (2000) Pivotal role of phosphoinositide-3 kinase in regulation of cytotoxicity in natural killer cells. Nat Immunol 1: 419–425.

Kawata, A., Hosokawa, M., Sawamura, Y., Ito, K., Une, Y., Shibata, T., Uchino, J., Kobayashi, H. (1990) Modification of lymphokine-activated killer cell accumulation into tumor sites by chemotherapy, local irradiation, or splenectomy. Mol Biother 2: 221–227.

Kawata, A., Une, Y., Hosokawa, M., Wakizaka, Y., Namieno, T., Uchino, J., Kobayashi, H. (1995) Adjuvant chemoimmunotherapy for hepatocellular carcinoma patients. Adriamycin, interleukin-2, and lymphokine-activated killer cells versus adriamycin alone. Am J Clin Oncol 18: 257–262.

Kershaw, M.H., Westwood, J.A., Zhu, Z., Witte, L., Libutti, S.K., Hwu, P. (2000) Generation of gene-modified T cells reactive against the angiogenic kinase insert domain-containing receptor (KDR) found on tumor vasculature. Hum Gene Ther 11: 2445–2452.

Kershaw, M.H., Westwood, J.A., Hwu, P. (2002) Dual-specific T cells combine proliferation and antitumor activity. Nat Biotechnol 20: 1221–1227.

Kershaw, M.H., Jackson, J.T., Haynes, N.M., Teng, M.W., Moeller, M., Hayakawa, Y., Street, S.E., Cameron, R., Tanner, J.E., Trapani, J.A., Smyth, M.J., Darcy, P.K. (2004) Gene-engineered T cells as a superior adjuvant therapy for metastatic cancer. J Immunol 173: 2143–2150.

Kipriyanov, S.M., Cochlovius, B., Schafer, H.J., Moldenhauer, G., Bahre, A., Le Gall, F., Knackmuss, S., Little, M. (2002) Synergistic antitumor effect of bispecific CD19 × CD3 and CD19 × CD16 diabodies in a preclinical model of non-Hodgkin's lymphoma. J Immunol 169: 137–144.

Kogure, K., Urabe, M., Mizukami, H., Kume, A., Sato, Y., Monahan, J., Ozawa, K. (2001) Targeted integration of foreign DNA into a defined locus on chromosome 19 in K562 cells using AAV-derived components. Int J Hematol 73: 469–475.

Kolanus, W., Romeo, C., Seed, B. (1993) T cell activation by clustered tyrosine kinases. Cell 74: 171–183.

Kornacker, M., Moldenhauer, G., Herbst, M., Weilguni, E., Tita-Nwa, F., Harter, C., Hensel, M., Ho, A.D. (2006) Cytokine-induced killer cells against autologous CLL: Direct cytotoxic effects and induction of immune accessory molecules by interferon-gamma. Int J Cancer 119: 1377–1382

Kuwana, Y., Asakura, Y., Utsunomiya, N., Nakanishi, M., Arata, Y., Itoh, S., Nagase, F., Kurosawa, Y. (1987) Expression of chimeric receptor composed of immunoglobulin-derived V regions and T-cell receptor-derived C regions. Biochem Biophys Res Commun 149: 960–968.

Lal, S., Lauer, U.M., Niethammer, D., Beck, J.F., Schlegel, P.G. (2000) Suicide genes: past, present and future perspectives. Immunol Today 21: 48–54.

Lamers, C.H., Gratama, J.W., Pouw, N.M., Langeveld, S.C., Krimpen, B.A., Kraan, J., Stoter, G., Debets, R. (2005) Parallel detection of transduced T lymphocytes after immunogene therapy of renal cell cancer by flow cytometry and real-time polymerase chain reaction: implications for loss of transgene expression. Hum Gene Ther 16: 1452–1462.

Lamers, C.H., Sleijfer, S., Vulto, A.G., Kruit, W.H., Kliffen, M., Debets, R., Gratama, J.W., Stoter, G., Oosterwijk, E. (2006) Treatment of metastatic renal cell carcinoma with autologous T-lymphocytes genetically retargeted against carbonic anhydrase IX: first clinical experience. J Clin Oncol 24: e20–22.

Li, J., Baker, M.D. (2000) Mechanisms involved in targeted gene replacement in mammalian cells. Genetics 156: 809–821.

Li, Q.S., Tanaka, S., Kisenge, R.R., Toyoda, H., Azuma, E., Komada, Y. (2000) Activation-induced T cell death occurs at G1A phase of the cell cycle. Eur J Immunol 30: 3329–3337.

Liberatore, C., Capanni, M., Albi, N., Volpi, I., Urbani, E., Ruggeri, L., Mencarelli, A., Grignani, F., Velardi, A. (1999) Natural killer cell-mediated lysis of autologous cells modified by gene therapy. *J Exp Med* 189: 1855–1862.

Lipshy, K.A., Kostuchenko, P.J., Hamad, G.G., Bland, C.E., Barrett, S.K., Bear, H.D. (1997) Sensitizing T-lymphocytes for adoptive immunotherapy by vaccination with wild-type or cytokine gene-transduced melanoma. *Ann Surg Oncol* 4: 334–341.

Lister, J., Rybka, W.B., Donnenberg, A.D., deMagalhaes Silverman, M., Pincus, S.M., Bloom, E.J., Elder, E.M., Ball, E.D., Whiteside, T.L. (1995) Autologous peripheral blood stem cell transplantation and adoptive immunotherapy with activated natural killer cells in the immediate posttransplant period. *Clin Cancer Res* 1: 607–614.

Lustgarten, J., Eshhar, Z. (1995) Specific elimination of IgE production using T cell lines expressing chimeric T cell receptor genes. *Eur J Immunol* 25: 2985–2991.

Ma, Q., Gonzalo-Daganzo, R.M., Junghans, R.P. (2002) Genetically engineered T cells as adoptive immunotherapy of cancer. *Cancer Chemother Biol Response Modif* 20: 315–341.

Maasho, K., Marusina, A., Reynolds, N.M., Coligan, J.E., Borrego, F. (2004) Efficient gene transfer into the human natural killer cell line, NKL, using the Amaxa nucleofection system. *J Immunol Methods* 284: 133–140.

Maki, G., Tam, Y.K., Berkahn, L., Klingemann, H.G. (2003) Ex vivo purging with NK-92 prior to autografting for chronic myelogenous leukemia. *Bone Marrow Transplant* 31: 1119–1125.

McGuinness, R.P., Ge, Y., Patel, S.D., Kashmiri, S.V., Lee, H.S., Hand, P.H., Schlom, J., Finer, M.H., McArthur, J.G. (1999) Anti-tumor activity of human T cells expressing the CC49-zeta chimeric immune receptor. *Hum Gene Ther* 10: 165–173.

McKnight, R.A., Wall, R.J., Hennighausen, L. (1995) Expression of genomic and cDNA transgenes after co-integration in transgenic mice. *Transgenic Res* 4: 39–43.

Mellado, M., Rodriguez-Frade, J.M., Manes, S., Martinez, A.C. (2001) Chemokine signaling and functional responses: the role of receptor dimerization and TK pathway activation. *Annu Rev Immunol* 19: 397–421.

Mitsuyasu, R.T., Anton, P.A., Deeks, S.G., Scadden, D.T., Connick, E., Downs, M.T., Bakker, A., Roberts, M.R., June, C.H., Jalali, S., Lin, A.A., Pennathur-Das, R., Hege, K.M. (2000) Prolonged survival and tissue trafficking following adoptive transfer of CD4zeta gene-modified autologous CD4(+) and CD8(+) T cells in human immunodeficiency virus-infected subjects. *Blood* 96: 785–793.

Mizoguchi, H., O'Shea, J.J., Longo, D.L., Loeffler, C.M., McVicar, D.W., Ochoa, A.C. (1992) Alterations in signal transduction molecules in T lymphocytes from tumor-bearing mice. *Science* 258: 1795–1798.

Moeller, M., Haynes, N.M., Trapani, J.A., Teng, M.W., Jackson, J.T., Tanner, J.E., Cerutti, L., Jane, S.M., Kershaw, M.H., Smyth, M.J., Darcy, P.K. (2004) A functional role for CD28 costimulation in tumor recognition by single-chain receptor-modified T cells. *Cancer Gene Ther* 11: 371–379.

Moertel, C.G., O'Fallon, J.R., Go, V.L., O'Connell, M.J., Thynne, G.S. (1986) The preoperative carcinoembryonic antigen test in the diagnosis, staging, and prognosis of colorectal cancer. *Cancer* 58: 603–610.

Moretta, A., Bottino, C., Vitale, M., Pende, D., Cantoni, C., Mingari, M.C., Biassoni, R., Moretta, L. (2001) Activating receptors and coreceptors involved in human natural killer cell-mediated cytolysis. *Annu Rev Immunol* 19: 197–223.

Moritz, D., Wels, W., Mattern, J., Groner, B. (1994) Cytotoxic T lymphocytes with a grafted recognition specificity for ERBB2-expressing tumor cells. *Proc Natl Acad Sci USA* 91: 4318–4322.

Moritz, D., Groner, B. (1995) A spacer region between the single chain antibody- and the CD3 zeta-chain domain of chimeric T cell receptor components is required for efficient ligand binding and signaling activity. *Gene Ther* 2: 539–546.

Muhlebach, M.D., Schmitt, I., Steidl, S., Stitz, J., Schweizer, M., Blankenstein, T., Cichutek, K., Uckert, W. (2003) Transduction efficiency of MLV but not of HIV-1 vectors is pseudotype dependent on human primary T lymphocytes. *J Mol Med* 81: 801–810.

Nalesnik, M.A., Rao, A.S., Furukawa, H., Pham, S., Zeevi, A., Fung, J.J., Klein, G., Gritsch, H.A., Elder, E., Whiteside, T.L., Starzl, T.E. (1997) Autologous lymphokine-activated killer cell therapy of Epstein-Barr virus-positive and -negative lymphoproliferative disorders arising in organ transplant recipients. *Transplantation* 63: 1200–1205.

Nolan, K.F., Yun, C.O., Akamatsu, Y., Murphy, J.C., Leung, S.O., Beecham, E.J., Junghans, R.P. (1999) Bypassing immunization: optimized design of "designer T cells" against carcinoembryonic antigen (CEA)-expressing tumors, and lack of suppression by soluble CEA. *Clin Cancer Res* 5: 3928–3941.

North, R.J. (1982) Cyclophosphamide-facilitated adoptive immunotherapy of an established tumor depends on elimination of tumor-induced suppressor T cells. *J Exp Med* 155: 1063–1074.

Ortaldo, J.R., Mason, A.T., O'Shea, J.J. (1995) Receptor-induced death in human natural killer cells: involvement of CD16. *J Exp Med* 181: 339–344.

Patel, S.D., Moskalenko, M., Smith, D., Maske, B., Finer, M.H., McArthur, J.G. (1999) Impact of chimeric immune receptor extracellular protein domains on T cell function. *Gene Ther* 6: 412–419.

Patel, S.D., Moskalenko, M., Tian, T., Smith, D., McGuinness, R., Chen, L., Winslow, G.A., Kashmiri, S., Schlom, J., Stanners, C.P., Finer, M.H., McArthur, J.G. (2000) T-cell killing of heterogenous tumor or viral targets with bispecific chimeric immune receptors. *Cancer Gene Ther* 7: 1127–1134.

Payne, A.S., Cornelius, L.A. (2002) The role of chemokines in melanoma tumor growth and metastasis. *J Invest Dermatol* 118: 915–922.

Perez, P., Hoffman, R.W., Shaw, S., Bluestone, J.A., Segal, D.M. (1985) Specific targeting of cytotoxic T cells by anti-T3 linked to anti-target cell antibody. *Nature* 316: 354–356.

Pule, M.A., Straathof, K.C., Dotti, G., Heslop, H.E., Rooney, C.M., Brenner, M.K. (2005) A chimeric T cell antigen receptor that augments cytokine release and supports clonal expansion of primary human T cells. *Mol Ther* 12: 933–941.

Reinhold, U., Liu, L., Ludtke Handjery, H.C., Heuser, C., Hombach, A., Wang, X., Tilgen, W., Ferrone, S., Abken, H. (1999) Specific lysis of melanoma cells by receptor grafted T cells is enhanced by anti-idiotypic monoclonal antibodies directed to the scFv domain of the receptor. *J Invest Dermatol* 112: 744–750.

Ren Heidenreich, L., Hayman, G.T., Trevor, K.T. (2000) Specific targeting of EGP-2+ tumor cells by primary lymphocytes modified with chimeric T cell receptors. *Hum Gene Ther* 11: 9–19.

Ren-Heidenreich, L., Mordini, R., Hayman, G.T., Siebenlist, R., LeFever, A. (2002) Comparison of the TCR zeta-chain with the FcR gamma-chain in chimeric TCR constructs for T cell activation and apoptosis. *Cancer Immunol Immunother* 51: 417–423.

Riddell, S.R., Elliott, M., Lewinsohn, D.A., Gilbert, M.J., Wilson, L., Manley, S.A., Lupton, S.D., Overell, R.W., Reynolds, T.C., Corey, L., Greenberg, P.D. (1996) T-cell mediated rejection of gene-modified HIV-specific cytotoxic T lymphocytes in HIV-infected patients. *Nat Med* 2: 216–223.

Roberts, M.R., Qin, L., Zhang, D., Smith, D.H., Tran, A.C., Dull, T.J., Groopman, J.E., Capon, D.J., Byrn, R.A., Finer, M.H. (1994) Targeting of human immunodeficiency virus-infected cells by CD8+ T lymphocytes armed with universal T-cell receptors. *Blood* 84: 2878–2889.

Roberts, M.R., Cooke, K.S., Tran, A.C., Smith, K.A., Lin, W.Y., Wang, M., Dull, T.J., Farson, D., Zsebo, K.M., Finer, M.H. (1998) Antigen-specific cytolysis by neutrophils and NK cells expressing chimeric immune receptors bearing zeta or gamma signaling domains. *J Immunol* 161: 375–384.

Roessig, C., Scherer, S.P., Baer, A., Vormoor, J., Rooney, C.M., Brenner, M.K., Juergens, H. (2002) Targeting CD19 with genetically modified EBV-specific human T lymphocytes. *Ann Hematol* 81 (Suppl 2): S42–43.

Romeo, C., Amiot, M., Seed, B. (1992) Sequence requirements for induction of cytolysis by the T cell antigen/Fc receptor zeta chain. *Cell* 68: 889–897.

Rooney, C.M., Smith, C.A., Ng, C.Y., Loftin, S., Li, C., Krance, R.A., Brenner, M.K.,

Heslop, H.E. (1995) Use of gene-modified virus-specific T lymphocytes to control Epstein-Barr-virus-related lymphoproliferation. *Lancet* 345: 9–13.

Rooney, C.M., Smith, C.A., Ng, C.Y., Loftin, S.K., Sixbey, J.W., Gan, Y., Srivastava, D.K., Bowman, L.C., Krance, R.A., Brenner, M.K., Heslop, H.E. (1998) Infusion of cytotoxic T cells for the prevention and treatment of Epstein-Barr virus-induced lymphoma in allogeneic transplant recipients. *Blood* 92: 1549–1555.

Rosenberg, S.A. (1991) Immunotherapy and gene therapy of cancer. *Cancer Res* 51: 5074–5079.

Rosenberg, S.A., Aebersold, P., Cornetta, K., Kasid, A., Morgan, R.A., Moen, R., Karson, E.M., Lotze, M.T., Yang, J.C., Topalian, S.L. et al. (1990) Gene transfer into humans—immunotherapy of patients with advanced melanoma, using tumor-infiltrating lymphocytes modified by retroviral gene transduction. *N Engl J Med* 323: 570–578.

Rosenstein, M., Eberlein, T.J., Rosenberg, S.A. (1984) Adoptive immunotherapy of established syngeneic solid tumors: role of T lymphoid subpopulations. *J Immunol* 132: 2117–2122.

Rossig, C., Bollard, C.M., Nuchtern, J.G., Merchant, D.A., Brenner, M.K. (2001) Targeting of G(D2)-positive tumor cells by human T lymphocytes engineered to express chimeric T-cell receptor genes. *Int J Cancer* 94: 228–236.

Rossig, C., Bollard, C.M., Nuchtern, J.G., Rooney, C.M., Brenner, M.K. (2002) Epstein-Barr virus-specific human T lymphocytes expressing antitumor chimeric T-cell receptors: potential for improved immunotherapy. *Blood* 99: 2009–2016.

Ruggeri, L., Capanni, M., Martelli, M.F., Velardi, A. (2001) Cellular therapy: exploiting NK cell alloreactivity in transplantation. *Curr Opin Hematol* 8: 355–359.

Ruggeri, L., Capanni, M., Urbani, E., Perruccio, K., Shlomchik, W.D., Tosti, A., Posati, S., Rogaia, D., Frassoni, F., Aversa, F., Martelli, M.F., Velardi, A. (2002) Effectiveness of donor natural killer cell alloreactivity in mismatched hematopoietic transplants. *Science* 295: 2097–2100.

Salvadori, S., Gansbacher, B., Zier, K. (1994) Functional defects are associated with abnormal signal transduction in T cells of mice inoculated with parental but not IL-2 secreting tumor cells. *Cancer Gene Ther* 1: 165–170.

Schirrmann, T., Pecher, G. (2001) Tumor-specific targeting of a cell line with natural killer cell activity by asialoglycoprotein receptor gene transfer. *Cancer Immunol Immunother* 50: 549–556.

Schirrmann, T., Pecher, G. (2002) Human natural killer cell line modified with a chimeric immunoglobulin T-cell receptor gene leads to tumor growth inhibition in vivo. *Cancer Gene Ther* 9: 390–398.

Schirrmann, T., Pecher, G. (2005) Specific targeting of CD33(+) leukemia cells by a natural killer cell line modified with a chimeric receptor. *Leuk Res* 29: 301–306.

Schmidt-Wolf, I.G., Lefterova, P., Mehta, B.A., Fernandez, L.P., Huhn, D., Blume, K.G., Weissman, I.L., Negrin, R.S. (1993) Phenotypic characterization and identification of effector cells involved in tumor cell recognition of cytokine-induced killer cells. *Exp Hematol* 21: 1673–1679.

Schubeler, D., Lorincz, M.C., Cimbora, D.M., Telling, A., Feng, Y.Q., Bouhassira, E.E., Groudine, M. (2000) Genomic targeting of methylated DNA: influence of methylation on transcription, replication, chromatin structure, and histone acetylation. *Mol Cell Biol* 20: 9103–9112.

Shores, E., Flamand, V., Tran, T., Grinberg, A., Kinet, J.P., Love, P.E. (1997) Fc epsilonRI gamma can support T cell development and function in mice lacking endogenous TCR zeta-chain. *J Immunol* 159: 222–230.

Stancovski, I., Schindler, D.G., Waks, T., Yarden, Y., Sela, M., Eshhar, Z. (1993) Targeting of T lymphocytes to Neu/HER2-expressing cells using chimeric single chain Fv receptors. *J Immunol* 151: 6577–6582.

Stitz, J., Buchholz, C.J., Engelstadter, M., Uckert, W., Bloemer, U., Schmitt, I., Cichutek, K. (2000) Lentiviral vectors pseudotyped with envelope glycoproteins derived from gibbon ape leukemia virus and murine leukemia virus 10A1. *Virology* 273: 16–20.

Tam, Y.K., Martinson, J.A., Doligosa, K., Klingemann, H.G. (2003) Ex vivo expansion of the highly cytotoxic human natural killer-92 cell-line under current good manufacturing practice conditions for clinical adoptive cellular immunotherapy. *Cytotherapy* 5: 259–272.

Tonn, T., Becker, S., Esser, R., Schwabe, D., Seifried, E. (2001) Cellular immunotherapy of malignancies using the clonal natural killer cell line NK-92. *J Hematother Stem Cell Res* 10: 535–544.

Tran, A.C., Zhang, D., Byrn, R., Roberts, M.R. (1995) Chimeric zeta-receptors direct human natural killer (NK) effector function to permit killing of NK-resistant tumor cells and HIV-infected T lymphocytes. *J Immunol* 155: 1000–1009.

Uckert, W., Willimsky, G., Pedersen, F.S., Blankenstein, T., Pedersen, L. (1998) RNA levels of human retrovirus receptors Pit1 and Pit2 do not correlate with infectibility by three retroviral vector pseudotypes. *Hum Gene Ther* 9: 2619–2627.

Uckert, W., Becker, C., Gladow, M., Klein, D., Kammertoens, T., Pedersen, L., Blankenstein, T. (2000) Efficient gene transfer into primary human CD8+ T lymphocytes by MuLV-10A1 retrovirus pseudotype. *Hum Gene Ther* 11: 1005–1014.

Uharek, L., Zeis, M., Glass, B., Steinmann, J., Dreger, P., Gassmann, W., Schmitz, N., Muller Ruchholtz, W. (1996) High lytic activity against human leukemia cells after activation of allogeneic NK cells by IL-12 and IL-2. *Leukemia* 10: 1758–1764.

Uherek, C., Tonn, T., Uherek, B., Becker, S., Schnierle, B., Klingemann, H.G., Wels, W. (2002) Retargeting of natural killer-cell cytolytic activity to ErbB2- expressing cancer cells results in efficient and selective tumor cell destruction. *Blood* 100: 1265–1273.

van Oers, N.S., Love, P.E., Shores, E.W., Weiss, A. (1998) Regulation of TCR signal transduction in murine thymocytes by multiple TCR zeta-chain signaling motifs. *J Immunol* 160: 163–170.

Visonneau, S., Cesano, A., Jeglum, K.A., Santoli, D. (1999) Adoptive therapy of canine metastatic mammary carcinoma with the human MHC non-restricted cytotoxic T-cell line TALL-104. *Oncol Rep* 6: 1181–1188.

Visonneau, S., Cesano, A., Porter, D.L., Luger, S.L., Schuchter, L., Kamoun, M., Torosian, M.H., Duffy, K., Sickles, C., Stadtmauer, E.A., Santoli, D. (2000) Phase I trial of TALL-104 cells in patients with refractory metastatic breast cancer. *Clin Cancer Res* 6: 1744–1754.

Vujanovic, N.L., Yasumura, S., Hirabayashi, H., Lin, W.C., Watkins, S., Herberman, R.B., Whiteside, T.L. (1995) Antitumor activities of subsets of human IL-2-activated natural killer cells in solid tissues. *J Immunol* 154: 281–289.

Walker, R.E., Bechtel, C.M., Natarajan, V., Baseler, M., Hege, K.M., Metcalf, J.A., Stevens, R., Hazen, A., Blaese, R.M., Chen, C.C., Leitman, S.F., Palensky, J., Wittes, J., Davey, R.T., Jr., Falloon, J., Polis, M.A., Kovacs, J.A., Broad, D.F., Levine, B.L., Roberts, M.R., Masur, H., Lane, H.C. (2000) Long-term in vivo survival of receptor-modified syngeneic T cells in patients with human immunodeficiency virus infection. *Blood* 96: 467–474.

Weijtens, M.E., Willemsen, R.A., Valerio, D., Stam, K., Bolhuis, R.L. (1996) Single chain Ig/gamma gene-redirected human T lymphocytes produce cytokines, specifically lyse tumor cells, and recycle lytic capacity. *J Immunol* 157: 836–843.

Weijtens, M.E., Willemsen, R.A., Hart, E.H., Bolhuis, R.L. (1998a) A retroviral vector system 'STITCH' in combination with an optimized single chain antibody chimeric receptor gene structure allows efficient gene transduction and expression in human T lymphocytes. *Gene Ther* 5: 1195–1203.

Weijtens, M.E., Willemsen, R.A., van Krimpen, B.A., Bolhuis, R.L. (1998b) Chimeric scFv/gamma receptor-mediated T-cell lysis of tumor cells is coregulated by adhesion and accessory molecules. *Int J Cancer.* 77: 181–187.

Weijtens, M.E., Hart, E.H., Bolhuis, R.L. (2000) Functional balance between T cell chimeric receptor density and tumor associated antigen density: CTL mediated cytolysis and lymphokine production. *Gene Ther* 7: 35–42.

Westwood, J.A., Smyth, M.J., Teng, M.W., Moeller, M., Trapani, J.A., Scott, A.M.,

Smyth, F.E., Cartwright, G.A., Power, B.E., Honemann, D., Prince, H.M., Darcy, P.K., Kershaw, M.H. (2005) Adoptive transfer of T cells modified with a humanized chimeric receptor gene inhibits growth of Lewis-Y-expressing tumors in mice. *Proc Natl Acad Sci USA* 102: 19051–19056.

Whiteside, T.L. (1999) Signaling defects in T lymphocytes of patients with malignancy. *Cancer Immunol Immunother* 48: 346–352.

Wiedle, G., Johnson-Leger, C., Imhof, B.A. (1999) A chimeric cell adhesion molecule mediates homing of lymphocytes to vascularized tumors. *Cancer Res* 59: 5255–5263.

Willemsen, R.A., Debets, R., Chames, P., Bolhuis, R.L. (2003) Genetic engineering of T cell specificity for immunotherapy of cancer. *Hum Immunol* 64: 56–68.

Williams, D.A., Baum, C. (2003) Medicine. Gene therapy – new challenges ahead. *Science* 302: 400–401.

Yannelli, J.R., Hyatt, C., McConnell, S., Hines, K., Jacknin, L., Parker, L., Sanders, M., Rosenberg, S.A. (1996) Growth of tumor-infiltrating lymphocytes from human solid cancers: summary of a 5-year experience. *Int J Cancer.* 65: 413–421.

Yew, N.S., Wysokenski, D.M., Wang, K.X., Ziegler, R.J., Marshall, J., McNeilly, D., Cherry, M., Osburn, W., Cheng, S.H. (1997) Optimization of plasmid vectors for high-level expression in lung epithelial cells. *Hum Gene Ther* 8: 575–584.

Yun, C.O., Nolan, K.F., Beecham, E.J., Reisfeld, R.A., Junghans, R.P. (2000) Targeting of T lymphocytes to melanoma cells through chimeric anti-GD3 immunoglobulin T-cell receptors. *Neoplasia* 2: 449–459.

Zenner, G., Vorherr, T., Mustelin, T., Burn, P. (1996) Differential and multiple binding of signal transducing molecules to the ITAMs of the TCR-zeta chain. *J Cell Biochem* 63: 94–103.

11
Emerging Therapeutic Concepts IV: Anti-idiotypic Antibodies

Peter Fischer and Martina M. Uttenreuther-Fischer

11.1
Introduction

Over 30 years ago the "network theory of the immune system" was published by Nils K. Jerne (Jerne 1974), for which, together with other "theories concerning the specificity in development and control of the immune system," he was awarded the Nobel Prize in physiology/medicine in 1984. Jerne's Nobel Prize was shared with Georges Köhler and César Milstein (for development of the hybridoma technique). As mentioned in the Nobel Prize announcement (http://nobelprize.org/), Jerne's theory had practical consequences for using anti-antibodies as a tool in medicine. Anti-idiotypic antibodies may be developed as vaccines against infectious diseases or tumors on the one hand and for the induction of tolerance by inhibition of (1) antipollen antibodies in allergy, (2) autoantibodies in autoimmune disease, and (3) graft rejection in transplantation as well. However, anti-idiotypic antibodes may themselves cause disease (e.g. autoimmunity) (Abu-Shakra et al. 1996) or suppression of the immune response (Metlas and Veljkovic 2004).

11.2
Definition of Anti-idiotypic Antibodies

The term "idiotype," which corresponds to the ensemble of idiotopes (i.e. the individual, unique variable regions of an antibody) was first introduced by Oudin, Michel, and Kunkel in 1963 (Abu-Shakra et al. 1996). The principle of Jerne's network theory was the observation that antibodies (Ab1) induce anti-antibodies (Ab2) directed against the idiotype of the first. In addition, those Ab2 will lead to anti-anti-antibodies (Ab3) and so on, resulting in a self-regulation of the antibody repertoire by stimulation and supression of B cells (Fig. 11.1).

Handbook of Therapeutic Antibodies. Edited by Stefan Dübel
Copyright © 2007 WILEY-VCH Verlag GmbH & Co. KGaA, Weinheim
ISBN 978-3-527-31453-9

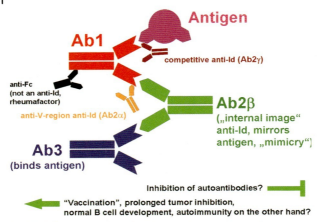

Fig. 11.1 Definition of anti-idiotypic antibodies. Antibodies (Ab1) induce anti-antibodies (Ab2) directed against the idiotype (unique variable region) of the first. In addition, those Ab2 will lead to anti-anti-antibodies (Ab3) and so on, resulting in a self-regulation of the antibody repertoire by stimulation and suppression of B cells. For details see text.

Importantly, some of the Ab2 may mirror the internal image of the original antigen recognized by Ab1. This type of Ab2, designated Ab2β, may mimic the antigen, eliciting Ab3 that bind to the original antigen similar to Ab1 (Abu-Shakra et al. 1996). In this way, Ab2β may be used as a vaccine instead of the antigen itself (Fig. 11.1). This may be advantageous when the antigen is not readily available (e.g. in the case of a dangerous pathogen or toxin, or a nonproteinous antigen such as a carbohydrate). This specific kind of anti-idiotypic interaction with the complementarity determining regions (CDRs) has been confirmed in crystallography and comparison of the complex between antigen (lysozyme) and Ab1 (anti-lysozyme antibody D1.3) to the complex of Ab2 (anti-D1.3 antibody E5.2) and Ab1. D1.3 contacted the antigen lysozyme and the anti-idiotope E5.2 through essentially the same combining-site residues, mimicking the noncovalent binding interactions (Fields et al. 1995).

Another variant of anti-idiotypic antibodies, Ab2γ, also inhibits the binding of Ab1 to its antigen. Ab2γ are also directed against the antigen-binding region (paratope) of Ab1, but they do not mirror the antigen and thus are not useful as a vaccine (Fig. 11.1). In contrast, Ab2α anti-idiotypic antibodies bind outside of the paratope and thus do not interfere with antigen binding. They may recognize parts of both CDR and framework regions (Fig. 11.1). Anti-constant region antibodies (anti-Fc) are known as rheumatoid factors which might enhance the destructive function of autoantibodies (Yang et al. 1999).

11.3
Anti-idiotypic Antibodies as Autoantigens

Anti-idiotypic antibodies against anti-pathogen antibodies (e.g. against mycobacteria or *Klebsiella*) may be the cause or trigger of autoimmune diesease by continuously inducing Ab3 against autoantigens (Shoenfeld 1994; Galeazzi et al. 1998). On the other hand, autoantibodies may be helpful in the treatment of autoimmune disease by blocking autoantigens or downregulation of cytokines (Shoenfeld 2005). Autoimmunity is not the focus of this chapter because there are excellent reviews and books available elsewhere (Peter and Shohat 1996; Shoenfeld et al. 1997; Sherer and Shoenfeld 2000).

11.4
Anti-idiotypes as a Tool for Generating Specific Antibodies

In phage display, the direct biopanning of specific antibody-presenting phage may be difficult or impossible, when the specificity or affinity for the antigen is too low compared with the background phage, resulting in nonspecific adherence of the phage. In many cases, the antigen is not available in a suitable form, because the expression or purification of the biomolecules, such as membrane proteins or carbohydrates, is not possible in sufficient quantities.

In those situations it may be advantageous to use anti-idiotypic antibodies for the panning procedure. These have been generated, for example, by immunization of rabbits with human IgG Fab (idiotype) that were affinity purified with the target antigen (Ishida et al. 1995). Alternatively, anti-idiotypic antibodies mimicking the target antigen were used to isolate the specific phage (Hombach et al. 1998; Fischer et al. 1999). In addition, the cloning of the anti-idiotypic antibodies themselves may be required to isolate antigen-mimicking antibody fragments (e.g. for vaccination, or to derive a tool for targeting clonal myeloma cells) (Willems et al. 1998).

In a different setting, the cloning of unknown antibodies that are the target of regulatory anti-idiotypic antibodies may be the primary goal. This was the case when autoantibodies from patients with autoimmune disease were selected by anti-idiotypic panning using intravenous IgG preparations (IVIG). This technique not only enabled the determination of the genetic origin of those antibodies (Jendreyko et al. 1998) but also allowed platelet-reactive IgG antibodies to be cloned from patients with autoimmune thrombocytopenia (AITP) for the first time (Fischer et al. 1999).

11.5
Intravenous Immunoglobulin Preparations (IVIG)

IVIG prepared by purification of serum IgG from several thousand healthy donors have shown positive effects in a variety of immuneological disorders.

IVIG have been successfully used in the treatment of autoimmune diseases such as autoimmune thrombocytopenia (Imbach et al. 1985; Imholz et al. 1988), acute myocarditis (Drucker et al. 1994), Kawasaki disease (Fischer et al. 1996b), and certain cases of systemic lupus erythematosus (SLE) (Schroeder et al. 1996; Sherer et al. 1999). Variable results and sometimes worsening of disease symptoms in SLE have been partially attributed to the existence of pathogenic anti-DNA idiotype-reactive IgG in IVIG preparations (Silvestris et al. 1994).

The diagnosis of Kawasaki disease, a mucocutaneous lymphadenopathy syndrome (Kawasaki 1967), mainly depends on clinical symptoms such as fever persisting for more than 5 consecutive days, since no specific laboratory criteria are available (Leung 1993; Fischer et al. 1996b). The etiology of Kawasaki disease is still unknown, although epidemiological and clinical data indicate an infectious origin (Kotzin et al. 1993; Burns and Matsubara 1995; Kawasaki 1995). Compiling the information from the literature, we proposed that the disease may be caused by simultaneous or successive dual infection with superantigen-producing and other pathogens (Fischer et al. 1996a,b). Without early diagnosis and treatment, up to 40% of all patients with Kawasaki disease will ultimately develop severe cardiac complications (Dajani et al. 1994). Although a specific therapy for Kawasaki disease is not available, coronary complications can be significantly reduced with high doses of IVIG combined with aspirin (acetylsalicylic acid) (Dajani et al. 1994).

The specific mechanisms of IVIG in autoimmune diseases are not clear. Various studies have suggested the inhibition of effector cells through blockade of Fc receptors (Debre et al. 1993), healing of persisting infections, induction of cytokines (Lacroix-Desmazes et al. 1996), inhibition of autoantibodies by *anti*-idiotypic antibodies (Berchtold et al. 1989; Levy et al. 1999), B cell modulation by *anti*-B cell antigen receptor antibodies and the reconstruction of the "idiotypic" or "V-region" connected (Lacroix-Desmazes et al. 1996) network (Jerne 1974) by IVIG.

However, the documented increase of serum IgM, decrease of certain autoantibodies and long-term therapeutic effects far beyond the half-life of infused IVIG often observed in children treated for AITP (Imholz et al. 1988) cannot be explained by nonspecific mechanisms such as Fc receptor blockade alone. For this effect on antibodies and B cells, the additional anti-idiotypic involvement of variable regions from IVIG molecules is required (Lacroix-Desmazes et al. 1996; Levy et al. 1999). Anti-idiotypic IVIG molecules may fit the antigen-binding site, particularly the complementarity determining regions (CDR) of antibodies, thus mimicking the original antigen (Ab2 beta, "internal image"). Other subsets of IVIG may be directed against idiotypes distinct from the paratope, recognizing mainly the framework regions (Ab2 alpha) (Fig. 1) (Abu-Shakra et al. 1996). The above-mentioned functions of IVIG may act synergistically in the regulation of T cell as well as B cell-dependent immune responses.

11.6
Anti-idiotypic Antibodies as Possible Superantigens?

To elucidate possible specific interactions of IVIG with (auto)antibodies, we recently investigated patient-derived monoclonal IgGs that were bound by IVIG in an anti-idiotypic manner, applying the combinatorial antibody phage display system (Barbas et al. 1991; Breitling et al. 1991; Duchosal et al. 1992; Yang et al. 1997, 1999; Hoet et al. 1998). From three different patients with AITP, a large number of clones specifically reacting with IVIG molecules were enriched. Many IVIG-selected Fab-phage from AITP strongly reacted with platelets in ELISA and fluorescent-activated cell sorting (FACS), in contrast to IVIG-selected Fab-phage derived from a healthy individual (Fischer et al. 1999) or a patient with SLE (Osei et al. 2000).

Sequencing revealed that the most frequently used germline gene loci of all IVIG-bound, platelet-negative Fabs from the three AITP libraries were 3–23 or 3–30/3–30.5 in the case of heavy chains and 3l, 2a2, and 3r for the light chains (Jendreyko et al. 1998; Fischer et al. 1999), while most platelet-reactive Fab were from V_H4 germline origin.

We observed the same favorite selection of 3–23 or 3–30/3–30.5-derived IgG and IgM Fabs by IVIG from libraries of healthy individuals (Hoffmann et al. 2000) and patients with SLE (Osei et al. 2000). Light chains, antigen specificity, and the high variation in mutation rates and CDR3 composition had little influence on this selection by IVIG. The observed interaction of IVIG with Fabs was characteristic for the binding of a B-cell superantigen (SAg).

Several potential B-cell SAgs defined by their ability to bind to the B-cell receptor (BCR) of certain V_H genes (mostly V_H3) outside of the antigen-binding groove have been characterized recently (reviewed in: Silverman 1997). B-cell SAgs include *Staphylococcus*-derived SpA (Sasano et al. 1993; Domiati-Saad and Lipsky 1998; Potter et al. 1998; Graille et al. 2000), mouse monoclonal "superantibody" D12 (Potter et al. 1998), CD5 (Pospisil et al. 1996), HIV gp120 (Karray et al. 1998; Neshat et al. 2000), and a novel clan III-restricted chicken monoclonal antibody (Cary et al. 2000). So far, only one internal human protein with SAg properties has been identified, the 175-kDa gut-associated sialoprotein Fv (pFv) (Silverman et al. 1995). Comparing the sequences, our IVIG-selected Fabs shared most of its critical amino acids with target antibodies of SpA, but comparative ELISAs revealed that at least some of the contact residues on Fabs for IVIG must be different from those for SpA.

The combined results suggested a specific interaction of a subset of IVIGs (normal immunoglobulin repertoires) with B cells that present BCRs derived from these two germline genes. Because 3–23 and 3–30/3–30.5 are the most frequently rearranged V_H germline gene segments among human B cells (Huang et al. 1996; Dewildt et al. 1999), this restricted anti-idiotypic interaction may have an important role for the development and control of the normal B-cell repertoire in health and disease (Fischer et al. 1998; Hoffmann et al. 2000). To investigate this further, we cloned and selected Fab fragments from a patient with Kawasaki disease before and after IVIG therapy. Again, a favored selection of antibodies derived from both the 3–23 or 3–30/3–30.5 germline gene segments was observed. Importantly, the reactivity with IVIG was significantly higher for clones from the library prepared after the IVIG treatment, providing the first *in vivo* functional evidence that a subset of IVIG may selectively activate B cells of this germline origin. This SAg-like, anti-idiotypic mechanism may add to the therapeutic effect of IVIG in the treatment of Kawasaki disease (Leucht et al. 2001).

11.7
Anti-idiotypic Antibodies in Cancer Therapy

The first report of successful use of a monoclonal antibody for the treatment of human cancer appeared in 1982 (Miller et al. 1982), describing the treatment of B-cell lymphoma with a patient-specific, monoclonal anti-idiotype antibody (reviewed in Waldmann 2003). The principle of anti-idiotypic vaccination and animal models are reviewed in detail elsewere (Veelken 2001; Corthay et al. 2004). A clinical benefit of idiotypic vaccination in patients with follicular lymphoma was proven recently (Inogès et al. 2006).

Numerous clinical trials using Ab2β as an antitumor vaccine have been performed or are currently underway (Hurvitz and Timmerman 2005; Maloney 2005), investigating Ab3 development and clinical remissions (Wettendorff et al. 1989; Mittelman et al. 1995; Reinartz et al. 1999; Foon et al. 2000; Chapman et al. 2004; Reinartz et al. 2004; Pritchard-Jones et al. 2005). For example, when human anti-idiotypic antibody 105AD7, produced by fusion of a human/mouse heteromyeloma and B cells of a patient treated with anti-791T/36, a mouse antibody directed against gp72, was applied to osteosarcoma patients, 11/28 patients had an Ab3 response (Austin et al. 1989; Pritchard-Jones et al. 2005).

Searching the National Institutes of Health (NIH) clinical trials database (www.clinicaltrials.gov/) in December 2005 for "idiotypic or idiotype" reveiled 32 studies, with nine of them still recruiting. However, not all of these directly investigate anti-idiotypic antibodies as a therapy. More specific results were observed from directly searching for cancer trials (www.cancer.gov/clinicaltrials/findtrials), searching for "idio" in the option for specific drugs. This revealed 20 closed and four open trials explicitly with idio- or anti-idiotypic antibodies. The targets in many of these studies were gangliosides, complex glycosphingolipids overexpressed in tumors of neuroectodermal origin, such as neuroblastoma or melanoma.

Despite continuing therapeutic efforts, advanced or relapsed stage IV neuroblastoma has a poor prognosis (Matthay et al. 1994; Matthay et al. 1999; Berthold et al. 2003). Passive immunotherapy with murine or human/mouse chimeric antibodies directed against GD2, a disialoganglioside overexpressed on neuroblastoma cells, in phase I/II trials demonstrated complete remissions and prolonged event-free survival in some neuroblastoma patients (Cheung et al. 1987, 1992, 1998, 2001; Frost et al. 1997; Handgretinger et al. 1992, 1995; Huang et al.

1992; Murray et al. 1992, 1994; Ziegler et al. 1997; Yu et al. 1998). Initially, the clinical efficacy of native anti-GD2 antibody therapy was mainly attributed to their ability to cause tumor cell killing by antibody-dependent cell-mediated cytotoxicity (ADCC) and complement-dependent cytotoxicity (CDC) (Kushner and Cheung 1989).

Although the infused antibodies were cleared from circulation after six halflives, clinical remissions were of longer duration, implying that additional antitumor defense mechanisms must have been triggered (Uttenreuther et al. 1992b, 1995b). The potential role of a "vaccination-like" effect in long-term neuroblastoma survivors, suggested by Cheung et al. and others, indicated that patients with an immune response benefited from passive immunotherapy (Cheung et al. 1987, 1993, 1994; Uttenreuther et al. 1992a; Handgretinger et al. 1995). Similar results were confirmed for other tumors in the adult patient population (Wagner et al. 1992; Baum et al. 1994; Madiyalakan et al. 1995).

Human anti-mouse antibodies (HAMAs) seemed to preclude repetitive therapeutic use of murine antibodies by diminished tumor targeting, accelerated clearance, and reduction of direct antitumor effects (Handgretinger et al. 1992; Saleh et al. 1992b; Khazaeli et al. 1993; Cheung et al. 1994; Uttenreuther-Fischer et al. 1995a). Some of these problems were addressed by decreasing the size and the xenogenic protein parts of antibodies (Gillies et al. 1989; Winter and Harris 1993). However, even with chimerized or humanized antibodies, (anti-idiotypic) immunogenicity was still observed (Fig. 11.2), although it neither limited treatment nor made patients prone to increased toxicity (Meredith et al. 1992; Saleh et al. 1992a; Uttenreuther-Fischer et al. 1996a; Yu et al. 1998).

In an analysis of a larger patient population of neuroblastoma survivors, Cheung et al. put up the hypothesis that transient levels of HAMAs, which are mainly directed against the constant regions of murine antibodies, were positively correlated with patient survival (Cheung et al. 1998). The immunogenicity of passively applied mAbs was not disadvantageous for the patient, but appeared beneficial by triggering an activation of the idiotypic network (Jerne 1974).

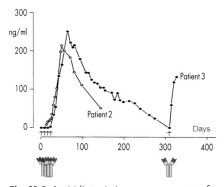

Fig. 11.2 Anti-idiotypic immune response of two ch14.18-treated neuroblastoma patients. Arrows indicate time-points of treatment.

Interestingly, in contrast to some studies (Koprowski et al. 1984; Herlyn et al. 1985; Wettendorff et al. 1989; Baum et al. 1994; Madiyalakan et al. 1995; Fagerberg et al. 1996; Schultes et al. 1998) Cheung et al. found that high HAMA levels prevented Ab3 formation in neuroblastoma patients treated with anti-GD2-MAbs (Cheung et al. 2000b).

Compiling results from previous studies they found that (1) patients with more intensive chemotherapy pretreatment before immunotherapy had lower and usually only transient HAMA/Ab2-levels than others (Cheung et al. 1994); (2) Ab3 production, starting around 6–14 months after MAb-therapy and persisting over years, was positively correlated with improved outcome (i.e. prolonged event-free survival) and more pronounced in patients with lower and only transient HAMA/Ab2-levels (Cheung et al. 2000b); (3) high HAMA concentrations were counterproductive for survival as they limited efficacy of further treatment cycles; (4) heavy chemotherapy eradicated lymphoid structures, reflected by low HAMA/Ab2 responses, and eliminated suppressor T-cell pathways and B cells. Ab2 might also bias the recovering immune repertoire towards the GD2 network (Cheung et al. 1994, 2000a,b; Cheung 2000). Moreover, Cheung et al. assumed that high Ab2 concentrations were not inductive for Ab3. In the murine model only IgM led to Ab3 production, while IgG was rather suppressive. In a healthy immune system dominant B-cell clones (Ab2) may prevent an anti-id response (Ab3) against themselves by early immunoglobulin class switch from immunogenic IgM to suppressive IgG class before anti-id B-cells (anti-Ab2) are activated (Reitan and Hannestad 1995; Cheung et al. 2000b).

11.8
Ab2β Vaccine Trials Mimicking GD$_2$

Currently two Ab2β vaccines are used for the treatment of GD2-positive malignancies: 1A7, a murine Ab2β against 14G2a, and 4B5 created by murine/human heteromyeloma technology, also directed against 14G2a (Saleh et al. 1993; Foon et al. 1998, 2000; Lutzky et al. 2002; Yu et al. 2002). Proof of principle was provided by all studies in the form of clinical responses and Ab3 serum levels in melanoma and neuroblastoma patients treated with 1A7, and positive Ab3 levels in animals treated with 4B5 (Saleh et al. 1993; Foon et al. 1998; Yu et al. 2002).

However, although Foon et al. did find Ab3-IgM in some of their melanoma patients, in general there was a positive correlation between IgM and IgG levels of Ab3, and isolated high Ab3-IgM levels were accompanied by disease progression (Foon et al. 1998, 2000). As the hypotheses and experimental findings presented above are partially contradictory, Cheung et al. suggested early on that more insight into the activation of the idiotypic network should be obtained by cloning such anti-idiotypic antibodies directly at the B-cell level (Cheung et al. 1994). Most clinical studies instead speculated on an activation of the idiotypic network driven by Ab2 and Ab3 levels detected in sera of patients treated with either Ab1 or Ab2.

Fusion cell lines of human/mouse heteromyelomas, however, frequently are not stable, and they still contain murine glycosylation (Harris et al. 1990; Lewis et al. 1992; Borrebaeck et al. 1993; Uttenreuther-Fischer et al. 1996b; Barnes et al. 2003). Phage display techniques provide an ideal method to clone and analyze the anti-idiotypic repertoire of patients after passive immunotherapy.

11.9
Molecular Characterization of the Anti-idiotypic Immune Response of a Relapse-free Cancer Patient

Assessing anti-ch14.18 levels in 65 serum samples of nine neuroblastoma patients after passive immunotherapy, only three patients showed a significant immune response against the variable region of ch14.18 (Ab2). These three patients are still alive without signs of disease 4.5–7 years later and have been off treatment for 4–5.5 years.

In a recent study, Uttenreuther-Fischer et al. (2004, 2006) reported a set of 40 human anti-idiotypic antibodies against monoclonal antibody ch14.18, cloned directly by antibody phage display technology from B cells of one of the neuroblastoma patients after ch14.18 treatment (Figs 11.3 and 11.4). Upon repetitive selection of lambda and kappa Fab-phage display libraries on target antigens ch14.18 or the murine equivalent 14G2a, positive binders were enriched. Selected Ab2 clones GK2 and GK8 as well as another 38 Fab phage clones demonstrated strong reactivity with both ch14.18 and 14G2a. Specificity and selectivity of binding was confirmed in Western blot analysis. Both anti-idiotypic clones GK2 and GK8 inhibited binding of ch14.18 to tumor-associated antigen GD2. Surprisingly, GK8-Fab alone was able to inhibit binding of the patient's serum to 14G2a by 80%, suggesting that GK8 it is identical to the majority of the patient's "original" anti-idiotypic antibodies. Immunization of rabbits with either purified GK2-

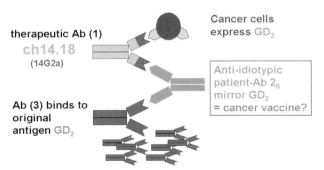

Fig. 11.3 Overview how patients develop anti-idiotypic antibodies against therapeutic ch14.18 which may function as an internal anticancer vaccine, eliciting Ab3 against the tumor-associated antigen GD2.

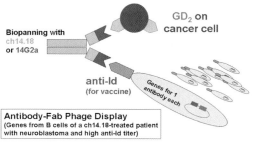

Fig. 11.4 Principle of phage display-cloning of anti-idiotypic antibodies (anti-Id) against ch14.18 as a possible vaccine for GD2-positive tumors. Therapeutic antibodies ch14.18 or the equivalent murine 14G2a are used in biopanning for selection of specific antibody-Fab phages.

and GK8-Fab or the complete Fab phage clearly produced a continuous rise in Ab3 serum levels (Ab3′) in these animals, as indicated by increased GD_2 binding. GK8 (and GK2) may be suitable as a fully human anti-idiotypic vaccine against GD_2-positive tumors.

Sequence analysis revealed at least 10 clones with different immunoglobulin genes. Homologies to putative V_H germline genes ranged between 94.90% and 100%; light chain homologies between 93.90% and 99.60%. An analysis of the R/S-ratio, giving the relation of replacement to silent mutations, suggested an antigen-driven selection of anti-idiotypic antibodies, triggered by ch14.18 treatment of our B-cell donor. Six out of 10 (60%) different clones showed an R/S-ratio above 2.9 for CDRs of their heavy chains and 6/10 (60%) for CDRs of their light chains. Taking into account that some clones appeared several times and giving a numeric relation, 32/36 (89%) clones showed somatic mutations in CDRs 1 and 2 of V_H regions and 11/36 (31%) for CDRs of their V_L regions (Uttenreuther-Fischer et al. 2006).

This was quite interesting, because earlier studies on bone marrow transplant (BMT) recipients had demonstrated that rearrangements in BMT recipients exhibited much less somatic mutations than did rearrangements from healthy subjects (Suzuki et al. 1996; Glas et al. 2000). Although our B-cell donor did not undergo BMT, intensive chemotherapy is also known to destroy lymphoid tissue. As the failure to accumulate somatic mutations in rearranged V_H genes is consistent with a maturational arrest at a very late state of B-cell differentiation, and as somatic mutations and affinity maturation are thought to take place in lymph node germinal centers (GC), it is a popular hypothesis that failure of germinal center processes prevents normal accumulation of somatic mutations following immunization in BMT recipients (Suzuki et al. 1996; Glas et al. 2000).

But unlike studies in BMT recipients, antibody clones picked from our "immunized" library from a heavily pretreated neuroblastoma patient exhibit a proportion of somatic mutations, which is comparable to what we and others found in B-cell libraries of healthy subjects (Suzuki et al. 1996; Hoffmann et al. 2000). In summary, the immunological capacity of our patient to respond to foreign antigens (i.e. idiotypes) seems comparable to that of a healthy subject.

11.10
Conclusion

Since their description approximately 30 years ago by Jerne, the recognition of the network theory and anti-idiotypic antibodies within the scientific community have had their ups and downs. Almost ignored by immunologists for more than a decade, new techniques such as phage display, DNA-based vaccines, and the recent demand for new tools in generating specific antibodies and effective vaccines against therapy-resistant tumors and old as well as emerging pathogens, for example by "retrovaccinology" (Check 2003), and the problem of breaking tolerance (Saha et al. 2004) has brought anti-idiotypic antibodies back to their attention. This is reflected by an increase of relevant publications (Shoenfeld et al. 1997; Weng et al. 2004; Park et al. 2005) and clinical studies (www.clinicaltrials.gov/) (Hurvitz and Timmerman 2005). Using fully human anti-idiotypic vaccines instead of animal-derived proteins should help to overcome problems of non-specific immunoreactivity or tolerance (Hernandez et al. 2005).

References

Abu-Shakra, M., Buskila, D., Shoenfeld, Y. (1996) Idiotypes and anti-idiotypic antibodies. In: Peter J.B., Shoenfeld, Y eds. *Autoantibodies*. Amsterdam: Elsevier, pp. 408–416.

Austin, E.B., Robins, R.A., Durrant, L.G., Price, M.R., Baldwin, R.W. (1989) Human monoclonal anti-idiotypic antibody to the tumour-associated antibody 791T/36. *Immunology* 67: 525–530.

Barbas, C.F.I., Kang, A.S., Lerner, R.A., Benkovic, S.J. (1991) Assembly of combinatorial antibody libraries on phage surfaces: the gene III site. *Proc Natl Acad Sci USA* 88: 7978–7982.

Barnes, L.M., Bentley, C.M., Dickson, A.J. (2003) Stability of protein production from recombinant mammalian cells, *Biotechnol Bioeng* 81: 631–639.

Baum, R.P., Niesen, A., Hertel, A., Nancy, A., Hess, H., Donnerstag, B., Sykes, T.R., Sykes, C.J., Suresh, M.R., Noujaim, A.A. et al. (1994) Activating anti-idiotypic human anti-mouse antibodies for immunotherapy of ovarian carcinoma. *Cancer* 73: 1121–1125.

Berchtold, P., Dale, G.L., Tani, P., McMillan, R. (1989) Inhibition of autoantibody binding to platelet glycoprotein IIb/IIIa by anti-idiotypic antibodies in intravenous gammaglobulin. *Blood* 74: 2414–2417.

Berthold, F., Hero, B., Kremens, B., Handgretinger, R., Henze, G., Schilling, F.H., Schrappe, M., Simon, T., Spix, C. (2003) Long-term results and risk profiles of patients in five consecutive trials (1979–1997) with stage 4 neuroblastoma over 1 year of age. *Cancer Lett* 197: 11–17.

Borrebaeck, C.K., Malmborg, A.C., Ohlin, M. (1993) Does endogenous glycosylation prevent the use of mouse monoclonal antibodies as cancer therapeutics? *Immunol Today* 14: 477–479.

Breitling, F., Dübel, S., Seehaus, T., Klewinghaus, I., Little, M. (1991) A surface expression vector for antibody screening. *Gene* 104: 147–153.

Burns, J.C., Matsubara, T. (1995) Kawasaki disease: Recent advances. *Trend Cardiovasc Med* 5: 3–7.

Cary, S.P., Lee, J., Wagenknecht, R., Silverman, G.J. (2000) Characterization of superantigen-induced clonal deletion with a novel clan III-restricted avian monoclonal antibody: exploiting evolutionary distance to create antibodies specific for a conserved VH region surface. *J Immunol* 164: 4730–4741.

Chapman, P.B., Williams, L., Salibi, N., Hwu, W.J., Krown, S.E., Livingston, P.O. (2004) A phase II trial comparing five dose levels of BEC2 anti-idiotypic monoclonal antibody vaccine that mimics GD3 ganglioside. *Vaccine* 22: 2904–2909.

Check, E. (2003) AIDS vaccines: back to "plan A", *Nature* 423: 912–914.

Cheung, N.K. (2000) Monoclonal antibody-based therapy for neuroblastoma, *Curr Oncol Rep* 2: 547–553.

Cheung, N.K., Lazarus, H., Miraldi, F.D., Abramowsky, C.R., Kallick, S., Saarinen, U.M., Spitzer, T., Strandjord, S.E., Coccia, P.F., Berger, N.A. (1987) Ganglioside GD2 specific monoclonal antibody 3F8: a phase I study in patients with neuroblastoma and malignant melanoma. *J Clin Oncol* 5: 1430–1440.

Cheung, N.K., Canete, A., Cheung, I.Y., Ye, J.N., Liu, C. (1993) Disialoganglioside GD2 anti-idiotypic monoclonal antibodies. *Int J Cancer* 54: 499–505.

Cheung, N.K., Cheung, I.Y., Canete, A., Yeh, S.J., Kushner, B., Bonilla, M.A., Heller, G., Larson, S.M. (1994) Antibody response to murine anti-GD2 monoclonal antibodies: correlation with patient survival, *Cancer Res* 54: 2228–2233.

Cheung, N.K.V., Kushner, B.H., Cheung, I.Y., Kramer, K., Canete, A., Gerald, W., Bonilla, M.A., Finn, R., Yeh, S.J., Larson, S.M. (1998) Anti-GD(2) antibody treatment of minimal residual stage 4 neuroblastoma diagnosed at more than 1 year of age. *J Clin Oncol* 16: 3053–3060.

Cheung, N.K., Guo, H.F., Cheung, I.Y. (2000a) Correlation of anti-idiotype network with survival following anti-G(D2) monoclonal antibody 3F8 therapy of stage 4 neuroblastoma. *Med Pediatr Oncol* 35: 635–637.

Cheung, N.K., Guo, H.F., Heller, G., Cheung, I.Y. (2000b) Induction of Ab3 and Ab3' antibody was associated with long-term survival after anti-G(D2) antibody therapy of stage 4 neuroblastoma. *Clin Cancer Res* 6: 2653–2660.

Cheung, N.K., Kushner, B.H., Laquaglia, M., Kramer, K., Gollamudi, S., Heller, G., Gerald, W., Yeh, S., Finn, R., Larson, S.M., Wuest, D., Byrnes, M., Dantis, E., Mora, J., Cheung, I.Y., Rosenfield, N., Abramson, S., O'Reilly, R.J. (2001) N7: a novel multi-modality therapy of high risk neuroblastoma (NB) in children diagnosed over 1 year of age. *Med Pediatr Oncol* 36: 227–230.

Cheung, N.V., Lazarus, H., Miraldi, F.D., Berger, N.A., Abramowsky, C.R., Saarinen, U.M., Spitzer, T., Strandjord, S.E., Coccia, P.F. (1992) Reassessment of patient response to monoclonal antibody 3F8 – letter. *J Clin Oncol* 10: 671–672.

Corthay, A., Lundin, K.U., Munthe, L.A., Froyland, M., Gedde-Dahl, T., Dembic, Z., Bogen, B. (2004) Immunotherapy in multiple myeloma: Id-specific strategies suggested by studies in animal models. *Cancer Immunol Immunother* 53: 759–769.

Dajani, A.S., Taubert, K.A., Takahashi, M., Bierman, F.Z., Freed, M.D., Ferrieri, P., Gerber, M., Shulman, S.T., Karchmer, A.W., Wilson, W. et al. (1994) Guidelines for long-term management of patients with Kawasaki disease. Report from the Committee on Rheumatic Fever, Endocarditis, Kawasaki Disease, Council on Cardiovascular Disease in the Young, American Heart Association. *Circulation* 89: 916–922.

Debre, M., Bonnet, M.C., Fridman, W.H., Carosella, E., Philippe, N., Reinert, P., Vilmer, E., Kaplan, C., Teillaud, J.L., Griscelli, C. (1993) Infusion of Fc gamma fragments for treatment of children with acute immune thrombocytopenic purpura. *Lancet* 342: 945–949.

Dewildt, R.M.T., Hoet, R.M.A., Vanvenrooij, W.J., Tomlinson, I.M., Winter, G. (1999) Analysis of heavy and light chain pairings indicates that receptor editing shapes the human antibody repertoire. *J Mol Biol* 285: 895–901.

Domiati-Saad, R., Lipsky, P.E. (1998) Staphylococcal enterotoxin A induces survival of V(H)3- expressing human B cells by binding to the V-H region with low affinity. *J Immunol* 161: 1257–1266.

Drucker, N.A., Colan, S.D., Lewis, A.B., Beiser, A.S., Wessel, D.L., Takahashi, M., Baker, A.L., Perez Atayde, A.R., Newburger, J.W. (1994) Gamma-globulin treatment of acute myocarditis in the pediatric population. *Circulation* 89: 252–257.

Duchosal, M.A., Eming, S.A., Fischer, P., Leturcq, D., Barbas, C.F.I., McConahey, P.J., Caothien, R.H., Thornton, G.B., Dixon, F.J., Burton, D.R. (1992) Immunization of hu-PBL-SCID mice and the rescue of human monoclonal Fab fragments through combinatorial libraries. *Nature* 355: 258–262.

Fagerberg, J., Ragnhammar, P., Liljefors, M., Hjelm, A.L., Mellstedt, H., Frodin, J.E. (1996) Humoral anti-idiotypic and anti-anti-idiotypic immune response in cancer patients treated with monoclonal antibody 17-1A. *Cancer Immunol Immunother* 42: 81–87.

Fields, B.A., Goldbaum, F.A., Ysern, X., Poljak, R.J., Mariuzza, R.A. (1995) Molecular basis of antigen mimicry by an anti-idiotope. *Nature* 374: 739–742.

Fischer, P., Uttenreuther-Fischer, M.M., Gaedicke, G. (1996a) Superantigens in the aetiology of Kawasaki disease. *Lancet* 348: 202.

Fischer, P., Uttenreuther-Fischer, M.M., Naoe, S., Gaedicke, G. (1996b) Kawasaki disease: update on diagnosis, treatment and a still controversial etiology. *Pediatr Hematol Oncol* 13: 487–501.

Fischer, P., Jendreyko, N., Lerch, H., Uttenreuther-Fischer, M.M., Gaedicke, G. (1998) Genetic origin of IgG antibodies cloned by phage display and anti-idiotypic panning from three patients with autoimmune thrombocytopenia. *Immunobiology* 199: 510–511.

Fischer, P., Jendreyko, N., Hoffmann, M., Lerch, H., Uttenreuther-Fischer, M.M., Chen, P.P., Gaedicke, G. (1999) Platelet reactive IgG antibodies cloned by phage display and panning with IVIG from three patients with autoimmune thrombocytopenia. *Br J Haematol* 105: 626–640.

Foon, K.A., Sen, G., Hutchins, L., Kashala, O.L., Baral, R., Banerjee, M., Chakraborty, M., Garrison, J., Reisfeld, R.A., Bhattacharya-Chatterjee, M. (1998) Antibody responses in melanoma patients immunized with an anti-idiotype antibody mimicking disialoganglioside GD2. *Clin Cancer Res* 4: 1117–1124.

Foon, K.A., Lutzky, J., Baral, R.N., Yannelli, J.R., Hutchins, L., Teitelbaum, A., Kashala, O.L., Das, R., Garrison, J., Reisfeld, R.A., Bhattacharyachatterjee, M. (2000) Clinical and immune responses in advanced melanoma patients immunized with an anti-idiotype antibody mimicking disialoganglioside GD2. *J Clin Oncol* 18: 376–384.

Frost, J.D., Hank, J.A., Reaman, G.H., Frierdich, S., Seeger, R.C., Gan, J., Anderson, P.M., Ettinger, L.J., Cairo, M.S., Blazar, B.R., Krailo, M.D., Matthay, K.K., Reisfeld, R.A., Sondel, P.M. (1997) A phase I/IB trial of murine monoclonal anti-GD2 antibody 14.G2a plus interleukin-2 in children with refractory neuroblastoma: A report of the Children's Cancer Group. *Cancer* 80: 317–333.

Galeazzi, M., Bellisai, F., Sebastiani, G.D., Morozzi, G., Marcolongo, R., Houssiau, F., Cervera, R., Levy, Y., George, J., Sherer, Y., Shoenfeld, Y. (1998) Association of 16/6 and SA1 anti-DNA idiotypes with anticardiolipin antibodies and clinical manifestations in a large cohort of SLE patients, *Clin Exp Rheumatol* 16: 717–720.

Gillies, S.D., Lo, K.M., Wesolowski, J. (1989) High-level expression of chimeric antibodies using adapted cDNA variable region cassettes. *J Immunol Methods* 125: 191–202.

Glas, A.M., van Montfort, E.H., Storek, J., Green, E.G., Drissen, R.P., Bechtold, V.J., Reilly, J.Z., Dawson, M.A., Milner, E.C. (2000) B-cell-autonomous somatic mutation deficit following bone marrow transplant. *Blood* 96: 1064–1069.

Graille, M., Stura, E.A., Corper, A.L., Sutton, B.J., Taussig, M.J., Charbonnier, J.B., Silverman, G.J. (2000) Crystal structure of a Staphylococcus aureus protein A domain complexed with the Fab fragment of a human IgM antibody: Structural basis for recognition of B-cell receptors and superantigen activity, *Proc Natl Acad Sci USA* 97: 5399–5404.

Handgretinger, R., Baader, P., Dopfer, R., Klingebiel, T., Reuland, P., Treuner, J., Reisfeld, R.A., Niethammer, D. (1992) A phase I study of neuroblastoma with the anti-ganglioside GD2 antibody 14.G2a. *Cancer Immunol Immunother* 35: 199–204.

Handgretinger, R., Anderson, K., Lang, P., Dopfer, R., Klingebiel, T., Schrappe, M., Reuland, P., Gillies, S.D., Reisfeld, R.A., Niethammer, D. (1995) A phase I study of human/mouse chimeric antiganglioside GD2 antibody ch14.18 in patients with neuroblastoma. *Eur J Cancer* 31A: 261–267.

Harris, J.F., Koropatnick, J., Pearson, J. (1990) Spontaneous and radiation-induced genetic instability of heteromyeloma hybridoma cells. *Mol Biol Med* 7: 485–493.

Herlyn, D., Lubeck, M., Sears, H., Koprowski, H. (1985) Specific detection of anti-idiotypic immune responses in cancer patients treated with murine monoclonal antibody. *J Immunol Methods* 85: 27–38.

Hernandez, A.M., Rodriguez, M., Lopez-Requena, A., Beausoleil, I., Perez, R., Vazquez, A.M. (2005) Generation of anti-Neu-glycolyl-ganglioside antibodies by immunization with an anti-idiotype monoclonal antibody: A self versus non-self-matter. *Immunobiology* 210: 11–21.

Hoet, R.M.A., Raats, J.M.H., de Wildt, R., Dumortier, H., Muller, S., Van den Hoogen, F., van Venrooij, W.J. (1998) Human monoclonal autoantibody fragments from combinatorial antibody libraries directed to the U1snRNP associated U1C protein; epitope mapping, immunolocalization and V-gene usage. *Mol Immunol* 35: 1045–1055.

Hoffmann, M., Uttenreuther-Fischer, M.M., Lerch, H., Gaedicke, G., Fischer, P. (2000) IVIG-bound IgG and IgM cloned by phage display from a healthy individual reveal the same restricted germ-line gene origin as in autoimmune thrombocytopenia. *Clin Exp Immunol* 121: 37–46.

Hombach, A., Pohl, C., Heuser, C., Sircar, R., Diehl, V., Abken, H. (1998) Isolation of single chain antibody fragments with specificity for cell surface antigens by phage display utilizing internal image antiidiotypic antibodies. *J Immunol Methods* 218: 53–61.

Huang, C.S., Uttenreuther, M., Reisfeld, R.A., Yu, A.L. (1992) Immunotherapy of GD2(+) tumors with a murine monoclonal antibody (MAB) 14G2a: A phase I study. *Proc ASCO* 11: 364.

Huang, S.C., Jiang, R.H., Glas, A.M., Milner, E.C.B. (1996) Non-stochastic utilization of Ig V region genes in unselected human peripheral B cells. *Mol Immunol* 33: 553–560.

Hurvitz, S.A., Timmerman, J.M. (2005) Current status of therapeutic vaccines for non-Hodgkin's lymphoma. *Curr Opin Oncol* 17: 432–440.

Imbach, P., Wagner, H.P., Berchtold, W., Gaedicke, G., Hirt, A., Joller, P., Mueller-Eckhardt, C., Muller, B., Rossi, E., Barandun, S. (1985) Intravenous immunoglobulin versus oral corticosteroids in acute immune thrombocytopenic purpura in childhood. *Lancet* 2: 464–468.

Imholz, B., Imbach, P., Baumgartner, C., Berchtold, W., Gaedicke, G., Gugler, E., Hirt, A., Hitzig, W., Mueller-Eckhardt, C., Wagner, H.P. (1988) Intravenous immunoglobulin (i.v. IgG) for previously treated acute or for chronic idiopathic thrombocytopenic purpura (ITP) in childhood: a prospective multicenter study. *Blut* 56: 63–68.

Inogès, S., Rodrìguez-Calvillo, M., Zabalegui, N., Lòpez-Dìaz de Cerio, A., Villanueva, H., Soria, E., Suárez, L., Rodríguez-Caballero, A., Pastor, F., García-Muñóz, R., Panizo, C., Pèrez-Calvo, J., Melero, I., Rocha, E., Orfao, A., Bendandi, M. (2006) Clinical benefit associated with idiotypic vaccination in patients with follicular lymphoma. *J Natl Cancer Inst* 98: 1292–1301.

Ishida, F., Gruel, Y., Brojer, E., Nugent, D.J., Kunicki, T.J. (1995) Repertoire cloning of a human IgG inhibitor of alpha(IIb)beta(3) function. The OG idiotype, *Mol Immunol* 32: 613–622.

Jendreyko, N., Uttenreuther-Fischer, M.M., Lerch, H., Gaedicke, G., Fischer, P. (1998) Genetic origin of IgG antibodies cloned by phage display and anti-idiotypic panning from three patients with autoimmune thrombocytopenia, *Eur J Immunol* 28: 4236–4247.

Jerne, N.K. (1974) Towards a network theory of the immune system. *Ann Immunol (Paris)* 125C: 373–389.

Karray, S., Juompan, L., Maroun, R.C., Isenberg, D., Silverman, G.J., Zouali, M. (1998) Structural basis of the gp120 superantigen-binding site on human immunoglobulins. *J Immunol* 161: 6681–6688.

Kawasaki, T. (1967) Acute febrile mucocutaneous syndrome: Clinical observation of 50 cases. *Jpn J Allergy* 16: 178–222.

Kawasaki, T. (1995) Kawasaki disease. *Acta Paediat* 84: 713–715.

Khazaeli, M.B., Saleh, M., Liu, T., Reisfeld, R.A., LoBuglio, A.F. (1993) Murine and chimeric antibodies to GD2 antigen in melanoma patients: pharmacokinetics and immune response. In: Epenetos, A.A. (ed.) *Monoclonal Antibodies 2. Applications in Clinical Oncology*, 1st edn. London: Chapman and Hall, pp. 413–419.

Koprowski, H., Herlyn, D., Lubeck, M., DeFreitas, E., Sears, H.F. (1984) Human anti-idiotype antibodies in cancer patients: Is the modulation of the immune response beneficial for the patient? *Proc Natl Acad Sci USA* 81: 216–219.

Kotzin, B.L., Leung, D.Y., Kappler, J., Marrack, P. (1993) Superantigens and their potential role in human disease. *Adv Immunol* 54: 99–166.

Kushner, B.H., Cheung, N.K. (1989) GM-CSF enhances 3F8 monoclonal antibody-dependent cellular cytotoxicity against human melanoma and neuroblastoma. *Blood* 73: 1936–1941.

Lacroix-Desmazes, S., Mouthon, L., Spalter, S.H., Kaveri, S., Kazatchkine, M.D. (1996) Immunoglobulins and the regulation of autoimmunity through the immune network. *Clin Exp Rheumatol* 14: S9–S15.

Leucht, S., Uttenreuther-Fischer, M.M., Gaedicke, G., Fischer, P. (2001) The B cell superantigen-like interaction of IVIG with Fab fragments of V_H 3–23 and 3–30/3–30.5 germ-line gene origin cloned from a patient with Kawasaki disease is enhanced after IVIG therapy. *Clin Immunol* 99: 18–29.

Leung, D.Y. (1993) Kawasaki disease. *Curr Opin Rheumatol* 5: 41–50.

Levy, Y., Sherer, Y., George, J., Langevitz, P., Ahmed, A., Bar-Dayan, Y., Fabbrizzi, F., Terryberry, J., Peter, J., Shoenfeld, Y. (1999) Serologic and clinical response to treatment of systemic vasculitis and associated autoimmune disease with intravenous immunoglobulin. *Int Arch Allergy Immunol* 119: 231–238.

Lewis, A.P., Parry, N., Peakman, T.C., Crowe, J.S. (1992) Rescue and expression of human immunoglobulin genes to generate functional human monoclonal antibodies. *Hum Antibodies Hybridomas* 3: 146–152.

Lutzky, J., Gonzalez-Angulo, A.M., Orzano, J.A. (2002) Antibody-based vaccines for the treatment of melanoma. *Semin Oncol* 29: 462–470.

Madiyalakan, R., Sykes, T.R., Dharampaul, S., Sykes, C.J., Baum, R.P., Hor, G., Noujaim, A.A. (1995) Anti-idiotype induction therapy. Evidence for the induction of immune response through the idiotype network in patients with ovarian cancer after administration of anti-CA125 murine monoclonal antibody B43.13. *Hybridoma* 14: 199–203.

Maloney, D.G. (2005) Immunotherapy for non-Hodgkin's lymphoma: monoclonal antibodies and vaccines. *J Clin Oncol* 23: 6421–6428.

Matthay, K.K., Seeger, R.C., Reynolds, C.P., Stram, D.O., Oleary, M.C., Harris, R.E., Selch, M., Atkinson, J.B., Haase, G.M., Ramsay, N.K. (1994) Allogeneic versus autologous purged bone marrow transplantation for neuroblastoma: A report from the Childrens Cancer Group. *J Clin Oncol* 12: 2382–2389.

Matthay, K.K., Villablanca, J.G., Seeger, R.C., Stram, D.O., Harris, R.E., Ramsay, N.K., Swift, P., Shimada, H., Black, C.T., Brodeur, G.M., Gerbing, R.B., Reynolds, C.P. (1999) Treatment of high-risk neuroblastoma with intensive chemotherapy, radiotherapy, autologous bone marrow transplantation, and 13-cis-retinoic acid. *N Engl J Med* 341: 1165–1173.

Meredith, R.F., Khazaeli, M.B., Plott, W.E., Saleh, M.N., Liu, T., Allen, L.F., Russell, C.D., Orr, R.A., Colcher, D., Schlom, J. (1992) Phase I trial of iodine-131-chimeric B72.3 (human IgG4) in metastatic colorectal cancer. *J Nucl Med* 33: 23–29.

Metlas, R., Veljkovic, V. (2004) HIV-1 gp120 and immune network. *Int Rev Immunol* 23: 413–422.

Miller, R.A., Maloney, D.G., Warnke, R., Levy, R. (1982) Treatment of B-cell lymphoma with monoclonal anti-idiotype antibody. *N Engl J Med* 306: 517–522.

Mittelman, A., Wang, X.H., Matsumoto, K., Ferrone, S. (1995) Antianti-idiotypic response and clinical course of the disease in patients with malignant melanoma immunized with mouse anti-idiotypic monoclonal antibody MK2-23. *Hybridoma* 14: 175–181.

Murray, J.L., Cunningham, J.E., Brewer, H.M., Janus, M.H., Podoloff, D.A., Bhadkamkar, V.P., Kasi, L.P., Shah, R.S., Benjamin, R.S., Legha, S.S., Plager, C., Papadopoulos, N.E., Jaffee, N., Ater, J.L., Mujoo, K., Itoh, K., Ross, M., Bucana, C.D., Rosenblum, M.G. (1992) Phase I trial of murine anti-ganglioside (GD2) monoclonal antibody (Mab) 14G2a in cancer patients. *J Immunother* 11: 135–136.

Murray, J.L., Cunningham, J.E., Brewer, H., Mujoo, K., Zukiwski, A.A., Podoloff, D.A., Kasi, L.P., Bhadkamkar, V., Fritsche, H.A., Benjamin, R.S. et al. (1994) Phase I trial of murine monoclonal antibody 14G2a administered by prolonged intravenous infusion in patients with neuroectodermal tumors. *J Clin Oncol* 12: 184–193.

Neshat, M.N., Goodglick, L., Lim, K., Braun, J. (2000) Mapping the B cell superantigen binding site for HIV-1 gp120 on a V(H)3 Ig. *Int Immunol* 12: 305–312.

Osei, A., Uttenreuther-Fischer, M.M., Lerch, H., Gaedicke, G., Fischer, P. (2000) Restricted V_H3 gene use in phage displayed Fab that are selected by IVIG. *Arthritis Rheum* 43: 2722–2732.

Park, I.H., Youn, J.H., Choi, I.H., Nahm, M.H., Kim, S.J., Shin, J.S. (2005) Anti-idiotypic antibody as a potential candidate vaccine for *Neisseria meningitidis* serogroup B. *Infect Immun* 73: 6399–6406.

Peter, J.B., Shohat, L., eds (1996) *Autoantibodies*. Amsterdam: Elsevier.

Pospisil, R., Fitts, M.G., Mage, R.G. (1996) CD5 is a potential selecting ligand for B cell surface immunoglobulin framework region sequences. *J Exp Med* 184: 1279–1284.

Potter, K.N., Li, Y., Mageed, R.A., Jefferis, R., Capra, J.D. (1998) Anti-idiotypic antibody D12 and superantigen SPA both interact with human V(H)3-encoded antibodies on the external face of the heavy chain involving FR1, CDR2 and FR3. *Mol Immunol* 35: 1179–1187.

Pritchard-Jones, K., Spendlove, I., Wilton, C., Whelan, J., Weeden, S., Lewis, I., Hale, J., Douglas, C., Pagonis, C., Campbell, B., Alvarez, P., Halbert, G., Durrant, L.G. (2005) Immune responses to the 105AD7 human anti-idiotypic vaccine after intensive chemotherapy, for osteosarcoma. *Br J Cancer* 92: 1358–1365.

Reinartz, S., Boerner, H., Koehler, S., Vonruecker, A., Schlebusch, H., Wagner, U. (1999) Evaluation of immunological responses in patients with ovarian cancer treated with the anti-idiotype vaccine ACA125 by determination of intracellular cytokines – a preliminary report. *Hybridoma* 18: 41–45.

Reinartz, S., Kohler, S., Schlebusch, H., Krista, K., Giffels, P., Renke, K., Huober, J., Mobus, V., Kreienberg, R., DuBois, A., Sabbatini, P., Wagner, U. (2004) Vaccination of patients with advanced ovarian carcinoma with the anti-idiotype ACA125: immunological response and survival (phase Ib/II). *Clin Cancer Res*, vol. 10, no. 5: 1580–1587.

Reitan, S.K., Hannestad, K. (1995) A syngeneic idiotype is immunogenic when borne by IgM but tolerogenic when joined to IgG. *Eur J Immunol* 25: 1601–1608.

Saha, A., Chatterjee, S.K., Foon, K.A., Primus, F.J., Sreedharan, S., Mohanty, K., Bhattacharya-Chatterjee, M. (2004) Dendritic cells pulsed with an anti-idiotype antibody mimicking carcinoembryonic antigen (CEA) can reverse immunological tolerance to CEA and induce antitumor immunity in CEA transgenic mice. *Cancer Res* 64: 4995–5003.

Saleh, M.N., Khazaeli, M.B., Wheeler, R.H., Allen, L., Tilden, A.B., Grizzle, W., Reisfeld, R.A., Yu, A.L., Gillies, S.D., LoBuglio, A.F. (1992a) Phase I trial of the chimeric anti-GD2 monoclonal antibody ch14.18 in patients with malignant melanoma. *Hum Antibodies Hybridomas* 3: 19–24.

Saleh, M.N., Khazaeli, M.B., Wheeler, R.H., Dropcho, E., Liu, T., Urist, M., Miller, D.M., Lawson, S., Dixon, P., Russell, C.H. (1992b) Phase I trial of the murine monoclonal anti-GD2 antibody 14G2a in metastatic melanoma. *Cancer Res* 52: 4342–4347.

Saleh, M.N., Stapleton, J.D., Khazaeli, M.B., LoBuglio, A.F. (1993) Generation of a human anti-idiotypic antibody that mimics the GD2 antigen. *J Immunol* 151: 3390–3398.

Sasano, M., Burton, D.R., Silverman, G.J. (1993) Molecular selection of human antibodies with an unconventional bacterial B cell antigen. *J Immunol* 151: 5822–5839.

Schroeder, J.O., Zeuner, R.A., Euler, H.H., Loffler, H. (1996) High dose intravenous immunoglobulins in systemic lupus erythematosus: Clinical and serological results of a pilot study. *J Rheumatol* 23: 71–75.

Schultes, B.C., Baum, R.P., Niesen, A., Noujaim, A.A., Madiyalakan, R. (1998) Anti-idiotype induction therapy: anti-CA125 antibodies (Ab(3)) mediated tumor killing in patients treated with Ovarex mAb B43.13 (Ab(1)). *Cancer Immunol Immunother* 46: 201–212.

Sherer, Y., Shoenfeld, Y. (2000) Idiotypic network dysregulation – A common etiopathogenesis of diverse autoimmune diseases. *Appl Biochem Biotechnol* 83: 155–162.

Sherer, Y., Levy, Y., Langevitz, P., Lorber, M., Fabrizzi, F., Shoenfeld, Y. (1999) Successful treatment of systemic lupus erythematosus cerebritis with intravenous immunoglobulin. *Clin Rheumatol* 18: 170–173.

Shoenfeld, Y. (1994) Idiotypic induction of autoimmunity: A new aspect of the idiotypic network. *FASEB J* 8: 1296–1301.

Shoenfeld, Y. (2005) Anti-DNA idiotypes: from induction of disease to novel therapeutical approaches. *Immunol Lett* 100: 73–77.

Shoenfeld, Y., Kennedy, R.C., Ferrone, S. (eds) (1997) *Idiotypes in Medicine: Autoimmunity, Infection and Cancer.* Amsterdam: Elsevier.

Silverman, G.J. (1997) B-cell superantigens. *Immunol Today* 18: 379–386.

Silverman, G.J., Roben, P., Bouvet, J.P., Sasano, M. (1995) Superantigen properties of a human sialoprotein involved in gut-associated immunity. *J Clin Invest* 96: 417–426.

Silvestris, F., Cafforio, P., Dammacco, F. (1994) Pathogenic anti-DNA idiotype-reactive IgG in intravenous immunoglobulin preparations. *Clin Exp Immunol* 97: 19–25.

Suzuki, I., Milner, E.C.B., Glas, A.M., Hufnagle, W.O., Rao, S.P., Pfister, L., Nottenburg, C. (1996) Immunoglobulin heavy chain variable region gene usage in bone marrow transplant recipients: Lack of somatic mutation indicates a maturational arrest. *Blood* 87: 1873–1880.

Uttenreuther, M., Huang, C.S., Tsui, C., Reisfeld, R.A., Gillies, S.D., Yu, A. (1992a) Comparison of biological activity and pharmacokinetics between a mouse and a human-mouse anti-GD2 chimeric monoclonal antibody (MAB). *Proc AACR* 33: 244.

Uttenreuther, M.M., Huang, Ch.S., Reisfeld, R.A., Yu, A.L. (1992b) Pharmacokinetics of chimeric human/mouse antibody ch14.18 in a phase I clinical trial with neuroblastoma patients. *J Cancer Res Clin Oncol* 118(Suppl.): 81.

Uttenreuther-Fischer, M.M., Huang, C.S., Reisfeld, R.A., Yu, A.L. (1995a) Pharmacokinetics of anti-ganglioside GD2 mAb 14G2a in a phase I trial in pediatric cancer patients. *Cancer Immunol Immunother* 41: 29–36.

Uttenreuther-Fischer, M.M., Huang, C.S., Yu, A.L. (1995b) Pharmacokinetics of human-mouse chimeric anti-GD2 mAb ch14.18 in a phase I trial in neuroblastoma patients. *Cancer Immunol Immunother* 41: 331–338.

Uttenreuther-Fischer, M.M., Fischer, P., Reisfeld, R.A., Gaedicke, G., Yu, A.L. (1996a) Monoclonal anti-disialoganglioside (GD2) antibodies in neuroblastoma-therapy: Summary of three phase I/II trials. *Proc ISHage-Meeting.*

Uttenreuther-Fischer, M.M., Yu, A., Gaedicke, G., Fischer, P. (1996b) A human anti-idiotypic monoclonal antibody against anti-GD2 chimeric antibody ch14.18 from a patient treated for neuroblastoma. *Hum Antibodies Hybridomas* 7: 55.

Uttenreuther-Fischer, M.M., Krueger, J.A., Fischer, P. (2004) Anti-idiotypic antibodies against anti-GD2 antibody ch14.18 – a potential tumor vaccine? *Monatsschr Kinderheilkd* 152(Suppl. 1): 229.

Uttenreuther-Fischer, M.M., Krueger, J.A., Fischer, P. (2006) Molecular characterization of the anti-idiotypic immune response of a relapse-free neuroblastoma patient following antibody therapy. A possible vaccine against tumors of neuroectodermal origin? *J Immunol* 176: 7775–7786.

Veelken, H. (2001) Tumor vaccination using monoclonal antibodies. *Internist (Berl)* 42: 874–883.

Wagner, U.A., Oehr, P.F., Reinsberg, J., Schmidt, S.C., Schlebusch, H.W., Schultes, B., Werner, A., Prietl, G., Krebs, D. (1992) Immunotherapy of advanced ovarian carcinomas by activation of the idiotypic network. *Biotechnol Ther* 3: 81–89.

Waldmann, T.A. (2003) Immunotherapy: past, present and future. *Nat Med* 9: 269–277.

Weng, W.K., Czerwinski, D., Timmerman, J., Hsu, F.J., Levy, R. (2004) Clinical outcome of lymphoma patients after idiotype vaccination is correlated with humoral immune response and immunoglobulin G Fc receptor genotype. *J Clin Oncol* 22: 4717–4724.

Wettendorff, M., Iliopoulos, D., Tempero, M., Kay, D., DeFreitas, E., Koprowski, H., Herlyn, D. (1989) Idiotypic cascades in cancer patients treated with monoclonal antibody CO17–1A. *Proc Natl Acad Sci USA* 86: 3787–3791.

Willems, P.M.W., Hoet, R.M.A., Huys, E.L.P.G., Raats, J.M.H., Mensink, E.J.B.M., Raymakers, R.A.P. (1998) Specific detection of myeloma plasma cells using anti-idiotypic single chain antibody fragments selected from a phage display library. *Leukemia* 12: 1295–1302.

Winter, G., Harris, W.J. (1993) Humanized antibodies. *Immunol Today* 14: 243–246.

Yang, Y.Y., Fischer, P., Leu, S.J.C., Olee, T., Carson, D.A., Chen, P.P. (1997) IgG rheumatoid factors isolated by surface displaying phage library technique. *Immunogenetics* 45: 301–310.

Yang, Y.Y., Fischer, P., Leu, S.J., Zhu, M., Woods, V.L., Jr., Chen, P.P. (1999) Possible presence of enhancing antibodies in idiopathic thrombocytopenic purpura. *Br J Haematol* 104: 69–80.

Yu, A.L., Uttenreuther-Fischer, M.M., Huang, C.S., Tsui, C.C., Gillies, S.D., Reisfeld, R.A., Kung, F. (1998) Phase I trial of a human-mouse chimeric anti-disialoganglioside monoclonal antibody ch14.18 in patients with refractory neuroblastoma and osteosarcoma. *J Clin Oncol* 16: 2169–2180.

Yu, A.L., Batova, A., Strother, D., Eskenazi, U., Castleberry, R.P., Bash, R., Matthay, K.K., Diller, L. (2002) A pilot trial of a GD-2 directed anti-idiotypic antibody as a vaccine for high risk neuroblastoma. *Proc Adv Neuroblastoma Res* 44: 17–6.

Ziegler, A., Macintosh, S.M., Torrance, L., Simon, W., Slabas, A.R. (1997) Recombinant antibody fragments that detect enoyl acyl carrier protein reductase in *Brassica napus*. *Lipids* 32: 805–809.

Part V
Ongoing Clinical Studies

12
Antibodies in Phase I/II/III: Cancer Therapy

P. Markus Deckert

12.1
Introduction

The promise of Paul Ehrlich's "magic bullet" was soon attached to Köhler and Milstein's – thankfully unpatented – invention of monoclonal antibody technology (Köhler and Milstein 1975) and then went unfulfilled for more than two decades. Only during the past 9 years have antibodies become an established modality of cancer treatment.

Eight antibodies are currently approved in the USA and Europe for the therapy of malignant diseases. Together, they exhibit many of the features to be expected from future antibody therapies. Today, more than 400 antibody varieties are in clinical trials for treatment of malignant diseases. Two hundred and five of these are listed in Tables 12.1–12.15. Although objective criteria have been employed, this implies that the selection will be arbitrary to a certain degree.

In large part, the clinical success of antibodies has been made possible through progress in recombinant technology, as chimerization and humanization helped to overcome the limitations of applying rodent proteins in humans. Furthermore, it has advanced the process of generating a desired binding specificity – from murine hybridoma towards *in vitro* systems such as phage display – and the capacity to design and tailor antigen-binding proteins to specific needs (e.g. by further reducing their immunogenicity or by creating bi- and even trifunctional molecules).

For practical purposes, four generations of monoclonal antibodies may be proposed:

1. Murine monoclonal antibodies generated by vaccination and hybridoma technology (m-mAb).
2. Variable region-grafted chimeric (ch-mAb) and complementarity determining region (CDR)-grafted humanized antibodies (hu-mAb), often collectively referred to as humanized antibodies.

Handbook of Therapeutic Antibodies. Edited by Stefan Dübel
Copyright © 2007 WILEY-VCH Verlag GmbH & Co. KGaA, Weinheim
ISBN 978-3-527-31453-9

3. Fully human antibodies generated either by applying the classical vaccination and hybridoma technology on transgenic animals bearing a "human" immune system or by phage display technology (h-mAb).
4. Recombinant antibody constructs such as single-chain variable fragments and diabodies or minibodies as well as recombinant fusion constructs of such fragments with proteins carrying various effector functions, such as enzymes or cytokines.

This outline is certainly not all-encompassing (e.g. it does not include chemical conjugates with drug molecules or radioisotopes, because they have been produced with antibodies of all generations) nor is it generational in a strict chronological sense, but rather marks the distance from the original murine hybridoma.

The future of antibody applications in clinical oncology will comprise members of all four generations. Thus, while most of the antibodies approved for cancer therapy today have been derived from classical hybridomas and improved by humanization, progress in recombinant antibody design is currently moving from bench to bedside as fourth-generation antibodies have entered clinical trials. On the other hand, there are still serious candidates of first-generation murine antibodies being evaluated in clinical studies today.

Along with this development, the discovery of new pathomechanisms in cell signaling and angiogenesis has led to new targets for therapeutic intervention, which are exploited both by antibodies and small molecule pharmaceuticals. Here, too, the focus of antibody research is shifting from the original concept of a cancer-specific surface marker allowing us to label a tumor cell for destruction by the immune system towards employing antibodies to activate or inhibit specific cellular mechanisms.

An up-to-date review of antibodies under clinical investigation has some intrinsic difficulties the reader should be aware of. First, clinical trials may follow years after the initial publication of a new candidate substance, while on the other hand successful trials may be carried out on substances not yet published at all. Second, the inception of a clinical trial is not normally published, nor, necessarily, are its results. Hence, the peer-reviewed literature will not cover the full spectrum of current clinical developments. All other sources, however, are biased in one way or another: the personal insight of a reviewer tends to cover only parts of the field, while company websites and business reviews rarely hold up to the standards of scientific communication. The best source for the start of a comprehensive search would be a central database of clinical trials. With ClinicalTrials.gov, there is such a registry for the United States, and it may be fairly complete for cancer and other diseases deemed life-threatening. However, although the US is certainly a strong motor in this field, there are no comparable registries yet for Europe, Asia, or the Commonwealth. Initiatives to fill this gap are underway, but as of now have not been successful (Haug et al. 2005).

The aim of this chapter is to provide a reasonably complete list of the monoclonal antibodies or antibody-derived constructs currently under clinical investigation (presented mainly in the form of tables) and to review critically the resulting trends and developments.

To this end, the peer-reviewed literature and registered clinical trials were scanned via MedLine and ClinicalTrials.gov, respectively, using various search algorithms, and the retrieved articles and trial protocols reviewed. Antibodies or antibody constructs that appeared in either database, but not in both, were then searched for by various internet search engines to retrieve information from company websites or business reviews. Only where this secondary search yielded reliable, scientifically relevant information were such antibodies included and the sources of information identified.

Following the perspective of this book as a textbook on antibodies, not on clinical oncology, and the author's view of future clinical antibody applications outlined above, this chapter is organized along antibody structure, function, and targets. Thus, bifunctional antibody constructs in clinical trials are presented first, then antibodies with specific targeting or effector mechanisms. The remainder of antibodies in the clinical study phase are grouped by categories of targeted antigen. Only where the antigen is not well described are they finally grouped by the disease studied.

Two important aspects are not mirrored in the outline of this chapter: (1) the original species of an antibody (i.e. whether it is the rodent, humanized, or "fully human" version) – antibody humanization may now be considered standard technology, and where murine antibodies are in clinical trial this is usually not for lack of a humanized version; and (2) the type of therapy (i.e. antibody monotherapy, radioimmunotherapy, combination chemotherapy etc.) – while the integration of antibodies in combination protocols was a major advance in their clinical application, too many antibodies are used in several types of therapy for this criterion to be helpful.

Finally, once an antibody is approved for clinical use in cancer, it leaves the scope of this article.

12.2
Novel Antibody Constructs

Recombinant and bifunctional antibody constructs are not new as such. Almost all currently approved antibodies are recombinantly chimerized or humanized. Early on in the development of antibodies for cancer, radioisotope conjugates have been produced, and two of these, the CD20-specific ^{90}Y-ibritumomab tiuxetan (Zevalin) and ^{131}I-tositumomab (Bexxar) are approved drugs today. The same applies to antibody–drug conjugates such as the CD33-specific chalicheamicin conjugate gemtuzumab-ozogamicin (Mylotarg).

This section, thus, deals with mostly recombinant antibody constructs of a function and structure not predefined by nature, but obtained by adding or subtracting functional domains.

12.2.1
Bispecific Antibodies

The first antibodies combining the binding regions for two different antigens were generated by heterohybridoma technology and were described as early as 1983 (Milstein and Cuello 1983). In cancer therapy, the original idea of bispecific antibodies was to bridge cellular antigens of cancer and immune cells and thus trigger a targeted immune reaction against the tumor (for an overview, see Table 12.1). A newer concept is the use of bispecific antibodies for pretargeting strategies – these are described in Section 12.4.3.

Based on conventional hybridoma technology is the chemically crosslinked dual Fab fragment MDX-210 and its semi-humanized version, MDX-H210 (humanized Fab anti-CD64::Fab anti-HER2/neu): The product of the protooncogene Her-2/neu (or erbB-2) is a member of the epithelial growth factor receptor superfamily that is overexpressed in about a third of breast cancers and is also the target of the approved antibody trastuzumab. CD64, synonymous for the Fc γ receptor I (FCγRI) of monocytes, macrophages, dendritic cells, and activated granulocytes, is the central mediator of antibody-dependent cellular cytotoxicity (ADCC). An interesting aspect of the approach chosen here and in other antibodies against effector receptors is the use of an FCγRI-specific idiotope for affinity reasons rather than its natural ligand, the IgG Fc fragment.

Clinical studies have been performed with both bispecific conjugates and have progressed to phase II with MDX-H210, which, interestingly, was first published for Her-2/neu-expressing prostate, not breast cancer (James et al. 2001). In summary, these studies have demonstrated good tolerance of the antibody, and immunological activity defined as binding to circulating monocytes and increases in plasma cytokine levels, saturation of circulating Fc receptors on myeloid cells, and invasion of monocytes into tumor tissue (Repp et al. 2003) was detectable, suggesting functioning of the therapeutic principle. However, no clinical effect was observed when the bispecific antibody was given alone, but only when granulocyte and/or monocyte-activating cytokines such as G-CSF, GM-CSF, or interferon γ (IFNγ), were added to the regimens. The figures generated from these studies are not yet sufficient to estimate response and survival rates, but between one-fifth and one-third of (usually extensively pretreated) patients in these studies showed some degree of response (Valone et al. 1995; James et al. 2001; Repp et al. 2003).

Comparable results were reported earlier in a phase I study of 2B1, an anti-Her-2neu bispecific with FcγRIII as the immunologic effector target (Weiner et al. 1995), whereas H22xKi-4 and MDX-447 are based on the same CD64-binding antibody H22 as MDX-210, but are directed against the CD30 antigen of Hodgkin's lymphoma (Borchmann et al. 2002) and the EGFR (Wallace et al. 2000), respectively. Similarly, with HRS-3/A9, binding FcγRIII and the CD30 antigen, in small phase I studies immunological responses were seen which demonstrated proof of principle, but were not predictive for clinical outcome

Table 12.1 Bispecific antibodies.

Antibody name	Characteristics	Target antigen	Target cancer	Phase	NCI study ID	Status	Source	Publ. phase	Publ. year	Comment
2B1	520C9 + 3G8, murine	p185/HER-2 + CD16	Breast	II	–		National Institutes of Health, Bethesda, MD, USA; Fox Chase Cancer Center, Philadelphia, USA	I	1995/1999	
BIS-1	F(ab')$_2$ of RIV-9 + MOC-31, murine	CD3 + EGP-2 (pan-carcinoma associated)	Lung cancer	I	–		Rijksuniversiteit Groningen, Groningen, Netherlands	I	1994/2000	Recruits T lymphocytes to tumor
Catumaxomab (Removab)	Trifunctional (trAb), murine/rat	EpCAM + CD3	Ovarian	I/II	NCT00189345	Closed 2005	Fresenius AG, Bad Homburg, Germany; Trion Pharma, München, Germany	–	2005	
H22 × Ki-4	Anti-CD30 + anti-CD64, human/murine	CD64 + CD30	Reed–Sternberg cells: Hodgkin's lymphoma	I	–		Medarex, Princeton, NJ, USA	I	2002	
HRS-3/A9	HRS-3 + A9, murine	CD30 + FCRIII (CD16)	Reed–Sternberg cells: Hodgkin's lymphoma	I/II	–		Universität des Saarlandes, Homburg, Germany; Biotest Pharma, Dreieich, Germany	I/II	2001	

Table 12.1 (Continued)

Antibody name	Characteristics	Target antigen	Target cancer	Phase	NCI study ID	Status	Source	Publ. phase	Publ. year	Comment
MDX-210	Chemically linked chimeric F(ab)′ of M22 + 520C9 murine	CD64 = FcγR1 + p185/HER-2	Leukemia, head-and-neck, breast, skin, pancreatic, renal cell, colon, bladder, and prostatic cancers	III	–		Medarex, Princeton, NJ, USA	I	1997/ 2004	
MDX-H210	Chemically linked humanized F(ab)′, based on H22 + 520C9	CD64 = FcγR1 + p185/HER-2	Head-and-neck, breast, skin, renal cell, ovarian, bladder, and prostatic cancers	II/III	–		Medarex, Princeton, NJ, USA	II	2003/ 2004	
MDX-214	Recombinant bispecific	CD89 (FcαR) + EGFR	Not specified	I/II	–		Medarex, Princeton, NJ, USA	–	–	
MDX-447	Humanized bispecific of H22 + mAb 425	CD64 = FcγR1 + EGFR	Head-and-neck, skin, renal cell, ovarian, bladder, and prostatic cancers	II	NCT00005813	Closed	Medarex, Princeton, NJ, USA; Merck KGaA, Darmstadt, Germany	–	2000	Combination with LAK cells
MT103 (BscCD19 × CD3)	scFv construct, murine	CD19 + CD3	Leukemia, non-Hodgkin's lymphoma	I	–		Micromet AG, München, Germany	–	2005	Derived from mAb HD37 and 145-2C11 (anti-CD3)
OC/TR (OC-TR)	F(ab′)$_2$ of Mov18 + anti-CD3, murine	CD3 + folate-binding protein	Ovarian cancer	I	–		Istituto Nazionale per lo Studio e la Cura dei Tumori, Milano, Italy; Universiteit Leiden, Leiden, Netherlands	I	1999	Recruits T lymphocytes to tumor

(Renner et al. 2000). Here, too, clinical trials employing costimulation by cytokines were initiated, but the results are not yet available (Hartmann et al. 2001).

A complete recombinant approach has been realized in MT103, a bispecific double scFv construct connecting the CD19 receptor of B lymphocytes as a tumor target of B-cellular non-Hodgkin lymphomas and acute or chronic leukemias with the CD3 part of the T-cell receptor defining the effector cells to be activated. Preclinical observations suggest that this bispecific antibody can induce repeated target cell lysis by activated T cells, while CD19-negative "bystanders" were spared (Hoffmann et al. 2005). *In vivo* T-cell activation and B-cell depletion have been demonstrated in chimpanzees, but data from an ongoing phase I clinical trial are not yet available (Schlereth et al. 2005).

Catumaxomab (Removab) is unique in that it has the structure of a "conventional" quadroma-derived bispecific complete IgG, but acts as a trifunctional agent that binds to tumor cells and harnesses two immunological pathways against the tumor. The trick here is the opposite approach to that introduced above with MDX-210: the natural Fc receptor-binding capacity of the IgG constant region is utilized for ADCC activation, leaving the bispecific variable regions for binding the epithelial cell adhesion molecule (EpCAM) overexpressed in various cancers and the CD3 portion of the T-cell receptor as (second) immunological effector. Thus, both T lymphocytes and antigen-presenting cells are activated, which should lead to a synergistic effect. In addition, the activation of T cells and granulocytes/monocytes means the parallel activation of acquired and innnate immunity. Preclinical data demonstrate activation of $CD83^+$ antigen-presenting cells, secretion of IFNγ, and granzyme B-mediated lysis of targeted cells by EpCAM-specific $CD8^+$ T cells after tumor cell opsonization with catumaxomab, and an *in vitro* cytotoxicity comparable to cisplatin (Gronau et al. 2005). As the secretion of cytokines such as TNFα, IFNγ, and IL-2b subsided within 24 h, the authors conclude that this system may be safe without provoking severe adverse events by a "cytokine storm" (Schmitt et al. 2004). Again, data from an ongoing clinical trial are not yet available.

12.2.2
Antibody Fusion Constructs

The idea of attaching an effector function other than its natural Fc fragment to an antibody has arisen early in the development of antibodies for cancer treatment. Again, the advent of recombinant technology has given this concept a boost in helping to overcome the drawbacks of chemical conjugation, namely heterogeneity of products with varying conjugation ratios and molecular attachment sites, problems of stability of the chemical linker, and often a low final yield of the desired product.

There are basically three groups of bifunctional constructs currently in clinical trials: drug conjugates, immunotoxins, and cytokine fusion proteins. They are summarized in Table 12.2.

Table 12.2 Antibody fusion constructs.

Antibody name	Characteristics	Target antigen	Target cancer	Phase	NCI study ID	Status	Source	Publ. phase	Publ. year	Comment
CMB-401	Calicheamycin conjugate of CTM01	RPAP epitope of MUC1	Breast, ovarian cancer	II	–		Wyeth, Madison, NJ, USA	II	2003/2005	Drug conjugate
260F9-rRA	260F9::ricin A conjugate, murine	55 kDa breast cancer-associated antigen	Breast, ovarian cancer	I	–		Cetus Corp., Emeryville, CA, USA	I	1989/1996	Immunotoxin
Anti-B4-bR	Ricin-A immunotoxin, murine	CD19	B lymphocytes: myeloma, leukemia	II	–		Dana Farber Cancer Institute, Boston, MA, USA	II	2001	Immunotoxin
RFB4(dsFv)-PE38 (BL22)	RFB4 scFv–*Pseudomonas* exotoxin PE38, disulfide stabilized, murine	CD22	B-cell lymphoma	I/II	NCT00071318 NCT00126646 NCT00075309 NCT00114751	Recruiting	Royal Free Hospital, London, UK; National Institutes of Health, Bethesda, MD, USA	I	2005	Immunotoxin
RFB4-dgA	RFB4–ricin A–conjugate, murine	CD22	B-cell lymphoma	I	NCT00006423 NCT00001271	Closed 2003	National Cancer Institute, Rockville, MD, USA	I	2000	Immunotoxin
RFT5(scFv)-ETA'	Recombinant humanized scFv–*Pseudomonas* exotoxin A (ETA)	CD25	Reed–Sternberg cells: Hodgkin's lymphoma	?	–		Medizinische Klinik I, Universität Köln, Köln, Germany	–	2002	Immunotoxin

Name	Description	Target	Indication	Phase	NCT	Status	Institution	Phase	Year	Type
RFT5-Dga RFT5-SMTP-dGA	Ricin-A immunotoxin, murine	CD25	Reed–Sternberg cells: Hodgkin's lymphoma	I/II	–		Royal Free Hospital, London, UK	I/II	1998	Immunotoxin
LMB-2 (anti-Tac(Fv)-PE38)	Recombinant *Pseudomonas* exotoxin PE38 construct, murine	CD25 (IL-2R)	T lymphocytes leukemia	II	NCT00080535 NCT00077922 NCT00263510 NCT00082004 NCT00080821 NCT00104975	Recruiting	National Institutes of Health, Bethesda, MD, USA	–	2000	Immunotoxin
HuM195/rGel	Chemical conjugate of chimeric huM195 and recombinant gelonin plant toxin	CD33	Myeloic malignancies	I	NCT00038051	Recruiting	University of Texas M.D. Anderson Cancer Center, Houston, TX, USA	–	2003	Immunotoxin
4D5MOCB-ETA	Recombinant humanized scFv–*Pseudomonas* exotoxin A (ETA)	EpCAM	Head-and-neck cancer	I	–		Universität Zurich, Zurich, Switzerland	–	2003	Immunotoxin
LMB-1 (B3-LysPE38)	B3–*Pseudomonas* exotoxin PE38, murine	Le-y	Gastric, ovarian, and colon cancers	II	NCT00001805	Closed 2000	National Institutes of Health, Bethesda, MD, USA	II	2004	Immunotoxin
LMB-7 (B3(Fv)-PE38)	B3 scFv–*Pseudomonas* exotoxin PE38 fusion, murine	Le-y	Leptomeningeal metastases	I	NCT00003020	Closed 2000	National Institutes of Health, Bethesda, MD, USA; Duke University, Durham, NC, USA	I	1995/2003	Immunotoxin

Table 12.2 (Continued)

Antibody name	Characteristics	Target antigen	Target cancer	Phase	NCI study ID	Status	Source	Publ. phase	Publ. year	Comment
LMB-9 (B3(dsFv) PE38)	B3 scFv–*Pseudomonas* exotoxin PE38, disulfide stabilized, murine	Le-y	Breast, lung, gastric, ovarian, and colon cancers	I/II	NCT00001691	Closed 2003	National Institutes of Health, Bethesda, MD, USA	–	2004	Immunotoxin
SS1P (SS1(dsFv)-PE38)	SS1 scFv–*Pseudomonas* exotoxin PE38, disulfide stabilized, murine	Mesothelin	Mesothelioma, lung (NSCLC), and ovarian cancers	I	NCT00065481	Recruiting	NeoPharm, Waukegan, IL, USA; National Institutes of Health, Bethesda, MD, USA	I	2004	Immunotoxin
huKS-IL2 (EMD 273066)	Recombinant huKS1/4::IL2 fusion protein	EpCAM	Lung, ovarian, prostate, and colorectal cancers	I	–		Merck KgaA, Darmstadt, Germany; EMD Pharmaceuticals, Durham, NC, USA	I	2004	Cytokine fusion
EMD 273063	Recombinant hu14.18::IL2(2) fusion protein	GD2	Neuroblastoma, melanoma	II	NCT00082758 NCT00109863	Recruiting	Merck KgaA, Darmstadt, Germany; EMD Pharmaceuticals, Durham, NC, USA	II	2004	Cytokine fusion

12.2.2.1 Antibody–Drug Conjugates

Potent drugs that exert intolerable toxicity when applied directly are the most obvious candidates for antibody-targeted delivery. In this respect, two classes of antibiotics have recently attracted attention: calicheamicins and maytansinoids. Calicheamicins are naturally synthesized by a soil microorganism and bind to the minor groove of DNA, causing double-strand breaks and subsequent cell death by apoptosis (Damle and Frost 2003).

Maytansine and its relatives are members of the ansamycin group of natural products, which are produced by higher plants, mosses and microorganisms (Cassady et al. 2004). Their antitumor activity relies on tubulin binding, inhibiting the microtubule assembly and thus mitosis. First described in 1972, their cytotoxic potency raised great interest for chemotherapeutic application, but despite – or rather, because of – their high cytotoxicity the results of clinical studies were disappointing: In the end, 21 out of about 800 treated patients with various tumors showed partial or, in one case, complete responses.

Because of their cytotoxic potency, both substance classes have been further developed for targeted tumor therapy. While a number of antibody–maytansinoid conjugates are in preclinical development, so far only gemtuzumab-ozogamicin (Mylotarg), directed against the CD33 antigen of acute myeloic leukemia cells, has been approved. CMB-401, a calicheamicin conjugate of the anti-MUC1 antibody CTM01, is currently in phase II clinical trials. While it has shown promising preclinical properties (Hamann et al. 2005), clinical data are not yet available.

12.2.2.2 Immunotoxins

The plant toxin ricin A and *Pseudomonas* exotoxin A are two highly poisonous biologic substances that, like the potent antibiotics above, cannot be applied systemically without deleterious toxicity. Again, the idea here is that targeted delivery via antibodies unleashes their potential against tumor cells without harming the rest of the organism.

Ricin, an N-glycosidase from the bean of the castor plant (*Ricinus communis*), is regarded as a biological weapon and has recently gained attention as a potential bioterrorism threat. It consists of an A and B chain, the latter of which binds to cell surface lectins, allowing the A chain to enter the cytosol and exert an essentially toxic ribosome blockade. Clinical effects start with various respiratory or gastrointestinal symptoms, depending on the route of ingestion, and progress to multiorgan failure and finally death. Treatment is supportive as no causal therapy is known (for a review see Audi et al. 2005).

Clinical trials with ricin conjugates have focused on lymphomas, not least due to their high density of specific surface antigens.

Initial clinical trials have tended to be disappointing. Several phase II clinical trials of the blocked ricin immunotoxin anti-B4-bR have been attempted, but they demonstrated too low tumor penetration to bring about a clinical effect in various therapeutic settings in non-Hodgkin's lymphoma patients (Multani et al. 1998;

Grossbard et al. 1999). A combination therapy trial finally yielded durable remissions, but only in a minority of patients (Longo et al. 2000).

RFB4-dgA is a chemically linked conjugate of anti-CD22 with ricin A that was well tolerated in B cell non-Hodgkin's lymphoma patients with a large proportion of circulating tumor cells, whereas the clinical course of patients without circulating cells was unpredictable and included two, probably treatment-associated, deaths. As this has been linked with aggregate formation of the pharmaceutical formulation, no further trials were conducted with this conjugate (Messmann et al. 2000).

Out of 15 heavily pretreated Hodgkin's patients treated with RFT5-dgA, an anti-CD25::ricin A conjugate, three experienced some degree of response. The predominant toxicity was a reversible vascular leak syndrome, and grade III toxicity was observed in all patients receiving the highest dose of $20\,mg\,m^{-2}$. While phase II studies at the maximum tolerated dose of $15\,mg\,m^{-2}$ were proposed (Schnell et al. 1998), to date none have been published.

However, both antibodies have also been fused with *Pseudomonas* exotoxin A, a bacterial toxin, or a truncated version of it, PE 38. RFT5(scFv)-ETA and a derived bispecific double scFv, RFT5/Ki-4(ScFv)-ETA, directed against both CD25 and CD30 (see also Section 12.2.1) have been expressed, and RFT5(scFv)-ETA has been announced for clinical studies after demonstrating specific cytotoxicity against disseminated human Hodgkin's lymphoma in severe combined immunodeficiency (SCID) mice (Barth et al. 2000).

The disulfide-stabilized recombinant *Pseudomonas* exotoxin fusion of RFB4 – RFB4(dsFv)-PE38 or BL22 – did not share the galenic problems of the ricin A conjugate in a phase I study for non-Hodgkin's lymphoma, chronic lymphocytic leukemia (B-CLL) and hairy cell leukemia. Here, 61% complete remissions of a mean duration of 36 months and 19% partial responses were seen in hairy cell leukemia. Among patients who received the maximum tolerated dose established in this study, the rate of complete responses was 86%, prompting a current phase II trial of BL22 for this indication. A characteristic adverse event was transient hemolytic uremic syndrome, observed in 5 out of 31 patients with hairy cell leukemia, but in none of the patients with other conditions. The most common toxicities were again symptoms of capillary leak syndrome, elevated liver enzymes, and fatigue (Kreitman et al. 2005).

The same approach has been followed for LMB-2 (anti-Tac(Fv)-PE38), which has also shown promising results in phase I in patients with various hematologic malignancies. Toxicity in this study was moderate, with fever and transient liver enzyme elevations. Remarkably, only 6 out of 35 patients developed neutralizing antibodies against the construct. One complete and seven partial responses were observed in patients with various hematologic malignancies, including hairy cell leukemia, cutaneous T-cell lymphoma, chronic lymphocytic leukemia, Hodgkin's disease, and T-cell leukemia (Kreitman et al. 2000).

An interesting group of PE38 constructs has been generated based on the anti-Lewis-Y antibody B3, termed LMB-1, LMB-7, and LMB-9. While LMB-1 is a chemical B3-PE38 conjugate, the other two are single-chain Fv (scFv)-based recombinant constructs, with the scFv being disulfide-stabilized in LMB-9.

An intriguing aspect of these recombinant constructs is the separate expression of the component proteins, which contain polyionic adapter peptides, allowing their covalent coupling. This technology facilitates a modular assembly of various antibody–toxin combinations and may in addition offer a solution to a common problem in the design and production of heterogeneous fusion proteins – the refusal of their components to be expressed in the same organism (Kleinschmidt et al. 2003).

LMB-7 and LMB-9 are currently in clinical trials of phase I or II. With LMB-1, clinical responses have been reported from a phase I trial for Le^y-positive solid tumors unresponsive to conventional treatment (Pai et al. 1996). As with ricin A, the predominant toxicity was a reversible vascular leak syndrome. An ensuing phase II trial addressed the question whether the human anti-murine immune response against the construct could be prevented by pretreatment with the anti-CD20 antibody rituximab, but found that, despite total suppression of circulating B lymphocytes, all patients developed neutralizing antibodies against LMB-1 that prevented repeated administration (Hassan et al. 2004).

With the availability of the smaller and more stable recombinant constructs, the clinical development of LMB-1 was discontinued in favor of LMB-7 and LMB-9. LMB-7 was very efficient in animal models, but disappointing in phase I clinical trials due to low stability and aggregation observed at 37 °C. Hence, the disulfide-stabilized scFv underlying the LMB-9 construct was generated, which showed a broader therapeutic window and enhanced stability in preclinical studies, thus opening the opportunity for more intensive dosing schedules such as continuous infusion in order to obtain better tumor penetration (Pastan 2003). LMB-9 is in a clinical phase I study for breast, colon, lung, ovarian, and gastric cancer that completed recruitment in 2003.

Disulfide stabilization has now become almost standard in new scFv-based constructs for clinical applications. It has also been applied in two other recombinant fusion constructs of *Pseudomonas* exotoxin currently in phase I, the anti-EpCAM 4D5MOCB-ETA for head and neck cancer (Di Paolo et al. 2003) and SS1P (SSI(dsFv)-PE38) directed against a mesothelioma antigen (Li et al. 2004).

While these results give the appearance of recombinant PE38 constructs as a comparably safe and potentially efficacious concept, the story of Erb38, a disulfide-stabilized recombinant scFv construct based on the e23 antibody against the Her2/neu (erb2) antigen, tells a cautionary tale. Apart from high expression in 30% of breast cancers, this antigen is also expressed at very low levels by hepatocytes, and despite a huge difference in expression, hepatic toxicity at the lowest dose level led to termination of a phase I trial of Erb38 (Pastan 2003).

As a third toxin, the plant protein gelonin, which inhibits ribosome function, has recently entered the clinical stage. Hum195/rGel is a chemical conjugate of a chimerized version of the anti-CD33 antibody M195 and recombinant gelonin. After initial *ex vivo* bone marrow purging studies (Duzkale et al. 2003), it is currently being evaluated in a still recruiting phase I trial for acute myeloid leukemia.

12.2.2.3 Cytokine Fusion Proteins

Like bispecific antibodies, cytokine fusion antibodies are intended to localize an immune effector function to cancer cells. One of the most effective cytokines in this respect is interleukin 2 (IL-2). It activates both cellular and humoral responses against an antigen. However, it is expressed by T lymphocytes only after their activation following recognition of an HLA-restricted antigen. To bypass this autoregulatory protection against the self-attack of host structures, including cancer cells, IL-2 is being applied with varying success in notoriously chemotherapy-refractory cancers such as melanoma (off-label) and renal cell carcinoma. Its adverse effects are numerous and partly severe.

Thus, as in the concept of drug conjugates and immunotoxins, it is desirable to target this cytokine more specifically to tumor cells in order to intensify its local antitumor effects and to minimize adverse events.

Two such cytokine–antibody fusions, both recombinantly generated, are currently in clinical trials: EMD-273063 and huKS-IL2.

The huKS-IL2 construct is derived from a humanized version of the murine anti-EpCAM antibody KS1/4. While clinical studies have been proposed or are ongoing for ovarian, lung, and colorectal cancer, the results of a phase I study with prostate cancer patients have already been published, indicating a favorable toxicity profile compared with the direct application of IL-2. Neutralizing antibodies were detected, but no hypersensitivity reactions were observed. This study showed increased parameters of IL-2-dependent immune activation, demonstrating the principle, although clinical responses were not reported (Ko et al. 2004).

EMD-273063, generated by the same technology from the same manufacturer as huKS-IL2, is based on the humanized antibody 14.18, which recognizes the GD2 ganglioside. It is currently being investigated in phase II in neuroblastoma. A phase I study in melanoma patients found frequent and partly dose-limiting grade 3 adverse events, including metabolic deregulation, hypotension, hypoxia, and elevated liver enzymes or bilirubin as well as opioid-dependent arthralgia or myalgia, but no grade 4 toxicity. As with huKS-IL2, indicators of an immunological response increased during treatment, but also without clinical responses (King et al. 2004).

12.3
Specific Targeting and Effector Mechanisms

The conventional effect of therapeutic antibodies is the initiation of antibody-dependent cellular cytotoxicity (ADCC) or complement-dependent cytotoxicity (CDC) upon binding to their cognate antigen. The previous section has introduced antibody constructs that have been designed and constructed to exert additional molecular functions. Many therapeutic antibodies, however, exert nonimmunological effects without a specific molecular design by activating or blocking the function of their target receptor (which may, of course, still be part of an immunological effect).

As the usual way to discover a hitherto unknown cellular antigen is its initial definition by a new monoclonal antibody, it is not surprising that often the function of a receptor, and thus the mechanism of action of a therapeutic antibody, has only been elucidated after the antibody has entered clinical development. Hence a strict line between "conventional" immunologically and "novel" non-immunologically acting antibodies cannot be drawn, and many of the "novel" antibodies still owe a proportion of their efficacy to ADCC or CDC. Similarly, the different effector principles introduced in this section cannot always be strictly separated. Thus, the proliferation signal pathways of tumor cells and neovasculature share many receptors, so that, for example, EGFR targeting could be classified both as inhibition of tumor cell growth and as antiangiogenesis. Nevertheless, from a systematical and didactic point of view it appears sensible to maintain these distinctions, as long as they are not taken for rigid borders.

12.3.1
Antiangiogenesis

The formation of new microvasculature is a prerequisite for tumor growth, and the recent characterization of the signaling and growth factors involved in this process has fostered the development of antiangiogenesis into a new principle of cancer therapy (Folkman 2006).

A broad array of antiangiogenic substances are now being investigated for their clinical antitumor effects. Of these, antibodies form a large proportion, notably those following the two already approved antiangiogenic antibodies, bevacizumab and cetuximab, in targeting vascular endothelial growth factor (VEGF) or epithelial growth factor receptor (EGFR), respectively. Other targets of antiangiogenic antibodies are phosphatidyl serine, IL-8, fibronectins, and integrin $\alpha_v\beta_3$. Antiangiogenic antibodies are summarized in Table 12.3.

12.3.1.1 Vascular Endothelial Growth Factor
Bevacizumab (Avastin) was the first anti-VEGF antibody approved in combination with chemotherapy for the first-line treatment of colorectal cancer. A4.6.1 and MV833 are two newer antibodies entering or already in clinical phase I trials.

A4.6.1 has been humanized from its murine counterpart by phage display optimization of the framework (Baca et al. 1997). Preclinical studies in an orthotopic mouse model of pancreatic cancer demonstrated additive antitumor efficacy of this antibody in combination with the matrix metalloproteinase inhibitor BB-94, a small-molecule antiangiogenic agent (Hotz et al. 2003). Clinical results are not yet available for this antibody, whereas for MV833 the results of a phase I study have been published. Here, one partial response and disease stabilization in another 9 out of 20 patients were reported. Toxicity was mostly limited to WHO grade I and II, predominantly fatigue, dyspnea, and erythema (Jayson et al. 2005).

Table 12.3 Antibodies with antiangiogenic effect.

Antibody name	Characteristics	Target antigen	Target cancer	Phase	NCI study ID	Status	Source	Publ. phase	Publ. year	Comment
A.4.6.1 (A4.6.1)	Phage-display humanized	VEGF	Rhabdomyosarcoma, pancreatic cancer	I	–		Genentech, South San Francisco, CA, USA	–	2000	
MV833	Murine	VEGF	Epithelial solid tumors	I	–		Protein Design Labs, Fremont, CA, US; Tsukuba Res. Lab., Japan; Toagosei Co., Japan	I	2005	
1121B (IMC-1121)		VEGF-2 (KDR)	Epithelial solid tumors	I	–		Imclone Systems, New York, NY, USA	–	2005	
1C11 (IMC-1C11)	Chimeric from murine	VEGFR-2	Colorectal cancer	I	–		Imclone Systems, New York, NY, USA	I	2003	
2C6 (IMC-2C6)	Humanized	VEGFR-2	Acute myeloid leukemia	I/II	–		Imclone Systems, New York, NY, USA	–	2004	
CDP791	Humanized from murine mAb VR165, PEG conjugate	VEGFR	Lung (NSCLC) cancer	I	NCT00152477	Recruiting	UCB S.A., Bruxelles, Belgium	–	–	No peer-reviewed publications
CNTO 95	Transgenic	α_V integrins	Melanoma	II	NCT00246012	Recruiting	Centocor, Horsham, PA, USA; Medarex, Princeton, NJ, USA	–	2005	
Volociximab Eos200-4 (M200)	Chimeric from murine	Integrin $\alpha_5\beta_1$	Renal cell carcinoma, melanoma	I	NCT00100685 NCT00099970 NCT00103077	Recruiting	Protein Design Labs, Fremont, CA, USA	–	2004	No peer-reviewed publications

Name	Type	Target	Indication	Phase		Institution	Year	Notes
Vitaxin	Humanized LM 609	Integrin $\alpha_v\beta_3$	Melanoma, sarcoma, prostatic	I/II	–	Scripps Research Institute, La Jolla, CA, USA	2002	
3G4	Humanized	Phosphatidylserine	Breast cancer, sarcoma	I	–	University of Texas Southwestern Medical Center, Dallas, TX USA	2005	
ABX-IL8	Transgenic	IL-8	Melanoma	I	–	Abgenix, Fremont, CA, USA	2005	Phase II study on COPD closed in 2003
BC1	Humanized	Fibronectin	Glioma, meningeoma, breast, lung	I/II	–	Istituto Nazionale di Ricerca di Cancro, Genova, Italy; Eidgenossische Technische Hochschule, Zürich, Switzerland	2001	
HUIV26	Humanized	Collagen type IV	Melanoma	I	–	University of Southern California, Los Angeles, CA, USA	2002	
CDP860	Humanized, PEG conjugated	PDGFβ receptor	Ovarian and colorectal cancers	II	II	UCB S.A., Bruxelles, Belgium	2005	Principle: reduction of intratumoral pressure

12.3.1.2 VEGF Receptors

As with EGF and EGFR, targeting the cellular VEGF receptor instead of the soluble ligand appears to be an obvious approach. Although no VEGF-R antibodies have so far been approved for therapy, three of them have recently entered phase I or I/II clinical trials. IMC-1121 and IMC-2C6, high-affinity versions of IMC-1C11, are fully human antibodies raised by phage display against the VEGF-R2, also termed kinase insert domain-containing receptor (KDR). With IMC-2C6, it was first demonstrated that VEGF-R2 can also be expressed by acute myeloid leukemia cells, which also produce VEGF, constituting autocrine and, via stimulation of production of IL-6 and GM-CSF, paracrine activation loops (Zhang et al. 2004). In addition to epithelium-derived solid tumors, for which IMC-1121 and IMC-1C11 are now in clinical phase I, all three antibodies have shown efficacy against leukemia cells in a murine *in vivo* model, probably by inhibition of VEGF-induced migration (Zhu et al. 2003). This study found a correlation between survival time of leukemia-inoculated mice and antibody affinity. IMC-2C6 is being investigated in a clinical phase I/II study.

Another anti-VEGF-R antibody, CDP791, has not been published in peer-reviewed journals yet, but is registered at the National Institutes of Health (NIH) in a phase II study for combination chemotherapy (carboplatin and paclitaxel) against non-small cell lung cancer (ClinicalTrials.gov ID: NCT00152477). It is noteworthy here because it is the only PEG-conjugated antibody in clinical trials the author is aware of. PEGylation has been successfully used to increase the circulating half-life and reduce the immunogenicity of foreign proteins for medical applications – mostly enzymes such as asparaginase, whose comparably small substrate can easily reach the catalytic region through the PEG meshwork. Protein–protein interactions such as antibody binding, however, are a different matter; it has been shown that PEGylation can be optimized so that binding activity is maintained, but immunogenicity diminished (Deckert et al. 2000).

12.3.1.3 Integrins

Integrins are cell surface receptors that mediate interactions of the cell with the extracellular matrix, which are responsible for cell attachment and spatial orientation, but also have crucial effects on growth, differentiation, and apoptosis pathways (Maile et al. 2006).

Vitronectin binding to integrin $\alpha_v\beta_3$ is necessary for the reaction of smooth muscle cells to insulin-like growth factor-I (IGF-I). Vitaxin, the humanized version of the LM609 antibody directed against a conformational epitope of integrin $\alpha_v\beta_3$, blocks this interaction and thus IGF-I-induced cell proliferation and angiogenesis. As this molecular target is expressed not only on endothelial cells, but also on breast cancer, Kaposi's sarcoma, melanoma, and other cancer cells, anti-integrin $\alpha_v\beta_3$ antibodies provide a two-sided method of tumor attack (Rader et al. 2002).

The fully human antibody CNTO 95 recognizes a common epitope of the α_v family of integrins. *In vitro*, this antibody inhibited melanoma cell adhesion, migration, and invasion, and substantially reduced human melanoma growth in a murine *in vivo* xenograft model. Its specificity for human, but not mouse integrins also suggests additional efficacy independent of its antiangiogenic effect (Trikha et al. 2004). CNTO 95 is being investigated in a recently started phase I/II clinical trial for melanoma patients.

M200 (volociximab), an antibody directed against integrin $α_5β_1$, has been reported to induce apoptosis in proliferating endothelial cells and to inhibit cancer cell growth *in vitro* (Bhaskar et al. 2004; Ramakrishnan et al. 2004), but no peer-reviewed clinical information on it has been published. Phase II clinical studies on its use in melanoma and renal cell carcinoma are currently recruiting patients.

12.3.1.4 Other Targets

In addition to the VEGF system and integrins, molecular targets exploited for antiangiogenic antibody therapy include the proangiogenic cytokine IL-8 (fully human antibody ABX-IL8) (Huang et al. 2002), the endothelial surface phospholipid phosphatidylserine (humanized antibody 3G4), which probably becomes exposed preferentially in tumor neovasculature due to oxidative stress (Ran et al. 2005), and structural extracellular matrix molecules such as collagen IV (humanized antibody HUIV26), which is exposed as an early event of vascular budding (Hangai et al. 2002), and fibronectin (BC1), the ligand of integrin $α_5β_1$ (Ebbinghaus et al. 2004).

12.3.2
Growth and Differentiation Signaling

If neo-angiogenesis is a supporting prerequisite for tumor growth, aberrant signaling of cell cycle, differentiation, or apoptosis lies at its core. Hence, a main focus of cancer research in past decades has been the identification of these molecular signaling pathways and the development of pharmaceutical ways to influence them – mostly by blocking proliferation signaling. In this effort, ground-breaking small-molecule drugs such as imatinib and other tyrosine kinase inhibitors (-inibs) and monoclonal antibodies stand on a par (for a summary, see Table 12.4).

12.3.2.1 Epithelial Growth Factor Receptor

Many tumors overexpress EGFR (Her1), which enhances both tumor cell proliferation and neoangiogenesis and thus is associated with an unfavorable prognosis. The prototype of anti-EGFR antibodies is cetuximab, which has now been approved for the treatment of colorectal cancer in combination with irinotecan chemotherapy in the US, Europe, and many Asian countries. Its mechanism of action is complex, comprising immunologic activation and target-specific non-immunologic effects. Competing with both EGF and TGFα, thanks to its higher affinity it blocks both ligands from the receptor. This inhibits EGFR tyrosine kinase signaling, which in turn reduces cell proliferation as well as production of angiogenic factors, and sensitizes the cell to chemotherapy and radiation (Petit et al. 1997; Grunwald and Hidalgo 2003).

Table 12.4 Antibodies acting on targets of growth and differentiation signaling.

Antibody name	Characteristics	Target antigen	Target cancer	Phase	NCI study ID	Status	Source	Publ. phase	Publ. year	Comment
225	Murine	EGFR	Melanoma, lung cancer (NSCLC)	I	–		Bristol Myers Squibb, New York, NY, USA	–	2002	Cetuximab predecessor, probably no longer pursued
425	Murine	EGFR	Head-and-neck, laryngeal, and papillary thyroid cancers	II	–		Fraunhofer IME, Aachen, Germany; Merck KGaA, Darmstadt, Germany	–	2004	
528	Murine	EGFR	Glioma, lung, skin, and vulvar cancers	I	–		Oncogene Science, Cambridge, MA, USA	–	2000	
806	Murine	EGFR v III	Glioma, head-and-neck, breast, lung, ovarian, colorectal, and prostatic cancers	I	–		Ludwig Institute for Cancer Research – New York, NY, USA and Melbourne, Australia	–	2005	
11F8 (IMC-11F8)	Recombinant Fab	EGFR	Epithelial cancers	I	–		Imclone Systems, New York, NY, USA	–	2005	
EMD72000 (matuzumab)	Humanized mAb 425	EGFR	Glioma, head-and-neck, laryngeal, and papillary thyroid cancers	II	NCT00215644 NCT00113581 NCT00111839	Recruiting	Merck KGaA, Darmstadt, Germany; EMD Pharmaceuticals, Durham, NC, USA	II	2005	

Name	Type	Target	Indication	Phase	Trial ID	Status	Company		Year	Notes
h-R3 (nimotuzumab)	Humanized version of murine IOR-EGF/R3	EGFR	Glioma, head-and-neck, breast, lung, and colon cancers	II	–		Centro Immunologia Molecular, Habana, Cuba	–	2005	
Humax-EGFR 2F8	Transgenic	EGFR	Head-and-neck cancer	I/II	NCT00093041	Closed 2005	Genmab, København, Denmark; Medarex, Princeton, NJ, USA	–	2005	
CR62	Rat	EGFR	Head-and-neck cancer	I	–		Institute of Cancer Research, Sutton, UK	I	1996/2005	
rHuMAb-EGFr (panitumumab, ABX-EGF)	Transgenic	EGFR	Nonsmall cell lung, renal, prostatic, colon, and other cancers	III	NCT00115765 NCT00101894 NCT00101907 ...	Recruiting	Abgenix, Fremont, CA, USA	II	2004/2005	
rhuMab-2C4 (pertuzumab, Omnitarg)	Humanized	p185/HER-2	Breast, lung, ovarian, prostate	II	NCT00096993 NCT00263224	Recruiting	Genentech, South San Francisco, CA, USA	I	2005	
HGS-ETR1 (TRM-1)	Agonist	TRAIL receptor-1 peptides (= TRIAL-R1 or DR4)	Lymphoma (NHL), colon, breast, lung, myeloma	II	NCT00092924 NCT00094848	Closed 2004	Cambridge Antibody Technology, Cambridge, UK	–	2005	
HGS-ETR2 (TRM-2)	High-affinity agonist	TRAIL receptor-2 peptides (= TRIAL-R2 or DR5)	Non-Hodgkin's lymphoma	I	–		Cambridge Antibody Technology, Cambridge, UK	–	2005	
AMG-162 (denosumab)	Recombinant human	RANK ligand	Bone metastases	I/II	NCT00104650	Recruiting	Amgen, Thousand Oaks, CA, USA	–	2005	Clinical study published for osteoporosis

In addition, cetuximab triggers ADCC via its Fc fragment, which contributes significantly to, but is not a prerequisite for its antitumor efficacy, as $F(ab')_2$ fragments also inhibited tumor growth in mice (Aboud-Pirak et al. 1988).

As can be seen so far, these features are characteristic of most if not all anti-EGFR antibodies currently investigated.

Panitumumab (ABX-EGF, rHuMAb-EGF) is a fully human IgG2 antibody. Its efficacy depends on the EGFR density on tumor cells, so that it shows sizable effects with EGFR-overexpressing tumors such as colon cancer. In other aspects, too, it is very similar to cetuximab: it appears to have moderate efficacy as monotherapy in phase II with pretreated and refractory patients, leading to about 10% partial responses and 36% stabilized disease, and it is well tolerated, with reversible acneiform (90% of patients) and other skin symptoms as the most common adverse effects. Whether it also emulates cetuximab in the resensitization of the cancer to irinotecan therapy remains to be seen, but initial clinical data suggest a similar effect (Foon et al. 2004; Rowinsky et al. 2004).

The humanized IgG1 antibody matuzumab (EMD72000) also acts both via immunological and non-immunological mechanisms: while EGFR blockade is regarded its main mechanism of action, it is capable of initiating ADCC. Two clinical phase II trials were reported at the 2005 ASCO meeting, one including 37 heavily pretreated patients with platinum-resistant ovarian or primary peritoneal (Müllerian) malignancies, the other 38 patients with recurrent, platinum-resistent cervical cancer. The spectrum and intensity of adverse effects in both cases was similar to that known for the other anti-EGFR antibodies, except for apparently more marked hepatotoxicity, and one case each of potentially therapy-related pancreatitis. While 21% and 23% of patients in these progression-prone cohorts experienced stable disease, zero and two objective responses were observed in the ovarian cancer study (Seiden et al. 2005) and the cervical cancer study (Blohmer et al. 2005), respectively.

With comparable preclinical data demonstrating the same complex mechanisms of action, the fully human antibody Humax-EGFR (also, but not uniquely, termed 2F8) has entered phase I/II trials (Bleeker et al. 2004).

Another humanized IgG1 anti-EGFR antibody, h-R3 has been radiolabeled with 99m-technetium (^{99m}Tc) and examined in a diagnostic clinical phase I/II trial on 25 patients with suspected epithelial-derived tumours. The highest antibody deposition was seen in the liver (53% of injected dose), followed by the other parenchymous and then hollow abdominal organs. No relevant adverse effects were observed. Tumor localizations were detected with a reported sensitivity of 76.5% and a specificity of 100%.

12.3.2.2 ErbB-2

Signal transduction of ErbB-2 (Her-2) is blocked by the humanized version of monoclonal antibody 2C4, pertuzumab (rhuMab-2C4, Omnitarg), exploiting a different mechanism than the approved antibody trastuzumab (Herceptin).

ErbB receptors require dimerization to activate their tyrosine kinase. The ErbB-2 (Her-2) receptor, which does not have a ligand, can be activated either by overexpression, leading to spontaneous homodimerization, or by heterodimerization with one of the three other erbB receptors upon their binding of ligand. While trastuzumab, besides other effects such as receptor downregulation and shedding inhibition, selectively inhibits ErbB-2 homodimer-induced tyrosine kinase activation, pertuzumab blocks ErbB-2 dimerization with any ErbB receptor (Badache and Hynes 2004). This recently described mechanism makes an obvious case for synergism of both antibodies, or pertuzumab with EGFR receptors, which has indeed been demonstrated *in vitro* (Nahta et al. 2004) and *in vivo*, respectively (Friess et al. 2005). A phase I study of pertuzumab found a favorable toxicity profile and pharmacokinetics allowing a 3-week dosage schedule. Out of 21 patients with progressive tumor disease resistant to chemotherapy or radiation, two experienced partial response and six stable disease for more than 2.5 months (Agus et al. 2005). A phase II double-blind randomized trial of gemcitabine with or without pertuzumab in patients with advanced ovarian, primary peritoneal, or fallopian tube cancer resistant to platinum-based chemotherapy is currently recruiting patients (ClinicalTrials.gov ID: NCT00096993).

12.3.2.3 TRAIL Receptors

An interesting approach to inducing proapoptotic rather than suppressing proliferation signals is followed by the fully human antibodies HGS-ETR1 (TRM-1) and HGS-ETR2 (TRM-2), which mimick the TNF-related apoptosis-inducing ligand (TRAIL) in binding the two apoptosis-inducing of its four receptors, TRAIL-R1 and TRAIL-R2, respectively (Younes et al. 2006). *In vitro* studies have shown that cell death is induced by either antibody as seen both in reduced cell survival and the activation of the intrinsic and extrinsic apoptosis pathways in a number of cell lines from different cancer entities, and *in vivo* efficacy against established tumors was demonstrated with colon, renal, and non-small cell lung cancer xenografts. Synergism of HGS-ETR1 with chemotherapy was demonstrated by increased chemosensitivity of cell lines resistant to the antibody alone, when treated with camptothecin, cisplatin, carboplatin, or 5-fluorouracil plus HGS-ETR1, and by increased sensitivity of colon cancer xenograft models treated with both modalities rather than either one alone (Pukac et al. 2005). Similar results have been obtained for both antibodies with primary lymphoma cells and lymphoma cell lines (Georgakis et al. 2005). Both antibodies are in clinical investigation, with a phase II study of HGS-ETR1 in non-small cell lung cancer only recently closed (ClinicalTrials.gov ID: NCT00092924), so that no clinical results are yet available.

12.3.2.4 RANK-Ligand

The recombinant fully human antibody denosumab (AMG 162) inhibits bone resorption by shifting the regulatory balance between the ligand of receptor activator of nuclear factor-κB (RANKL), a member of the TNF family, and its opponent, osteoprotegerin (OPG). In this balance, RANKL acts pro-osteoclastic by a number of mechanisms, while osteoprotegerin inhibits osteoclast proliferation and function by blocking RANKL. Denosumab mimicks the function of osteoprotegerin, but with a longer half-life (Kostenuik 2005). So far, its main clinically investigated application has been the prevention or remediation of osteoporosis in postmenopausal women. Here, while clinical end-points or bone densitometry have not been evaluated, long-lasting reduction in bone turnover could be demonstrated by biochemical surrogate parameters, and the treatment was well tolerated with only temporary and mostly mild effects on calcium metabolism (Bekker et al. 2004). Given that osteoclast activation through cytokines plays a major role in the formation of bone metastases and osteolyses, this concept is of obvious interest for the treatment of many cancers. Hence, in the spring of 2005 a phase II study of denosumab has opened for patients treated with bisphosphonates for bone manifestations of cancer or myeloma (ClinicalTrials.gov ID: NCT00104650).

12.3.3
Pretargeting Strategies

Pretargeting strategies are two-step procedures using antibodies as antigen-specific "anchors," to which cytotoxic effectors bind in a second step, thus localizing their effect to tumor cells. This approach allows the combination of the specific, but slow localization of antibodies with the efficacy of faster to distribute, but more toxic effectors. In addition, it may allow multiple applications of the effector for the duration of antibody localization in the tumor tissue. To improve the desired pharmacokinetic differential, often a clearing agent removing unbound antibody – mostly another monoclonal antibody against the first one – needs to be applied to take full advantage of this approach. Two kinds of cytotoxic effectors are currently in clinical trials: radioisotopes and prodrugs (see Table 12.5).

12.3.3.1 Radioimmunotherapy
In radioimmunotherapeutic pretargeting, the radioisotope is usually bound to a protein or peptide. The peptide binds either to a peptide-specific idiotope of a bispecific antibody, or in biotin–avidin systems to avidin conjugated to the antibody. The radioactive isotope can thus be concentrated in the tumor without prolonged circulation (and thus exposure) of the rest of the body. A preclinical study illustrating the charm of this concept has been performed on a conjugate of IMMU-106 (anti-CD20, see below) with a murine antihistamine–succinyl–glycine (HSG) antibody and ^{111}In- or ^{90}Y-labeled HSG. Here, 10 times higher tumor-to-blood ratios were found, and the same maximum dose of HSG-radionuclide was reached at a much earlier time point compared with directly radiolabeled IMMU-106 (Sharkey et al. 2005b).

A streptavidin–biotin system was clinically tested with NR-LU-10/SA. The antibody NR-LU-10 recognizes a colon cancer-associated antigen. In a phase II study, the streptavidin conjugate was followed 48h later by biotin–galactose–human serum albumin as a clearing agent. The radioisotope ^{90}Y bound to biotin via the linker tetra-azacyclododecanetetra-acetic acid (DOTA) was administered another 24h later in a dose of 110mCi m^{-2}. While there was considerable toxicity

Table 12.5 Pretargeting strategies.

Antibody name	Characteristics	Target antigen	Target cancer	Phase	NCI study ID	Status	Source	Publ. phase	Publ. year	Comment
A5PC	Conjugate of murine A5B7 with carboxypeptidase G2	CEA	CEA-expressing epithelial tumors	I	–		Cancer Research UK Targeting and Imaging Group, UK	I	2002	
hMN-14 × m734	Colorectal, ovarian	CEA	CEA-expressing epithelial tumors	I	–		Nuclear Medicine Department, Rene Gauducheau Cancer Center, France	I	2003	
NR-LU-10/SA	Streptavidin conjugate, murine	EpCAM	Colon cancer	II	–		Stanford University, Stanford, CA, USA; University of Nebraska, Lincoln, NE, USA	II	2000	
96.5 (P96.5, 96.5-hCE2)	Irinotecan-activating ADEPT approach with carboxylesterase 2, murine	p97 melanotransferrin	Melanoma, glioma	I	–		Bristol Myers Squibb, New York, NY, USA	I	1985/ 2001	Past trials with 96.5 alone, current preclinical investigation into ADEPT

with grade 3 or 4 diarrhea experienced by about a third of patients and grade 3 or 4 hematologic toxicities by a cumulative 40%, and objective responses were seen in less than 10%, the feasibility of pretargeting radioimmunotherapy was demonstrated (Knox et al. 2000). A similar pretargeting system for yttrium-90, however, based on BC2, an anti-tenascin antibody in phase I/II clinical studies for glioblastoma, found a substantial survival advantage for treated over untreated patients (Grana et al. 2002). Another biotin-based approach to pretargeted radioimmunotherapy has been reported using an anti-TAG72 antibody fusion protein based on the CC49 antibody (Shen et al. 2005).

The bispecific antibody approach has been realized with hMN-14 × m734 and ^{131}I-labeled di-diethylenetriaminepentaacetic acid-indium. A clinical dose-optimization study highlights the difficulty of the exact dosing and timing necessary to apply such complex therapy regimens successfully. Radioactively trace-labeled antibody was first administered in dose escalation and repeated dose adjustments to measure its tumor uptake, then the uptake of a fixed dose of the therapeutic radioconjugate was monitored. The antibody dose generating the highest uptake also led to considerable toxicity. Reducing the antibody dose by about half, however, optimized the blood clearance of the radioconjugate so that good tumor accretion was achieved with more acceptable toxicity. (Kraeber-Bodere et al. 2003).

12.3.3.2 Antibody-directed Enzyme Prodrug Therapy

In antibody-directed enzyme prodrug therapy (ADEPT), the targeting construct is an antibody–enzyme fusion protein or conjugate. After its tumor localization and clearance from the bloodstream, an inert prodrug is administered which is then selectively cleaved by the enzyme component to deliver a toxic chemotherapeutic agent. A number of clinical studies have been performed which in aggregate demonstrate the feasibility of clinical ADEPT (Martin et al. 1997; Francis et al. 2002). While the general problem of antibody immunogenicity preventing repeated therapy cycles is shared by this approach, recombinant humanization could so far not solve, but only alleviate it for ADEPT: the enzyme must be essentially non-human, otherwise enzymatic prodrug activation would occur "wild" and no longer be controlled by the antibody construct.

Nevertheless, the elucidation of the catalytic regions of enzymes and their three-dimensional structures on the one hand and the identification of particularly immunogenic protein motifs on the other has spurred a similar deimmunization development for enzymes as seen before with the chimerization and then humanization of antibodies (Mayer et al. 2004).

Of the clinical studies published so far, a recent one focused on the need for a clearing agent. A5CP, a galactosylated conjugate of a murine F(ab')$_2$ fragment of A5B7 (an anti-CEA antibody) and carboxypeptidase G2 had been previously tested in a clinical trial, but showed low efficacy due to prolonged antibody conjugate circulation. Thus, a new prodrug bis-iodo phenol mustard (ZD2767P) was designed to avoid the necessity of a clearing antibody by means of its high potency and short half-life. However, although the regimen was comparably well tolerated, no clinical or radiological responses were seen, and a closer look at antibody–

enzyme conjugate localization revealed inadequate tumor localization (Francis et al. 2002), stressing once more the need not only for more potent drug–prodrug systems, but in particular for antibody–enzyme constructs with better tumor penetration and affinity. One current approach to this is the design of single-chain fragment-based recombinant antibody–enzyme fusion proteins (Svensson et al. 1992; Siemers et al. 1997; Deckert et al. 2003).

12.3.4
Immune Signaling

Strictly speaking, the classical immunological effect of antibody therapy via Fc-mediated ADCC and CDC is in itself a form of targeted immune signaling. For the purposes of this section, however, this term addresses antibodies whose cognate antigens are receptors or ligands of the immune system (see Table 12.6). As is so often the case, a strict line cannot be drawn without force, as the first example will show.

12.3.4.1 Cytokines and Cytokine Receptors

Interleukin-2 receptor α   The high-affinity receptor for IL-2, IL-2 receptor α, also termed CD25 or Tac, is the target of anti-Tac-H or daclizumab (Zenapax). Activation of the IL-2 receptor plays a central role in the activation of specific or acquired immunity in general and in several forms of autoimmune reactions. Hence, as a functional antibody that blocks immune signaling, daclizumab is applied in autoimmune conditions and for immunosuppression (it was originally approved in 1997 for use in kidney transplantation). Its role in oncology, however, appears to be more that of a conventional tumor-targeting antibody, which it plays on the predominantly CD25-positive cells of adult T-cell leukemia/lymphoma (ATL), hairy cell leukemia (HCL), cutaneous T-cell lymphoma (CTCL), chronic lymphocytic leukemia (CLL), Hodgkin's disease, non-Hodgkin's lymphoma (NHL), and other lymphoid leukemias or lymphomas. This is not true without exception, though: in their very early stages, malignant cells may grow IL-2-dependent, and thus this principle may be important e.g. in minimal residual disease (Staak et al. 2004).

In murine models, the efficacy of daclizumab in the treatment of hematologic malignancies has been demonstrated. Currently, this humanized antibody is in clinical studies for patients with bone marrow failure, Hodgkin's disease, non-Hodgkin's lymphoma, and lymphoid leukemia (NCT00001962, NCT00001575, NCT00001941).

Interleukin 6   Interleukin 6 (IL-6) is an important growth and survival factor in several cancers, contributing to drug resistance, cachexia and formation of bone metastases. A disease in which its role has been well characterized is multiple myeloma. Elsilimomab (B-E8) is a murine anti-IL-6 antibody and has been given in parallel with chemotherapy and autologous stem cell rescue in this disease.

Table 12.6 Antibodies acting on immune signaling.

Antibody name	Characteristics	Target antigen	Target cancer	Phase	NCI study ID	Status	Source	Publ. phase	Publ. year	Comment
Anti-Tac-H (daclizumab, Zenapax)	Humanized version of murine 1H4	CD25, IL-2 receptor	T-cell lymphoma, leukemia	I/II	NCT00001962 NCT00001575 NCT00001941	Recruiting	Protein Design Labs, Fremont, CA, USA; Stanford University, Stanford, CA, USA	I	2003/ 2005	Approved for transplant therapy
B-E8 (elsilimomab)	Murine	IL-6	Lymphoma (NHL), myeloma	I/II	–		Diaclone, Besançon, France	I/II	2005	
CNTO 328 (cCLB8)	Chimeric from murine, blocker of amyloid A	IL-6	Cancer cachexia, renal	I/II	NCT0026513S	Recruiting	Centocor, Horsham, PA, USA	I/II	2005	Clinical study: abstract
Humax-Lymphoma	Transgenic	IL-15 receptor	Lymphoma, myeloma	II	–		Amgen, Thousand Oaks, CA, USA; Genmab, København, Denmark	–	–	No peer-reviewed publ.
PM-1 (rsHu-PM1, atlizumab, tocilizumab)	Humanized version of murine PM-1	IL-6 receptor	Multiple myeloma	II	–		Chugai Pharmaceuticals Co. Ltd, Tokyo, Japan	–	2003	
G5/44	Human	CD22	B-cell lymphoma (NHL)	I	–		Wyeth, Madison, NJ, USA	–	2005	
HD37	Murine	CD19	B-cell lymphoma (NHL)	I	–		Deutsches Krebsforschungszentrum, Heidelberg, Germany	I	2000	

Name	Description	Target	Indication	Phase	NCT number	Status	Sponsor		Year	Notes
hLL2 (epratuzumab, LymphoCide)	Humanized version of LL2	CD22	B-lymphocytic non-Hodgkin's lymphoma (NHL)	III	NCT00113802 NCT00113971 NCT00098839 NCT00111306	Recruiting	Immunomedics, Morris Plains, NJ, USA	I	2005	
Humax-CD20	Transgenic	–	Lymphoma (NHL)	I/II	–		Genmab, København, Denmark	–	–	
IDEC-152 (lumiliximab)	Primatized chimeric	CD23	Leukemia (CLL)	I	NCT00103558	Recruiting	Biogen IDEC, Cambridge, MA, USA; Seikagaku Corp., Tokyo, Japan	I	2004	IgE Fc receptor
IMMU-106	Humanized, identical backbone with hLL2	CD20	Lymphoma	I/II	–		Immunomedics, Morris Plains, NJ, USA	–	2005	
LL2 bectumomab	F(ab′)₂, murine	CD22	B-cell lymphoma (NHL)	III	–		Immunomedics, Morris Plains, NJ, USA	I	2002/ 2003	
Lym-1	Murine	HLA-DR	B-cell lymphoma (NHL), leukemia	II/III	NCT00008021 NCT00009776 NCT00028613	Closed	University of Southern California, Los Angeles, CA, USA; Schering AG, Berlin, Germany	I/II	1999	RIT, CIT
MB-1	Murine	CD37	B-cell lymphoma (NHL)	I	–		Stanford University, Stanford, CA, USA	I	1997	
TACI-Ig	Recombinant fusion protein targeting TNF receptors (BlyS, APRIL)	–	B-cell leukemia (B-cellular autoimmune diseases)	I	–		Zymogenetics, Seattle, WA, USA	–	2005	Soluble TACI receptor

Table 12.6 (Continued)

Antibody name	Characteristics	Target antigen	Target cancer	Phase	NCI study ID	Status	Source	Publ. phase	Publ. year	Comment
Ber-Act8	Murine	CD103	T lymphocytes: hairy cell leukemia	I	–		Freie Universität Berlin, Berlin, Germany	–	1997	
CP-675,206	Transgenic	CTLA-4	T lymphocytes various cancers	I	NCT00090896 NCT00075192	Recruiting	Abgenix, Fremont, CA, USA	I	2005	
Humax-CD4 (zanolimumab)	Transgenic	CD4	Lymphoma	II	NCT00127881	Recruiting	Medarex, Princeton, NJ, USA; Genmab, København, Denmark	II	2004/ 2005	
IDEC-114 galiximab	Primatized chimeric	CD80	T-cell lymphoma (NHL)	II	NCT00117975	Recruiting	Biogen IDEC, Cambridge, MA, USA	I/II	2005	
MDX-010	Transgenic	CTLA-4	T lymphocytes: melanoma, prostatic cancer	III	NCT00094653 NCT00108888	Recruiting	Medarex, Princeton, NJ, USA	I	2005	(>10 studies)
MEDI-507 (siplizumab)	Humanized from rat	CD2	Lymphoma	II	NCT00071825 NCT00105313 NCT00123942 NCT00063817 NCT00113646	Recruiting	Medimmune, Gaithersburg, MD, USA	I	2005	
OKT3	Murine	CD3	Colorectal, melanoma, ovarian, renal, leukemia	I	NCT00091611 NCT00001832 NCT00080353	Recruiting	Ortho Biotech, Bridgewater, NJ, USA	I	2005	Approved
T101	Fab, murine	CD5	T-cell leukemia (CLL), lymphoma	I/II	–		Scripps Research Institute, La Jolla, CA, USA	I	1998/ 1999	RIT

Name	Type	Target	Indication	Phase	NCT number	Status	Institution	Phase	Year	Notes
Visilizumab (HuM291, Nuvion)	Humanized, competes with OKT3	CD3	T-cell leukemia, T-cellular lymphoma, myelodysplastic syndrome, transplant rejection	II	–		Protein Design Labs, Fremont, CA, USA	I/II	2005	Immunosuppressant, studied in GvHD, not leukemia itself
1D10 (apolizumab, Remitogen)	Humanized from murine 1D10 (IgG2a)	28/32 (HLA-DR)	B- and T-cell lymphoma (NHL), leukemia	II	NCT00089154 NCT00022971	Recruiting	Protein Design Labs, Fremont, CA, USA	II	2001	Phase II: abstract only
BC8	Murine	CD45	Leukemia	I	–		Fred Hutchinson Cancer Center, Seattle, WA, USA	I	1999/2004	
M195	Murine	CD33	leukemia (AML)	I/II	NCT00014495 NCT00016159	Recruiting	Memorial Sloan-Kettering Cancer Center, New York, NY, USA	I	1995	RIT
MDX-33 (H22)	Humanized M22	CD64 (Fc gamma receptor I)	Various cancers, idiopathic thrombocytopenic purpura	II	–		Medarex, Princeton, NJ, USA	II	2000/2003	Clinical study: abstract
MN-3 (lamelesomab, sulesomab)	Fab' fragment, murine	NCA-90 (CD66)	Granulocytes: osteomyelitis, bone metastasis	II	–		Immunomedics, Morris Plains, NJ, USA	–	2005	
SGN-40 PRO64553	Humanized from murine S2C6	CD40	Leukemia, multiple myeloma	I	NCT00079716 NCT00103779	Recruiting	Genentech, South San Francisco, CA, USA	I	2005	
YTH54.12	Rat	CD45	Leukemia	I	–		Addenbrooks Hospital and Cantab Research Ltd., Cambridge, UK	–	2003	
79IT/36	Murine	CD55 (gp72)	Osteosarcoma, colorectal cancer	I	–		Cancer Research UK, London, UK	–	2001	
SC-1	Heterohybridoma	CD55/SC-1	Gastric	I/II	–		Universität Würzburg, Würzburg, Germany	I	2000/2004	

RIT, radioimmunotherapy; CIT, chemoimmunotherapy; GvHD, graft versus host disease.

Evaluation of C-reactive protein production showed neutralization of IL-6 activity, and mucositis and fever were significantly reduced. Moreover, a median event-free survival of 35 months and an overall survival of 68.2% at 5 years were observed in this uncontrolled trial (Rossi et al. 2005). Another study focused on cachexia in multiple myeloma patients investigating CNTO 328, a chimeric anti-IL-6 antibody. A phase I study demonstrated good tolerability and a half-life of approximately 17 days, making it applicable in prolonged dosage intervals. Preclinical data indicated that IL-6 blockade may counteract cachexia (Zaki et al. 2004; Mulders et al. 2005).

12.3.4.2 B-cell Signaling

Judging from preclinical data, the humanized anti-CD20 antibody IMMU-106 appears to resemble the approved chimeric anti-CD20 antibody rituximab in almost every respect. Due to its completely human framework, it is expected to reveal improved pharmacokinetics, tolerability, and efficacy when applied to humans. Clinical trials have reportedly commenced, but no results are published as of this writing (Stein et al. 2004).

In response to anti-CD20 treatment, non-Hodgkin's lymphoma cells have been shown to increase their expression of CD22. Hence it has been suggested that IMMU-106 treatment should be combined with the humanized anti-CD22 antibody epratuzumab (hLL2). Epratuzumab has been examined in a phase I radio-immunotherapy trial of Re-186–epratuzumab with 15 patients, of whom five reached objective responses with only mild or moderate toxicity (Postema et al. 2003), and a phase I/II immunotherapy trial in non-Hodgkin's lymphoma patients revealed objective responses in about 10% of patients, one of whom was refractory to rituximab, without dose-limiting toxicity. This antibody is now in phase I or II studies for Waldenström's disease (NCT00113802), relapsed acute lymphoblastic leukemia (NCT00098839), and systemic lupus erythematodes (NCT00113971, NCT00111306).

12.3.4.3 T-cell Signaling

The anti-CD2 antibody siplizumab (MEDI-507) is designed to achieve T cell-depletion in various clinical settings (Dey et al. 2005). It is currently in clinical phase I for T-cellular lymphoproliferative disease (NCT00071825), for CD2-positive lymphoma/leukemia (NCT00105313, NCT00123942), as an immunosupressant in kidney and bone marrow transplantation (NCT00063817), and for *ex vivo* T-cell depletion (NCT00113646).

Humax CD4 or zanolimumab is a transgenic fully human antibody from transgenic mice, expressed in CHO cells. It prevents T-cell activation by blocking CD4 interaction with MHC-II (Skov et al. 2003). Mainly developed for psoriasis vulgaris and rheumatoid arthritis, it has been tested clinically in cutaneous T-cell lymphoma (Kim et al. 2004) with encouraging results, and a case report has been published on angioimmunoblastic T-cell lymphoma (Hagberg et al. 2005).

CD80, also named B7 or BB1, a member of the immunoglobulin superfamily that is involved in the alloactivation of T cells and plays a critical role in autoimmune, humoral, and transplant responses, is the cognate antigen of galiximab (IDEC-114). Apart from autoimmune diseases such as psoriasis (Gottlieb et al. 2002), it has been investigated clinically in patients with relapsed or refractory follicular lymphoma, most of stage III or IV disease, by weekly infusions. The therapy was well tolerated without dose-limiting toxicities. A response rate of 11% was reported, interestingly with a time to best response of up to 12 months, which makes a causal assignment difficult, but two responders were reported progression-free at 22 and 24.4 months (Czuczman et al. 2005).

12.3.4.4 Other Lymphocyte Signals

The anti-CD40 antibody SGN-40 is a humanized IgG1 antibody in phase I for the treatment of multiple myeloma that has also shown antilymphoma activity against other B-lineage lymphomas preclinically (Law et al. 2005). Further elucidating its mechanism of action, blockade of sCD40L-mediated phosphatidylinositol 3′-kinase and nuclear factor kappa B activation was demonstrated in myeloma cells, and proliferation triggered by IL-6 was inhibited (Tai et al. 2004). Clinical phase I trials on its use for multiple myeloma (NCT00079716) and non-Hodgkin's lymphoma (NCT00103779) are under way.

CTLA-4 has been implicated in peripheral immunologic tolerance, so that its blocking may lead to effective anticancer responses. The fully human MDX-010 has already been investigated in renal cancer (Yang et al. 2005) and in high-risk resected stage III and IV melanoma in combination with gp100, MART-1, and tyrosinase vaccines. Apparently, development of autoimmune responses correlated with a better clinical outcome (Sanderson et al. 2005). Similar results were found with CP675,206, also a fully human anti-CTLA4 monoclonal antibody (Ribas et al. 2005).

SC-1 is a human monoclonal antibody found to bind a stomach carcinoma-associated isoform of CD55 or decay-accelerating factor B. In a clinical phase I trial, a significant induction of apoptotic activity was reported in 90% of the cases, and a significant regression of tumor mass in 50% of the patients. The authors suggest this antibody for adjuvant therapy (Vollmers et al. 2000).

12.3.4.5 The Cautionary Tale of TGN1412

TGN1412, a T cell-activating antibody, made sad headlines in early 2006 because of an utterly failed phase I clinical trial. Since then "the strange case of TGN1412" has been thoroughly examined and reviewed (Farzaneh et al. 2006; Hansen and Leslie 2006; Schneider et al. 2006; Bhogal and Combes 2006a,b). Briefly, six healthy volunteers who received the antibody rather than placebo experienced a cytokine storm disease within minutes to hours after application, leading to multiorgan failure and prolonged intensive care in all of them. Although as of the writing of this chapter, all six have survived and left the hospital, albeit with serious sequelae, this event has raised general questions about the conduct of clinical studies and the secrecy in which the generated data are kept. Thus, although no regulations had been broken, with hindsight it appears absurd to apply the experimental drug to all six volunteers almost simultaneously, and the future development of similar drugs trying to avoid the risk of TGN1412 will be hindered by the fact that the data continue to be kept confidential.

While these aspects are independent of antibodies and T-cell stimulatory drugs, the more specific lessons from this trial are more complicated. The target of TGN1412, CD28, acts as a costimulatory signal in T cell activation: upon binding of the antigen-laden MHC-II complex to the T-cell receptor, the intracellular signal cascades for cytokine production and release as well as proliferation are actually released by binding of B7 to CD28. However, both receptors can be "superstimulated," thus bypassing the need for the other signal to activate the T cell. TGN1412 acts as such a superagonist on CD28.

At least two mechanisms normally act to counterbalance and regulate T-cell activation by CD28. First, there is a dose-dependent activation of regulatory T cells, which are preferentially activated at lower anti-CD28 doses in experimental models, thus suppressing an autoreactive T-cell response. In addition, the natural CD28 ligand B7 has a regulatory function in that it is also a ligand for CTLA-4, which is expressed by activated T lymphocytes and promotes apoptosis or anergy. Both mechanisms were outrun by the nonlocalized, systemic superactivation of CD28.

In addition, this cautionary tale highlights species specificity as another problem of drug development in general, exacerbated in antibody therapeutics. The distribution and quantitative reaction patterns of a human receptor molecule and its, say, avian homolog may very well differ substantially. This is also true for the targets of small molecule drugs. In antibodies, however, their species-specific sequence and structure add another twist to this conundrum: even with identical target function, a human (or humanized) antibody administered in a macaque may not reveal all its beneficial or adverse effects because of its immunogenicity in the animal. The other way round, using a nonhumanized version for preclinical tests may obscure molecular effects unique to the human sequence.

Thus, preclinical animal testing will continue to have its intrinsic limitations, and the first application of any new drug in a human being will always carry a risk that can at best be minimized, but never avoided. Which speaks against having six first human beings at once.

12.3.5
Anti-Idiotype Vaccines

If the tumor vaccines are still regarded as highly experimental by many, anti-idiotype vaccine concepts may at first glance appear to be outright esoteric. They relate to Niels K. Jerne's concept of idiotype networks and the observation that the idiotope of an antibody may itself be another antibody's cognate antigen. Thus, antibody 1 may mimick a tumor antigen and thus induce an anti-idiotypic antibody 2 that not only binds antibody 1, but also cross-reacts with the tumor antigen. Taken one step further, another antibody 1 may bind to a tumor antigen, leading to induction of a tumor-mimicking anti-idiotypic antibody 2 – a so-called internal image antibody of the tumor antigen – which in turn induces an anti-anti-idiotypic antibody 3, that recognizes again antibody 2 and the tumor antigen. The advantages of this approach are the potentially precise definition of the antigen used as vaccine, the possibility of vaccinating against antigens not yet fully characterized or sequenced, and the general advantage of active vaccination: a potentially sustained response of the host.

105AD7 is an example of both approaches characterized above, divided between an original recipient of antibody 1 who generated antibody 2, and the prospective recipients of this antibody 2, who may generate the antitumor antibody 3: the heterohybridoma antibody 105AD7 was originally isolated from a colorectal cancer patient receiving the antitumor antibody 791T/36. This antibody's binding site has been identified as a defined part of the complement regulatory protein CD55. Thus, 105AD7 is a spontaneous CD55-mimicking anti-idiotype to 791T/36. Indeed, amino acid and structural homology between the CDRs of 105AD7 and the "mimicked" regions of CD55 has been demonstrated (Spendlove et al. 2000). This vaccine has been tested in several clinical trials. Phase I studies confirmed that the antibody is well tolerated and that it generated T-cell responses. That result encouraged a prospective, randomized, double-blind, placebo-controlled phase II survival study of 162 patients with advanced colorectal cancer. The result was a nonsignificant survival disadvantage for the vaccine patients, which the authors attribute to high tumor burden or poor patient compliance (only 50% of patients completed at least three immunizations) (Maxwell-Armstrong et al. 2001).

Another study looked at myelosuppressively pretreated young patients with osteosarcoma, finding a mixed picture as to immune responses, but, remarkably, two patients with possible clinical responses who were allowed to continue immunization for 2 years had remained disease free for around 6 years at the time of reporting (Pritchard-Jones et al. 2005).

Clinical anti-idiotype vaccination trials have been reported on with several antibodies against mucins (see also Section 12.4.4), such as pemtumomab (HMFG1), an antibody against mucin 1 (Nicholson et al. 2004), and the anti-mucin 16 antibodies ACA125 (Reinartz et al. 2004) and oregovomab (B43.13) (Schultes et al. 1998). All have also been used in nonvaccination approaches, and the difference between the two kinds of approaches lies in the repeated administration of the antibody with an adjuvant that enhances the specific immune response. With all three, the induction of anti-idiotypic networks has been observed and partly described in great detail, the therapies were well tolerated, but clinical responses were rare at best. Thus, some authors suggest alternative delivery systems that may improve specific immunogenicity (Hann et al. 2005). Anti-idiotype vaccine strategies are summarized in Table 12.7.

12.4
Antigens Without Known Effector Function

Regardless of the advances in functional tumor targeting outlined so far, antigens selected only for their cancer specificity as "dumb" targets continue to play an

Table 12.7 Anti-idiotype vaccines.

Antibody name	Characteristics	Target antigen	Target cancer	Phase	NCI study ID	Status	Source	Publ. phase	Publ. year	Comment
105AD7 (Onyvax-105)	Heterohybridoma, p72/CD55 image	mAb 791T/36	Osteosarcoma, colorectal carcinoma	I/II	NCT00007826	Closed 2003	Cancer Research UK, London, UK	I/II	2005	
11D10 (TriAb)	HMFG image, murine	mAb BrE1	Breast, ovarian cancers	II	NCT00006470 NCT00033748 NCT00045617	Closed 2004	Titan Pharmaceuticals, South San Francisco, CA, USA	I	2000	Abstract only
1A7 (TriGem)	Image of ganglioside GD2, murine	mAb 14.G2a	Melanoma, neuroblastoma, sarcoma, small cell lung cancers	II	NCT00045617	Closed 2005	Titan Pharmaceuticals, South San Francisco, CA, USA	I/II	1998	
1E10	Image of ganglioside GM3, murine	mAb P3	Melanoma	I/II	–		Centro Immunologia Molecular, Habana, Cuba	–	2003	
3H1 (CeaVac)	CEA image, murine	mAb 8019	Colon cancer	II/III	NCT00006470 NCT00033748	Closed 2004	Roswell Park Memorial Institute, Buffalo, USA	–	2004	Combination vaccine with 11D10
ACA125	CA125 image, murine	mAb OC125, CA125 anti-id	Ovarian, breast cancers	I/II	NCT00103545 NCT00058435	Closed	Cell Control Biomedical, Martinsried, Germany	II	2005	

Name	Description	mAb	Indication	Phase	NCT	Status	Institution		Year	Notes
Mitumomab (BEC-2)	Image of ganglioside GD3, murine	mAb R24	Melanoma, small cell lung cancer	III	NCT00037713 NCT00003279 NCT00006352	Closed 2001	Memorial Sloan-Kettering Cancer Center, New York, NY, USA	–	2004	
BR3E4	Image of mAb CO17-1A, rat	mAb 17-1A	Colon cancer	I	–		Wistar Institute, Philadelphia, PA, USA	I	2003	
I-Mel(pg)-2 MEL-2	Murine	mAb MEM-136	Melanoma	I/II	–		Biogen IDEC, Cambridge, MA,	I	1998/ 2004	
IGN101	Murine mAb 17-1A adsorbed on aluminium hydroxide as adjuvant	EpCAM	Nonsmall cell lung cancer	II/III	–	Closed 2005	Igeneon, Vienna, Austria	–	–	Anti-anti-idiotype vaccine. Company information only.
MF11-30	Murine	mAb 225.28	Melanoma	I/II	–		Biogen IDEC, Cambridge, MA, USA	I	1990/ 2004	
MK2-23	Image of HMW-MAA, murine	mAb 763.74	Melanoma	I	–		New York Medical College, Valhalla, NY, USA	–	2005	

important role in the development of new antibody-based cancer therapies. They are introduced here in an attempt to categorize them. The author is aware of two caveats: this categorization is not possible without inconsistencies, and today's "dumb" target may be tomorrow's functional trigger, as happened in the past and will be pointed out where appropriate.

12.4.1
Adhesion Molecules

Adhesion molecules comprise a heterogeneous group of proteins involved in mechanical cell adhesion, but also in cell–cell interaction and signaling. Antibodies against adhesion molecules are listed in Table 12.8.

12.4.1.1 EpCAM

The epithlial cell adhesion molecule (EpCAM, also termed GA733-2, KSA, 17-1A antigen) is a cell surface glycoprotein expressed by the majority of epithelial malignancies. The antigen had been implicated in mechanisms allowing tumors to escape immune surveillance (Armstrong and Eck 2003), but the suggested underlying ligand interaction, intriguing though it was, turned out to be an artifact (Nechansky and Kircheis 2005). Recently, EpCAM has been found to colocalize with tight-junction proteins and to form complexes with CD44v and tetraspanins, which may play a role in apoptosis resistance (Ladwein et al. 2005). This finding may finally give a clue to its function and overexpression in carcinomas.

Antibodies directed against EpCAM are supposed to act by ADCC and CDC, and anti-idiotype network activity has been hypothesized.

Edrecolomab (17-1A, Panorex) was the first anti-EpCAM antibody, defining the antigen. It was also the first antibody approved for treatment of colorectal cancer (in Germany only), and probably the first approved antibody in clinical use to be withdrawn from the market. Its story is emblematic for the pitfalls on the way from an anticancer antibody to an approved drug (and back).

A chimeric version of the murine IgG2a antibody has been produced, and two pharmacokinetic clinical studies were published in the early 1990s (Trang et al. 1990; Meredith et al. 1991).

With the murine original, a number of clinical studies have been conducted, proving good tolerability, with anaphylactic reactions to the murine protein as the most critical ones. Adjuvant therapy in patients with resected Dukes' stage C colorectal cancer and minimal residual disease was found to improve survival and reduce tumor recurrence significantly. According to the pivotal study by Riethmueller et al., conducted on 189 patients, adjuvant edrecolomab treatment reduced the risk of distant metastases by 32% over a follow-up period of 7 years (Riethmuller et al. 1998). While the differences seen in this study were significant, their power was actually small, with broad overlap of the 95% confidence intervals.

Since then, a number of clinical trials in advanced stages of colorectal or pancreatic cancer found limited efficacy at best for antibody monotherapy or

Table 12.8 Antibodies targeting adhesion molecules.

Antibody name	Characteristics	Target antigen	Target cancer	Phase	NCI study ID	Status	Source	Publ. phase	Publ. year	Comment
17-1A (CO17-1A edrecolomab Panorex)	Murine	EpCAM	Breast, lung, gastric, colorectal cancers	II/III	NCT00002968 NCT00002664	Closed 2001	Wistar Institute, Philadelphia, PA, USA	III	2005	
17-1A chimeric	Chimeric from murine	EpCAM	Gastric, colorectal cancers	I	–		Centocor, Horsham, PA, USA	I	1991/1996	
323/A3	Murine	EpCAM	Breast, ovarian, and prostatic cancers	I/II	–		Centocor, Horsham, PA, USA	–	2003	
3622W94	Humanized	EpCAM	Prostatic cancer	I/II	–		Glaxo-SmithKline, Brentford, UK	–	1999	
GA73.3 (GA733-2)	Murine	EpCAM	Gastric, colorectal cancers	I	–		Wistar Institute, Philadelphia, PA, USA	–	2002	
ING-1	Recombinant chimeric	EpCAM	Breast cancer	II	NCT00051675	Closed 2003	Xoma LLC, Berkeley, CA, USA	I	2005	
MT201 adecatumumab	Recombinant ("human engineered")	FpCAM	Breast, prostatic cancers	II	–		Micromet AG, München, Germany	I	2004/2005	
A33	Murine	gpA33	Pancreatic and colorectal cancers	I/II	–		Ludwig Institute for Cancer Research, New York, NY, USA; Memorial Sloan-Kettering Cancer Center, New York, NY, USA	I/II	1996/2005	

Table 12.8 (Continued)

Antibody name	Characteristics	Target antigen	Target cancer	Phase	NCI study ID	Status	Source	Publ. phase	Publ. year	Comment
hu-A33	Humanized	gpA33	Gastric, colorectal cancers	I	NCT00199797 NCT00003360 NCT00003543 NCT00199862	Recruiting	Ludwig Institute for Cancer Research, New York, NY, USA; Memorial Sloan-Kettering Cancer Center, New York, NY, USA	I	2005	RIT, CIT
BIWA-1	Murine	CD44v6	Head-and-neck, lung, breast cancers, multiple myeloma	I	–		Boehringer Ingelheim, Ingelheim, Germany	I	2000/ 2002	
BIWA-4 (bivatuzumab)	Humanized	CD44v6	Head-and-neck, lung, breast cancers, multiple myeloma	I	–		Boehringer Ingelheim, Ingelheim, Germany	I	2004/ 2005	
E48	F(ab')$_2$, murine	E48 (GPI-linked Ly-6 antigen)	Head and neck, vulvar cancers	I/II	–		Vrije Universiteit Medisch Centrum, Amsterdam, Netherlands	I	1997/ 2000	
ERIC-1	Murine	CD56 (NCAM)	NK cells: retinoblastoma, glioma	I	–		University of Bristol, Bristol, UK	–	2001	

FC-2.15	Murine	CD15 (Le-x), Ag2.15	Leukemia (AML), Hodgkin's lymphoma, breast and colorectal cancers	I	–	Instituto de Investigaciones Bioquímicas, Fundación Campomar, Buenos Aires, Argentina	1995/1999
G-22	Murine	CD44 spliced form	Glioma, melanoma, lung cancer	I	–	Nagoya University, Nagoya, Japan	1997
Leu-M1	Murine	CD15 (Le-x)	Reed–Sternberg cells: Hodgkin's lymphoma, renal cell carcinoma	I	–	BD Biosciences, San Jose, CA, USA	1998
MOC-1	Murine	NCAM	Neuroectodermal mesothelioma	I	–	Rijksuniversiteit Groningen, Groningen, Netherlands	1999	Diagnostic clinical studies
hu-N901	Humanized	CD56 (NCAM)	NK-cells: neuroblastoma, lung cancer (SCLC)	II	–	Biogen IDEC, Cambridge, MA, USA	1997/2005
RS7-3G11	Humanized	EGP-1 (GA733-1)	Lung (NSCLC), breast, gastric, ovarian, prostatic, bladder cancer, etc.	I	–	Immunomedics, Morris Plains, NJ, USA	2001

combination regimens (Adkins and Spencer 1998). Combining the antibody with the cytokines GM-CSF and IL-2 showed no objective responses (Fiedler et al. 2001); another study combining edrecolomab with the oral fluoropyrimidine capecitabine (Xeloda) saw one complete and two partial remissions among 27 patients treated (Makower et al. 2003).

The relative clinical ineffectiveness of edrecolomab had become a given with the advent of new small-molecule drugs such as irinotecan and oxaliplatin, when one study found an increase in tumor-protective cytokines and immune reactivity in patients treated with edrecolomab compared with standard therapy only, leading the authors to conclude that edrecolomab could be helpful in postoperative restoration of immune function (Tsavaris et al. 2004).

A recent phase III study looked again at the adjuvant setting in advanced non-metastasized (UICC stage II or Dukes' stage B) colon cancer. A total of 377 patients were stratified according to tumor extension (T3 or T4) after resection and randomly assigned either to edrecolomab or observation. No significant difference in overall and disease-free survival was found after a median follow-up of 42 months (Hartung et al. 2005). This study, together with the body of existing evidence, overturned the very rationale on which edrecolomab had been approved in Germany, albeit to no regulatory consequences: this study itself had to be terminated in 2000 because edrecolomab was withdrawn from the market and is no longer produced.

In its time, however, the pivotal study by Riethmüller et al. was state of the art, and in a way, edrecolomab was a pioneer that paved the way for the successful modern antibodies bevacizumab and cetuximab now approved for colon cancer therapy.

A number of other anti-EpCAM antibodies are or have been in clinical studies, many of which appear to have gone "silent" in the literature during the past 5 years. Of two recombinant human antibodies, ING-1 and MT201, early clinical results have recently been presented at meetings. For both of these, good tolerability with only mild toxicity, preliminary evidence of tumor localization, comparably low or absent human anti-human antibody (HAHA) immunogenicity, and no objective responses were reported (Naundorf et al. 2002; Peters et al. 2004; de Bono et al. 2004; Kiner-Strachan et al. 2005). Whether these antibodies will open new clinical potential for EpCAM targeting remains to be seen.

12.4.1.2 gpA33

The history of antibodies against the A33 antigen (now termed gpA33) is another – and not yet decided – example of the complications encountered on the way from a tumor-selective antibody to controlled studies and a working drug. A33 antigen has been sequenced and identified as a member of the immunglobulin superfamily (Heath et al. 1997). It is now accepted as an adhesion molecule, although details of its function and possible natural ligand are still unclear. The antigen, recently termed gpA33, is expressed by gastrointestinal epithelia including pancreas and by more than 95% of colon cancers. The first clinical phase I/II radioimmunotherapy trials commenced in the early 1990s. Interestingly, in a trial with ^{131}I-A33, there was little or no bowel toxicity despite expression of the antigen in normal gastrointestinal tissue. Radioimmunoscintigrams confirmed the localization of the radioisotope to almost all known primary or metastatic tumor sites. Furthermore, while there was initial uptake in normal bowel, time-dependent specificity became visible with a maximum at about 2 weeks after injection, when normal tissue had almost completely cleared the antibody, while tumor sites still showed intensive activity (Welt et al. 1994). Another study with ^{125}I-labeled A33 found modest antitumor activity with improved tolerability in the absence of gastrointestinal or hematological toxicity (Welt et al. 1996). The controlled studies encouraged by these results had to wait for the CDR-grafted humanized version. Even with this huA33, however, immunogenicity remained a problem as HAHA developed in 8 out of 11 patients, going along with strong infusion reactions. Only patients with high HAHA titers developed significant toxicity in four cases, and the only radiographic objective response was observed in one of the HAHA-negative patients (Welt et al. 2003a).

Very similar results were observed in a parallel combination chemotherapy study with BCNU, vincristin, fluorouracil, streptozocin (BOF-Step) in a fixed dose and huA33 dose escalation. Apart from higher toxicity because of the chemotherapy component, the results were encouraging, with 3 of 12 patients reaching radiographic partial responses, but were almost identical with respect to HAHA reactions (7 of 12 patients) (Welt et al. 2003b).

More recent studies found a somewhat more favorable HAHA incidence (Chong et al. 2005). In a neoadjuvant setting, one study demonstrated selective and rapid localization of radiolabeled huA33 to colorectal carcinoma sites and penetration into the centers of large necrotic tumors by postsurgical analysis of tumors resected after radioimmunotherapy (Scott et al. 2005).

To overcome the problem of HAHA reactions, an scFv against gpA33 has been developed in a phage display system (Rader et al. 2000). Based on this A33scFv, recombinant fusion proteins for pretargeting enzymes that activate inert prodrugs in ADEPT are currently in preclinical development (Deckert et al. 2003).

12.4.1.3 CD44v6

A transmembrane glycoprotein expressed by cells of almost all tissues, CD44 is a receptor for hyaluronan which mediates interactions with other cells and with the extracellular matrix. It helps anchoring cells to basal membranes and thus to achieve and maintain a polar orientation. It is also involved in leukocyte aggregation and their adhesion to endothelia and in mechanisms of cancer proliferation and dissemination. The human CD44 gene in fact encodes a variety of proteins which differ in glycosylation, but essentially are splice variants of the 19 human exons. CD44 variants containing variant domain 6 (CD44v6), in particular, seem to play a role in tumor cell invasion into tissues and metastasis of carcinomas (Heider et al. 2004).

Clinical trials have been reported on for bivatuzumab (BIWA-4). Apart from safety of administration and low immunogenicity, antitumor effects in incurable head-and-neck cancer patients with bulky disease were reported from a radioimmunotherapy study (Borjesson et al. 2003). The same approach tested in early stage breast cancer, however, revealed disappointing tumor-to-blood and tumor-to-normal tissue ratios, which could not be correlated with CD44v6 expression or the other tumor characteristics investigated (Koppe et al. 2004).

12.4.2
Stromal Antigens and Molecules Interacting with Extracellular Matrix

Tumor stroma and the extracellular matrix form essential support structures of a tumor, but the cells involved are not part of the malignant process and are not transformed themselves. Like neo-angiogenesis, this makes them attractive targets for tumor therapy as they are less prone to develop escape mechanisms than tumor cells themselves. Antibodies of this category currently in clinical trials are listed in Table 12.9.

12.4.2.1 FAP Antigen

Fibroblast activation protein is a stromal antigen involved in normal growth during childhood, wound healing, and tumorigenesis. Sibrotuzumab (BIBH-1) is the CDR-grafted humanized version of F-19, an anti-FAP antibody that has shown good tolerability in early-phase clinical trials and in particular did not interfere significantly with wound healing. A phase I radioimmunotherapy trial of ^{131}I-sibrotuzumab in patients with colorectal and non-small cell lung cancer found no objective responses, whereas the toxicity profile again was very mild (Scott et al. 2003). Another study of the unconjugated antibody in patients with metastatic colorectal cancer receiving weekly intravenous infusions did not reveal any objective responses. In contrast, the disease was progressive during the study in all but two patients, who progressed post study (Hofheinz et al. 2003).

12.4.2.2 Extracellular Matrix

The interactions of cells with extracellular matrix (ECM) proteins play a crucial role in the establishment and proliferation of solid tumors. The underlying mechanisms and the extent of their contribution in various cancer entities is still under investigation, but a couple of interesting ECM proteins or their cellular ligands such as tenascin C and annexin II have emerged, and some anti-ECM antibodies have reached clinical trials. 81C6, an anti-tenascin antibody whose murine and chimeric versions have been published under the same name, has shown prolonged and thereby specific tumor uptake in tumor compared to visceral organs, and objective responses in 2 of 9 patients in a phase I study of ^{131}I-radioimmunotherapy for lymphoma (Rizzieri et al. 2004). Phase II results of intracavity radioimmunotherapy after tumor resection in patients with recurrent malignant glioma indicate prolonged median survival over historical controls, encouraging modified phase II studies and a multicenter phase III trial (Akabani et al. 2005; Reardon et al. 2006).

12.4 Antigens Without Known Effector Function

Table 12.9 Antibodies against stromal and extracellular matrix antigens.

Antibody name	Characteristics	Target antigen	Target cancer	Phase	NCI study ID	Status	Source	Publ. phase	Publ. year	Comment
BIBH-1 (sibrotuzumab)	Humanized F19	FAP (fibroblast activation protein)	Colorectal cancer	II	NCT00004042 NCT00005616	Closed 2000	Boehringer Ingelheim, Ingelheim, Germany	II	2004	
Po66	Murine	Po66-CBP	Lung cancer (NSCLC)	I	–		Centres de Lutte contre le Cancer, Rennes, France	–	2003	
81C6	Chimeric of murine G2b/k	Tenascin	Glioma, lymphoma	I/II	NCT00002752 NCT00002753 NCT00003461 NCT00003478 NCT00003484	Closed 2000	Duke University, Durham, NC, USA	II	2005	RIT
BC2 (BC2-biotin)	Conjugate for pretargeting approach, murine	Tenascin	Glioma	I/II	–		Istituto Nazionale per la Ricerca sul Cancro, Genova, Italy	–	2003	

12.4.3
Gangliosides

Gangliosides are cytoplasmic membrane glycolipids characteristic of the myelin sheaths of neurons. Their oligosaccharide chains extend from the cell surface into the extracellular space and thus play an important role in cell identification and molecular recognition. They are overexpressed in many tumors of neuroectodermal origin. Gangliosides D2 (GD2), D3 (GD3), and M2 (GM2), in particular, have been investigated as targets for immunotherapy (see Table 12.10) because antibodies directed against them are retained for a longer time than those recognizing other gangliosides (Hanai et al. 2000).

12.4.3.1 Gangliosides D2 and D3

Of phase I and II studies of the murine anti-GD2/3 antibody IgG 3 antibody 3F8 in patients with stage 4 neuroblastoma, objective responses, including a sizable proportion of complete responses, and, more importantly, long-term survival of a subset of patients were reported with tolerable acute adverse effects and no delayed neurological toxicity (Cheung et al. 1998). Stratified for resistance to induction therapy, recurrent disease and progressive disease, the majority of patients in the first group achieved complete remission, whereas the effect in patients with progressive disease was small. (Kushner et al. 2001). In a larger controlled study of stage 4 neuroblastoma, the chimeric antibody ch14.18 was studied in 334 patients older than 1 year who had completed initial treatment without event, 166 of whom received the antibody. The tolerability again was good with reversible acute symptoms. While no significant differences in event-free survival were seen between treatment and control groups, the result for overall survival was unclear (Simon et al. 2004). These results were confirmed in a study with infants younger than 1 year (Simon et al. 2005).

Combination therapy of ch14.18 with antibody R24 and IL-2 in patients with melanoma or sarcoma lowered the maximum tolerated dose of each antibody, but despite objective responses and induction of immunological reactions, a therapeutic advantage could not be established (Choi et al. 2005).

12.4.3.2 Gangliosides M2 and M3

The GM2-recognizing human IgG1 kappa chimeric antibody chL6 has been tested in phase I trials for non-small cell lung, colon, and breast cancer. Apart from fever, chills, nausea, and drops in blood cell counts, the antibody was well tolerated, and good tumor localization was demonstrated in biopsies. While human anti-mouse antibody (HAMA) responses in the vast majority of patients hindered the announced phase II trials after encouraging phase I results with the murine monoclonal antibody, antibodies against the chimeric version were detected in only 4 of 18 patients. However, as with the murine version, no objective responses were seen (Goodman et al. 1993).

Chimeric L6-based radioimmunotherapy of patients with chemotherapy-refractory metastatic breast cancer then led to objective tumor responses in about

Table 12.10 Antibodies against ganglioside antigens.

Antibody name	Characteristics	Target antigen	Target cancer	Phase	NCI study ID	Status	Source	Publ. phase	Publ. year	Comment
ch14.18	Humanized mAb 14.18 (derived IL-2 fusion protein: EMD-273063)	GD2	Neuroblastoma	II/III	NCT00026312 NCT00030719	Recruiting	EMD Pharmaceuticals, Durham, NC, USA	I	2005	
14.G2a	Isotype variant of mAb 14.18, murine	GD2	Neuroblastoma	I	–		Scripps Research Institute, La Jolla, CA, USA	I	1997/ 1999	
3F8	Murine	GD2	Neuroblastoma, melanoma, osteosarcoma	II	NCT00072358 NCT00089258 NCT00058370	Recruiting	Case Western Reserve University, Cleveland, OH, USA	II	2004	Clinical abstract
KM871 (ecromeximab, KW-2871, KW-2971)	Chimeric of mAb KM641	GD3	Melanoma	I	–		Kyowa Hakko, Tokyo, Japan	I	2001	
ME36.1	Murine	GD2 and GD3	Melanoma	I	–		Wistar Institute, Philadelphia, PA, USA	–	2000	
R24	Murine	GD3	Leukemia, lymphoma, melanoma, soft tissue sarcoma	I/II	–		Memorial Sloan-Kettering Cancer Center, New York, NY, USA	I	1998/ 2003	

Table 12.10 (Continued)

Antibody name	Characteristics	Target antigen	Target cancer	Phase	NCI study ID	Status	Source	Publ. phase	Publ. year	Comment
L55	EBV-transformed line	GM2 (OFA-I-1)	Breast, ovarian, prostatic, colon cancer, melanoma	I	–		John Wayne Cancer Institute, Santa Monica, CA, USA	I	1995/ 1999	
L6	Murine	GM2	Breast, lung (NSCLC), ovarian, prostatic, and colon cancers	I	–		Scripps Research Institute, La Jolla, CA, USA	I	1995/ 2005	
chL6	Chimeric	GM2	Breast, lung (SCLC), ovarian, and colon cancers	I/II	–		Bristol Myers Squibb, New York, NY, USA	I	1995/ 2005	
L612	Recombinant human	GM3, GM4	Melanoma	I	–		John Wayne Cancer Institute, Santa Monica, CA, USA	I	2004	

50%. Myelosuppression was the limiting toxicity. Hence, transfusions of G-CSF-mobilized peripheral blood progenitor cells were tested in three patients. This reduced hematologic toxicity, but antibodies against chL6 emerged as a problem in the first two patients, so that ciclosporin was administered in the third patient. She did not develop specific antibodies or significant toxicity and enjoyed decreased tumor parameters and improved performance status for a follow-up of 9 months (Richman et al. 1995). Another strategy to overcome the problem of emerging antibodies against the therapeutic antibody is the administration of deoxyspergualin, which was shown to suppress to some extent HAMA responses to L6 (Dhingra et al. 1995). While no other clinical studies following these have been published yet, the preclinical development of new therapeutic approaches with this antibody appears to continue (DeNardo et al. 2005a).

A fully human IgM antibody against ganglioside M3, L612 has been tested in melanoma patients. Two of nine patients with resected tumors had been followed up for more than 5 years without evidence of tumor (Irie et al. 2004).

12.4.4
Mucins and Mucin-like Proteins

Mucins are large O-glycoproteins with clustered oligosaccharides linked to threonine, serine, and proline-rich tandem repeats. They fall into two distinct classes: gel-forming mucins are the main component of the secretions produced by mucous glands and goblet cells, whereas transmembrane mucins are cell surface proteins that appear to act rather as adhesion molecules. The latter comprise mucins 1, 3a and b, 4, 12, and 17. Mucin 1 (MUC1) expression is increased in many colon cancers and is correlated with a worse prognosis. MUC16 is one of a number of mucins that do not fit into either class. It has a transmembrane domain, but the structure of its tandem repeats differs from that in other transmembrane mucins. Termed CA125, it has long been known as a tumor marker for ovarian cancer (Porchet and Aubert 2004; Byrd and Bresalier 2004).

As it shares many characteristics with mucins without belonging to the MUC gene family, the tumor-associated glycoprotein 72 (TAG72) has been termed a mucin-like protein (Pavlinkova et al. 1999). Antibodies of this category are compiled in Table 12.11.

12.4.4.1 **Mucin 1**
A considerable number of antibodies against MUC1 antigens have been in clinical studies. Recently, trials have been reported on for pintumomab and pemtumomab. Pintumomab (170H.82, m170), a murine antibody, has been investigated for hormone-refractory prostate and breast cancer looking at two aspects: combination chemo-radioimmunotherapy with paclitaxel and the prevention of anti-mouse antibodies by immunosuppression rather than deimmunization of the antibody. Targeting of bone and soft tissue metastases was documented in

Table 12.11 Antibodies against mucins and mucin-like antigens.

Antibody name	Characteristics	Target antigen	Target cancer	Phase	NCI study ID	Status	Source	Publ. phase	Publ. year	Comment
170H.82 (m170 pintumomab)	Murine	MUC1	Breast, pancreatic, ovarian cancers	III	–		Cross Cancer Institute, Edmonton, AL, Canada	I/II	2005	
AR20.5 (BrevaRex)	Murine	MUC1	Myeloma, breast, pancreatic cancers	I/II	–		AltaRex, Edmonton, AL, Canada	–	2001	Epithelial cell surface mucin, function see comment
BrE-3	Murine	MUC1 HFMG	Breast	I	NCT00007891	Closed 2002	Cancer Research Fund of Contra Costa, Walnut Creek, CA, USA	I	1998	
h-CTM0	Humanized	MUC1 (RPAP epitope)	Breast, lung, ovarian, endometrial, and colorectal cancers	II	–		UCB S.A., Bruxelles, Belgium	I	1998/ 2005	
HMFG1 (pemtumomab)	Murine	MUC1	Breast, lung (NSCLC), gastric, ovarian, and colorectal cancers	III	NCT00004115	Closed	Cancer Research UK, London, UK	I	2005	RIT. Also used in anti-idiotype approach
hu-HMFG1	Humanized from HMFG1	MUC1	Breast, lung (NSCLC), gastric, ovarian, and colorectal cancers	I	–		Cancer Research UK, London, UK	–	2005	RIT. Idiotypic vaccination

Antibody	Type	Antigen	Indication	Phase	NCT	Status	Institution		Year	Notes
HMFG2	Murine	MUC1	Breast and ovarian cancers	II	–		Cancer Research UK, London, UK; Scripps Research Institute, La Jolla, CA, USA	–	2004	
MA5	Murine	MUC1	Breast and colorectal cancers	I	–		McGill University, Montral, QC, Canada	–	2002	
Nd2	Murine	MUC1-associated antigen	Pancreatic cancer	I	–		Osaka City Univiversity, Osaka, Japan	I	1997/2002	Diagnostic study 1999
PAM-4	Murine	MUC1	Pancreatic and colorectal cancers	I/II	–		Garden State Cancer Center, Newark, NJ, USA	–	1997	
SM3	Murine	MUC1	Breast and ovarian cancers	I	–		Cancer Research UK, London, UK	–	1996	
145-9	Murine	CA125, CA130	Ovarian cancer	I	–		Kyoto University, Kyoto, Japan	I	1997	
B43.13 (oregovomab)	Murine	CA125	Ovarian cancer	III	NCT00086632 NCT00050375	Recruiting	Altarex, Waltham, MA, USA	III	2004	
Mu-9	Murine	CSA-p	Colorectal cancer	I	–		Garden State Cancer Center, Newark, NJ, USA	I	1994/2005	CSA-p is related to and possibly identical with CA125
B72.3 chimeric	Chimeric	TAG-72 (CA72-4)	Adenocarcinoma of various origin, mesothelioma	I	–		National Institutes of Health, Bethesda, MD, USA	I	2005	
CC49 (HUcc49y10)	Humanized	TAG-72 (CA72-4)	Adenocarcinoma: colorectal, gastric, ovarian	II	–		National Institutes of Health, Bethesda, MD, USA	–	2003	
CC49 (minretumomab, 90Y-CC49)	Murine	TAG-72 (CA72-4)	Adenocarcinoma: colorectal, gastric	III	NCT00002734 NCT00002532 NCT00023933	Closed	National Institutes of Health, Bethesda, MD, USA	I	2005	RIT: phase I

Table 12.11 (Continued)

Antibody name	Characteristics	Target antigen	Target cancer	Phase	NCI study ID	Status	Source	Publ. phase	Publ. year	Comment
scFvHu-CC49	scFv humanized	TAG-72 (CA72-4)	Adenocarcinoma: colorectal, gastric	I/II	–		National Institutes of Health, Bethesda, MD, USA	–	2001	
hu-CC49	Humanized	TAG-72 (CA72-4)	Adenocarcinoma: colorectal, gastric	I	–		National Institutes of Health, Bethesda, MD, USA	–	2003	
hu-CC49 delta CH2	Humanized, constant heavy 2 domain-deleted	TAG-72 (CA72-4)	Adenocarcinoma: colorectal, gastric	I	NCT00025532	Closed	Ohio University, Athens, OH, USA	I	2005	RIT
MDX-220	Bispecific humanized	TAG-72	Breast, prostate cancers	I/II	–		Medarex, Princeton, NJ, USA	–	–	
3E1.2	Murine	MSA (mammary serum antigen)	Lung, breast, and colon cancers	I	–		Melbourne University, Melbourne, Australia	–	1998	
ABX-MA1 (c3.19.1)	Transgenic	MUC18	melanoma	I	–		Abgenix, Fremont, CA, USA	–	2003	
MLS102	Murine	sialyl-Tn	Pancreatic and colorectal cancers	I	–		Kyoto University, Kyoto, Japan	–	1997	

almost all patients, and ciclosporin cotreatment limited human anti-mouse antibodies to one of 16 patients (compared with 12 of 17 patients in a previous trial without ciclosporin), encouraging the authors to test multidose fractionated radioimmunotherapy in future trials (Richman et al. 2005).

As the limiting toxicity was grade 4 neutropenia, requiring autologous peripheral blood stem cell support in the combination therapy cohort, another trial looked at increasing the tumor-to-normal tissue ratio by a cathepsin-cleavable linker between antibody and radioisotope and an scFv-based pretargeting approach. While targeting remained successful and clinical responses were observed in a subset of patients, myelosuppression still limited applicable dose and patient eligibility (DeNardo et al. 2005b).

Several radioimmunotherapy approaches for ovarian cancer have been evaluated with pemtumomab (HMFG1), which binds to an epitope of MUC1 also termed human milk fat globulin 1 (hence the antibody acronym). Peritoneal single administration of ^{90}Y-pemtumomab was tested in 52 patients with ovarian cancer at stage I C through IV who had completed platinum-containing standard chemotherapy, 21 of whom were in complete remission, while 31 had residual disease. For those patients in complete remission, this study found long-term survival with a 10-year rate of close to 80%, while no profit was seen for patients with residual disease (Epenetos et al. 2000).

Combining external radiotherapy and radioimmunotherapy as a means of intensifying radiotherapy but not adverse effects was evaluated in another trial including 23 patients with nonsmall cell lung cancer. It showed low tumor uptake, low tumor-to-normal tissue ratios, and a short residence time despite good radioimaging of tumor localizations (Garkavij et al. 2005).

An anti-idiotypic vaccination approach with this antibody has been mentioned above.

12.4.4.2 Mucin 16 or CA125

The murine monoclonal anti-MUC16 antibody oregovomab (B43.13) had initially been applied for diagnostic immunoscintigraphy, and the observation of a survival benefit for the patients involved led to the hypothesis of an anti-idiotypic network generating CA125-neutralizing antibodies as the therapeutic principle. Further scrutiny then revealed that the observed induction of anti-CA125 responses correlated with the concentration of CA125 circulating in the blood at the time of antibody injection. The epitopes recognized by the induced antibodies were not restricted to that bound by oregovomab. These observations speak against anti-idiotypic immunization and in favor of direct immunization by CA125, probably induced by rapid formation of antigen–antibody complexes after oregovomab injection. Detection of oregovomab-induced CA125-specific immune responses correlated with improved survival (Noujaim et al. 2001). A retrospective analysis of immune responses and survival of patients with recurrent ovarian carcinoma who had received ^{99m}Tc-labeled oregovomab confirmed the initial observation of a survival benefit conferred by oregovomab. According to this study, 56.8% of these heavily pretreated patients survived for longer than 12 months, 34.1% for more than 24 months, and 6 out of 44 patients were still alive 4–7.5 years after the initial antibody dose. The detection of HAMA and anti-CA125 responses was associated with 2 to 3 times prolonged average survival (Mobus et al. 2003). Based on these results, a randomized study of consolidation therapy for patients with stage III or IV ovarian cancer in complete response after primary treatment compared oregovomab against placebo. While only a marginal and nonsignificant survival benefit was reported for the oregovomab group in total (13.3 months for oregovomab versus 10.3 months for placebo), patients who did mount an anti-CA125 immune response tended to have a prolonged time until the disease relapsed. Subgroup analyses revealed prognostic factors for improvement of the time to relapse, which was more than doubled to 24 months in one group (Berek et al. 2004). A phase III study to assess this effect prospectively has been announced.

12.4.4.3 Tumor-associated Glycoprotein 72

The mucin-like TAG72 has also been classified as an oncofetal antigen. It is expressed in adenocarcinomas, but except for secretory endometrium, no appreciable presence in normal tissues has been found.

The murine original antibody CC49 (minretumomab) is parent to chimerized, humanized, single-chain and C_H2 domain-deleted versions (see Table 12.11) which have been investigated at several different institutions in numerous clinical phase I and II trials including patients with lung, colorectal, breast, ovarian, and prostate cancers. The predominant treatment modality has been radioimmunotherapy, with ^{131}I, ^{90}Y, and ^{177}Lu as therapeutic radioisotopes and ^{111}In as trace label. A thorough and recent review of the clinical trials performed with CC49 and its derivatives has been compiled by Meredith et al. (2003b). This review finds consistent pharmacokinetic and biodistribution characteristics for the individual antibody entities, independent of the tumor diagnosis or bone versus soft tissue metastases. Unsurprisingly, the longest biologic half-life was seen with the chimeric complete IgG antibody cB72.3, and the shortest with the heavy-chain constant region-deleted version HuCC49dCH2, with the "conventional" CC49 in the middle. The longest half-life was about twice as long as the shortest. Myelosuppression was the dose-limiting toxicity in radioimmunotherapy. Hence, localized application such as interaperitoneal administration increased the tolerated dose. Combination therapy with interferon increased expression of TAG72 and possibly tumor uptake of ^{131}I-CC49. A methodologic problem highlighted by this review was the lack of standardization in study design, monitoring, and reporting, so that dosimetry and other data are difficult to compare between studies and to translate into predictions of safe dose levels (Meredith et al. 2003a).

Recently, ^{90}Y-mCC49 was evaluated in patients with non-small cell lung cancer in combination with interferon α for TAG72 upregulation and paclitaxel. In addition, the chelators EDTA and DTPA were tested for their ability to reduce myelosuppression. The latter was found true to a modest extent for DTPA.

No objective tumor responses were observed (Forero et al. 2005), fitting this study into the pattern previously described.

Another recent study evaluated the suitability of the domain-deleted Hu-CC49DeltaC$_H$2 for a diagnostic rather than therapeutic role in radioimmuno-guided surgery. This is the intraoperative identification of tumor localizations by gamma-detecting probes after systemic application of a radiolabeled antibody. Exploratory laparotomy was performed between days 3 and 20 after antibody injection, and pharmacokinetic data monitored in the meantime. The results fitted a two-compartment pharmacokinetic model. After day 3, the antibody was preferentially localized in tumor tissue, especially in intestinal and metastatic liver lesions (Xiao et al. 2005).

12.4.5
Lineage- or Tissue-specific Antigens

Of many antigens targeted for cancer therapy, immunologic identity and expression specificity for a certain differentiation line or stage have been established although the molecule itself and its function have not yet been defined or are themselves lineage- or tissue-specific. For an overview of such antibodies in clinical trials see also Table 12.12.

12.4.5.1 **CD30**
The type 1 glycoprotein CD30 was originally identified as the Ki-1 antigen on Reed–Sternberg cells in Hodgkin's disease and non-Hodgkin's lymphomas such as diffuse large cell, anaplastic large cell, and immunoblastic lymphomas. It is a member of the TNF receptor family and activates nuclear factor κB (NFκB) via TNF receptor-associated factors (TRAF) 2 and 5, leading to pleiotropic effects *in vitro* ranging from cell proliferation to cell death. It is also expressed on lymphocytes infected with HIV, HTLV-1 or EBV, and crosslinking of HIV-infected T cells is involved in the activation of HIV production. Its selective expression in lymphomas gave rise to hopes of repeating the success of antibody therapy seen with B-cell non-Hodgkin's lymphomas.

So far, this expectation has yet to be met, but interesting interactions were observed that are encouraging for clinical trials. For example, 5F11 (MDX-060), which was well tolerated and had some clinical activity in a recently closed phase I/II trial, was shown to activate NFκB and the antiapoptotic protein c-flip in Hodgkin's lymphoma-derived cell lines. This limitation of the clinical use of 5F11 has been overcome *in vitro* and *in vivo* by combination with the NFκB-suppressing proteasome inhibitor bortezomib. The combination had a synergistic cytotoxic effect if 5F11 is followed by bortezomib, indicating a bortezomib-sensitizing function of the antibody (Boll et al. 2005). Similarly, a drug-sensitizing effect of the chimeric antibody SGN-30, currently in clinical phase II trials, was demonstrated for a number of chemotherapeutic drugs in Hodgkin's and anaplastic large cell lymphoma cell lines *in vitro* and partly *in vivo* (Cerveny et al. 2005).

Table 12.12 Antibodies against lineage- or tissue-specific antigens.

Antibody name	Characteristics	Target antigen	Target cancer	Phase	NCI study ID	Status	Source	Publ. phase	Publ. year	Comment
5F11 (MDX-060)	Transgenic, fully human	CD30	Lymphoma	I/II	NCT00059995	Closed 2005	Medarex, Princeton, NJ, USA	I/II	2004/2005	T and B cell activation, M. Hodgkin, embryonal tumors
Ber-H2	Murine	CD30	Reed–Sternberg cells: Hodgkin's lymphoma, testicular cancer	I/II	–		Freie Universität Berlin, Berlin, Germany	–	1999	T and B cell activation, M. Hodgkin, embryonal tumors
HeFi-1	Murine	CD30	Reed–Sternberg cells: Hodgkin's lymphoma	I	NCT00048880	Recruiting	National Institutes of Health, Bethesda, MD, USA	–	2004	T and B cell activation, M. Hodgkin, embryonal tumors
HRS-3	Murine	CD30	Reed–Sternberg cells: Hodgkin's lymphoma	I	–		Cancer Research UK, London, UK; Biotest AG, Dreieich, Germany	I/II	2001	T and B cell activation, M. Hodgkin, embryonal tumors
SGN-30	Chimeric of murine AC10	CD30	Hodgkin's lymphoma	II	–		Seattle Genetics, Bothell, WA, USA	–	2005	T and B cell activation, M. Hodgkin, embryonal tumors

Name	Type	Antigen	Indication	NCT	Status	Institution	Phase	Year	Notes
1F5	Murine	CD20	B lymphocytes: leukemia, lymphoma	–		Bristol Myers Squibb, New York, NY, USA	I	2000/2003	RIT. Intracellular accumulation
CAMPATH-1M YTH66.9	Studied for *ex vivo* applications, rat	CD52	leukemia	–		Cambridge University, Cambridge, UK	I	1998/2002	Campath history: −1M → 1G → 1H alemtuzumab
Lintuzumab (HuM195, Zamyl)	Humanized immunotoxin see Hum-195/rGel	CD33	Myelodysplastic syndrome, acute myeloid leukemia	–		Memorial Sloan-Kettering Cancer Center, New York, NY, USA	III	2003	Studied for *ex vivo* purging
G250	Murine	G250 (MN/CA9 or MN/CA IX)	Renal cell carcinoma	–		Universiteit Leiden, Leiden, Netherlands; Wilex, München, Germany	II	1998/2005	RIT
G250, c-WX-G250	Chimeric	G250 (MN/CA9 or MN/CA IX)	Renal cell carcinoma	NCT00199875 NCT00087022 NCT00199888	Recruiting	Universiteit Leiden, Leiden, Netherlands; Wilex, München, Germany	II/III	2004/2005	RIT
J591	Murine	PSMA	Prostatic cancer	NCT00040586 NCT00195039 NCT00081172	Recruiting	Ludwig Institute for Cancer Research, New York, NY, USA; Millenium Pharmaceuticals, Cambridge, MA, USA	II	2006	
huJ591	Humanized J591	PSMA	Prostatic cancer	NCT00195039	Recruiting	Millenium Pharmaceuticals, Cambridge, MA, USA	II		
3H11	Murine scFv	3H11Ag	Gastric cancer	–		Beijing Institute for Cancer Research, Beijing, China	I	2004	T-cell antigen?
C242	Murine	CA242 (CanAg)	Lung, pancreatic, and colorectal cancers	–		Göteborgs Universitet, Göteborg, Sweden; Glaxo-SmithKline, Brentford, UK	I/II	2003	

Table 12.12 (Continued)

Antibody name	Characteristics	Target antigen	Target cancer	Phase	NCI study ID	Status	Source	Publ. phase	Publ. year	Comment
H11	Humanized	C antigen	Breast, pancreatic, liver, and gastric cancers, lymphoma	II	NCT00058292	Recruiting	Shizuoka University, Shizuoka, Japan; Viventia Biotech, Toronto, ON, Canada	–	2003	RIT
HM1.24	Chimeric	HM1.24	Lymphoma	I	–		Chugai Pharmaceuticals Co. Ltd, Tokyo, Japan	I	1999/ 2005	B-cell differentiation. Phase I trial halted.
MBr1	Murine	CaMBr-1 (GL-6)	Breast and lung (SCLC) cancers	I	–		Istituto Nazionale per lo Studio e la Cura dei Tumori, Milano, Italy; Memorial Sloan-Kettering Cancer Center, New York, NY, USA	–	–	
Mel-14	Murine	Chondroitin sulfate (proteoglycan antigen of gliomas, melanomas	Glioma, melanoma, medulloblastoma	I	NCT00002751 NCT00002754	Closed 1999	Ludwig Institute for Cancer Research, Lausanne, Switzerland; Duke University, Durham, NC, USA	I	2001	Phase I study reported in case report. Ab also quoted as Me1-14
RAV12	High-affinity variant of mAb KID3, chimeric	RAAG12	Gastric, pancreatic, and colorectal cancers	I	NCT00101972	Recruiting	Raven Biotechnologies, South San Francisco, CA, USA	–	–	Glycoprotein antigen – ion pump-related?

12.4.5.2 G250 or Carbonic Anhydrase IX

Carbonic anhydrase IX (CA IX) is a membrane-associated enzyme implicated in cell proliferation under hypoxia and thus in oncogenesis and tumor progression. The antigen-defining antibody G250 was raised by immunization of mice with human renal cell carcinoma. CA IX is expressed in various malignancies, in particular renal cell carcinoma (RCC), but is absent in normal tissues except for gastric mucosa. In RCC patients, high expression of CA IX has been determined as a positive prognostic factor for response to IL-2 therapy and survival (Atkins et al. 2005). Selective uptake of G250 antibody by antigen-positive cells has been demonstrated immunohistochemically, and comparably low protein doses were required for effective tumor targeting (Lam et al. 2005).

These findings led to clinical studies with murine G250 in the mid 1990s, but as so often, anti-murine responses abrogated repeated treatment and thus, despite good tumor targeting and tolerable adverse effects, the clinical potential could not be evaluated until a chimeric version, cG250, was available (Divgi et al. 1998). A series of clinical trials have been conducted since then.

The first phase I study with ^{131}I-cG250 observed encouraging clinical effects. Myelosuppression was determined as the dose-limiting toxicity (Steffens et al. 1999). A subsequent escalation trial of whole-body absorbed dose tried to reduce myelotoxicity by fractionated administration. Concomitant monitoring of whole-body and serum clearance allowed rational treatment planning and dynamic dose adaptation, but there was no evidence for fractionation-induced sparing of the hematopoietic system (Divgi et al. 2004).

The unlabeled antibody was investigated in a multicenter phase II study of metastasized RCC. Here, no drug-related grade III or IV toxicity was observed, and only a small proportion of patients experienced grade II toxicity. Among the 36 patients entered, one complete response and one partial response were noted, and five patients with initially progressive disease experienced stabilization for more than 6 months. The overall median survival was 15 months (Bleumer et al. 2004). Given the limited therapeutic options and bleak prognosis of this disease, cG250 may thus be of clinical benefit in renal cell carcinoma.

12.4.5.3 Prostate-specific Membrane Antigen

The transmembrane glycoprotein prostate-specific membrane antigen (PSMA) is primarily expressed on prostatic epithelial cells, including those of prostate cancer, and has been successfully employed for immunoscintigraphy of prostate cancer metastases. The murine IgG1 monoclonal antibody J591 directed against the extracellular domain of PSMA has been epitope deimmunized (Milowsky et al. 2004). It could therefore be administered repeatedly with good tolerability and induced dose-correlated ADCC in a phase I trial of patients with androgen-independent prostate cancer (Morris et al. 2005). A humanized version has also been developed and is also currently in a phase II clinical trial. For radioimmunotherapy with repeated ^{177}Lu-J591, a phase I trial including 35 patients with progressive androgen-independent prostate cancer found dose-limiting myelosuppression after more than three doses of 30 mCi m^{-2}, but no serious nonhematologic toxicity. All known tumor sites in a subset of patients with positive preceding imaging studies were detected by the antibody, and biologic activity as determined by reduction of serum prostate-specific antigen to less than half the initial value was seen in four patients (11.4%) (Bander et al. 2005). Another study found prolonged tumor responses and reduced myelotoxicity as a result of dose fractionation (Vallabhajosula et al. 2005).

12.4.6
Oncofetal Antigens

Oncofetal antigens are a very heterogeneous group whose members are expressed physiologically only during embryonal or fetal development and – obviously associated with cell proliferation – again as a pathological feature of various malignancies (see Table 12.13).

12.4.6.1 **Lewis Y**

The Lewis Y (Le^Y) antigen was first described as a blood group-related antigen expressed on activated granulocytes. Its association with proteins of the CD66 family of oncofetal antigens led to its classification in this category, while its interaction with the ErbB family of EGF receptors might well justify listing it in Section 12.3.2.2 as interfering with growth signaling (Klinger et al. 2004). Le^Y is expressed in a large proportion of epithelial cancers such as lung, breast, ovarian, and colon carcinoma.

3S193 is a murine anti-Le^Y antibody that has been humanized by linking the murine variable region and human framework cDNA in various combinations and selecting for Le^Y binding. The avidity of the resultant hu3S193 matched that of the murine version, while its capacity to induce ADCC and CDC exceeded that of the parent antibody. *In vivo*, no tumor reduction could be seen in mice bearing established xenograft tumors of the MCF-7 breast cancer cell line against which 3S193 was originally raised, but tumor growth was decelerated in a preventive model compared with controls (Scott et al. 2000). In the same murine xenograft model, ^{131}I-labeled hu3S193 combined with taxol in subtherapeutic doses of each agent significantly inhibited tumor growth in 80% of mice as opposed to either substance alone (Clarke et al. 2000). A clinical phase I study is recruiting patients (NCT00084799), but no clinical results are available yet. Blocking of ErbB1 and ErbB2 signaling has been investigated with IgN311. This humanized anti-Le^Y antibody immunoprecipitated both EGF receptors from detergent lysates of human tumor cells and blocked EGF-stimulated downstream signals with an efficacy similar to that of trastuzumab (Klinger et al. 2004).

12.4.6.2 **Carcinoembryonal Antigen**

Carcinoembryonal antigen (CEA) is the prototypical oncofetal antigen and has long been used as a diagnostic tumor marker to monitor the course of adenocarcinomas, in particular colon cancer. Hence, despite the problem of circulating soluble CEA – which is the very rationale of its diagnostic use but limits tumor

Table 12.13 Antibodies against oncofetal antigens.

Antibody name	Characteristics	Target antigen	Target cancer	Phase	NCI study ID	Status	Source	Publ. phase	Publ. year	Comment
hu3S193	Humanized	Le-y	Breast, ovarian, and colon cancers	I	NCT00084799	Recruiting	Ludwig Institute for Cancer Research, New York, NY, USA; Biovation Ltd., Aberdeen, Scotland, UK	–	2001	
ABL364 (BR55-2)	Murine	Le-y	Breast, lung, gastric, and colorectal cancers	I	–		Wistar Institute, Philadelphia, PA, USA; Igeneon AG, Vienna, Austria	–	2004	
cBR96	Chimeric	Le-y	Lung, gastric, colon, and prostatic cancers	II	NCT00028483	Closed 2004	Bristol Myers Squibb, New York, NY, USA	I	2000/ 2001	
IGN311	Humanized ABL364	Le-y	Epithelial: carcinomas	I	–		Igeneon AG, Vienna, Austria; Protein Design Labs, Fremont, CA, USA	I	2004/ 2005	
mAb 35	Murine	CEA	Colorectal cancer	I/II	–		Ludwig Institute for Cancer Research, Lausanne, Switzerland	I	2001	
38S1	Murine	CEA	colorectal cancer	I	–		Wenner Gren Institute, Stockholms Universitet, Stockholm, Sweden	–	2000	
F6	Murine	CEA	Thyroid and colorectal cancers	I/II	–		Université de Lausanne, Lausanne, Switzerland; Intervet Danmark AS, Skovlund, Denmark	I/II	1998/ 2005	

Table 12.13 *(Continued)*

Antibody name	Characteristics	Target antigen	Target cancer	Phase	NCI study ID	Status	Source	Publ. phase	Publ. year	Comment
hMN14 (hMN-14, labetuzumab)	Humanized. Bispecific MN-14 × DTIn-1 in preclinical development publ 2004	CEA	Lung, pancreatic, ovarian, and colorectal cancers	I/II	NCT00040599 NCT00041639 NCT00041652 NCT00041691	Closed 2004	Immunomedics, Morris Plains, NJ, USA	II	2005	
IOR-CEA1	Murine	CEA	Colorectal cancer	II	—		Instituto Nacional di Oncologia y Radiobiologia, Habana, Cuba	II	2000	
MN14 (MN-14)	F(ab)$_2$ fragment, murine	CEA	Ovarian and colorectal cancers	I/II	NCT00004048 NCT00004085 ...	Closed 2004	Immunomedics, Morris Plains, NJ, USA	I/II	2000/ 2004	
PR1A3	Murine	CEA	Colorectal cancer	I	—		Cancer Research UK, London, UK	I	2005	
T84.66	Murine	CEA	Colorectal cancer	I	—		City of Hope Medical Center, Duarte, USA	—	2005	
cT84.66	Chimeric from murine	CEA	Colorectal cancer	I	—		City of Hope Medical Center, Duarte, USA	I	2004/ 2005	
5T4	Fab fragment, murine	5T4	Trophoblast cells: breast, lung, gastric, ovarian, and colorectal cancers	I	—		Cancer Research UK Immunology Group, Manchester, UK	I	2004/ 2005	trophoblast surface cell marker
A6H	Murine	CD26 (dipeptidyl peptidase IV)	T lymphocytes: renal cell cancer	I/II	—		University of Minnesota, Minneapolis, MN, USA; University of Washington, Seattle, WA, USA	I/II	1987/ 2001	
H17E2	Murine	Placental alkaline phosphatase	Lung, breast, ovarian, uterine, and testicular cancers	II	—		Cancer Research UK and Hammersmith Hospital, London, UK	—	2000	

targeting – it is not surprising that numerous approaches to its therapeutic exploitation are being investigated (see Table 12.13).

Among these, the humanized antibody labetuzumab (hMN-14) has been tested in a series of recently completed clinical studies, two of which, on radioimmunotherapy with ^{90}Y- and ^{131}I-labeled conjugates, have already been published. In a combination therapy protocol with doxorubicin and peripheral blood stem cell support, high-dose ^{90}Y-labetuzumab was employed in patients with advanced medullary thyroid cancer. The majority of known tumor lesions were detected, but 20% went insufficiently targeted. Most targeted lesions received more than 20 Gy with an administered dose of less than 1.5 MBq m^{-2}. The average tumor-to-marrow ratio appeared safe at 15.0, albeit with a confidence interval from 4.0 to 26, but a tumor-to-lung ratio of 6.9 ± 6.1 was mirrored in dose-limiting cardiopulmonary toxicity. One patient showed an objective response, two more had minor responses (Sharkey et al. 2005a). Salvage therapy after complete resection of liver metastases of colorectal cancer was the objective of a phase II trial including 23 patients. Doses of 40–60 mCi m^{-2} of ^{131}I-labetuzumab were administered. A median overall survival time of 68.0 months and a median disease-free survival time of 18.0 months with a 5-year survival rate of 51% were reported, compared with historical data giving a 5-year survival of about one-third. Based on these data demonstrating a real clinical benefit, a multicenter, randomized trial has been projected (Liersch et al. 2005).

Another anti-CEA antibody, PR1A3, is remarkable in that its target initially had not been identified as CEA for lack of reaction with circulating serum antigen. Its binding site has then been identified as a conformational epitope involving the glycosyl-phosphatidylinositol (GPI) anchor and the B3 domain at the site of membrane attachment. Putative conformation changes associated with antigen release from the cell surface apparently masked soluble CEA from this antibody (Durbin et al. 1994). A humanized version without a separate name has been constructed and shown to retain the affinity and cell surface specificity of the original (Stewart et al. 1999). Unfortunately, its unique binding pattern could not be translated into clinical benefit. While a couple of clinical imaging studies have been published, the only peer-reviewed therapy trial investigated an anti-idiotype vaccine approach. Despite interesting immunological findings, its result is best summarized by quoting the authors: "Progressive disease was observed in 14 of the 15 patients with minimal toxicity" (Zbar et al. 2005).

A more optimistic outlook can be drawn from reports on a chimeric high-affinity anti-CEA antibody, T84.66. Pretherapy imaging trials demonstrated little toxicity, little immunogenicity, and good tumor targeting, but rapid liver clearance in a subset of patients (Wong et al. 1995). A more recent study combined ^{90}Y-cT84.66 with continuous infusion 5-fluorouracil for patients with chemotherapy-refractory metastatic colorectal cancer. Of 21 heavily pretreated patients, none developed an objective response, but one mixed response was observed and 11 patients with progressive disease at study entry experienced disease stabilization for 3–8 months. An additional effect was reduced incidence of immune reactions against the chimeric antibody (Wong et al. 2003). A new cT84.66-based, ^{123}I-labeled minibody construct demonstrated successful tumor imaging in 7 of 8 patients, maintained tumor residence, and reduced circulation half-life without drug-related adverse reactions (Wong et al. 2004).

12.4.7
Individual or Noncategorized Antigens

A couple of antigens targeted by clinically investigated antibodies are defined and characterized, but do not fit into any of the above categories. These include detrimental products of malignant conditions such as amyloid (11–1F4), or parts of the cytoskeleton and other intracellular antigens that may be exposed as a result of malignant transformation and changes in cell morphology such as annexins (28A32) or cytokeratins (COU-1) or a cytokeratin-associated protein (174H.64). Finally, qualitative or quantitative metabolic markers of tumors such as parathyroid hormone-related protein (CAL) or the alpha folate receptor (MOv18) are falling into this category, which is summarized in Table 12.14.

12.5
Disease-specific Concepts of Unknown Antigen Function and Structure

There are a few antibodies in clinical investigation whose target antigens have not been characterized at all so far apart from their presumed specificity for a certain type of cancer (see Table 12.15). Little published information is available about most of these. An exception is BDI-1, which has so far only been characterized as an antibody against bladder cancer. Studies of this antibody have been published on radioimmunoimaging by intravesical administration, demonstrating immunoreactivity and tumor-specific localization (Zhang et al. 1998). Intravesical administration of BDI-1-RT immunotoxin as adjuvant therapy after tumor resection in 31 bladder cancer patients compared to mitomycin C in a control group of 36 patients revealed about half the recurrence rate in the immunotoxin group, albeit without statistical significance, and significantly lower toxicity. Interestingly, the immunoreactivity of BDI-1-RT was found to correlate with tumor grade (Zang et al. 2000). Recently, the production and *in vitro* characterization of a novel immunotoxin consisting of arsenic trioxide-loaded nanospheres linked to BDI-1 was reported (Zhou et al. 2005). Although the article is at times difficult to follow, the authors go to great lengths to demonstrate specific binding activity and cytotoxic effect, and this may be an interesting approach.

12.6
Summary

Antibodies as anticancer agents have come of age and are now a standard modality of both established and experimental therapies. Originally conceived as a strategy for the specific targeting of "passive" markers of malignant cells, their immediate effects have in fact evolved into a wide array of biological functions.

Table 12.14 Antibodies against individual antigens not otherwise categorized.

Antibody name	Characteristics	Target antigen	Target cancer	Phase	NCI study ID	Status	Source	Publ. phase	Publ. year	Comment
28A32	Heterohybridoma	CTAA 28A32-32K, annexin-related	Colorectal cancer	I/II	–		Intracel, Frederick, MD, USA	I/II	1990/ 1998	Intracellular
MOv18	Murine	Alpha folate receptor	Epithelial: ovarian cancer	I	–		Istituto Nazionale per lo Studio e la Cura dei Tumori, Milano, Italy	I	1995/ 2005	RIT
11-1F4	Chimeric	Amyloid	Primary amyloidosis	I/II	–		University of Tennessee, Knoxville, TN, USA	–	2003	
174H.64 (174H.64 × antibiotin)	Murine antibody (bispecific conjugate for pretargeting, not in clinical study)	Cytokeratin	Lung, breast, uterine, skin and other cancers	I/II	–		University of Alberta, Edmonton, AL, Canada; Abbott Labs, Abbott Pk, IL, USA	–	2000	
COU-1	Human	Cytokeratin 8 and 18	Lung, breast, ovarian, pancreatic, and colorectal cancers	I/II	–		Bristol Myers Squibb, New York, NY, USA	–	2002	
CAL (anti-PTHrP)	Humanized	Parathyroid hormone related protein (PTHrP)	Breast cancer, bone metastases	I/II	NCT00060138 NCT00051779	Closed 2004	Chugai Pharmaceuticals Co. Ltd. Tokyo, Japan	–	–	

Table 12.15 Antibodies in disease-specific use with undefined antigen.

Antibody name	Characteristics	Target antigen	Target cancer	Phase	NCI study ID	Status	Source	Publ. phase	Publ. year	Comment
1A3	Murine	–	Colorectal cancer	I/II	–		Washington University, St. Louis, MO, USA	I	1993/2001	Clinical study: abstract
chA7 (chA7Fab-NCS)	Chimeric A7	–	Gastric, pancreatic, and colorectal cancers, leukemia, lymphoma	I	–		Kyoto Prefectural University of Medicine, Kyoto, Japan	I	2003	RIT. Neocarcinostatin conjugate
BDI-1 (BDI-1-RT immunotoxin)	Murine	–	Bladder cancer	I	–		Beijing University, Beijing, China	I	2000/2005	Clinical study on immunotoxin
chSF-25	Chimeric	–	Head-and-neck, liver, and colon cancers	I/II	–		Massachusetts General Hospital, Boston, MA, USA; Centocor, Horsham, PA, USA	I/II	1994/2002	Radioimmunoscintigraphy trial
30.6 (c30.6)	Murine (c30.6: chimeric)	–	Colorectal cancer	I	–		Melbourne University, Melbourne, Australia	I	2000	
GAH	Heterohybridoma, Fab fragment	–	Gastrointestinal cancers	I	–		Mitsubishi Pharma, Osaka, Japan	I	2004	Internalized. rGAH-conjugated, doxorubicin encapsulated immunoliposomes
Hepama-1	Murine	–	Liver cancer	I/II	–		Hipple Cancer Research Center, Dayton, OH, USA	I	2004	

In reviewing which antibodies have made it from bench to bedside in terms of clinical trials, two trends become visible regarding antibody structure and therapeutic principle: on the one hand, sophisticated recombinant antibody constructs succeed "plain old" immunoglobulins, on the other, antibodies become increasingly important either as targeting vehicles for the delivery of secondary effector molecules or as specific interactors with cell signaling pathways. While they compete with small molecule drugs in the latter, their unique potential lies in the complex bi- and multifunctional approaches made possible by the former.

In interpreting results of clinical trials, however, one caveat must be kept in mind: It is tempting to judge the potential of the tested therapy by clinical response and survival rates, but this is not what early phase, and particular phase I trials, are designed for. Facing life-threatening diseases, for ethical reasons most investigational new drugs can only be applied in patients who have exhausted all established treatment options. At this point, these patients have been selected for a bad prognosis. Thus, while a significant clinical response in these trials is certainly a good sign, the lack thereof has little predictive value. Especially in solid tumors, the natural strength of antibody-based therapies must be expected from specific treatment of disseminated systemic disease of small total volume, as in early stage or minimal residual disease (Koppe et al. 2005). Successful antibody-based treatment of bulky disease with large tumor mass will probably remain the exception. This does not diminish the role of antibodies in future cancer therapy, as sufficient therapeutic options are often available to remove large tumors, but disseminated micrometastatic disease or minimal residual disease is the most ominous threat to the patient's survival in most malignancies.

References

Aboud-Pirak, E., Hurwitz, E., Pirak, M.E., Bellot, F., Schlessinger, J., Sela, M. 1988. Efficacy of antibodies to epidermal growth factor receptor against KB carcinoma in vitro and in nude mice. *J Natl Cancer Inst* 80: 1605–1611.

Adkins, J.C., Spencer, C.M. (1998) Edrecolomab (monoclonal antibody 17-1A) *Drugs* 56: 619–626.

Agus, D.B., Gordon, M.S., Taylor, C., Natale, R.B., Karlan, B., Mendelson, D.S., Press, M.F., Allison, D.E., Sliwkowski, M.X., Lieberman, G., Kelsey, S.M., Fyfe, G. (2005) Phase I clinical study of pertuzumab, a novel HER dimerization inhibitor, in patients with advanced cancer. *J Clin Oncol* 23: 2534–2543.

Akabani, G., Reardon, D.A., Coleman, R.E., Wong, T.Z., Metzler, S.D., Bowsher, J.E., Barboriak, D.P., Provenzale, J.M., Greer, K.L., DeLong, D., Friedman, H.S., Friedman, A.H., Zhao, X.G., Pegram, C. N., McLendon, R.E., Bigner, D.D., Zalutsky, M.R. (2005) Dosimetry and radiographic analysis of I-131-labeled anti-tenascin 81C6 murine monoclonal antibody in newly diagnosed patients with malignant gliomas: A phase II study. *J Nucl Med* 46: 1042–1051.

Armstrong, A., Eck, S.L. (2003) EpCAM – A new therapeutic target for an old cancer antigen. *Cancer Biol Ther* 2: 320–325.

Atkins, M., Regan, M., McDermott, D., Mier, J., Stanbridge, E., Youmans, A., Febbo, P., Upton, M., Lechpammer, M., Signoretti, S. (2005) Carbonic anhydrase IX expression predicts outcome of interleukin 2 therapy for renal cancer. *Clin Cancer Res* 11: 3714–3721.

Audi, J., Belson, M., Patel, M., Schier, J., Osterloh, J. (2005) Ricin poisoning: a comprehensive review. *JAMA* 294: 2342–2351.

Baca, M., Presta, L.G., O'Connor, S.J., Wells, J.A. (1997) Antibody humanization using monovalent phage display. *J Biol Chem* 272: 10678–10684.

Badache, A., Hynes, N.E. (2004) A new therapeutic antibody masks ErbB2 to its partners. *Cancer Cell* 5: 299–301.

Bander, N.H., Milowsky, M.I., Nanus, D.M., Kostakoglu, L., Vallabhajosula, S., Goldsmith, S.J. (2005) Phase I trial of (177)lutetium-labeled J591, a monoclonal antibody to prostate-specific membrane antigen, in patients with androgen-independent prostate cancer. *J Clin Oncol* 23: 4591–4601.

Barth, S., Huhn, M., Matthey, B., Schnell, R., Tawadros, S., Schinkothe, T., Lorenzen, J., Diehl, V., Engert, A. (2000) Recombinant anti-CD25 immunotoxin RFT5(scFv)-ETA′ demonstrates successful elimination of disseminated human Hodgkin lymphoma in SCID mice. *Int J Cancer* 86: 718–724.

Bekker, P.J., Holloway, D.L., Rasmussen, A.S., Murphy, R., Martin, S.W., Leese, P.T., Holmes, G.B., Dunstan, C.R., DePaoli, A.M. (2004) A single-dose placebo-controlled study of AMG 162, a fully human monoclonal antibody to RANKL, in postmenopausal women. *J Bone Miner Res* 19: 1059–1066.

Berek, J.S., Taylor, P.T., Gordon, A., Cunningham, M.J., Finkler, N., Orr, J., Jr., Rivkin, S., Schultes, B.C., Whiteside, T.L., Nicodemus, C.F. (2004) Randomized, placebo-controlled study of oregovomab for consolidation of clinical remission in patients with advanced ovarian cancer. *J Clin Oncol* 22: 3507–3516.

Bhaskar, V., Wales, P., Bremberg, D., Fox, M., Ho, S. (2004) M200, a chimeric antibody against integrin alpha5 beta1, inhibits cancer cell growth in vitro. *J Immunother* 27: S9.

Bhogal, N., Combes, R. (2006a) An Update on TGN1412. *Altern Lab Anim* 34: 351–356.

Bhogal, N., Combes, R. (2006b) TGN1412: time to change the paradigm for the testing of new pharmaceuticals. *Altern Lab Anim* 34: 225–239.

Bleeker, W.K., van Bueren, J.J.L., van Ojik, H.H., Gerritsen, A.F., Pluyter, M., Houtkamp, M., Halk, E., Goldstein, J., Schuurman, J., van Dijk, M.A., van de Winkel, J.G.J., Parren, P.W.H.I. (2004) Dual mode of action of a human anti-epidermal growth factor receptor monoclonal antibody for cancer therapy. *J Immunol* 173: 4699–4707.

Bleumer, I., Knuth, A., Oosterwijk, E., Hofmann, R., Varga, Z., Lamers, C., Kruit, W., Melchior, S., Mala, C., Ullrich, S., De Mulder, P., Mulders, P.F., Beck, J. (2004) A phase II trial of chimeric monoclonal antibody G250 for advanced renal cell carcinoma patients. *Br J Cancer* 90: 985–990.

Blohmer, J., Gore, M., Kuemmel, S., Verheijen, R.H., Kimmig, R., Massuger, L.F.A.G., Du Bois, A., Smit, W.M., Kaye, S., Deubelbeiss, C. (2005) Phase II study to determine response rate, pharmacokinetics (PK), pharmacodynamics (PD), safety, and tolerability of treatment with the humanized anti-epidermal growth factor receptor (EGFR) monoclonal antibody EMD 72000 (matuzumab) in patients with recurrent cervical cancer. *J Clin Oncol* 23: 174S.

Boll, B., Hansen, H., Heuck, F., Reiners, K., Borchmann, P., Rothe, A., Engert, A., von Strandmann, E.P. (2005) The fully human anti-CD30 antibody 5F11 activates NF-kappa B and sensitizes lymphoma cells to bortezomib-induced apoptosis. *Blood* 106: 1839–1842.

Borchmann, P., Schnell, R., Fuss, I., Manzke, O., Davis, T., Lewis, L.D., Behnke, D., Wickenhauser, C., Schiller, P., Diehl, V., Engert, A. (2002) Phase 1 trial of the novel bispecific molecule H22xKi-4 in patients with refractory Hodgkin lymphoma. *Blood* 100: 3101–3107.

Borjesson, P.K.E., Postema, E.J., Roos, J.C., Colnot, D.R., Marres, H.A.M., van Schie, M.H., Stehle, G., de Bree, R., Snow, G.B., Oyen, W.J.G., van Dongen, G.A.M.S. (2003) Phase I therapy study with Re-186-labeled humanized monoclonal antibody BIWA 4 (bivatuzumab) in patients with head and neck squamous cell carcinoma. *Clin Cancer Res* 9: 3961S–3972S.

Byrd, J.C., Bresalier, R.S. (2004) Mucins and mucin binding proteins in colorectal cancer. *Cancer Metastasis Rev* 23: 77–99.

Cassady, J.M., Chan, K.K., Floss, H.G., Leistner E. (2004) Recent developments in the maytansinoid antitumor agents. *Chem Pharm Bull (Tokyo)* 52: 1–26.

Cerveny, C.G., Law, C.L., McCormick, R.S., Lenox, J.S., Hamblett, K.J., Westendorf, L.E., Yamane, A.K., Petroziello, J.M., Francisco, J.A., Wahl, A.F. (2005) Signaling via the anti-CD30 mAb SGN-30 sensitizes Hodgkin's disease cells to conventional chemotherapeutics. *Leukemia* 19: 1648–1655.

Cheung, N.K.V., Kushner, B.H., Yeh, S.D.J., Larson, S.M. (1998) 3F8 monoclonal antibody treatment of patients with stage 4 neuroblastoma: a phase II study. *Int J Oncol* 12: 1299–1306.

Choi, B.S., Sondel, P.M., Hank, J.A., Schalch, H., Gan, J., King, D.M., Kendra, K., Mahvi, D., Lee, L.Y., Kim, K., Albertini, M.R. (2005) Phase I trial of combined treatment with ch14.18 and R24 monoclonal antibodies and interleukin-2 for patients with melanoma or sarcoma. *Cancer Immunol Immunother* 1–14.

Chong, G., Lee, F.T., Hopkins, W., Tebbutt, N., Cebon, J.S., Mountain, A.J., Chappell, B., Papenfuss, A., Schleyer, P., Murphy, R., Wirth, V., Smyth, F.E., Potasz, N., Poon, A., Davis, I.D., Saunder, T., O'keefe, G.J., Burgess, A.W., Hoffman, E.W., Old, L.J., Scott, A.M. (2005) Phase I trial of 131I-huA33 in patients with advanced colorectal carcinoma. *Clin Cancer Res* 11: 4818–4826.

Clarke, K., Lee, F.T., Brechbiel, M.W., Smyth, F.E., Old, L.J., Scott, A.M. (2000) Therapeutic efficacy of anti-Lewis(y) humanized 3S193 radioimmunotherapy in a breast cancer model: Enhanced activity when combined with taxol chemotherapy. *Clin Cancer Res* 6: 3621–3628.

Czuczman, M.S., Thall, A., Witzig, T.E., Vose, J.M., Younes, A., Emmanouilides, C., Miller, T.P., Moore, J.O., Leonard, J.P., Gordon, L.I., Sweetenham, J., Alkuzweny, B., Finucane, D.M., Leigh, B.R. (2005) Phase I/II study of galiximab, an anti-CD80 antibody, for relapsed or refractory follicular lymphoma. *J Clin Oncol* 23: 4390–4398.

Damle, N.K., Frost, P. (2003) Antibody-targeted chemotherapy with immunoconjugates of calicheamicin. *Curr Opin Pharmacol* 3: 386–390.

de Bono, J.S., Tolcher, A.W., Forero, A., Vanhove, G.F.A., Takimoto, C., Bauer, R.J., Hammond, L.A., Patnaik, A., White, M.L., Shen, S., Khazaeli, M.B., Rowinsky, E.K., LoBuglio, A.F. (2004) ING-1, a monoclonal antibody targeting Ep-CAM in patients with advanced adenocarcinomas. *Clin Cancer Res* 10: 7555–7565.

Deckert, P.M., Jungbluth, A., Montalto, N., Clark, M.A., Finn, R.D., Williams, C. Jr, Richards, E.C., Panageas, K.S., Old, L.J., Welt, S. 2000. Pharmacokinetics and microdistribution of polyethylene glycol-modified humanized A33 antibody targeting colon cancer xenografts. *Int J Cancer* 87: 382–390.

Deckert, P.M., Renner, C., Cohen, L.S., Jungbluth, A., Ritter, G., Bertino, J.R., Old, L.J., Welt, S. (2003) A33scFv-cytosine deaminase: a recombinant protein construct for antibody-directed enzyme-prodrug therapy. *Br J Cancer* 88: 937–939.

DeNardo, S.J., DeNardo, G.L., Miers, L.A., Natarajan, A., Foreman, A.R., Gruettner, C., Adamson, G.N., Ivkov, R. (2005a) Development of tumor targeting bioprobes ((111)In-chimeric L6 monoclonal antibody nanoparticles) for alternating magnetic field cancer therapy. *Clin Cancer Res* 11: 7087s–7092s.

DeNardo, S.J., Richman, C.M., Albrecht, H., Burke, P.A., Natarajan, A., Yuan, A., Gregg, J.P., O'Donnel, R.T., DeNardo, G.L. (2005b) Enhancement of the therapeutic index: From nonmyeloablative and myeloablative toward pretargeted radioimmunotherapy for metastatic prostate cancer. *Clin Cancer Res* 11: 7187S–7194S.

Dey, B.R., McAfee, S., Colby, C., Cieply, K., Caron, M., Saidman, S., Preffer, F., Shaffer, J., Tarbell, N., Sackstein, R., Sachs, D., Sykes, M., Spitzer, T.R. (2005) Anti-tumour response despite loss of donor chimaerism in patients treated with non-myeloablative conditioning and allogeneic stem cell transplantation. *Br J Haematol* 128: 351–359.

Dhingra, K., Fritsche, H., Murray, J.L., LoBuglio, A.F., Khazaeli, M.B., Kelley, S., Tepper, M.A., Grasela, D., Buzdar, A., Valero, V. (1995) Phase I clinical and pharmacological study of suppression of human antimouse antibody response to monoclonal antibody L6 by deoxyspergualin. *Cancer Res* 55: 3060–3067.

Di Paolo, C., Willuda, J., Kubetzko, S., Lauffer, I., Tschudi, D., Waibel, R., Pluckthun, A., Stahel, R.A., Zangemeister-Wittke, U. (2003) A recombinant immunotoxin derived from a humanized epithelial cell adhesion molecule-specific single-chain antibody fragment has potent and selective antitumor activity. *Clin Cancer Res* 9: 2837–2848.

Divgi, C.R., Bander, N.H., Scott, A.M., O'Donoghue, J.A., Sgouros, G., Welt, S., Finn, R.D., Morrissey, F., Capitelli, P., Williams, J.M., Deland, D., Nakhre, A., Oosterwijk, E., Gulec, S., Graham, M.C., Larson, S.M., Old, L.J. (1998) Phase I/II radioimmunotherapy trial with iodine-131-labeled monoclonal antibody G250 in metastatic renal cell carcinoma. *Clin Cancer Res* 4: 2729–2739.

Divgi, C.R., O'Donoghue, J.A., Welt, S., O'Neel, J., Finn, R., Motzer, R.J., Jungbluth, A., Hoffman, E., Ritter, G., Larson, S.M., Old, L.J. (2004) Phase I clinical trial with fractionated radioimmunotherapy using I-13-labeled chimeric G250 in metastatic renal cancer. *J Nucl Med* 45: 1412–1421.

Durbin, H., Young, S., Stewart, L.M., Wrba, F., Rowan, A.J., Snary, D., Bodmer, W.F. (1994) An epitope an carcinoembryonic antigen defined by the clinically relevant antibody Pr1A3. *Proc Natl Acad Sci USA* 91: 4313–4317.

Duzkale, H., Pagliaro, L.C., Rosenblum, M.G., Varan, A., Liu, B.S., Reuben, J., Wierda, W.G., Korbling, M., McMannis, J.D., Glassman, A.B., Scheinberg, D.A., Freireich, E.J. (2003) Bone marrow purging studies in acute myelogenous leukemia using the recombinant anti-CD33 immunotoxin HuM195/rGel. *Biol Blood Marrow Transplant* 9: 364–372.

Ebbinghaus, C., Scheuermann, J., Neri, D., Elia, G. (2004) Diagnostic and therapeutic applications of recombinant antibodies: targeting the extra-domain B of fibronectin, a marker of tumor angiogenesis. *Curr Pharm Des* 10: 1537–1549.

Epenetos, A.A., Hird, V., Lambert, H., Mason, P., Coulter, C. (2000) Long term survival of patients with advanced ovarian cancer treated with intraperitoneal radioimmunotherapy. *Int J Gynecol Cancer* 10(S1): 44–46.

Farzaneh, L., Kasahara, N., Farzaneh, F. (2006) The strange case of TGN1412. *Cancer Immunol Immunother* (in press).

Fiedler, W., Kruger, W., Laack, E., Mende, T., Vohwinkel, G., Hossfeld DK. (2001) A clinical trial of edrecolomab, interleukin-2 and GM-CSF in patients with advanced colorectal cancer. *Oncol Rep* 8: 225–231.

Folkman, J. (2006) Angiogenesis. *Annu Rev Med* 57: 1–18.

Foon, K.A., Yang, X.D., Weiner, L.M., Belldegrun, A.S., Figlin, R.A., Crawford, J., Rowinsky, E.K., Dutcher, J.P., Vogelzang, N.J., Gollub, J., Thompson, J.A., Schwartz, G., Bukowski, R.M., Roskos, L.K., Schwab, G.M. (2004) Preclinical and clinical evaluations of ABX-EGF, a fully human anti-epidermal growth factor receptor antibody. *Int J Radiat Oncol Biol Phys* 58: 984–990.

Forero, A., Meredith, R.F., Khazaeli, M.B., Shen, S., Grizzle, W.E., Carey, D., Busby, E., LoBuglio, A.F., Robert, F. (2005) Phase I study of 90Y-CC49 monoclonal antibody therapy in patients with advanced non-small cell lung cancer: effect of chelating agents and paclitaxel co-administration. *Cancer Biother Radiopharm* 20: 467–478.

Francis, R.J., Sharma, S.K., Springer, C., Green, A.J., Hope-Stone, L.D., Sena, L., Martin, J., Adamson, K.L., Robbins, A., Gumbrell, L., O'Malley, D., Tsiompanou, E., Shahbakhti, H., Webley, S., Hochhauser, D., Hilson, A.J., Blakey, D., Begent, R.H. (2002) A phase I trial of antibody directed enzyme prodrug therapy (ADEPT) in patients with advanced colorectal carcinoma or other CEA producing tumours. *Br J Cancer* 87: 600–607.

Friess, T., Scheuer, W., Hasmann, M. (2005) Combination treatment with erlotinib and pertuzumab against human tumor xenografts is superior to monotherapy. *Clin Cancer Res* 11: 5300–5309.

Garkavij, M., Samarzija, M., Ewers, S.B., Jakopovic, M., Tezak, S., Tennvall, J. (2005) Concurrent radiotherapy and tumor targeting with 111In-HMFG1-F(ab')$_2$ in patients with MUC1-positive non-small cell lung cancer. *Anticancer Res* 25: 4663–4671.

Georgakis, G.V., Li, Y., Humphreys, R., Andreeff, M., O'Brien, S., Younes, M., Carbone, A., Albert, V., Younes, A. (2005) Activity of selective fully human agonistic antibodies to the TRAIL death receptors TRAIL-R1 and TRAIL-R2 in primary and cultured lymphoma cells: induction of apoptosis and enhancement of doxorubicin- and bortezomib-induced cell death. *Br J Haematol* 130: 501–510.

Goodman, G.E., Hellstrom, I., Yelton, D.E., Murray, J.L., O'Hara, S., Meaker, E., Zeigler, L., Palazzolo, P., Nicaise, C., Usakewicz J, . (1993) Phase I trial of chimeric (human-mouse) monoclonal antibody L6 in patients with non-small-cell lung, colon, and breast cancer. *Cancer Immunol Immunother* 36: 267–273.

Gottlieb, A.B., Lebwohl, M., Totoritis, M.C., Abdulghani, A.A., Shuey, S.R., Romano, P., Chaudhari, U., Allen, R.S., Lizambri, R.G. (2002) Clinical and histologic response to single-dose treatment of moderate to severe psoriasis with an anti-CD80 monoclonal antibody. *J Am Acad Dermatol* 47: 692–700.

Grana, C., Chinol, M., Robertson, C., Mazzetta, C., Bartolomei, M., De Cicco, C., Fiorenza, M., Gatti, M., Caliceti, P., Paganelli, G. (2002) Pretargeted adjuvant radioimmunotherapy with yttrium-90-biotin in malignant glioma patients: a pilot study. *Br J Cancer* 86: 207–212.

Gronau, S.S., Schmitt, M., Thess, B., Reinhardt, P., Wiesneth, M., Schmitt, A., Riechelmann, H. (2005) Trifunctional bispecific antibody-induced tumor cell lysis of squamous cell carcinomas of the upper aerodigestive tract. *Head Neck* 27: 376–382.

Grossbard, M.L., Multani, P.S., Freedman, A.S., O'Day, S., Gribben, J.G., Rhuda, C., Neuberg, D., Nadler, L.M. (1999) A Phase II study of adjuvant therapy with anti-B4-blocked ricin after autologous bone marrow transplantation for patients with relapsed B-cell non-Hodgkin's lymphoma. *Clin Cancer Res* 5: 2392–2398.

Grunwald, V., Hidalgo, M. (2003) Developing inhibitors of the epidermal growth factor receptor for cancer treatment. *J Natl Cancer Inst* 95: 851–867.

Hagberg, H., Pettersson, M., Bjerner, T., Enblad, G. (2005) Treatment of a patient with a nodal peripheral T-cell lymphoma (angioimmunoblastic T-cell lymphoma) with a human monoclonal antibody against the CD4 antigen (HuMax-CD4). *Med Oncol* 22: 191–194.

Hamann, P.R., Hinman, L.M., Beyer, C.F., Lindh, D., Upeslacis, J., Shochat, D., Mountain, A. (2005) A calicheamicin conjugate with a fully humanized anti-MUC1 antibody shows potent antitumor effects in breast and ovarian tumor xenografts. *Bioconjug Chem* 16: 354–360.

Hanai, N., Nakamura, K., Shitara, K. (2000) Recombinant antibodies against ganglioside expressed on tumor cells. *Cancer Chemother Pharmacol* 46: S13–S17.

Hangai, M., Kitaya, N., Xu, J.S., Chan, C.K., Kim, J.J., Werb, Z., Ryan, S.J., Brooks, P.C. (2002) Matrix metalloproteinase-9-dependent exposure of a cryptic migratory control site in collagen is required before retinal angiogenesis. *Am J Pathol* 161: 1429–1437.

Hann, E., Reinartz, S., Clare, S.E., Passow, S., Kissel, T., Wagner, U. (2005) Development of a delivery system for the continuous endogenous release of an anti-idiotypic antibody against ovarian carcinoma. *Hybridoma* 24: 133–140.

Hansen, S., Leslie, R.G. (2006) TGN1412: scrutinizing preclinical trials of antibody-based medicines. *Nature* 441: 282.

Hartmann, F., Renner, C., Jung, W., da Costa, L., Tembrink, S., Held, G., Sek, A., Konig, J., Bauer, S., Kloft, M., Pfreundschuh, M. (2001) Anti-CD16/CD30 bispecific antibody treatment for Hodgkin's disease: Role of infusion schedule and costimulation with cytokines. *Clin Cancer Res* 7: 1873–1881.

Hartung, G., Hofheinz, R.D., Dencausse, Y., Sturm, J., Kopp-Schneider, A., Dietrich, G., Fackler-Schwalbe, I., Bornbusch, D., Gonnermann, M., Wojatschek, C., Lindemann, W., Eschenburg, H., Jost, K., Edler, L., Hochhaus, A., Queisser, W. (2005) Adjuvant therapy with edrecolomab versus observation in stage II colon cancer: A multicenter randomized phase III study. *Onkologie* 28: 347–350.

Hassan, R., Williams-Gould, J., Watson, T., Pai-Scherf, L., Pastan, I. (2004)

Pretreatment with rituximab does not inhibit the human immune response against the immunogenic protein LMB-1. *Clin Cancer Res* 10(1 Pt 1): 16–18.

Haug, C., Gotzsche, P.C., Schroeder, T.V. (2005) Registries and registration of clinical trials. *N Engl J Med* 353: 2811–2812.

Heath, J.K., White, S.J., Johnstone, C.N., Catimel, B., Simpson, R.J., Moritz, R.L., Tu, G.F., Ji, H., Whitehead, R.H., Groenen, L.C., Scott, A.M., Ritter, G., Cohen, L., Welt, S., Old, L.J., Nice, E.C., Burgess, A.W. (1997) The human A33 antigen is a transmembrane glycoprotein and a novel member of the immunoglobulin superfamily. *Proc Natl Acad Sci USA* 94: 469–474.

Heider, K.H., Kuthan, H., Stehle, G., Munzert, G. (2004) CD44v6: a target for antibody-based cancer therapy. *Cancer Immunol Immunother* 53: 567–579.

Hoffmann, P., Hofmeister, R., Brischwein, K., Brandl, C., Crommer, S., Bargou, R., Itin, C., Prang, N., Baeuerle, P.A. (2005) Serial killing of tumor cells by cytotoxic T cells redirected with a CD19-/CD3-bispecific single-chain antibody construct. *Int J Cancer* 115: 98–104.

Hofheinz, R.D., al Batran, S.E., Hartmann, F., Hartung, G., Jager, D., Renner, C., Tanswell, P., Kunz, U., Amelsberg, A., Kuthan, H., Stehle, G. (2003) Stromal antigen targeting by a humanised monoclonal antibody: An early phase II trial of sibrotuzumab in patients with metastatic colorectal cancer. *Onkologie* 26: 44–48.

Hotz, H.G., Hines, O.J., Hotz, B., Foitzik, T., Buhr, H.J., Reber, H.A. 2003. Evaluation of vascular endothelial growth factor blockade and matrix metalloproteinase inhibition as a combination therapy for experimental human pancreatic cancer. *J Gastrointest Surg* 7: 220–227.

Huang, S.Y., Mills, L., Mian, B., Tellez, C., McCarty, M., Yang, X.D., Gudas, J.M., Bar-Eli, M. (2002) Fully humanized neutralizing antibodies to interleukin-8 (ABX-IL8) inhibit angiogenesis, tumor growth, and metastasis of human melanoma. *Am J Pathol* 161: 125–134.

Irie, R.F., Ollila, D.W., O'Day, S., Morton, D.L. (2004) Phase I pilot clinical trial of human IgM monoclonal antibody to ganglioside GM3 in patients with metastatic melanoma. *Cancer Immunol Immunother* 53: 110–117.

James, N.D., Atherton, P.J., Jones, J., Howie, A.J., Tchekmedyian, S., Curnow, R.T. (2001) A phase II study of the bispecific antibody MDX-H210 (anti-HER2 x CD64) with GM-CSF in HER2+advanced prostate cancer. *Br J Cancer* 85: 152–156.

Jayson, G.C., Mulatero, C., Ranson, M., Zweit, J., Jackson, A., Broughton, L., Wagstaff, J., Hakansson, L., Groenewegen, G., Lawrance, J., Tang, M., Wauk, L., Levitt, D., Marreaud, S., Lehmann, F.F., Herold, M., Zwierzina, H. (2005) Phase I investigation of recombinant anti-human vascular endothelial growth factor antibody in patients with advanced cancer. *Eur J Cancer* 41: 555–563.

Kim, Y.H., Obitz, E., Iversen, L., Oesterborg, A., Whittaker, S., Illidge, T.M., Schwarz, T., Kaufmann, R., Gniadecki, R., Duvic, M., Cooper, K., Jensen, P., Baadsgaard, O., Knox, S.J. (2004) HuMax-CD4, fully human monoclonal antibody: Phase II trial in cutaneous T cell lymphoma. *J Invest Dermatol* 122: A57.

Kiner-Strachan, B., Goel, S., Vanhove, G., Bauer, R.J., Verdier-Pinard, D., Karri, S., Desai, K., Bulgaru, A., Macapinlac, M., Mani, S. (2005) Phase I and pharmacokinetic study of a subcutaneously administered human-engineered monoclonal antibody ING-1 in patients with advanced adenocarcinomas. *J Clin Oncol* 23: 180S.

King, D.M., Albertini, M.R., Schalch, H., Hank, J.A., Gan, J., Surfus, J., Mahvi, D., Schiller, J.H., Warner, T., Kim, K., Eickhoff, J., Kendra, K., Reisfeld, R., Gillies, S.D., Sondel, P. (2004) Phase I clinical trial of the immunocytokine EMD 273063 in melanoma patients. *J Clin Oncol* 22: 4463–4473.

Kleinschmidt, M., Rudolph, R., Lilie, H. (2003) Design of a modular immunotoxin connected by polyionic adapter peptides. *J Mol Biol* 327: 445–452.

Klinger, M., Farhan, H., Just, H., Drobny, H., Himmler, G., Loibner, H., Mudde, G.C., Freissmuth, M., Sexl, V. (2004) Antibodies directed against Lewis-Y antigen inhibit signaling of Lewis-Y modified ErbB receptors. *Cancer Res* 64: 1087–1093.

Knox, S.J., Goris, M.L., Tempero, M., Weiden, P.L., Gentner, L., Breitz, H., Adams, G.P., Axworthy, D., Gaffigan, S., Bryan, K., Fisher, D.R., Colcher, D., Horak, I.D., Weiner, L.M. (2000) Phase II trial of yttrium-90-DOTA-biotin pretargeted by NR-LU-10 antibody/streptavidin in patients with metastatic colon cancer. *Clin Cancer Res* 6: 406–414.

Ko, Y.J., Bubley, G.J., Weber, R., Redfern, C., Gold, D.P., Finke, L., Kovar, A., Dahl, T., Gillies, S.D. (2004) Safety, pharmacokinetics, and biological pharmacodynamics of the immunocytokine EMD 273066 (huKS-IL2) – Results of a phase 1 trial in patients with prostate cancer. *J Immunother* 27: 232–239.

Köhler, G., Milstein, C. (1975) Continuous cultures of fused cells secreting antibody of predefined specificity. *Nature* 256: 495–497.

Koppe, M., van Schaijk, F., Roos, J., van Leeuwen, P., Heider, K.H., Kuthan, H., Bleichrodt, R. (2004) Safety, pharmacokinetics, immunogenicity, and biodistribution of Re-186-labeled humanized monoclonal antibody BIWA 4 (bivatuzumab) in patients with early-stage breast cancer. *Cancer Biother Radiopharm* 19: 720–729.

Koppe, M.J., Bleichrodt, R.P., Oyen, W.J., Boerman, O.C. (2005) Radioimmunotherapy and colorectal cancer. *Br J Surg* 92: 264–276.

Kostenuik, P.J. (2005) Osteoprotegerin and RANKL regulate bone resorption, density, geometry and strength. *Curr Opin Pharmacol* 5: 618–625.

Kraeber-Bodere, F., Faivre-Chauvet, A., Ferrer, L., Vuillez, J.P., Brard, P.Y., Rousseau, C., Resche, I., Devillers, A., Laffont, S., Bardies, M., Chang, K., Sharkey, R.M., Goldenberg, D.M., Chatal, J.F., Barbet, J. (2003) Pharmacokinetics and dosimetry studies for optimization of anti-carcinoembryonic antigen x anti-hapten bispecific antibody-mediated pretargeting of Iodine-131-labeled hapten in a phase I radioimmunotherapy trial. *Clin Cancer Res* 9: 3973S–3981S.

Kreitman, R.J., Wilson, W.H., White, J.D., Stetler-Stevenson, M., Jaffe, E.S., Giardina, S., Waldmann, T.A., Pastan, I. (2000) Phase I trial of recombinant immunotoxin anti-Tac(Fv)PE38 (LMB-2) in patients with hematologic malignancies. *J Clin Oncol* 18: 1622–1636.

Kreitman, R.J., Squires, D.R., Stetler-Stevenson, M., Noel, P., Fitzgerald, D.J.P., Wilson, W.H., Pastan, I. (2005) Phase I trial of recombinant immunotoxin RFB4(dsFv)-PE38 (BL22) in patients with B-cell malignancies. *J Clin Oncol* 23: 6719–6729.

Kushner, B.H., Kramer, K., Cheung, N.K.V. (2001) Phase II trial of the anti-G(D2) monoclonal antibody 3F8 and granulocyte-macrophage colony-stimulating factor for neuroblastoma. *J Clin Oncol* 19: 4189–4194.

Ladwein, M., Pape, U.F., Schmidt, D.S., Schnolzer, M., Fiedler, S., Langbein, L., Franke, W.W., Moldenhauer, G., Zoller, M. (2005) The cell-cell adhesion molecule EpCAM interacts directly with the tight junction protein claudin-7. *Exp Cell Res* 309: 345–357.

Lam, J.S., Pantuck, A.J., Belldegrun, A.S., Figlin, R.A. (2005) G250: a carbonic anhydrase IX monoclonal antibody. *Curr Oncol Rep* 7: 109–115.

Law, C.L., Gordon, K.A., Collier, J., Klussman, K., McEarchern, J.A., Cerveny, C.G., Mixan, B.J., Lee, W.P., Lin, Z.H., Valdez, P., Wahl, A.F., Grewal, I.S. (2005) Preclinical antilymphoma activity of a humanized anti-CD40 monoclonal antibody, SGN-40. *Cancer Res* 65: 8331–8338.

Li, Q., Verschraegen, C.F., Mendoza, J., Hassan, R. (2004) Cytotoxic activity of the recombinant anti-mesothelin immunotoxin, SS1(dsFv)PE38, towards tumor cell lines established from ascites of patients with peritoneal mesotheliomas. *AntiCancer Res* 24: 1327–1335.

Liersch, T., Meller, J., Kulle, B., Behr, T.M., Markus, P., Langer, C., Ghadimi, B.M., Wegener, W.A., Kovacs, J., Horak, I.D., Becker, H., Goldenberg, D.M. (2005) Phase II trial of carcinoembryonic antigen radioimmunotherapy with ^{131}I-labetuzumab after salvage resection of colorectal metastases in the liver: five-year safety and efficacy results. *J Clin Oncol* 23: 6763–6770.

Longo, D.L., Duffey, P.L., Gribben, J.G., Jaffe, E.S., Curti, B.D., Gause, B.L., Janik, J.E., Braman, V.M., Esseltine, D., Wilson, W.H., Kaufman, D., Wittes, R.E., Nadler, L.M., Urba, W.J. (2000) Combination chemotherapy followed by an immunotoxin (anti-B4-blocked ricin) in patients with indolent lymphoma: results of a phase II study. *Cancer J* 6: 146–150.

Maile, L.A., Busby, W.H., Sitko, K., Capps, B.E., Sergent, T., Badley-Clarke, J., Clemmons, D.R. (2006) Insulin-like growth factor-i signaling in smooth muscle cells is regulated by ligand binding to the 177CYDMKTTC184 Sequence of the β_3-subunit of $\alpha_v\beta_3$. *Mol Endocrinol* 20: 405–413.

Makower, D., Sparano, J.A., Wadler, S., Fehn, K., Landau, L., Wissel, P., Versola, M., Mani, S. (2003) A pilot study of edrecolomab (Panorex, 17-1A antibody) and capecitabine in patients with advanced or metastatic adenocarcinoma. *Cancer Invest* 21: 177–184.

Martin, J., Stribbling, S.M., Poon, G.K., Begent, R.H., Napier, M., Sharma, S.K., Springer, C.J. (1997) Antibody-directed enzyme prodrug therapy: pharmacokinetics and plasma levels of prodrug and drug in a phase I clinical trial. *Cancer Chemother Pharmacol* 40: 189–201.

Maxwell-Armstrong, C.A., Durrant, L.G., Buckley, T.J.D., Scholefield, J.H., Robins, R.A., Fielding, K., Monson, J.R.T., Guillou, P., Calvert, H., Carmichael, J., Hardcastle, J.D. (2001) Randomized double-blind phase II survival study comparing immunization with the anti idiotypic monoclonal antibody 105AD7 against placebo in advanced colorectal cancer. *Br J Cancer* 84: 1443–1446.

Mayer, A., Sharma, S.K., Tolner, B., Minton, N.P., Purdy, D., Amlot, P., Tharakan, G., Begent, R.H., Chester, K.A. (2004) Modifying an immunogenic epitope on a therapeutic protein: a step towards an improved system for antibody-directed enzyme prodrug therapy (ADEPT). *Br J Cancer* 90: 2402–2410.

Meredith, R.F., LoBuglio, A.F., Plott, W.E., Orr, R.A., Brezovich, I.A., Russell, C.D., Harvey, E.B., Yester, M.V., Wagner, A.J., Spencer, S.A., Wheeler, R.H., Saleh, M.N., Rogers, K.J., Polansky, A., Salter, M.M., Khazaeli, M.B. (1991) Pharmacokinetics, immune-response, and biodistribution of iodine-131-labeled chimeric mouse human Igg1,K 17-1A monoclonal-antibody. *J Nucl Med* 32: 1162–1168.

Meredith, R., Shen, S., Macey, D., Khazaeli, M.B., Carey, D., Robert, F., LoBuglio, A. (2003a) Comparison of biodistribution, dosimetry, and outcome from clinical trials of radionuclide-CC49 antibody therapy. *Cancer Biother Radiopharm* 18: 393–404.

Meredith, R., Shen, S., Macey, D., Khazaeli, M.B., Carey, D., Robert, F., LoBuglio, A. (2003b) Comparison of biodistribution, dosimetry, and outcome from clinical trials of radionuclide-CC49 antibody therapy. *Cancer Biother Radiopharm* 18: 393–404.

Messmann, R.A., Vitetta, E.S., Headlee, D., Senderowicz, A.M., Figg, W.D., Schindler, J., Michiel, D.F., Creekmore, S., Steinberg, S.M., Kohler, D., Jaffe, E.S., Stetler-Stevenson, M., Chen, H.C., Ghetie, V., Sausville, E.A. (2000) A phase I study of combination therapy with immunotoxins IgG-HD37-deglycosylated ricin A chain (dgA) and IgG-RFB4-dgA (Combotox) in patients with refractory CD19(+), CD22(+) B cell lymphoma. *Clin Cancer Res* 6: 1302–1313.

Milowsky, M.I., Nanus, D.M., Kostakoglu, L., Vallabhajosula, S., Goldsmith, S.J., Bander, N.H. (2004) Phase I trial of yttrium-90-labeled anti-prostate-specific membrane antigen monoclonal antibody J591 for androgen-independent prostate cancer. *J Clin Oncol* 22: 2522–2531.

Milstein, C., Cuello, A.C. (1983) Hybrid hybridomas and their use in immunohistochemistry. *Nature* 305: 537–540.

Mobus, V.J., Baum, R.P., Bolle, M., Kreienberg, R., Noujaim, A.A., Schultes, B.C., Nicodemus, C.F. (2003) Immune responses to murine monoclonal antibody-B43.13 correlate with prolonged survival of women with recurrent ovarian cancer. *Am J Obstet Gynecol* 189: 28–36.

Morris, M.J., Divgi, C.R., Pandit-Taskar, N., Batraki, M., Warren, N., Nacca, A., Smith-Jones, P., Schwartz, L., Kelly, W.K., Slovin, S., Solit, D., Halpern, J., Delacruz, A.,

Curley, T., Finn, R., O'Donoghue, J.A., Livingston, P., Larson, S., Scher, H.I. (2005) Pilot trial of unlabeled and indium-111-labeled anti-prostate-specific membrane antigen antibody J591 for castrate metastatic prostate cancer. *Clin Cancer Res* 11: 7454–7461.

Mulders, P.F., Debruyne, F.M.J., de Weijer, K., Rossi, J.F., Patil, S., Cooper, J., Prabhakar, U., Garay, C., Corringham, R. (2005) A phase I/II study of a chimeric antibody against interleukin-6 (CNTO 328) in subjects with metastatic renal cell carcinoma. *J Urol* 173: 380.

Multani, P.S., O'Day, S., Nadler, L.M., Grossbard, M.L. (1998) Phase II clinical trial of bolus infusion anti-B4 blocked ricin immunoconjugate in patients with relapsed B-cell non-Hodgkin's lymphoma. *Clin Cancer Res* 4: 2599–2604.

Nahta, R., Hung, M.C., Esteva, F.J. (2004) The HER-2-targeting antibodies trastuzumab and pertuzumab synergistically inhibit the survival of breast cancer cells. *Cancer Res* 64: 2343–2346.

Naundorf, S., Preithner, S., Mayer, P., Lippold, S., Wolf, A., Hanakam, F., Fichtner, I., Kufer, P., Raum, T., Riethmuller, G., Baeuerle, P.A., Dreier, T. (2002) In vitro and in vivo activity of MT201, a fully human monoclonal antibody for pancarcinoma treatment. *Int J Cancer* 100: 101–110.

Nechansky, A., Kircheis, R. (2005) EpCAM is not a LAIR-1 ligand! *Cancer Biol Ther* 4: 357.

Nicholson, S., Bomphray, C.C., Thomas, H., McIndoe, A., Barton, D., Gore, M., George, A.J. (2004) A phase I trial of idiotypic vaccination with HMFG1 in ovarian cancer. *Cancer Immunol Immunother* 53: 809–816.

Noujaim, A.A., Schultes, B.C., Baum, R.P., Madiyalakan, R. (2001) Induction of CA125-specific B and T cell responses in patients injected with MAb-B43.13– evidence for antibody-mediated antigen-processing and presentation of CA125 in vivo. *Cancer Biother Radiopharm* 16: 187–203.

Pai, L.H., Wittes, R., Setser, A., Willingham, M.C., Pastan, I. (1996) Treatment of advanced solid tumors with immunotoxin LMB-1: an antibody linked to *Pseudomonas* exotoxin. *Nat Med* 2: 350–353.

Pastan, I. (2003) Immunotoxins containing *Pseudomonas* exotoxin A: a short history. *Cancer Immunol Immunother* 52: 338–341.

Pavlinkova, G., Beresford, G.W., Booth, B.J., Batra, S.K., Colcher, D. (1999) Pharmacokinetics and biodistribution of engineered single-chain antibody constructs of MAb CC49 in colon carcinoma xenografts. *J Nucl Med* 40: 1536–1546.

Peters, M., Dasilva, A., Weckermann, D., Oberneder, R., Ebner, B., Kirchinger, P., Fetter, A., Kohne-Volland, R., Baeuerle, P., Gjorstrup, P. (2004) Phase I study of the novel fully human monoclonal antibody MT201, directed against epithelial cellular adhesion molecule (Ep-CAM), in patients with hormone-refractory prostate cancer (HRPC). *J Clin Oncol* 22: 188S.

Petit, A.M., Rak, J., Hung, M.C., Rockwell, P., Goldstein, N., Fendly, B., Kerbel, R.S. (1997) Neutralizing antibodies against epidermal growth factor and ErbB-2/neu receptor tyrosine kinases down-regulate vascular endothelial growth factor production by tumor cells in vitro and in vivo: angiogenic implications for signal transduction therapy of solid tumors. *Am J Pathol* 151: 1523–1530.

Porchet, N., Aubert, J.P. (2004) [MUC genes: mucin or not mucin? That is the question]. *Med Sci (Paris)* 20: 569–574.

Postema, E.J., Raemaekers, J.M.M., Oyen, W.J.G., Boerman, O.C., Mandigers, C.M.P.W., Goldenberg, D.M., van Dongen, G.A.M.S., Corstens, F.H.M. (2003) Final results of a phase I radioimmunotherapy trial using Re-186-epratuzumab for the treatment of patients with non-Hodgkin's lymphoma. *Clin Cancer Res* 9: 3995S–4002S.

Pritchard-Jones, K., Spendlove, I., Wilton, C., Whelan, J., Weeden, S., Lewis, I., Hale, J., Douglas, C., Pagonis, C., Campbell, B., Alvarez, P., Halbert, G., Durrant, L.G. (2005) Immune responses to the 105AD7 human anti-idiotypic vaccine after intensive chemotherapy, for osteosarcoma. *Br J Cancer* 92: 1358–1365.

Pukac, L., Kanakaraj, P., Humphreys, R., Alderson, R., Bloom, M., Sung, C.,

Riccobene, T., Johnson, R., Fiscella, M., Mahoney, A., Carrell, J., Boyd, E., Yao, X.T., Zhang, L., Zhong, L., von Kerczek, A., Shepard, L., Vaughan, T., Edwards, B., Dobson, C., Salcedo, T., Albert, V. (2005) HGS-ETR1, a fully human TRAIL-receptor 1 monoclonal antibody, induces cell death in multiple tumour types in vitro and in vivo. *Br J Cancer* 92: 1430–1441.

Rader, C., Ritter, G., Nathan, S., Elia, M., Gout, I., Jungbluth, A.A., Cohen, L.S., Welt, S., Old, L.J., Barbas, C.F., III. (2000) The rabbit antibody repertoire as a novel source for the generation of therapeutic human antibodies. *J Biol Chem* 275: 13668–13676.

Rader, C., Popkov, M., Neves, J.A., Barbas, C.F. (2002) Integrin alpha v beta 3-targeted therapy for Kaposi's sarcoma with an in vitro-evolved antibody. *FASEB J* 16.

Ramakrishnan, V., Bhaskar, V., Law, D.A., Wong, M.H.L., Green, J., Tso, J., Jeffry, U., Finck, B., Murray, R. (2004) M200, an integrin alpha 5 beta 1 antibody, promotes apoptosis in proliferating endothelial cells. *J Clin Oncol* 22: 240S.

Ran, S., He, J., Huang, X., Soares, M., Scothorn, D., Thorpe, P.E. (2005) Antitumor effects of a monoclonal antibody that binds anionic phospholipids on the surface of tumor blood vessels in mice. *Clin Cancer Res* 11: 1551–1562.

Reardon, D.A., Akabani, G., Coleman, R.E., Friedman, A.H., Friedman, H.S., Herndon, J.E., McLendon, R.E., Pegram, C.N., Provenzale, J.M., Quinn, J.A., Rich, J.N., Vredenburgh, J.J., Desjardins, A., Guruangan, S., Badruddoja, M., Dowell, J.M., Wong, T.Z., Zhao, X.G., Zalutsky, M.R., Bigner, D.D. (2006) Salvage radioimmunotherapy with murine iodine-131-labeled antitenascin monoclonal antibody 81C6 for patients with recurrent primary and metastatic malignant brain tumors: phase II study results. *J Clin Oncol* 24: 115–122.

Reinartz, S., Kohler, S., Schlebusch, H., Krista, K., Giffels, P., Renke, K., Huober, J., Mobus, V., Kreienberg, R., duBois, A., Sabbatini, P., Wagner, U. (2004) Vaccination of patients with advanced ovarian carcinoma with the anti-idiotype ACA125: Immunological response and survival (Phase Ib/II) *Clin Cancer Res* 10: 1580–1587.

Renner, C., Hartmann, F., Jung, W., Deisting, C., Juwana, M., Pfreundschuh, M. (2000) Initiation of humoral and cellular immune responses in patients with refractory Hodgkin's disease by treatment with an anti-CD16/CD30 bispecific antibody. *Cancer Immunol Immunother* 49: 173–180.

Repp, R., van Ojik, H.H., Valerius, T., Groenewegen, G., Wieland, G., Oetzel, C., Stockmeyer, B., Becker, W., Eisenhut, M., Steininger, H., Deo, Y.M., Blijham, G.H., Kalden, J.R., van de Winkel, J.G., Gramatzki, M. (2003) Phase I clinical trial of the bispecific antibody MDX-H210 (anti-FcgammaRI × anti-HER-2/neu) in combination with Filgrastim (G-CSF) for treatment of advanced breast cancer. *Br J Cancer* 89: 2234–2243.

Ribas, A., Camacho, L.H., Lopez-Berestein, G., Pavlov, D., Bulanhagui, C.A., Millham, R., Comin-Anduix, B., Reuben, J.M., Seja, E., Parker, C.A., Sharma, A., Glaspy, J.A., Gomez-Navarro, J. (2005) Antitumor activity in melanoma and anti-self responses in a phase I trial with the anti-cytotoxic T lymphocyte-associated antigen 4 monoclonal antibody CP-675,206. *J Clin Oncol* 23: 8968–8977.

Richman, C.M., DeNardo, S.J., O'Grady, L.F., DeNardo, G.L. (1995) Radioimmunotherapy for breast cancer using escalating fractionated doses of ^{131}I-labeled chimeric L6 antibody with peripheral blood progenitor cell transfusions. *Cancer Res* 55(23 Suppl): 5916s–5920s.

Richman, C.M., DeNardo, S.J., O'Donnell, R.T., Yuan, A., Shen, S., Goldstein, D.S., Tuscano, J.M., Wun, T., Chew, H.K., Lara, P.N., Kukis, D.L., Natarajan, A., Meares, C.F., Lamborn, K.R., DeNardo, G.L. (2005) High-dose radioimmunotherapy combined with fixed, low-dose paclitaxel in metastatic prostate and breast cancer by using a MUC-1 monoclonal antibody, m170, linked to indium-111/yttrium-90 via a cathepsin cleavable linker with cyclosporine to prevent human anti-mouse antibody. *Clin Cancer Res* 11: 5920–6927.

Riethmuller, G., Holz, E., Schlimok, G., Schmiegel, W., Raab, R., Hoffken, K., Gruber, R., Funke, I., Pichlmaier, H., Hirche, H., Buggisch, P., Witte, J., Pichlmayr, R. (1998) Monoclonal antibody therapy for resected Dukes' C colorectal cancer: seven-year outcome of a multicenter randomized trial. *J Clin Oncol* 16: 1788–1794.

Rizzieri, D.A., Akabani, G., Zalutsky, M.R., Coleman, R.E., Metzler, S.D., Bowsher, J.E., Toaso, B., Anderson, E., Lagoo, A., Clayton, S., Pegram, C.N., Moore, J.O., Gockerman, J.P., DeCastro, C., Gasparetto, C., Chao, N.J., Bigner, D.D. (2004) Phase 1 trial study of I-131-labeled chimeric 81C6 monoclonal antibody for the treatment of patients with non-Hodgkin lymphoma. *Blood* 104: 642–648.

Rossi, J.F., Fegueux, N., Lu, Z.Y., Legouffe, E., Exbrayat, C., Bozonnat, M.C., Navarro, R., Lopez, E., Quittet, P., Daures, J.P., Rouille, V., Kanouni, T., Widjenes, J., Klein, B. (2005) Optimizing the use of anti-interleukin-6 monoclonal antibody with dexamethasone and 140 mg/m^2 of melphalan in multiple myeloma: results of a pilot study including biological aspects. *Bone Marrow Transplant* 36: 771–779.

Rowinsky, E.K., Schwartz, G.H., Gollob, J.A., Thompson, J.A., Vogelzang, N.J., Figlin, R., Bukowski, R., Haas, N., Lockbaum, P., Li, Y.P., Arends, R., Foon, K.A., Schwab, G., Dutcher, J. (2004) Safety, pharmacokinetics, and activity of ABX-EGF, a fully human anti-epidermal growth factor receptor monoclonal antibody in patients with metastatic renal cell cancer. *J Clin Oncol* 22: 3003–3015.

Sanderson, K., Scotland, R., Lee, P., Liu, D.X., Groshen, S., Snively, J., Sian, S., Nichol, G., Davis, T., Keler, T., Yellin, M., Weber, J. (2005) Autoimmunity in a phase I trial of a fully human anti-cytotoxic T-lymphocyte antigen-4 monoclonal antibody with multiple melanoma peptides and montanide ISA 51 for patients with resected stages III and IV melanoma. *J Clin Oncol* 23: 741–750.

Schlereth, B., Quadt, C., Dreier, T., Kufer, P., Lorenczewski, G., Prang, N., Brandl, C., Lippold, S., Cobb, K., Brasky, K., Leo, E., Bargou, R., Murthy, K., Baeuerle, P.A. (2005) T-cell activation and B-cell depletion in chimpanzees treated with a bispecific anti-CD19/anti-CD3 single-chain antibody construct. *Cancer Immunol Immunother* 1–12.

Schmitt, M., Schmitt, A., Reinhardt, P., Thess, B., Manfras, B., Lindhofer, H., Riechelmann, H., Wiesneth, M., Gronau, S. (2004) Opsonization with a trifunctional bispecific (alphaCD3 x alphaEpCAM) antibody results in efficient lysis in vitro and in vivo of EpCAM positive tumor cells by cytotoxic T lymphocytes. *Int J Oncol* 25: 841–848.

Schneider, C.K., Kalinke, U., Lower, J. (2006) TGN1412 – a regulator's perspective. *Nat Biotechnol* 24: 493–496.

Schnell, R., Vitetta, E., Schindler, J., Barth, S., Winkler, U., Borchmann, P., Hansmann, M.L., Diehl, V., Ghetie, V., Engert, A. (1998) Clinical trials with an anti-CD25 ricin A-chain experimental and immunotoxin (RFT5-SMPT-dgA) in Hodgkin's lymphoma. *Leukemia Lymphoma* 30: 525.

Schultes, B.C., Baum, R.P., Niesen, A., Noujaim, A.A., Madiyalakan, R. (1998) Anti-idiotype induction therapy: anti-CA125 antibodies (Ab3) mediated tumor killing in patients treated with Ovarex mAb B43.13 (Ab1). *Cancer Immunol Immunother* 46: 201–212.

Scott, A.M., Geleick, D., Rubira, M., Clarke, K., Nice, E.C., Smyth, F.E., Stockert, E., Richards, E.C., Carr, F.J., Harris, W.J., Armour, K.L., Rood, J., Kypridis, A., Kronina, V., Murphy, R., Lee, F.T., Liu, Z.Q., Kitamura, K., Ritter, G., Laughton, K., Hoffman, E., Burgess, A.W., Old, L.J. (2000) Construction, production, and characterization of humanized anti-Lewis Y monoclonal antibody 3S193 for targeted immunotherapy of solid tumors. *Cancer Res* 60: 3254–3261.

Scott, A.M., Wiseman, G., Welt, S., Adjei, A., Lee, F.T., Hopkins, W., Divgi, C.R., Hanson, L.H., Mitchell, P., Gansen, D.N., Larson, S.M., Ingle, J.N., Hoffman, E.W., Tanswell, P., Ritter, G., Cohen, L.S., Bette, P., Arvay, L., Amelsberg, A., Vlock, D., Rettig, W.J., Old, L.J. (2003) A phase I dose-escalation study of sibrotuzumab in patients with advanced or metastatic fibroblast activation protein-positive cancer. *Clin Cancer Res* 9: 1639–1647.

Scott, A.M., Lee, F.T., Jones, R., Hopkins, W., MacGregor, D., Cebon, J.S., Hannah, A., Chong, G., U, P., Papenfuss, A., Rigopoulos, A., Sturrock, S., Murphy, R., Wirth, V., Murone, C., Smyth, F.E., Knight, S., Welt, S., Ritter, G., Richards, E., Nice, E.C., Burgess, A.W., Old, L.J. (2005) A phase I trial of humanized monoclonal antibody A33 in patients with colorectal carcinoma: biodistribution, pharmacokinetics, and quantitative tumor uptake. *Clin Cancer Res* 11: 4810–4817.

Seiden, M., Burris, H.A., Matulonis, U., Hall, J., Armstrong, D., Speyer, J., Tillner, J., Weber, D., Muggia, F. (2005) A phase II trial of EMD7(2000) (matuzumab), a humanized anti-EGFR monoclonal antibody in subjects with heavily treated and platinum-resistant advanced Mullerian malignancies. *J Clin Oncol* 23: 229S.

Sharkey, R.M., Hajjar, G., Yeldell, D., Brenner, A., Burton, J., Rubin, A., Goldenberg, D.M. (2005a) A phase I trial combining high-dose 90Y-labeled humanized anti-CEA monoclonal antibody with doxorubicin and peripheral blood stem cell rescue in advanced medullary thyroid cancer. *J Nucl Med* 46: 620–633.

Sharkey, R.M., Karacay, H., Chang, C.H., McBride, W.J., Horak, I.D., Goldenberg, D.M. (2005b) Improved therapy of non-Hodgkin's lymphoma xenografts using radionuclides pretargeted with a new anti-CD20 bispecific antibody. *Leukemia* 19: 1064–1069.

Shen, S., Forero, A., LoBuglio, A.F., Breitz, H., Khazaeli, M.B., Fisher, D.R., Wang, W., Meredith, R.F. (2005) Patient-specific dosimetry of pretargeted radioimmunotherapy using CC49 fusion protein in patients with gastrointestinal malignancies. *J Nucl Med* 46: 642–651.

Siemers, N.O., Kerr, D.E., Yarnold, S., Stebbins, M.R., Vrudhula, V.M., Hellstrom, I., Hellstrom, K.E., Senter, P.D. (1997) Construction, expression, and activities of L49-sFv-beta-lactamase, a single-chain antibody fusion protein for anticancer prodrug activation. *Bioconjug Chem* 8: 510–519.

Simon, T., Hero, B., Faldum, A., Handgretinger, R., Schrappe, M., Niethammer, D., Berthold, F. (2004) Consolidation treatment with chimeric anti-GD2-antibody ch14.18 in children older than 1 year with metastatic neuroblastoma. *J Clin Oncol* 22: 3549–3557.

Simon, T., Hero, B., Faldum, A., Handgretinger, R., Schrappe, M., Niethammer, D., Berthold, F. (2005) Infants with stage 4 neuroblastoma: the impact of the chimeric anti-GD2-antibody ch14.18 consolidation therapy. *Klin Padiatr* 217: 147–152.

Skov, L., Kragballe, K., Zachariae, C., Obitz, E.R., Holm, E.A., Jemec, G.B.E., Solvsten, H., Ibsen, H.H., Knudsen, L., Jensen, P., Petersen, J.H., Menne, T., Baadsgaard, O. (2003) HuMax-CD4 – A fully human monoclonal anti-CD4 antibody for the treatment of psoriasis vulgaris. *Arch Dermatol* 139: 1433–1439.

Spendlove, I., Li, L., Potter, V., Christiansen, D., Loveland, B.E., Durrant, L.G. (2000) A therapeutic human anti-idiotypic antibody mimics CD55 in three distinct regions. *Eur J Immunol* 30: 2944–2953.

Staak, J.O., Colcher, D., Wang, J.Y., Engert, A., Raubitschek, A.A. (2004) Radioimmunotherapy utilizing the humanized anti-CD25 MAb daclizumab is highly effective in mice bearing human Hodgkin's lymphoma xenografts. *Blood* 104: 241B–242B.

Steffens, M.G., Boerman, O.C., De Mulder, P.H., Oyen, W.J., Buijs, W.C., Witjes, J.A., van den Broek, W.J., Oosterwijk-Wakka, J.C., Debruyne, F.M., Corstens, F.H., Oosterwijk, E. (1999) Phase I radioimmunotherapy of metastatic renal cell carcinoma with ^{131}I-labeled chimeric monoclonal antibody G250. *Clin Cancer Res* 5(10 Suppl): 3268s–3274s.

Stein, R., Qu, Z.X., Chen, S., Rosario, A., Shi, V., Hayes, M., Horak, I.D., Hansen, H.J., Goldenberg, D.M. (2004) Characterization of a new humanized anti-CD20 monoclonal antibody, IMMU-106, and its use in combination with the humanized anti-CD22 antibody, epratuzumab, for the therapy of non-Hodgkin's lymphoma. *Clin Cancer Res* 10: 2868–2878.

Stewart, L.M., Young, S., Watson, G., Mather, S.J., Bates, P.A., Band, H.A.,

Wilkinson, R.W., Ross, E.L., Snary, D. (1999) Humanisation and characterisation of PR1A3, a monoclonal antibody specific for cell-bound carcinoembryonic antigen. *Cancer Immunol Immunother* 47: 299–306.

Svensson, H.P., Kadow, J.F., Vrudhula, V.M., Wallace, P.M., Senter, P.D. (1992) Monoclonal antibody-beta-lactamase conjugates for the activation of a cephalosporin mustard prodrug. *Bioconjug Chem* 3: 176–181.

Tai, Y.T., Catley, L.P., Mitsiades, C.S., Burger, R., Podar, K., Shringpaure, R., Hideshima, T., Chauhan, D., Hamasaki, M., Ishitsuka, K., Richardson, P., Treon, S.P., Munshi, N.C., Anderson, K.C. (2004) Mechanisms by which SGN-40, a humanized anti-CD40 antibody, induces cytotoxicity in human multiple myeloma cells: Clinical implications. *Cancer Res* 64: 2846–2852.

Trang, J.M., LoBuglio, A.F., Wheeler, R.H., Harvey, E.B., Sun, L., Ghrayeb, J., Khazaeli, M.B. (1990) Pharmacokinetics of a mouse/human chimeric monoclonal antibody (C-17-1A) in metastatic adenocarcinoma patients. *Pharm Res* 7: 587–592.

Trikha, M., Zhou, Z., Nemeth, J.A., Chen, Q., Sharp, C., Emmell, E., Giles-Komar, J., Nakada, M.T. (2004) CNTO 95, a fully human monoclonal antibody that inhibits alphav integrins, has antitumor and antiangiogenic activity in vivo. *Int J Cancer* 110: 326–335.

Tsavaris, N.B., Katsoulas, H.L., Kosmas, C., Papalambros, E., Gouveris, P., Papantoniou, N., Rokana, S., Kosmas, N., Skopeliti, M., Tsitsilonis, O.E. (2004) The effect of Edrecolomab (Mo17-1A) or fluorouracil-based chemotherapy on specific immune parameters in patients with colorectal cancer – A comparative study. *Oncology* 67: 403–410.

Vallabhajosula, S., Goldsmith, S.J., Kostakoglu, L., Milowsky, M.I., Nanus, D.M., Bander, N.H. (2005) Radioimmunotherapy of prostate cancer using Y-90- and Lu-177-labeled J591 monoclonal antibodies: Effect of multiple treatments on myelotoxicity. *Clin Cancer Res* 11: 7195S–7200S.

Valone, F.H., Kaufman, P.A., Guyre, P.M., Lewis, L.D., Memoli, V., Deo, Y., Graziano, R., Fisher, J.L., Meyer, L., Mrozekorlowski, M., Wardwell, K., Guyre, V., Morley, T.L., Arvizu, C., Fanger, M.W. (1995) Phase Ia/Ib trial of bispecific antibody Mdx-210 in patients with advanced breast or ovarian-cancer that overexpresses the protooncogene Her-2/Neu. *J Clin Oncol* 13: 2281–2292.

Vollmers, H.P., Timmermann, W., Hensel, F., Illert, B., Thiede, A., Muller-Hermelink, H.K. (2000) Adjuvant immunotherapy for stomach carcinoma with the apoptosis-inducing human monoclonal antibody SC-1. *Zentralbl Chir* 125: 37–40.

Wallace, P.K., Romet-Lemonne, J.L., Chokri, M., Kasper, L.H., Fanger, M.W., Fadul, C.E. (2000) Production of macrophage-activated killer cells for targeting of glioblastoma cells with bispecific antibody to Fc gamma RI and the epidermal growth factor receptor. *Cancer Immunol Immunother* 49: 493–503.

Weiner, L.M., Clark, J.I., Davey, M., Li, W.S., Garcia, dP., I., Ring, D.B., Alpaugh, R.K. (1995) Phase I trial of 2B1, a bispecific monoclonal antibody targeting c-erbB-2 and Fc gamma RIII. *Cancer Res* 55: 4586–4593.

Welt, S., Divgi, C.R., Kemeny, N., Finn, R.D., Scott, A.M., Graham, M., Germain, J.S., Richards, E.C., Larson, S.M., Oettgen, H.F. (1994) Phase I/II study of iodine 131-labeled monoclonal antibody A33 in patients with advanced colon cancer. *J Clin Oncol* 12: 1561–1571.

Welt, S., Scott, A.M., Divgi, C.R., Kemeny, N.E., Finn, R.D., Daghighian, F., Germain, J.S., Richards, E.C., Larson, S.M., Old, L.J. (1996) Phase I/II study of iodine 125-labeled monoclonal antibody A33 in patients with advanced colon cancer. *J Clin Oncol* 14: 1787–1797.

Welt, S., Ritter, G., Williams, C., Jr., Cohen, L.S., John, M., Jungbluth, A., Richards, E.A., Old, L.J., Kemeny, N.E. (2003a) Phase I study of anticolon cancer humanized antibody A33. *Clin Cancer Res* 9: 1338–1346.

Welt, S., Ritter, G., Williams, C., Jr., Cohen, L.S., Jungbluth, A., Richards, E.A., Old, L.J., Kemeny, N.E. (2003b) Preliminary report of a phase I study of combination chemotherapy and humanized A33 antibody immunotherapy in patients with advanced colorectal cancer. *Clin Cancer Res* 9: 1347–1353.

Wong, J.Y.C., Williams, L.E., Yamauchi, D.M., OdomMaryon, T., Esteban, J.M., Neumaier, M., Wu, A.M., Johnson, D.K., Primus, F.J., Shively, J.E., Raubitschek, A.A. (1995) Initial experience evaluating (90)yttrium-radiolabeled anticarcinoembryonic antigen chimeric T84.66 in A phase-I radioimmunotherapy trial. *Cancer Res* 55: S5929–S5934.

Wong, J.Y.C., Shibata, S., Williams, L.E., Kwok, C.S., Liu, A., Chu, D.Z., Yamauchi, D.M., Wilczynski, S., Ikle, D.N., Wu, A.M., Yazaki, P.J., Shively, J.E., Doroshow, J.H., Raubitschek, A.A. (2003) A phase I trial of Y-90-anti-carcinoembryonic antigen chimeric T84.66 radioimmunotherapy with 5-fluorouracil in patients with metastatic colorectal cancer. *Clin Cancer Res* 9: 5842–5852.

Wong, J.Y.C., Chu, D.Z., Williams, L.E., Yamauchi, D.M., Ikle, D.N., Kwok, C.S., Liu, A., Wilczynski, S., Colcher, D., Yazaki, P.J., Shively, J.E., Wu, A.M., Raubitschek, A.A. (2004) Pilot trial evaluating an I-123-labeled 80-kilodalton engineered anticarcinoembryonic antigen antibody fragment (cT84.66 minibody) in patients with colorectal cancer. *Clin Cancer Res* 10: 5014–5021.

Xiao, J., Horst, S., Hinkle, G., Cao, X., Kocak, E., Fang, J., Young, D., Khazaeli, M., Agnese, D., Sun, D., Martin, E. Jr. (2005) Pharmacokinetics and clinical evaluation of 125I-radiolabeled humanized CC49 monoclonal antibody (HuCC49deltaC(H)2) in recurrent and metastatic colorectal cancer patients. *Cancer Biother Radiopharm* 20: 16–26.

Yang, J.C., Beck, K.E., Blansfield, J.A., Tran, K.Q., Lowy, I., Rosenberg, S.A. (2005) Tumor regression in patients with metastatic renal cancer treated with a monoclonal antibody to CTLA4 (MDX-010) *J Clin Oncol* 23: 166S.

Younes, M., Georgakis, G.V., Rahmani, M., Beer, D., Younes, A. (2006) Functional expression of TRAIL receptors TRAIL-R1 and TRAIL-R2 in esophageal adenocarcinoma. *Eur J Cancer* 42: 542–547.

Zaki, M.H., Nemeth, J.A., Trikha, M. (2004) CNTO 328, a monoclonal antibody to IL-6, inhibits human tumor-induced cachexia in nude mice. *Int J Cancer* 111: 592–595.

Zang, Z., Xu, H., Yu, L., Yang, D., Xie, S., Shi, Y., Li, Z., Li, J., Wang, J., Li, M., Guo, Y., Gu, F. (2000) Intravesical immunotoxin as adjuvant therapy to prevent the recurrence of bladder cancer. *Chin Med J (Engl)* 113: 1002–1006.

Zbar, A.P., Thomas, H., Wilkinson, R.W., Wadhwa, M., Syrigos, K.N., Ross, E.L., Dilger, P., Allen-Mersh, T.G., Kmiot, W.A., Epenetos, A.A., Snary, D., Bodmer, W.F. (2005) Immune responses in advanced colorectal cancer following repeated intradermal vaccination with the anti-CEA murine monoclonal antibody, PR1A3: results of a phase I study. *Int J Colorectal Dis* 20: 403–414.

Zhang, C.L., Yu, L.Z., Gu, F.L., Buka, S.D., Zu, S.L., Xie, S.S., Pan, Z.Y. (1998) Targeted diagnosis of bladder and ureteral carcinoma using radiolabelled BDI-1. *Urol Res* 26: 343–348.

Zhang, H.F., Li, Y.W., Li, H.L., Bassi, R., Jimenez, X., Witte, L., Bohlen, P., Hicklin, D.J., Zhu, Z.P. (2004) Inhibition of both the autocrine and the paracrine growth of human leukemia with a fully human antibody directed against vascular endothelial growth factor receptor 2. *Leuk Lymphoma* 45: 1887–1897.

Zhou, J., Zeng, F.Q., Li, C., Tong, Q.S., Gao, X., Xie, S.S., Yu, L.Z. (2005) Preparation of arsenic trioxide-loaded albuminutes immuno-nanospheres and its specific killing effect on bladder cancer cell in vitro. *Chin Med J (Engl)* 118: 50–55.

Zhu, Z., Hattori, K., Zhang, H., Jimenez, X., Ludwig, D.L., Dias, S., Kussie, P., Koo, H., Kim, H.J., Lu, D., Liu, M., Tejada, R., Friedrich, M., Bohlen, P., Witte, L., Rafii, S. (2003) Inhibition of human leukemia in an animal model with human antibodies directed against vascular endothelial growth factor receptor 2. Correlation between antibody affinity and biological activity. *Leukemia* 17: 604–611.

13
Antibodies in Phase I/II/III: Targeting TNF

Martin H. Holtmann and Markus F. Neurath

13.1
Introduction

The past decades have dramatically increased our understanding of the pathophysiology of chronic inflammatory bowel disease (IBD). This has mainly been facilitated by the advent of molecular biological and recombinant techniques. The focus on immunological aspects, in particular on T-cell immunology with the analysis of regulatory cytokine signaling pathways, has provided the basis for innovative treatment strategies. Many of these novel strategies are still experimental and subject to clinical studies, while others have already been incorporated into the treatment armamentarium of routine clinical management in IBD. The availability of an increasing number of recombinant proteins in the treatment of IBD is a unique example of translational research from the bedside to the bench and back to the bedside.

Interestingly, while the elucidation of the immunoregulatory pathomechanisms provided the basis for the development of these novel strategies, ongoing studies on the mechanisms of action of these substances in the clinical setting further increase our pathophysiological understanding considerably. This will further promote the development of more effective and more specific treatment strategies.

It is the purpose of this chapter to outline our current pathoimmunological understanding of IBD with special focus on the characterization of those proinflammatory players that have become targets for the rational development of recombinant antagonistic proteins. In particular, in trying to understand the mechanisms of action of the various different antitumor necrosis factor (anti-TNF) strategies, important insights in TNF signaling pathways in IBD and other diseases have been made.

Handbook of Therapeutic Antibodies. Edited by Stefan Dübel
Copyright © 2007 WILEY-VCH Verlag GmbH & Co. KGaA, Weinheim
ISBN 978-3-527-31453-9

13.2
Inflammatory Bowel Disease

Crohn's disease and ulcerative colitis are the two most common forms of IBD and have gained increasing medical and health economic importance during recent decades, especially in Western countries due to their early manifestion in life between 15 and 35 years of age and their chronic character. The etiology of IBD is still unclear and should be considered as multifactorial according to recent studies (Podolsky 2002). Genetic factors seem to play a pathogenetic role as well as environmental, infectious, and immunological factors (Fig. 13.1) (Fiocchi 1998).

Mucosal surfaces such as the intestinal mucosa are special interfaces for the interaction between the organism and the environment and possess an especially adapted immune system (gut-associated lymphoid tissue – GALT) (Nagler-Anderson 2000). Due to its large surface the gut can be regarded as the biggest immune organ of the human body. In the gut the organism is physiologically exposed to a large amount of antigens from the natural flora and the food. Whereas a systemic immune response to these foreign antigens should be prevented in order not to damage the organism, potentially pathogenic antigens have to be recognized and eliminated at a local level. The intestinal epithelium with the mucus represents a mechanical barrier, but it is mainly the role of the GALT to keep this fragile immunological balance between hyporesponsiveness and efficient immune defense.

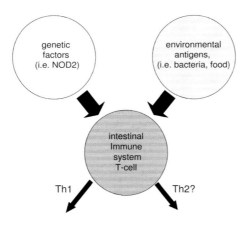

Fig. 13.1 Multifactorial pathogenesis of inflammatory bowel disease (IBD). According to the current paradigm the etiology of IBD is multifactorial. It evolves in genetically susceptibel patients (i.e. NOD2/CARD15), who are exposed to yet unknown environmental antigens (i.e. bacterial or nutritional antigens). This leads to an activation of the intestinal immune system with a pivotal role of T cell-mediated immune responses. While Crohn's disease is prototypic of a Th1-mediated disease, data supporting a Th2-driven pathogenesis of ulcerative colitis are less convincing.

Substantial progress has been made in our understanding of the pathogenesis of IBD in recent years, pursuing the view that IBD could result from disturbances of the intestinal barrier and a pathologic activation of the intestinal immune response towards luminal bacterial and nutritional antigens in a genetically susceptible individual. The main focus of immunologic research in IBD has been on the role and behavior of the T cell and its interaction with other cell populations (Neurath et al. 2002).

13.3
Pathophysiologic Role of T Cells

The acute and chronic inflammation of the bowel goes along with an activation of lamina propria T cells, leading to increased cytokine production (Strober et al. 1998). The cytokine patterns in Crohn's disease and ulcerative colitis show significant differences. In Crohn's disease, $CD4^+$ T cells produce increased amounts of interleukin 12 (IL-12), TNF-α, and interferon gamma (IFN-γ), which would fit to a Th1 cytokine profile. The production of the Th2 cytokines IL-4 and IL-5 in Crohn's disease is decreased. In ulcerative colitis, IL-5 production is increased while IFN-γ production by anti-CD2/CD28-stimulated lamina propria T cells is unchanged compared with control patients (Fuss et al. 1996). IL-13 has recently been identified as a key mediator in experimental ulcerative colitis-like disease (Heller et al. 2005). However, the cytokine profile in ulcerative colitis cannot be classified as a Th2 phenotype without restrictions, since the Th2 cytokine IL-4 is decreased in ulcerative colitis.

The relevance of the Th1–Th2 paradigm has been questioned additionally by the emergence of a third type of $CD4^+$ helper cells termed Th3 cells, mainly producing transforming growth factor beta (TGF-β). In an experimental model, blockade of endogenous TGF-β aggravates colitis, suggesting an anti-inflammatory, protective role of TGF-β (Fuss et al. 2002). $CD4^+CD25^+$ regulatory T cells have also recently been identified as a distinct T-cell population characterized by the production of the anti-inflammatory cytokine IL-10 (Maul et al. 2005).

Numerous experimental animal models, especially knockout or transgenic models, have documented the pathophysiological relevance of distinct cytokine dysregulations (i.e. TNF-α, IL-2, IL-6, IL-10, IL-12, and INF-γ) (Wirtz and Neurath 2000).

13.4
Tumor Necrosis Factor-α

TNF-α is one of the best characterized cytokines in IBD (Holtmann et al. 2002a). The primary translational product of the human TNF-α is the membrane-bound TNF-α (mTNF-α) of 233 amino acids length (26 kDA). Metalloproteinases such as TACE and ADAM 10 cleave the extracellular domain of mTNF-α and thus release soluble TNF-α (sTNF-α) of 157 amino acids (17 kDa). A probably crucial role of mTNF-α signaling via TNF receptor 2 in chronic inflammatory disease states was not recognized until recently, as will be discussed below. Both mTNF-α and sTNF-α form noncovalently linked homotrimers, which occur intracellularly. Membrane-bound TNF-α is capable of inducing signaling in a receptor-independent manner (reverse signaling) (Eissner et al. 2000).

The pathogenic role of TNF-α in inflammatory bowel disease is well established by clinical and experimental studies (Holtmann et al. 2002a). Although serum levels of TNF-α are not elevated significantly in patients with IBD (Nielsen et al. 2000), lamina propria mononuclear cells from colon biopsies of untreated patients with Crohn's disease and ulcerative colitis spontanously produce more TNF-α than cells from controls when cultured *in vitro* (Reinecker et al. 1993).

An important pathogenic mechanism of TNF-α in the mucosa seems to be the stimulation of a Th1 T-cell response. Lamina propria T lymphocytes from colon biopsies of patients with Crohn's disease incubated with TNF-α, produce increased amounts of the Th1 cytokines IL-2, IFNγ, and TNF-α itself (Plevy et al. 1997). The stimulation of TNF-α secretion by TNF-α itself suggests a possible positive feedback mechanism, which could potentially contribute to the perpetuation of inflammation.

Yet another possible mechanism of TNF-α action could be the activation of endogenous matrix metalloproteinases (MMP), which results in damage of the extracellular matrix of the mucosa (Pender et al. 1997).

The critical role of TNF-α for the development of colitis has been reproduced in various established experimental animal models of colitis including the TNBS model (2,4,6-trinitrobenzene sulfonic acid), the DSS model (dextrane sulfate sodium), the IL-10 knockout mouse and the $CD4^+CD45RB^{high}$ respectively $CD4^+CD62L^+$ adoptive transfer model (Kuhn et al. 1993; Atreva et al. 2000; Neurath et al. 2000; Singh et al. 2001).

13.5
TNF Receptors and Signaling

TNF-α exerts its effects via two TNF-specific membrane-bound receptors with a molecular weight of approximately 55–60 kDA (TNF-R1, p55) and 75–80 kDA (TNF-R2, p75) respectively, which belong to the TNF/nerve growth factor (NGF) receptor family (Fig. 13.2) (Loetscher et al. 1991). Both TNF-R1 and TNF-R2 are composed of two identical subunits and are glycosylated. While TNF-R1 and TNF-R2 are quite similar in their extracellular regions (both receptors possess multiple cysteine-rich motifs), their intracellular domains exhibit striking structural differences, most likely reflecting different signaling pathways.

Extracellular ligand binding elicits complex intracellular signaling cascades. Important signaling cascades of TNF-R1 include the activation of kinases, the induction of apoptosis, and the activation of the proinflammatory transcription factor nuclear factor kappa B (NFκB). NFκB is found in the cytosol of nonacti-

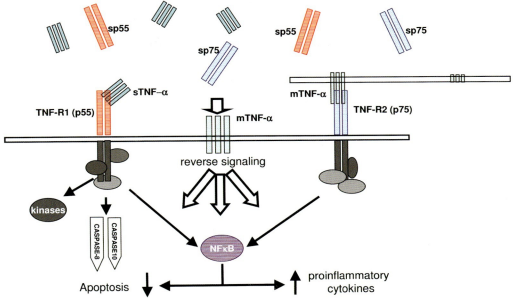

Fig. 13.2 Tumor necrosis factor signaling. TNF-α mediates its effects via two specific cell surface receptors, TNF-R1 (p55) and TNF-R2 (p75). TNF-R1 is mainly activated via soluble TNF-α (sTNF-α), while membrane-bound TNF-α (mTNF-α) is the principal ligand of TNF-R2. Recent data suggest that signaling via mTNF-α/TNF-R2 plays a critical role in the pathogenesis of Crohn's disease. Signaling via TNF-R2 leads to activation of NFκB, a strong proinflammatory transcription factor. Concurrent inhibition of proapoptotic factor may futher promote inflammation. TNF receptors exist in a soluble form generated by shedding of the extracellular portions. These soluble receptor have preserved ligand binding capacity and might thus contribute to the regulation of ligand availability. The finding that mTNF-α is capable of eliciting intracellular signaling without receptor interaction (reverse signaling) adds yet another level of complexity to TNF signaling. Novel recombinant protein-based anti-TNF strategies target different factors of this pathway, some by imitating the mechanism of action of endogenous components, such as soluble receptor p55.

vated monocytes and T cells as an inactive heterodimer of a p50 and p65 subunit, bound to the inhibitory IκB protein. Upon stimulation of the cell IκB is phosphorylated by IκB kinase and consecutively releases NFκB (Auphan et al. 1995). The active subunit of NFκB, p65, moves into the nucleus and directly interacts with the promoter region of several proinflammatory genes such as TNF-α, IL-1, IL-6, and IL-12 (Neurath et al. 2005). Inhibition of NFκB is the underlying mechanism of many established anti-inflammatory treatment strategies. While salicylic acids inhibit IκB kinase (IKK), corticosteroids stimulate IκB synthesis and additionally inhibit p65 in the nucleus by direct complexation (Yin et al. 1998).

For a long time TNF-R1 was considered the principal mediator of TNF signal transduction. This view was mainly based on the fact that TNF-R2 binds sTNF-α with a 20-fold lower binding affinity than TNF-R1 (Grell et al. 1995). However, it was then found that TNF-R2 is preferentially activated by mTNF-α with high affinity in a paracrine fashion and probably in an autocrine loop, too (Haas et al. 1999). The main cellular response of TNF-R2 is the activation of NFκB. These observations make it possible that the mTNF-α/TNF-R2 system could play an important immunoregulatory role at the local level.

In order to assess the net cellular response of TNF signaling it is important to consider that the two principal effects of TNF-α signal transduction, activation of NFκB and induction of apoptosis, can be antagonistic to each other. NFκB inhibits apoptosis through upregulation of antiapoptotic signaling factors (Roth et al. 1995; Wang et al. 1998). It is conceivable that concurrent inhibition of apoptosis in addition to the upregulation of proinflammatory cytokines is an important proinflammatory mechanism of NFκB.

The hypothesis that TNF-α signaling via TNF-R2 may play an independent pathogenic role can be demonstrated impressively in a transgenic mouse model overexpressing two alleles of the human TNF-R2, which can be activated by murine TNF-α just as efficiently as the mouse receptor (Douni and Kollias 1998), These mice spontanously developed a severe general inflammatory syndrome involving pancreas, liver, kidneys, and lungs. NFκB was constitutively increased in mononuclear cells of the peripheral blood.

Strong direct evidence for a crucial role of TNF-R2 signaling in the pathogenesis of Crohn's disease has been provided by recent experimental studies (Holtmann et al. 2002b). In this work, it could be shown that TNF-R2 expression is significantly increased on mononuclear cells in peripheral blood and in the lamina propria of patients with active Crohn's disease. In a murine model system of experimental colitis overexpression of TNF-R2 led to severe aggravation of colitis. This was mediated by induction of a Th1-like cytokine profile of mononuclear cells and by inhibition of apoptosis in the lamina propria. Both pathomechanisms are relevant in Crohn's disease, too. These data contribute to our understanding of the differential clinical efficacy of different anti-TNF strategies (see below).

Elevated levels of soluble TNF-R1 and TNF-R2 can be detected in the urine of patients with Crohn's disease and ulcerative colitis and correlate with high disease indices (Hadziselimovic et al. 1995). Elevated levels of soluble TNF-R2 can also be detected in various severe inflammatory or autoimmune disease states (i.e. sepsis, chronic viral hepatitis, acute pancreatitis, systemic lupus erythematosus, rheumatoid arthritis, and AIDS (Godfried et al. 1993; Schroder et al. 1995; Marinos et al. 1995; de Beaux et al. 1996; Gabay et al. 1997). It is unclear, however, how these elevated levels of soluble TNF-R2 can be explained. They could represent a regulatory mechanism to bind and inactivate soluble TNF-α ligand.

TNF signaling thus represents a complex network of ligands in a membrane-bound and soluble form, two receptors with different ligand affinities and intracellular signaling cascades, soluble forms of receptors that modify the availability of ligand and reverse signaling of the membrane-bound ligand without receptor

13.6
Anti-TNF Antibodies and Fusion Proteins in Clinical Testing

Currently there are four recombinant anti-TNF antibodies and two recombinant fusion proteins available or under clinical testing in IBD, respectively: infliximab, adalimumab, certolizumab, CDP571, etanercept, and onercept. Although they share TNF-antagonizing properties, these drugs display striking differences in clinical efficacy in distinct inflammatory disorders. The elucidation of their different mechanisms of action in distinct clinical settings helps us to learn more about the details and peculiarities of TNF signal transduction in distinct inflammatory disorders.

The different features of TNF-α antagonists are summarized in Table 13.1 and Fig. 13.3.

Table 13.1 Mechanisms of action of anti-TNF strategies.

Substance	Construct	Binding properties	Complement fixation	Antibody-dependent cytotoxicity	Induction of apoptosis	Efficacy in IBD	Efficacy in rheumatological disorders
Infliximab	Chimeric mouse-human IgG1 mAb	sTNF, mTNF	Yes	Yes	Yes	Yes	Yes
Adalimumab	Human IgG1 mAb	sTNF, mTNF	Yes	Yes	Yes	Yes	Yes
Certolizumab (CDP870)	Humanized Fab fragment linked to PEG	sTNF, mTNF	No	No	?	(Yes)	(Yes)
CDP571	Humanized IgG4 mAb	sTNF, mTNF	No	No	?	No	No
Etanercept	IgG1 Fc fragment linked to 2 p75	sTNF	No	No	No	No	Yes
Onercept	Soluble p55	sTNF, mTNF	No	No	?	?	?

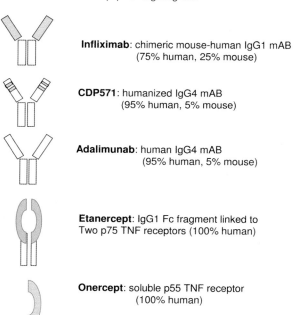

Fig. 13.3 Recombinant anti-TNF strategies. Infliximab was the first recombinant anti-TNF antibody. The residual murine proportions have been considered a major drawback in clinical practise because the mouse epitopes were thought to be responsible for the development of human antichimeric antibodies (HACAs), causing allergic side effects. In an attempt to reduce immunogenicity, humanized antibodies were generated, further reducing the proportions of murine origin (CDP571). Adalimumab is a fully human monoclonal antibody. However, antibodies to CDP571 and adalimumab can still be detected (human anti-human antibodies). Etanercept and onercept represent extracellular portions of p75 and p55 with preserved ligand-binding capacity, thus probably imitating the regulatory role of endogenous soluble receptor. CDP870 distinguished itself by the attachment of two polyethylene glycol molecules, ensuring prolonged continuous release after subcutanous application.

13.6.1
Infliximab

Most clinical experience exists for infliximab (cA2). This is a genetically engineered IgG1 murine–human monoclonal antibody with a constant region of human IgG1κ-immunoglobulin representing 75% of the molecule and a variable region of a monoclonal mouse anti-human antibody representing 25%. Infliximab binds both sTNF-α and mTNF-α and most likely blocks the interaction of TNF-α with the TNF receptors this way (Siegel et al. 1995; Agnholt et al. 2003).

Clinical studies were first performed in rheumatoid arthritis, where good efficacy could be shown (Maini 2004). Other rheumatological disorders followed, including psoriatic arthritis (Antoni et al. 2005), juvenile chronic arthritis (Gerloni et al. 2005), psoriasis (Reich et al. 2005), and ankylosing spondylitis (Marzo-Ortega et al. 2005).

The efficacy of infliximab in the induction and maintenance of remission in patients with both luminal and fistulizing Crohn's disease has been established in several controlled studies (Rutgeerts et al. 1999; Hanauer et al. 2002; Sands et al. 2004). Recent data support the utilization of infliximab in severe refractory ulcerative colitis, too (Rutgeerts et al. 2005; Sandborn et al. 2005a).

The general tolerability of infliximab is good. Most adverse side effects are probably related to the immunogenicity of infliximab that leads to the formation of antibodies to infliximab (ATI) (human antichimeric antibodies, HACA) in 3–17% of the cases (Hanauer et al. 2004) and include infusion reactions (in 4–16%), delayed hypersensitivity-like reactions (in 1%), and autoimmune phenomena. A rare, but severe side effect is the exacerbation of latent infections, especially tuberculosis (Keane et al. 2001; Myers et al. 2002). This reflects the important role of TNF-α in the antimicrobial defense of the immune system. Available long-term safety data provide no evidence for an increased risk for lymphoproliferative disorders or other malignancies (Lichtenstein et al. 2005).

13.6.2
Adalimumab

Adalimumab is a recombinant human IgG1 monoclonal antibody that binds to sTNF-α and mTNF-α and induces apoptosis in monocytes (Shen et al. 2005). A recent, large, phase III study has shown that adalimumab is efficacious and well tolerated in the long-term treatment of rheumatoid arthritis (Weinblatt et al. 2006). Short-term efficacy could also be shown for psoriatic arthritis (Mease et al. 2005). The role of adalimumab in IBD is currently under investigation. Uncontrolled studies showed the clinical efficacy of adalimumab in cases of infliximab intolerability and refractoriness (Sandborn et al. 2004a). This efficacy could be confirmed in a larger controlled study in infliximab-naive patients (Sandborn et al. 2005b,c).

The theoretical advantage of adalimumab with regard to adverse side effects is reduced immunogenicity due to the lack of mouse epitopes. However, development of human anti-human antibodies (anti-adalimumab antibodies, AAAs) is still observed. Side effects include injection-site reactions and infections.

13.6.3
Certolizumab

Certolizumab (CDP870) is a Fab fragment of a humanized monoclonal anti-TNF antibody attached to polyethylene glycol molecules. Although it binds to both sTNF-α and mTNF-α, certolizumab does not induce apoptosis in peripheral blood lymphocytes and monocytes (Fossati and Nesbitt 2005). The clinical efficacy of certolizumab has not yet been established. In a phase II trial, certolizumab showed clinical response after 12 weeks only in a subgroup of patients with elevated C-reactive protein level as sign of systemic inflammation (Schreiber et al. 2005). Results of larger phase III trials have not been published, yet. Adverse events include injection-site reactions, exacerbations of Crohn's disease, and infections.

13.6.4
CDP571

CDP571 (nerelimomab) is a humanized anti-TNF antibody derived from a mouse anti-human TNF-α monoclonal antibody. The complementarity determining region of this antibody has been linked to a human IgG4 antibody. Binding occurs to both sTNF-α and mTNF-α, but no data have been reported regarding induction of apoptosis. CDP571 has shown only temporary marginal efficacy 2 weeks after treatment and in a subgroup of patients with elevated C-reactive protein (Sandborn et al. 2004b). Tolerability and safety data are limited and are comparable to infliximab.

13.6.5
Etanercept

Etanercept is a fusion protein consisting of two identical chains of recombinant human anti-TNF receptor p75 monomers fused to the Fc domain of human IgG1. Etanercept binds only to sTNF-α, but not to mTNF-α. Unlike infliximab and adulimumab, etanercept fails to induce apoptosis in lamina propria T lymphocytes (Van den Brande et al. 2003). Etanercept has shown good efficacy in rheumatoid arthritis and the indication has been extended to many other rheumatoid disorders, including psoriatic arthritis, juvenile chronic arthritis, plaque psoriasis, and ankylosing spondylitis (Moreland et al. 1999; Bathon et al. 2000; Takei et al. 2001; Leonardi et al. 2003; Mease et al. 2004; Davis et al. 2005; Tyring et al. 2006). Interestingly, in Crohn's disease etanercept has failed to show any efficacy (Sandborn et al. 2001).

13.6.6
Onercept

Onercept is a recombinant form of human soluble TNF-R1 (p55). In an initial pilot study on 12 patients with Crohn's disease, onercept showed clinical response beyond placebo (Rutgeerts et al. 2003). This result could not be confirmed in a phase II study (Rugeerts et al. 2004). More importantly, however, phase III trials on onercept in psoriasis had to be discontinued because of two cases of severe sepsis, one of which was lethal.

13.7
Mechanisms of Action

Soon after clinical efficacy of infliximab was shown, the mechanisms of action *in vivo* were investigated. In preclinical studies using a transfection model system with expression of uncleavable mTNF-α, it had been shown that infliximab has a higher binding affinity to mTNF-α than etanercept, while both agents bind sTNF-α (Scallon et al. 1995). These results could be confirmed on activated lamina propria T lymphocytes (Van den Brande et al. 2003). The potential of infliximab and etanercept to neutralize biologically active sTNF-α was similar (Van den Brande et al. 2005).

Since defective apoptosis of mononuclear cells seems to play a role in the pathogenesis of IBD (Boirivant et al. 1999), it was soon hypothesized that the rapid clinical effect of infliximab might be due to induction of apoptosis. And in fact, in Crohn's disease patients treated with infliximab, dose-dependent induction of peripheral blood monocyte apoptosis by a CD95/CD95L-dependent pathway within a few hours could be shown (Lugering et al. 2001). This effect could be imitated by F(ab)$_2$ fragments of infliximab which lack the Fc domain. The finding that infliximab induces increased apoptosis of lamina propria T cells in patients with Crohn's disease is probably even more relevant (ten Hove et al. 2002).

The ability to bind mTNF-α correlates with the inducibility of apoptosis, since only infliximab, but not etanercept induces apoptosis in lamina propria T lymphocytes. Interestingly, induction of apoptosis in cultured peripheral blood monocytes could also be shown for adalimumab, which shares the isotype class IgG1 with infliximab (Shen et al. 2005). In this experimental setting, adalimumab, infliximab and etanercept reduced the levels of sTNF-α in the culture, but only adalimumab and infliximab reduced the production of IL-10 and IL-12.

For infliximab and adalimumab, additional relevant mechanisms of action may be the ability to fix complement and antibody-dependent cytotoxicity. These are features that may be referred to their common subtype class IgG1.

However, binding to mTNF-α does not always lead to blockade of TNF signaling, but can elicit reverse trans-signaling. In an intriguing experimental approach, wildtype and serin-replaced mutant forms of mTNF-α were stably transfected in human Jurkat T cells. Treatment with infliximab, but not with etanercept led to IL-10 production, apoptosis, and G0/G1 cell cycle arrest. These effects were abolished by substitution of all three cytoplasmic serine residues of mTNF-α by alanine residues (Mitoma et al. 2005). This work does not only reveal an additional mechanism of action of infliximab, but also provides insight into a so far unrecognized role of mTNF-α in chronic inflammation.

13.8
Other Anti-TNF Biologicals

CNTO-148 (golimumab) is another fully human anti-TNF-antibody, that can be administered subcutaneously. Clinical testing has been restricted to rheumatoid arthritis so far and preliminary unpublished data suggest statistically significant reductions of signs and symptoms of rheumatoid arthritis. Afelimomab is a monoclonal anti-TNF antibody F(ab')$_2$ fragment. Published data exist from controlled clinical trials in patients with severe sepsis syndrome. Afelimomab slightly reduces 28-day all-cause mortality in a subgroup of patients with sepsis with elevated IL-6 levels (Reinhart et al. 2001; Panecek et al. 2004). No data or press releases, however, are available on these agents in IBD.

13.9
Other Cytokine-based and Anti-CD4$^+$ T-Cell Approaches

Many of the proinflammatory cytokines considered as pathogenetically relevant in IBD have become targets of novel recombinant treatment approaches. Fontolizumab is a humanized anti-IFNγ antibody developed by Protein Design Labs and currently under investigation in phase I and II studies (Harmony I and II) in patients with Crohn's disease. The human IgG1 monoclonal anti-IL-12p40 antibody ABT-874 from Abbott Laboratories has been successfully tested in a controlled phase II study in 76 patients with CD and shows efficacy in remission induction (Mannon et al. 2004). The anti-inflammatory IL-10 (from Schering Plough) has been utilized directly as recombinant protein delivered subcutaneously in three large phase III trials comprising 800 patients, but has failed to show any effect (van Deventer et al. 1997; Fedorak et al. 2000; Schreiber et al. 2005). Phase II trials with IL-11 (from Genetics Institute) were discontinued, too. MRA is a humanized anti-IL-6 receptor antibody from Roche currently tested in a phase II trial in Crohn's disease.

The pivotal role of the CD4$^+$ T-helper cell in the pathogenesis of IBD is also taken into account in treatment strategies that attempt to downregulate T-cell recruitment to the inflamed focus by blocking adhesion molecules. Early studies blocking the intercellular adhesion molecule 1 (ICAM-1) with an antisense oligonucleotide showed no clinical efficacy (Schreiber et al. 2001).

A monoclonal antibody approach was pursued with natalizumab, a humanized IgG4 antibody to α_4 integrin (Fig. 13.4). While effective in multiple sclerosis, which is T-cell mediated, too, natalizumab failed to show any clinically relevant efficacy in several large studies in active Crohn's disease (Gordon et al. 2001; Ghosh et al. 2003; Miller et al. 2003; Sandborn et al. 2005d). Of note, three cases of JC virus-associated progressive multifocal leukoencephalopathy under natalizumab treatment have been reported (Kleinschmidt-DeMasters and Tyler 2005; Van Assche et al. 2005).

MLN-02 is another humanized IgG1 monoclonal anti-integrin antibody directed against the heterodimeric epitope $\alpha_4\beta_7$ (Fig. 13.4). This antibody showed efficacy in ulcerative colitis (Feagan et al. 2005). Concerns regarding the risk of polymorphonuclear leukocytes under MLN-02 treatment have been allayed by the following considerations. Natalizumab blocks specifically the α_4 subunit of integrins that is incorporated in both the $\alpha_4\beta_1$ heterodimer that interacts with VCAM-1 in

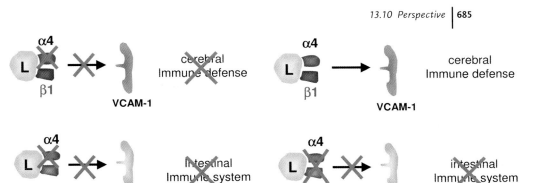

Natalizumab = anti-α4

Fig. 13.4 Differential mechanisms of action of anti-integrins. Inhibition of T-cell homing is a therapeutic approach to decrease inflammation. Migration of T cells is mediated by distinct adhesion molecules which interact specifically. T cells can pass the blood–brain barrier through interaction of the $\alpha_4\beta_1$ integrin heterodimer with the adhesion molecule VCAM-1. For homing into the gut, T cells expressing the heterodimer

MLN02 = anti-α4β7

$\alpha_4\beta_7$ interact with the adhesion molecule MAdCAM-1, which is specifically expressed in the gut. The monoclonal anti-integrin antibody natalizumab blocks the α subunit and thus inhibits homing of T cells to both the brain and the gut. The antigenetic epitope of MLN-02 is the heterodimer $\alpha_4\beta_7$ as a whole. MLN-02 thus blocks homing of T cells to the gut, but not to the brain.

the brain and the $\alpha_4\beta_7$ heterodimer that interacts with MAdCAM-1, specifically expressed in the gut. Natalizumab thus interferes with T cell homing in both the brain and the gut. Being directed against the heterodimer $\alpha_4\beta_7$ as a whole, MLN-02 only blocks MAdCAM-1-mediated T-cell homing in the gut, thus preventing the negative impact on immune defense in the brain. It remains to be seen if these theoretical considerations really apply in clinical practice in the long run.

13.10
Perspective

It is striking that only 50–70% of all patients respond clinically to distinct anti-cytokine strategies such as anti-TNF-α or anti-IL-12 antibodies given the pivotal role of these cytokines in murine models of colitis. In fact, there is increasing evidence that the classic Th1-Th2 paradigm derived from the focus on T-cell immunology may be too simplistic to explain IBD.

Two considerations may be relevant. Crohn's disease and ulcerative colitis may represent common final pathophysiological pathways of different etiological triggers. These different pathomechanisms may involve different signaling cascades. In addition, individual cytokines may play different roles in different phases of the disease. In particular Crohn's disease is characterized by two distinct phases, an initial, inductive phase and a chronic, effector phase.

Increasing evidence suggests that for the inductive phase, cell components of the innate immune system, such as monocytes/macrophages, play a predominant role (Kelsall and Strober 1999; Berrebi et al. 2003). Monocytes/macrophages are important antigen presentation cells and produce proinflammatory cytokines such as IL-1, TNF-α, IL-6, and possibly IL-12 and IL-18. The first Crohn's disease susceptibility gene codes for the intracellular protein NOD2/CARD15, which represents an intracellular receptor for bacterial products, in particular muramyl dipeptide, strongly expressed in monocytes/macrophages and epithelial cells (Hugot et al. 2001; Ogura et al. 2001; Hisamatsu et al. 2003). *In vitro* studies on loss of function mutations of NOD2 revealed that wildtype NOD2/CARD15 activates NFκB.

This apparently paradoxical finding has led to the view that in Crohn's disease the initial response of the innate immune system to antigens of the physiological intestinal flora or the food may be defective, resulting in a detrimental activation of the acquired immune system with chronic inflammation. In this phase dysregulation of T cells and T cell-associated cytokine regulatory networks play the crucial role. And in fact, in a murine model of a loss-of-function mutation of NOD2, elevated NFκB activation in response to muramyl dipeptide could be shown with increased susceptibility to bacterial-induced inflammation (Kobayashi et al. 2005; Maeda et al. 2005).

It is thus conceivable that the blockade of a certain cytokine – although a key proinflammatory cytokine in colitis models – could be beneficial under certain circumstances, but ineffective or even detrimental otherwise. This is likely to apply not only to the targets of anticytokine strategies currently available, but also for all potential future targets such as IL-13, IL-18, IL-23, IL-27, and others. For medical and cost-effectiveness reasons it will be challenging to identify predictive markers for the responsiveness of an individual patient to a distinct cytokine-targeted treatment strategy.

With regard to safety, the serious adverse events following natalizumab treatment and the risk of exacerbation of silent infections and sepsis under anti-TNF-α treatment should alert to the two faces of Janus inherent in T cell and T cell-associated cytokine-targeted strategies.

References

Agnholt, J., Dahlerup, J.F., Kaltoft, K. (2003) The effect of etanercept and infliximab on the production of tumour necrosis factor alpha, interferon-gamma and GM-CSF in in vivo activated intestinal T lymphocyte cultures. *Cytokine* 23: 76–85.

Antoni, C.E., Kavanaugh, A., Kirkham, B., Tutuncu, Z., Burmester, G.R., Schneider, U., Furst, D.E., Molitor, J., Keystone, E., Gladman, D., Manger, B., Wassenberg, S., Weier, R., Wallace, D.J., Weisman, M.H., Kalden, J.R., Smolen, J. (2005) Sustained benefits of infliximab therapy for dermatologic and articular manifestations of psoriatic arthritis: results from the infliximab multinational psoriatic arthritis controlled trial (IMPACT). *Arthritis Rheum* 52: 1227–1236.

Atreya, R., Mudter, J., Finotto, S., Mullberg, J., Jostock, T., Wirtz, S., Schutz, M., Bartsch, B., Holtmann, M., Becker, C., Strand, D., Czaja, J., Schlaak, J.F., Lehr, H.A., Autschbach, F., Schurmann, G., Nishimoto, N., Yoshizaki, K., Ito, H., Kishimoto, T., Galle, P.R., Rose-John, S., Neurath, M.F. (2000) Blockade of interleukin 6 trans signaling suppresses T-cell resistance against apoptosis in chronic intestinal inflammation: evidence in Crohn disease and experimental colitis in vivo. *Nat Med* 6: 583–588.

Auphan, N., DiDonato, J.A., Rosette, C., Helmberg, A., Karin, M. (1995) Immunosuppression by glucocorticoids: inhibition of NF-kappa B activity through induction of I kappa B synthesis. *Science* 270: 286–290.

Bathon, J.M., Martin, R.W., Fleischmann, R.M., Tesser, J.R., Schiff, M.H., Keystone, E.C., Genovese, M.C., Wasko, M.C., Moreland, L.W., Weaver, A.L., Markenson, J., Finck, B.K. (2000) A comparison of etanercept and methotrexate in patients with early rheumatoid arthritis. *N Engl J Med* 343: 1586–1593.

Berrebi, D., Maudinas, R., Hugot, J.P., Chamaillard, M., Chareyre, F., De Lagausie, P., Yang, C., Desreumaux, P., Giovannini, M., Cezard, J.P., Zouali, H., Emilie, D., Peuchmaur, M. (2003) Card15 gene overexpression in mononuclear and epithelial cells of the inflamed Crohn's disease colon. *Gut* 52: 840–846.

Boirivant, M., Marini, M., Di Felice, G., Pronio, A.M., Montesani, C., Tersigni, R., Strober, W. (1999) Lamina propria T cells in Crohn's disease and other gastrointestinal inflammation show defective CD2 pathway-induced apoptosis. *Gastroenterology* 116: 557–565.

Davis, J.C., van der Heijde, D.M., Braun, J., Dougados, M., Cush, J., Clegg, D., Inman, R.D., Kivitz, A., Zhou, L., Solinger, A., Tsuji, W. (2005) Sustained durability and tolerability of etanercept in ankylosing spondylitis for 96 weeks. *Ann Rheum Dis* 64: 1557–1562.

de Beaux, A.C., Goldie, A.S., Ross, J.A., Carter, D.C., Fearon, K.C. (1996) Serum concentrations of inflammatory mediators related to organ failure in patients with acute pancreatitis. *Br J Surg* 83: 349–353.

Douni, E., Kollias, G. (1998) A critical role of the p75 tumor necrosis factor receptor (p75TNF-R) in organ inflammation independent of TNF, lymphotoxin alpha, or the p55TNF-R. *J Exp Med* 188: 1343–1352.

Eissner, G., Kirchner, S., Lindner, H., Kolch, W., Janosch, P., Grell, M., Scheurich, P., Andreesen, R., Holler, E. (2000) Reverse signaling through transmembrane TNF confers resistance to lipopolysaccharide in human monocytes and macrophages. *J Immunol* 164: 6193–6198.

Feagan, B.G., Greenberg, G.R., Wild, G., Fedorak, R.N., Pare, P., McDonald, J.W., Dube, R., Cohen, A., Steinhart, A.H., Landau, S., Aguzzi, R.A., Fox, I.H., Vandervoort, M.K. (2005) Treatment of ulcerative colitis with a humanized antibody to the alpha4beta7 integrin. *N Engl J Med* 352: 2499–2507.

Fedorak, R.N., Gangl, A., Elson, C.O., Rutgeerts, P., Schreiber, S., Wild, G., Hanauer, S.B., Kilian, A., Cohard, M., LeBeaut, A., Feagan, B. (2000) Recombinant human interleukin 10 in the treatment of patients with mild to moderately active Crohn's disease. The Interleukin 10 Inflammatory Bowel Disease Cooperative Study Group. *Gastroenterology* 119: 1473–1482.

Fiocchi, D. (1998) Inflammatory bowel disease: etiology and pathogenesis. *Gastroenterology* 115: 182–205.

Fossati, G., Nesbitt, A.M. (2005) Effect of the anti-TNF agents, adalimumab, etanercept, infliximab, and certolizumab PEGOL (CDP870) on the induction of apoptosis in activated peripheral blood lymphocytes and monocytes (abstract). *Am J Gastroenterol* 100: S298.

Fuss, I.J., Neurath, M., Boirivant, M., Klein, J.S., de la Motte, C., Strong, S.A., Fiocchi, C., Strober, W. (1996) Disparate CD4+ lamina propria (LP) lymphokine secretion profiles in inflammatory bowel disease. Crohn's disease LP cells manifest increased secretion of IFN-gamma, whereas ulcerative colitis LP cells manifest increased secretion of IL-5. *J Immunol* 157: 1261–1270.

Fuss, I.J., Boirivant, M., Lacy, B., Strober, W. (2002) The interrelated roles of TGF-beta and IL-10 in the regulation of experimental colitis. *J Immunol* 168: 900–908.

Gabay, C., Cakir, N., Moral, F., Roux-Lombard, P., Meyer, O., Dayer, J.M., Vischer, T., Yazici, H., Guerne, P.A. (1997) Circulating levels of tumor necrosis factor soluble receptors in systemic lupus erythematosus are significantly higher than in other rheumatic diseases and correlate with disease activity. *J Rheumatol* 24: 303–308.

Gerloni, V., Pontikaki, I., Gattinara, M., Desiati, F., Lupi, E., Lurati, A., Salmaso, A., Fantini, F. (2005) Efficacy of repeated intravenous infusions of an anti-tumor necrosis factor alpha monoclonal antibody, infliximab, in persistently active, refractory juvenile idiopathic arthritis: results of an open-label prospective study. *Arthritis Rheum* 52: 548–553.

Ghosh, S., Goldin, E., Gordon, F.H., Malchow, H.A., Rask-Madsen, J., Rutgeerts, P., Vyhnalek, P., Zadorova, Z., Palmer, T., Donoghue, S. (2003) Natalizumab for active Crohn's disease. *N Engl J Med* 348: 24–32.

Godfried, M.H., van der Poll, T., Jansen, J., Romijn, J.A., Schattenkerk, J.K., Endert, E., van Deventer, S.J., Sauerwein, H.P. (1993) Soluble receptors for tumour necrosis factor: a putative marker of disease progression in HIV infection. *Aids* 7: 33–36.

Gordon, F.H., Lai, C.W., Hamilton, M.I., Allison, M.C., Srivastava, E.D., Fouweather, M.G., Donoghue, S., Greenlees, C., Subhani, J., Amlot, P.L., Pounder, R.E. (2001) A randomized placebo-controlled trial of a humanized monoclonal antibody to alpha4 integrin in active Crohn's disease. *Gastroenterology* 121: 268–274.

Grell, M., Douni, E., Wajant, H., Lohden, M., Clauss, M., Maxeiner, B., Georgopoulos, S., Lesslauer, W., Kollias, G., Pfizenmaier, K., Scheurich, P. (1995) The transmembrane form of tumor necrosis factor is the prime activating ligand of the 80 kDA tumor necrosis factor receptor. *Cell* 83: 793–802.

Haas, E., Grell, M., Wajant, H., Scheurich, P. (1999) Continuous autotropic signaling by membrane-expressed tumor necrosis factor. *J Biol Chem* 274: 18107–18112.

Hadziselimovic, F., Emmons, L.R., Gallati, H. (1995) Soluble tumour necrosis factor receptors p55 and p75 in the urine monitor disease activity and the efficacy of treatment of inflammatory bowel disease. *Gut* 37: 260–263.

Hanauer, S.B., Feagan, B.G., Lichtenstein, G.R., Mayer, L.F., Schreiber, S., Colombel, J.F., Rachmilewitz, D., Wolf, D.C., Olson, A., Bao, W., Rutgeerts, P. (2002) Maintenance infliximab for Crohn's disease: the ACCENT I randomised trial. *Lancet* 359: 1541–1549.

Hanauer, S.B., Wagner, C.L., Bala, M., Mayer, L., Travers, S., Diamond, R.H., Olson, A., Bao, W., Rutgeerts, P. (2004) Incidence and importance of antibody responses to infliximab after maintenance or episodic treatment in Crohn's disease. *Clin Gastroenterol Hepatol* 2: 542–553.

Heller, F., Florian, P., Bojarski, C., Richter, J., Christ, M., Hillenbrand, B., Mankertz, J., Gitter, A.H., Burgel, N., Fromm, M., Zeitz, M., Fuss, I., Strober, W., Schulzke, J.D. (2005) Interleukin-13 is the key effector Th2 cytokine in ulcerative colitis that affects epithelial tight junctions, apoptosis, and cell restitution. *Gastroenterology* 129: 550–564.

Hisamatsu, T., Suzuki, M., Reinecker, H.C., Nadeau, W.J., McCormick, B.A., Podolsky, D. (2003) CARD15/NOD2 functions as an anti-bacterial factor i human intestinal epithelial cells. *Gastroenterology* 124: 993–1000.

Holtmann, M.H., Schutz, M., Galle, P.R., Neurath, M.F. (2002a) Functional relevance of soluble TNF-alpha, transmembrane TNF-alpha and TNF-signal transduction in gastrointestinal diseases with special reference to inflammatory bowel diseases. *Z Gastroenterol* 40: 587–600.

Holtmann, M.H., Douni, E., Schutz, M., Zeller, G., Mudter, J., Lehr, H.A., Gerspach, J., Scheurich, P., Galle, P.R., Kollias, G., Neurath, M.F. (2002b) Tumor necrosis factor-receptor 2 is up-regulated on lamina propria T cells in Crohn's disease and promotes experimental colitis in vivo. *Eur J Immunol* 32: 3142–3151.

Hugot, J.P., Chamaillard, M., Zouali, H., Lesage, S., Cezard, J.P., Belaiche, J., Almer, S., Tysk, C., O'Morain, C.A., Gassull, M.,

Binder, V., Finkel, Y., Cortot, A., Modigliani, R., Laurent-Puig, P., Gower-Rousseau, C., Macry, J., Colombel, J. F., Sahbatou, M., Thomas, G. (2001) Association of NOD2 leucine-rich repeat variants with susceptibility to Crohn's disease. *Nature* 411: 599–603.

Keane, J., Gershon, S., Wise, R.P., Mirabile-Levens, E., Kasznica, J., Schwieterman, W.D., Siegel, J.N., Braun, M.M. (2001) Tuberculosis associated with infliximab, a tumor necrosis factor alpha-neutralizing agent. *N Engl J Med* 345: 1098–1104.

Kelsall, B., Strober, W. (1999) *Gut-Associated Lymphoid Tissue: Antigen Handling and T Lymphocyte Responses.* San Diego: Academic Press, 1999.

Kleinschmidt-DeMasters, B.K., Tyler, K.L. (2005) Progressive multifocal leukoencephalopathy complicating treatment with natalizumab and interferon beta-1a for multiple sclerosis. *N Engl J Med* 353: 369–374.

Kobayashi, K.S., Chamaillard, M., Ogura, Y., Henegariu, O., Inohara, N., Nunez, G., Flavell, R.A. (2005) Nod2-dependent regulation of innate and adaptive immunity in the intestinal tract. *Science* 307: 731–734.

Kuhn, R., Lohler, J., Rennick, D., Rajewsky, K., Muller, W. (1993) Interleukin-10-deficient mice develop chronic enterocolitis. *Cell* 75: 263–274.

Leonardi, C.L., Powers, J.L., Matheson, R.T., Goffe, B.S., Zitnik, R., Wang, A., Gottlieb, A.B. (2003) Etanercept as monotherapy in patients with psoriasis. *N Engl J Med* 349: 2014–2022.

Lichtenstein, G.R., Feagan, B.G., Cohen, R.D., Salzberg, B.A., Diamond, R.H., Chen, D.M., Pritchard, M.L., Sandborn, W.J. (2005) Safety of infliximab and other Crohn's disease therapies – updated TREAT registry data with over 10,000 patient-years of follow-up (abstract). *Gastroenterology* 128: A-580.

Loetscher, H., Brockhaus, M., Dembic, Z., Gentz, R., Gubler, U., Hohmann, H.P., Lahm, H.W., Van Loon, A.P., Pan, Y.C., Schlaeger, E.J., et al. (1991) Two distinct tumour necrosis factor receptors – members of a new cytokine receptor gene family. *Oxf Surv Eukaryot Genes* 7: 119–142.

Lugering, A., Schmidt, M., Lugering, N., Pauels, H.G., Domschke, W., Kucharzik, T. (2001) Infliximab induces apoptosis in monocytes from patients with chronic active Crohn's disease by using a caspase-dependent pathway. *Gastroenterology* 121: 1145–1157.

Maeda, S., Hsu, L.C., Liu, H., Bankston, L.A., Iimura, M., Kagnoff, M.F., Eckmann, L., Karin, M. (2005) Nod2 mutation in Crohn's disease potentiates NF-kappaB activity and IL-1beta processing. *Science* 307: 734–738.

Maini, S.R. (2004) Infliximab treatment of rheumatoid arthritis. *Rheum Dis Clin North Am* 30: 329–347, vii.

Mannon, P.J., Fuss, I.J., Mayer, L., Elson, C.O., Sandborn, W.J., Present, D., Dolin, B., Goodman, N., Groden, C., Hornung, R.L., Quezado, M., Yang, Z., Neurath, M.F., Salfeld, J., Veldman, G.M., Schwertschlag, U., Strober, W. (2004) Anti-interleukin-12 antibody for active Crohn's disease. *N Engl J Med* 351: 2069–2079.

Marinos, G., Naoumov, N.V., Rossol, S., Torre, F., Wong, P.Y., Gallati, H., Portmann, B., Williams, R. (1995) Tumor necrosis factor receptors in patients with chronic hepatitis B virus infection. *Gastroenterology* 108: 1453–1463.

Marzo-Ortega, H., McGonagle, D., Jarrett, S., Haugeberg, G., Hensor, E., O'Connor, P., Tan, A.L., Conaghan, P.G., Greenstein, A., Emery, P. (2005) Infliximab in combination with methotrexate in active ankylosing spondylitis: a clinical and imaging study. *Ann Rheum Dis* 64: 1568–1575.

Maul, J., Loddenkemper, C., Mundt, P., Berg, E., Giese, T., Stallmach, A., Zeitz, M., Duchmann, R. (2005) Peripheral and intestinal regulatory $CD4^+$ $CD25^{high}$ T cells in inflammatory bowel disease. *Gastroenterology* 128: 1868–1878.

Mease, P.J., Kivitz, A.J., Burch, F.X., Siegel, E.L., Cohen, S.B., Ory, P., Salonen, D., Rubenstein, J., Sharp, J.T., Tsuji, W. (2004) Etanercept treatment of psoriatic arthritis: safety, efficacy, and effect on disease progression. *Arthritis Rheum* 50: 2264–2272.

Mease, P.J., Gladman, D.D., Ritchlin, C.T., Ruderman, E.M., Steinfeld, S.D., Choy, E.H., Sharp, J.T., Ory, P.A., Perdok, R.J.,

Weinberg, M.A. (2005) Adalimumab for the treatment of patients with moderately to severely active psoriatic arthritis: results of a double-blind, randomized, placebo-controlled trial. *Arthritis Rheum* 52: 3279–3289.

Miller, D.H., Khan, O.A., Sheremata, W.A., Blumhardt, L.D., Rice, G.P., Libonati, M.A., Willmer-Hulme, A.J., Dalton, C.M., Miszkiel, K.A., O'Connor, P.W. (2003) A controlled trial of natalizumab for relapsing multiple sclerosis. *N Engl J Med* 348: 15–23.

Mitoma, H., Horiuchi, T., Hatta, N., Tsukamoto, H., Harashima, S., Kikuchi, Y., Otsuka, J., Okamura, S., Fujita, S., Harada, M. (2005) Infliximab induces potent anti-inflammatory responses by outside-to-inside signals through transmembrane TNF-alpha. *Gastroenterology* 128: 376–392.

Moreland, L.W., Schiff, M.H., Baumgartner, S.W., Tindall, E.A., Fleischmann, R.M., Bulpitt, K.J., Weaver, A.L., Keystone, E.C., Furst, D.E., Mease, P.J., Ruderman, E.M., Horwitz, D.A., Arkfeld, D.G., Garrison, L., Burge, D.J., Blosch, C.M., Lange, M.L., McDonnell, N.D., Weinblatt, M.E. (1999) Etanercept therapy in rheumatoid arthritis. A randomized, controlled trial. *Ann Intern Med* 130: 478–486.

Myers, A., Clark, J., Foster, H. (2002) Tuberculosis and treatment with infliximab. *N Engl J Med* 346: 623–626.

Nagler-Anderson, C. (2000) Tolerance and immunity in the intestinal immune system. *Crit Rev Immunol* 20: 103–120.

Neurath, A.R., Strick, N., Li, Y.Y., Debnath, A.K. (2005) *Punica granatum* (pomegranate) juice provides an HIV-1 entry inhibitor and candidate topical microbicide. *Ann NY Acad Sci* 1056: 311–327.

Neurath, M., Fuss, I., Strober, W. (2000) TNBS-colitis. *Int Rev Immunol* 19: 51–62.

Neurath, M.F., Finotto, S., Glimcher, L.H. (2002) The role of Th1/Th2 polarization in mucosal immunity. *Nat Med* 8: 567–573.

Nielsen, O.H., Vainer, B., Madsen, S.M., Seidelin, J.B., Heegaard, N.H. (2000) Established and emerging biological activity markers of inflammatory bowel disease. *Am J Gastroenterol* 95: 359–367.

Ogura, Y., Bonen, D.K., Inohara, N., Nicolae, D.L., Chen, F.F., Ramos, R., Britton, H., Moran, T., Karaliuskas, R., Duerr, R.H., Achkar, J.P., Brant, S.R., Bayless, T.M., Kirschner, B.S., Hanauer, S.B., Nunez, G., Cho, J.H. (2001) A frameshift mutation in NOD2 associated with susceptibility to Crohn's disease. *Nature* 411: 603–606.

Panacek, E.A., Marshall, J.C., Albertson, T.E., Johnson, D.H., Johnson, S., MacArthur, R.D., Miller, M., Barchuk, W.T., Fischkoff, S., Kaul, M., Teoh, L., Van Meter, L., Daum, L., Lemeshow, S., Hicklin, G., Doig, C. (2004) Efficacy and safety of the monoclonal anti-tumor necrosis factor antibody F(ab')$_2$ fragment afelimomab in patients with severe sepsis and elevated interleukin-6 levels. *Crit Care Med* 32: 2173–2182.

Pender, S.L., Tickle, S.P., Docherty, A.J., Howie, D., Wathen, N.C., MacDonald, T.T. (1997) A major role for matrix metalloproteinases in T cell injury in the gut. *J Immunol* 158: 1582–1590.

Plevy, S.E., Landers, C.J., Prehn, J., Carramanzana, N.M., Deem, R.L., Shealy, D., Targan, S.R. (1997) A role for TNF-alpha and mucosal T helper-1 cytokines in the pathogenesis of Crohn's disease. *J Immunol* 159: 6276–6282.

Podolsky, D. (2002) Inflammatory bowel disease. *N Engl J Med* 347: 417–429.

Reich, K., Nestle, F.O., Papp, K., Ortonne, J.P., Evans, R., Guzzo, C., Li, S., Dooley, L.T., Griffiths, C.E. (2005) Infliximab induction and maintenance therapy for moderate-to-severe psoriasis: a phase III, multicentre, double-blind trial. *Lancet* 366: 1367–1374.

Reinecker, H.C., Steffen, M., Witthoeft, T., Pflueger, I., Schreiber, S., MacDermott, R.P., Raedler, A. (1993) Enhanced secretion of tumour necrosis factor-alpha, IL-6, and IL-1 beta by isolated lamina propria mononuclear cells from patients with ulcerative colitis and Crohn's disease. *Clin Exp Immunol* 94: 174–181.

Reinhart, K., Menges, T., Gardlund, B., Harm Zwaveling, J., Smithes, M., Vincent, J.L., Tellado, J.M., Salgado-Remigio, A., Zimlichman, R., Withington, S., Tschaikowsky, K., Brase, R., Damas, P., Kupper, H., Kempeni, J., Eiselstein, J.,

Kaul, M. (2001) Randomized, placebo-controlled trial of the anti-tumor necrosis factor antibody fragment afelimomab in hyperinflammatory response during severe sepsis: The RAMSES Study. *Crit Care Med* 29: 765–769.

Rothe, M., Sarma, V., Dixit, V.M., Goeddel, D.V. (1995) TRAF2-mediated activation of NF-kappa B by TNF receptor 2 and CD40. *Science* 269: 1424–1427.

Rutgeerts, P., D'Haens, G., Targan, S., Vasiliauskas, E., Hanauer, S.B., Present, D.H., Mayer, L., Van Hogezand, R.A., Braakman, T., DeWoody, K.L., Schaible, T.F., Van Deventer, S.J. (1999) Efficacy and safety of retreatment with anti-tumor necrosis factor antibody (infliximab) to maintain remission in Crohn's disease. *Gastroenterology* 117: 761–769.

Rutgeerts, P., Lemmens, L., Van Assche, G., Noman, M., Borghini-Fuhrer, I., Goedkoop, R. (2003) Treatment of active Crohn's disease with onercept (recombinant human soluble p55 tumour necrosis factor receptor): results of a randomized, open-label, pilot study. *Aliment Pharmacol Ther* 17: 185–192.

Rutgeerts, P., Fedorak, R.N., Rachmilewitz, D., Tarabar, D., Gibson, P., Nielsen, O.H., Wild, G., Schreiber, S., Zignani, M. (2004) Onercept (recombinant human soluble p55 tumour necrosis factor receptor) treatment in patients with active Crohn's disease: randomized, placebo-controlled, dose-finding phase II study (abstract). *Gut* 53: A47.

Rutgeerts, P., Feagan, B.G., Olson, A., Johanns, J., Travers, S.B., Present, D.H., Sands, B.E., Sandborn, W.J. (2005) A randomized placebo-controlled trial of infliximab therapy for active ulcerative colitis: act i trial (abstract). *Gastroenterology* 128: A-105.

Sandborn, W.J., Hanauer, S.B., Katz, S., Safdi, M., Wolf, D.G., Baerg, R.D., Tremaine, W.J., Johnson, T., Diehl, N.N., Zinsmeister, A.R. (2001) Etanercept for active Crohn's disease: a randomized, double-blind, placebo-controlled trial. *Gastroenterology* 121: 1088–1094.

Sandborn, W.J., Hanauer, S., Loftus, E.V., Jr., Tremaine, W.J., Kane, S., Cohen, R., Hanson, K., Johnson, T., Schmitt, D., Jeche, R. (2004a) An open-label study of the human anti-TNF monoclonal antibody adalimumab in subjects with prior loss of response or intolerance to infliximab for Crohn's disease. *Am J Gastroenterol* 99: 1984–1989.

Sandborn, W.J., Feagan, B.G., Radford-Smith, G., Kovacs, A., Enns, R., Innes, A., Patel, J. (2004b) CDP571, a humanised monoclonal antibody to tumour necrosis factor alpha, for moderate to severe Crohn's disease: a randomised, double blind, placebo controlled trial. *Gut* 53: 1485–1493.

Sandborn, W.J., Rachmilewitz, D., Hanauer, S.B., Lichtenstein, G.R., Villiers, W.J., Olson, A., Johanns, J., Travers, S.B., Colombel, J.F. (2005a) Infliximab induction and maintenance therapy for ulcerative colitis: the act 2 trial (abstract). *Gastroenterology* 128: A-104.

Sandborn, W.J., Hanauer, S.B., Lukas, M., Wolf, D.C., Isaacs, K.L., MacIntosh, D., Panaccione, R. (2005b) Maintenance of remission over 1 year in patients with active Crohn's disease treated with adalimumab: results of a blinded, placebo-controlled study (abstract). *Am J Gastroenterol* 100: S311.

Sandborn, W.J., Hanauer, S.B., Lukas, M., Wolf, D.C., Isaacs, K.L., MacIntosh, D., Panaccione, R., Rutgeerts, P., Pollack, P. (2005c) Remission and clinical response induced and maintained in patients with active Crohn's disease treated for 1-year open-label with adalimumab. *Am J Gastroenterol* 100: S316.

Sandborn, W.J., Colombel, J.F., Enns, R., Feagan, B.G., Hanauer, S.B., Lawrance, I.C., Panaccione, R., Sanders, M., Schreiber, S., Targan, S., van Deventer, S., Goldblum, R., Despain, D., Hogge, G.S., Rutgeerts, P. (2005d) Natalizumab induction and maintenance therapy for Crohn's disease. *N Engl J Med* 353: 1912–1925.

Sands, B.E., Anderson, F.H., Bernstein, C.N., Chey, W.Y., Feagan, B.G., Fedorak, R.N., Kamm, M.A., Korzenik, J.R., Lashner, B.A., Onken, J.E., Rachmilewitz, D., Rutgeerts, P., Wild, G., Wolf, D.C., Marsters, P.A., Travers, S.B., Blank, M.A., van Deventer, S.J. (2004) Infliximab maintenance therapy for fistulizing Crohn's disease. *N Engl J Med* 350: 876–885.

Scallon, B.J., Moore, M.A., Trinh, H., Knight, D.M., Ghrayeb, J. (1995) Chimeric anti-TNF-alpha monoclonal antibody cA2 binds recombinant transmembrane TNF-alpha and activates immune effector functions. *Cytokine* 7: 251–259.

Schreiber, S., Nikolaus, S., Malchow, H., Kruis, W., Lochs, H., Raedler, A., Hahn, E. G., Krummenerl, T., Steinmann, G. (2001) Absence of efficacy of subcutaneous antisense ICAM-1 treatment of chronic active Crohn's disease. *Gastroenterology* 120: 1339–1346.

Schreiber, S., Rutgeerts, P., Fedorak, R.N., Khaliq-Kareemi, M., Kamm, M.A., Boivin, M., Bernstein, C.N., Staun, M., Thomsen, O.O., Innes, A. (2005) A randomized, placebo-controlled trial of certolizumab pegol (CDP870) for treatment of Crohn's disease. *Gastroenterology* 129: 807–818.

Schroder, J., Stuber, F., Gallati, H., Schade, F.U., Kremer, B. (1995) Pattern of soluble TNF receptors I and II in sepsis. *Infection* 23: 143–148.

Shen, C., Assche, G.V., Colpaert, S., Maerten, P., Geboes, K., Rutgeerts, P., Ceuppens, J.L. (2005) Adalimumab induces apoptosis of human monocytes: a comparative study with infliximab and etanercept. *Aliment Pharmacol Ther* 21: 251–258.

Siegel, S.A., Shealy, D.J., Nakada, M.T., Le, J., Woulfe, D.S., Probert, L., Kollias, G., Ghrayeb, J., Vilcek, J., Daddona, P.E. (1995) The mouse/human chimeric monoclonal antibody cA2 neutralizes TNF in vitro and protects transgenic mice from cachexia and TNF lethality in vivo. *Cytokine* 7: 15–25.

Singh, B., Read, S., Asseman, C., Malmstrom, V., Mottet, C., Stephens, L.A., Stepankova, R., Tlaskalova, H., Powrie, F. (2001) Control of intestinal inflammation by regulatory T cells. *Immunol Rev* 182: 190–200.

Strober, W., Ludviksson, B.R., Fuss, I.J. (1998) The pathogenesis of mucosal inflammation in murine models of inflammatory bowel disease and Crohn disease. *Ann Intern Med* 128: 848–856.

Takei, S., Groh, D., Bernstein, B., Shaham, B., Gallagher, K., Reiff, A. (2001) Safety and efficacy of high dose etanercept in treatment of juvenile rheumatoid arthritis. *J Rheumatol* 28: 1677–1680.

ten Hove, T., van Montfrans, C., Peppelenbosch, M.P., van Deventer, S.J. (2002) Infliximab treatment induces apoptosis of lamina propria T lymphocytes in Crohn's disease. *Gut* 50: 206–211.

Tyring, S., Gottlieb, A., Papp, K., Gordon, K., Leonardi, C., Wang, A., Lalla, D., Woolley, M., Jahreis, A., Zitnik, R., Cella, D., Krishnan, R. (2006) Etanercept and clinical outcomes, fatigue, and depression in psoriasis: double-blind placebo-controlled randomised phase III trial. *Lancet* 367: 29–35.

Van Assche, G., Van Ranst, M., Sciot, R., Dubois, B., Vermeire, S., Noman, M., Verbeeck, J., Geboes, K., Robberecht, W., Rutgeerts, P. (2005) Progressive multifocal leukoencephalopathy after natalizumab therapy for Crohn's disease. *N Engl J Med* 353: 362–368.

Van den Brande, J.M., Braat, H., van den Brink, G.R., Versteeg, H.H., Bauer, C.A., Hoedemaeker, I., van Montfrans, C., Hommes, D.W., Peppelenbosch, M.P., van Deventer, S.J. (2003) Infliximab but not etanercept induces apoptosis in lamina propria T-lymphocytes from patients with Crohn's disease. *Gastroenterology* 124: 1774–1785.

Van den Brande, J., Hommes, D.W., Peppelenbosch, M.P. (2005) Infliximab induced T lymphocyte apoptosis in Crohn's disease. *J Rheumatol Suppl* 74: 26–30.

van Deventer, S.J., Elson, C.O., Fedorak, R.N. (1997) Multiple doses of intravenous interleukin 10 in steroid-refractory Crohn's disease. Crohn's Disease Study Group. *Gastroenterology* 113: 383–389.

Wang, C.Y., Mayo, M.W., Korneluk, R.G., Goeddel, D.V., Baldwin, A.S., Jr. (1998) NF-kappaB antiapoptosis: induction of TRAF1 and TRAF2 and c-IAP1 and c-IAP2 to suppress caspase-8 activation. *Science* 281: 1680–1683.

Weinblatt, M.E., Keystone, E.C., Furst, D.E., Kavanaugh, A.F., Chartash, E.K., Segurado, O.G. (2006) Long-term efficacy and safety of adalimumab plus methotrexate in patients with rheumatoid arthritis: ARMADA 4-year extended study. *Ann Rheum Dis* 65: 753-759.

Wirtz, S., Neurath, M.F. (2000) Animal models of intestinal inflammation: new insights into the molecular pathogenesis and immunotherapy of inflammatory bowel disease. *Int J Colorectal Dis* 15: 144–160.

Yin, M.J., Yamamoto, Y., Gaynor, R.B. (1998) The anti-inflammatory agents aspirin and salicylate inhibit the activity of I(kappa)B kinase-beta. *Nature* 396: 77–80.